Surfactants
in
Tribology

VOLUME 3

Edited by
Girma Biresaw
Kashmiri Lal Mittal

Surfactants
in
Tribology

VOLUME 3

CRC Press
Taylor & Francis Group
Boca Raton London New York

CRC Press is an imprint of the
Taylor & Francis Group, an **informa** business

CRC Press
Taylor & Francis Group
6000 Broken Sound Parkway NW, Suite 300
Boca Raton, FL 33487-2742

First issued in paperback 2019

© 2013 by Taylor & Francis Group, LLC
CRC Press is an imprint of Taylor & Francis Group, an Informa business

No claim to original U.S. Government works

ISBN-13: 978-1-4398-8958-9 (hbk)
ISBN-13: 978-0-367-38026-7 (pbk)

Visit the Taylor & Francis Web site at
http://www.taylorandfrancis.com

and the CRC Press Web site at
http://www.crcpress.com

Contents

PART I Nanotribology and Polymeric Systems

PART II Biobased and Environmentally Friendly Lubricants and Additives

PART III Tribological Properties of Aqueous and Nonaqueous Systems

PART IV   Advanced Tribological Concepts

Preface

Surfactants perform a wide variety of functions in tribology. These range from the basic lubrication function to control friction and wear to controlling a wide range of lubricant properties such as emulsification/demulsification, bioresistance, oxidation resistance, rust/corrosion prevention, and so on. Surfactants also spontaneously form a variety of organized structures in solution that have interesting tribological properties. These include monolayers, normal/reverse micelles, o/w and w/o microemulsions, hexagonal and lamellar lyotropic liquid crystals, and uni- and multilamellar vesicles. Recently, another group of organized assemblies has become of great interest in lubrication. These are called self-assembled monolayers (SAMs) and play critical roles in the lubrication of a wide range of products, including microelectromechanical systems (MEMS) and nanoelectromechanical systems (NEMS).

Although there is a great deal of literature on the topics of surfactants and tribology individually, there is not much information on the subject of surfactants and tribology together, in spite of the fact that surfactants play many critical roles in tribology. In order to fill this lacuna in the literature linking surfactants and tribology, we decided to organize the premier symposium on "Surfactants in Tribology." This occurred as a part of the 16th International Symposium on Surfactants in Solution (SIS) in Seoul, South Korea, June 4–9, 2006.

The SIS series of biennial events started in 1976 and since then these meetings have been held in many corners of the globe, and have been attended by "who is who" in surfactant community. These meetings are recognized by the international community as the premier forum for discussing the latest research findings on surfactants in solution. In keeping with the SIS tradition, leading researchers from around the world engaged in unraveling the importance and relevance of surfactants in tribological phenomena were invited to present their latest findings.

The first "Surfactants in Tribology" symposium was a huge success, and it elicited a high level of interest in the subject. Thus, we decided to invite leading scientists working in this area, who may or may not have participated in the symposium, to submit written accounts (chapters) of their recent research findings. This culminated in the publication of the first book *Surfactants in Tribology* in 2008.

Since the first symposium, interest in the relevance of surfactants in tribology has continued to grow unabated among scientists and engineers working in the areas of both surfactants and tribology. So we decided to organize a follow-up symposium on the subject, which took place during the 17th International Symposium on Surfactants in Solution (SIS-2008), August 17–22, 2008, in Berlin, Germany. Again, because of the high tempo of interest and enthusiasm displayed during the second symposium, we decided to publish a follow-up book on the subject. So we invited researchers in both surfactants and tribology, who may or may not have participated in the second symposium, to submit chapters for *Surfactants in Tribology*, volume 2. We received an overwhelming response from the contributors to submit high-quality chapters, and volume 2 materialized in 2011.

Owing to the continuing strong interest in the field of surfactants in tribology, the third symposium on the subject was organized at the 18th International Symposium on Surfactants in Solution (SIS-2010), November 14–19, 2010, in Melbourne, Australia. The symposium was a smashing success; so, the decision was made to publish a follow-up book, *Surfactants in Tribology*, volume 3, based on the topics discussed at the symposium. As before, invitations were sent out to experts in the fields of both surfactants and tribology, who may or may not have attended the Melbourne symposium. We were pleased with the excellent response from the contributors and the quality of chapters submitted to the current compilation.

Volume 3 comprises a total of 19 chapters dealing with various aspects of surfactants and tribology, some of which had not been covered at all in the first two volumes in this series. These 19 chapters have been logically grouped into four parts as follows. Part I consists of six chapters dealing with nanotribology and polymeric systems. Topics covered in Part I include superlubric C_{60} molecular bearings; tribological properties of nanoparticles, ionic polymer brushes, ionic liquid polymer brushes; friction dynamics of confined semi-dilute polymer solutions; surfactant modification of biopolymer friction. Part II consists of five chapters dealing with biobased and environmentally friendly lubricants and additives. Topics discussed in Part II include biobased nitrogen containing lubricant additives; environmentally friendly surface-active agents; functional fluids from lesquerella oil; green tannic acid surfactants and metal complexes; and elastohydrodynamics of biobased binary blends. Part III comprises five chapters dealing with tribological properties of aqueous and nonaqueous systems. Topics discussed in Part III include ionic liquids, alkyl carbonates, aqueous solutions of silicone polyethers, aqueous mixed surfactant systems, and cutting oil emulsifier formulations. Part IV deals with advanced tribological concepts. Topics discussed in Part IV include the moving contact line problem in electrowetting, adsorption of surfactants on hematite weighting material in water-based petroleum drilling fluids, and electric charge and evaporation control of motor bearing oil for magnetic recording disk drive.

Surface science and tribology play very critical roles in a host of industries. Manufacture and use of almost every consumer and industrial product rely on application of advanced knowledge of surface science and tribology. Examples of the major economic sectors where these two disciplines are of critical importance include mining, agriculture, manufacturing (metals, plastics, wood, automotives, computers, MEMS, NEMS, appliances, planes, rails, etc.), construction (homes, roads, bridges, etc.), transportation (cars, boats, rails, airplanes), medical (instruments and diagnostic devices, and transplants for knee, hips, and other body parts). The chapters in *Surfactants in Tribology*, vol. 3 discuss some of the underlying tribological and surface science issues relevant to many situations in diverse industries. We believe that the information compiled in this book will be a valuable resource to scientists working in or those entering the fields of tribology and surface science.

This volume and its predecessors (vols. 1 and 2) contain bountiful information and reflect the latest developments highlighting the relevance of surfactants in various tribological phenomena pertaining to many and different situations. As we learn more about the connection between surfactants and tribology, new and improved ways to control lubrication, friction, and wear utilizing surfactants will emerge.

Now it is our pleasant duty to thank all those who helped in creating this book. First and foremost, we are very thankful to the contributors for their interest, enthusiasm, and cooperation as well as for sharing their findings, without which this book could not be born. Also we would like to extend our appreciation to Barbara Glunn (Taylor & Francis, publisher) for her continued interest in and support of this book project and the staff at Taylor & Francis for giving this book its form.

Girma Biresaw
Bio-Oils Research Unit
NCAUR-MWA-ARS-USDA
Peoria, Illinois

Kashmiri Lal Mittal
Hopwell Jct., New York

Editors

Girma Biresaw received a PhD in physical-organic chemistry from the University of California, Davis, and spent 4 years as a postdoctoral research fellow at the University of California, Santa Barbara, investigating reaction kinetics and products in surfactant-based organized assemblies. Dr. Biresaw then joined the Aluminum Company of America as a scientist and conducted research in tribology, surface/colloid science, and adhesion for 12 years. Dr. Biresaw joined the Agricultural Research Service (ARS) of the U.S. Department of Agriculture in Peoria, Illinois, in 1998 as a research chemist, and became a lead scientist in 2002. At ARS, Dr. Biresaw conducts research in tribology, adhesion, and surface/colloid science in support of programs aimed at developing biobased products from farm-based raw materials. Dr. Biresaw has received more than 150 national and international invitations including requests to participate in and/or conduct training, workshops, advisory, and consulting activities. Dr. Biresaw is a member of the editorial board of the *Journal of Biobased Materials and Bioenergy*. Dr. Biresaw has authored/coauthored more than 225 invited and contributed scientific publications, including more than 60 peer-reviewed articles, 6 patents, 4 edited books, more than 35 proceedings and book chapters, and more than 120 scientific abstracts.

Kashmiri Lal Mittal received a PhD in chemistry from the University of Southern California in 1970 and was associated with IBM Corp. as a research scientist from 1972 to 1994. He is currently teaching and consulting worldwide in the areas of adhesion and surface cleaning. He is the editor of 106 published books, as well as others that are in the process of publication, within the realms of surface and colloid science and of adhesion. He has received many awards and honors and is listed in many biographical reference works. Mittal was a founding editor of the *Journal of Adhesion Science and Technology* and was its editor-in-chief until April 2012. He has served on the editorial boards of a number of scientific and technical journals. He was recognized for his contributions and accomplishments by the international adhesion community that organized the First International Congress on Adhesion Science and Technology in Amsterdam in 1995 on the occasion of his 50th birthday (235 papers from 38 countries were presented). In 2002, he was honored by the global surfactant community, which instituted the Kash Mittal Award in the surfactant field in his honor. In 2003, he was honored by the Maria Curie-Sklodowska University, Lublin, Poland, which awarded him the title of doctor *honoris causa*. In 2010, he was honored by both adhesion and surfactant communities on the occasion of publication of his 100th edited book. More recently, he has started the new journal titled *Reviews of Adhesion and Adhesives*.

Contributors

Ahmed M. Al-Sabagh
Egyptian Petroleum Research
    Institute
Cairo, Egypt

Grigor B. Bantchev
Bio-Oils Research Unit
National Center for Agricultural
    Utilization Research
USDA-Agricultural Research
    Service
Peoria, Illinois

Girma Biresaw
Bio-Oils Research Unit
National Center for Agricultural
    Utilization Research
USDA-Agricultural Research Service
Peoria, Illinois

Sophie Bistac
Laboratoire de Photochimie
    et d'Ingénierie Macromoléculaires
    (LPIM)
Université de Haute Alsace (UHA)
Mulhouse, France

Atanu Biswas
Plant Polymer Research Unit
National Center for Agricultural
    Utilization Research
USDA-Agricultural Research Service
Peoria, Illinois

Maurice Brogly
Laboratoire de Photochimie et
    d'Ingénierie Macromoléculaires
    (LPIM)
Université de Haute Alsace (UHA)
Mulhouse, France

J. Cayer-Barrioz
LTDS–UMR CNRS 5513–Ecole
    Centrale de Lyon
Ecully, France

Steven C. Cermak
Bio-Oils Research Unit
National Center for Agricultural
    Utilization Research
USDA-Agricultural Research Service
Peoria, Illinois

Kenneth M. Doll
Bio-Oils Research Unit
National Center for Agricultural
    Utilization Research
USDA-Agricultural Research Service
Peoria, Illinois

Mahmoud R. Noor El-Din
Egyptian Petroleum Research Institute
Cairo, Egypt

Mohamed M. El Sukkary
Petrochemicals Department
Egyptian Petroleum Research Institute
Cairo, Egypt

Roque L. Evangelista
Bio-Oils Research Unit
National Center for Agricultural
    Utilization Research
USDA-Agricultural Research Service
Peoria, Illinois

Ahmad Fahs
Laboratoire de Photochimie et
    d'Ingénierie Macromoléculaires
    (LPIM)
Université de Haute Alsace (UHA)
Mulhouse, France

J. M. González
Production Department (PRIE)
PDVSA Intevep
Estado Miranda, Venezuela

David R. K. Harding
Institute of Fundamental Science
Massey University
Palmerston North, New Zealand

Tatsuya Ishikawa
Graduate School of Engineering
Kyushu University
Fukuoka, Japan

Dina A. Ismail
Petrochemicals Department
Egyptian Petroleum Research Institute
Cairo, Egypt

Nadia G. Kandile
Ain Shams University
Cairo, Egypt

T. E. Karis
San Jose Research Center
HGST, a Western Digital Company
San Jose, California

Motoyasu Kobayashi
Japan Science and Technology Agency
ERATO
Kyushu University
Fukuoka, Japan

D. Mazuyer
LTDS–UMR CNRS 5513–Ecole
  Centrale de Lyon
Ecully, France

Kouji Miura
Department of Physics
Aichi University of Education
Aichi, Japan

Nabel A. Negm
Petrochemicals Department
Egyptian Petroleum Research Institute
Cairo, Egypt

Marta Ogorzalek
Department of Chemistry
Technical University of Radom
Radom, Poland

F. Quintero
Production Department (PRIE)
PDVSA Intevep
Estado Miranda, Venezuela

Leslie R. Rudnick
Designed Materials Group
Wilmington, Delaware

Naruo Sasaki
Department of Materials and
  Life Science
Seikei University
Tokyo, Japan

Marian W. Sulek
Department of Chemistry
Technical University of Radom
Radom, Poland

Lei Sun
Key Laboratory for Special Functional
  Materials of Ministry of Education
Henan University
Kaifeng, People's Republic of China

Atsushi Takahara
Japan Science and Technology Agency
ERATO
and
Graduate School of Engineering
and
Institute for Materials Chemistry
  and Engineering
Kyushu University
Fukuoka, Japan

Salah M. Tawfik
Department of Petrochemicals
Egyptian Petroleum Research Institute
Cairo, Egypt

Masami Terada
Japan Science and Technology Agency
ERATO
Kyushu University
Fukuoka, Japan

A. Tonck
LTDS–UMR CNRS 5513–Ecole
  Centrale de Lyon
Ecully, France

Ying Wang
State Key Laboratory of Nonlinear
  Mechanics
Institute of Mechanics
Beijing, People's Republic of China

Zhishen Wu
Key Laboratory for Special
  Functional Materials of Ministry
  of Education
Henan University
Kaifeng, People's Republic of China

Carlo Zecchini
Organic Carbonates Marketing
  Consultant
Lodi, Italy

Zhijun Zhang
Key Laboratory for Special
  Functional Materials of
  Ministry of Education
Henan University
Kaifeng, People's Republic of China

Ya-Pu Zhao
State Key Laboratory of Nonlinear
  Mechanics
Institute of Mechanics
Beijing, People's Republic of China

Yanbao Zhao
Key Laboratory for Special Functional
  Materials of Ministry of Education
Henan University
Kaifeng, People's Republic of China

Malgorzata Zieba
Department of Chemistry
Technical University of Radom
Radom, Poland

Part I

Nanotribology and Polymeric Systems

1 Nano-Scale Friction and Superlubricity at Carbonic Interfaces

Naruo Sasaki and Kouji Miura

CONTENTS

The nanoscale superlubricity or ultralow friction on carbon hybrid interfaces is investigated. The following interfaces are considered: atomic force microscopy (AFM) tip on graphite surface (tip/graphite); AFM tip on C_{60}/graphite (tip/C_{60}/graphite); graphite on graphite surface (graphite/graphite); and graphite on C_{60}/graphite (graphite/C_{60}/graphite). For the tip/graphite and graphite/graphite interfaces, frictional force maps are compared between simulations and experiments, which can be explained by stick–slip motion of the tip apex atom and graphite flake. For the graphite/C_{60}/graphite interface, C_{60} molecular bearing, superlubricity appears, where the (mean) lateral force becomes nearly zero within the atomic resolution of the frictional force microscopy. The C_{60} intercalated graphite film is one of the most successful systems we have developed as graphite/C_{60}/graphite interface, and it exhibits an extremely low friction coefficient, $\mu < 0.001$. Simulated superlubricity of graphite/C_{60}/graphite interface shows a marked anisotropy, which reflects the symmetry of the six-membered rings of the C_{60} molecule and graphene sheet. The physical origin of the maximum peak and near-zero minimum is numerically clarified. Controlling the superlubricity of graphite/C_{60}/graphite interface will contribute to solving energy and environmental problems.

1.1 INTRODUCTION

It is the ultimate goal of tribology researchers to realize an ideal friction-free machinery system with zero energy consumption. Since the first observation of atomic-scale friction [1] and the proposal of the concept of an ideal frictionless sliding [2], fundamental studies on superlubricity have been carried out to date based mainly on two different mechanisms: incommensurate contact [3–16] and weak interfacial interaction [17–21]. Control of superlubricity leads to energetically effective control of the motion of nano- or micro-scale objects at the interface.

As efficient solid lubricants, lamellar solids, such as graphite, MoS_2, and boron nitride, are widely used because flakes cleaved from these substrates can contribute to superlubricity. This is due to the incommensurate stacking of the lattice between the flakes and the surface. Such a structure is observed between mica flakes on a mica surface [3], MoS_2 flakes on a MoS_2 surface [5,6], MoO_3 nanocrystals on a MoS_2 surface [7], $MoS_2/MoO_3/MoS_2$ interface [8], and graphite flakes on a graphite surface [9,10].

Although it can be easily expected that the graphite-intercalated compounds (GICs) are suitable for superlubricants, little work has been done to use GICs as practical lubricants. Recently, as a promising artificial superlubricant system, we have developed several types of GICs, C_{60} molecular bearing system, and have studied the superlubrication mechanism of the graphite/fullerene interface [13–16]. An important point to note in this system is that the weak chemical- or van der Waals-type bonds between C_{60} molecules and graphite sheets are not only weak enough to move C_{60} molecules smoothly but also strong enough to hold the structures of the graphite/C_{60}/graphite interface firmly. This feature does not appear in the case of the C_{60} molecules adsorbed on the Si surface since strong chemical bonds are formed between the C_{60} molecules and the Si surface. Therefore, the graphite/C_{60}/graphite interface can be considered as one of the best candidates to achieve superlubricity. First, we have developed a C_{60} monolayer system confined by graphite walls and have shown that this system exhibits superlubricity or ultralow mean lateral force [13,14]. Next, we have developed a C_{60} intercalated graphite film prepared by chemical and thermal treatments [15,16], and have shown that this film exhibits excellent superlubricity with an ultralow friction coefficient $\mu < 0.001$, which is smaller than $\mu < 0.002$ for MoS_2 [7] and $\mu \cong 0.001$ for graphite [9]. We anticipate our novel lubrication system to be a starting point for developing more practical superlubricants using intercalated graphite, which will contribute to solving the energy and environmental problems.

Therefore, in this chapter, we discuss frictional properties and superlubricity of carbon hybrid interfaces. First, in Sections 1.3.1 through 1.3.5, the following interfaces are studied mainly from an experimental standpoint [9,12–16]: atomic-force microscopy (AFM) tip on graphite (tip/graphite), AFM tip on C_{60}/graphite (tip/C_{60}/graphite), graphite flake on graphite (graphite/graphite), and graphite on C_{60}/graphite (graphite/C_{60}/graphite). As mentioned above, graphite/C_{60}/graphite interface is experimentally achieved by fabricating the C_{60} sandwiched system [13,14] and the C_{60} intercalated graphite film [15,16]. In particular, in Sections 1.3.1 and 1.3.3, comparison between simulation and experiment is carried out [9,14]. Next, in Section 1.3.6, graphite/C_{60}/graphite interfaces are studied from a theoretical point of view [22–26]. In Section 1.3.6.1, the simulated structure of the graphite/C_{60}/graphite interface is investigated using static molecular mechanics simulation [22,23], which confirms the validity of our simulated results. Then, in Section 1.3.6.2, the effect of the scan direction of the graphene sheet on superlubricity at the graphite/C_{60}/graphite interface is studied [24]. Marked anisotropy and load dependence are observed. The physical origin of the maximum peak [24] and minimum [25] is discussed in Sections 1.3.6.3 and 1.3.6.4, respectively.

1.2 EXPERIMENTAL

1.2.1 MATERIALS

1.2.1.1 C_{60} Sandwiched Graphite

A graphite substrate was prepared by cleaving highly oriented pyrolytic graphite (HOPG) [13,14]. The C_{60} film on HOPG was prepared by evaporation from a BN crucible as shown in Figure 1.1. The substrate temperature during evaporation was maintained at 150°C to 200°C. A cleavaged graphite flake with an area of 1 mm²

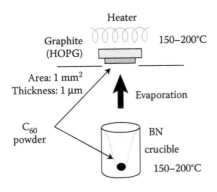

FIGURE 1.1 Method of preparation of C_{60} sandwiched system.

and a thickness of about 5 μm was used. A manipulator was used to place a graphite flake on a suitable location on the fixed C_{60} film/graphite, and thereafter, a rect-
angular silicon cantilever with a normal spring constant of 0.75 N/m was placed
on the graphite flake on the C_{60} film/graphite using the position data of C_{60} film/
graphite within an experimental error of 0.1 μm [6]. Normal and lateral forces were
measured simultaneously under humidity-controlled conditions at room temperature using commercially available instruments (Seiko Instruments Inc., SPI-3700
and Digital Instruments Inc., Multi-mode SPM). The scan speed was 0.13 μm/s.
Zero normal force was defined as the position at which the cantilever was not bent.
The frictional forces were calibrated using the method presented in Refs. [27–29].

1.2.1.2 C_{60} Intercalated Graphite

C_{60} intercalated graphite films were prepared as illustrated in Figure 1.2 [15,16,30–32].
Graphite (HOPG) for frictional force measurement and natural graphite powder for
high-resolution transmission electron microscopy (HRTEM) were stirred for 16 h in
a reaction mixture of concentrated sulfuric acid and nitric acid (4:1, v/v). The acid-
treated natural graphite was washed with water until neutralized and then dried at
100°C to remove any remaining water. The dried graphite particles were heat-treated
at 1050°C for 15 s to obtain exfoliated graphite particles, which were then immersed
in 70% ethyl alcohol solution in an ultrasonic bath. C_{60} powder and the exfoliated
graphite enclosed in a vacuum-sealed quartz tube were placed in a furnace at 600°C
for 15 days. The structure of the C_{60} intercalated graphite film was investigated using
HRTEM, JEM-2000EX, on very thin sections of an intercalated graphite film prepared from natural graphite powder, which is not representative of the entire sample.
Frictional forces were measured at room temperature using the same instruments
mentioned above for an intercalated graphite block prepared from the HOPG.

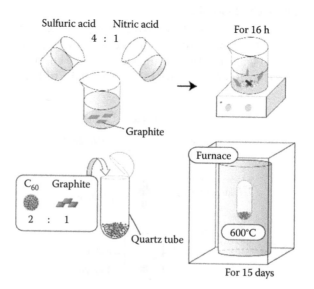

FIGURE 1.2   Method of preparation of C_{60} intercalated graphite film.

1.2.2 SIMULATION

1.2.2.1 Tip/Graphite Interface

Simulation of single-atom friction was performed using the single-atom tip connected to the three-dimensional cantilever spring scanned on a rigid monolayer graphite surface as shown in Figure 1.3 [18–20]. The model of the graphite surface consists of 600 carbon atoms and 271 hexagons. The lattice constant of graphite is 1.421Å. The total potential energy V is assumed to consist of the elastic energy of the cantilever V_T and the microscopic tip–surface interaction energy V_{TS} such as $V = V_T + V_{TS}$.

V_T is assumed to be a harmonic potential as follows:

$$V_T = \frac{1}{2}(k_x(x - x_s)^2 + k_y(y - y_s)^2 + k_z(z - z_s)^2), \tag{1.1}$$

where $k_i (i = x, y, z)$ is an elastic constant of the effective lateral spring including the cantilever parallel to the i-th ($i = x, y, z$) direction and the microscopic interatomic bonds of the tip. (x, y, z) denotes the tip atom position, and (x_s, y_s, z_s) denotes the equilibrium tip support position for the system without having interaction with the surface.

V_{TS} is assumed to be the sum of all the pairwise Lennard–Jones interaction energies between the single-atom tip and the substrate surface atoms as follows:

$$V_{TS} = \sum_i 4\epsilon \left[\left(\frac{\sigma}{r_{0i}} \right)^{12} - \left(\frac{\sigma}{r_{0i}} \right)^6 \right], \tag{1.2}$$

where r_{0i} is the distance between the tip atom and the i-th atom in the graphite surface, and the parameters are assumed to be $\epsilon = 0.87381 \times 10^{-2}$ eV, $\sigma = 2.4945$ Å. This

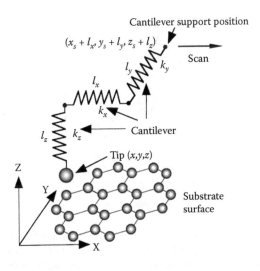

Cantilever support position

$(x_s + l_x, y_s + l_y, z_s + l_z)$

Scan

l_y　k_y

l_x

k_x

Cantilever

Tip (x, y, z)

Z

Y

Substrate surface

X

FIGURE 1.3 Schematic illustration of the model of the tip/graphite interface in frictional force microscopy. Single-atom tip is connected to a three-dimensional cantilever.

interaction potential with these parameters can excellently reproduce corrugation amplitude of AFM images of the graphite [33].

In the simulation, a static Tomlinson model is employed. It is assumed that the tip–surface system is in the limit of absolute zero temperature, $T \rightarrow 0$ K, where thermal activated processes can be completely neglected. Since the scanning velocity of the FFM tip v is much smaller than the characteristic velocity of the lattice vibration, v can be assumed to be in the limit of zero, $v \rightarrow 0$. Note that the adiabatic total potential energy surface itself is changed very slowly over time by the scan in the tip support position. As shown in Figure 1.4, the tip atom is always located at a stable equilibrium position of the potential surface for each tip support position. During some period of the cantilever scan, the tip atom continuously moves trapped in the local minimum. But it discretely jumps from one minimum to another deeper minimum and the stored elastic energy is dissipated instantaneously when the barrier

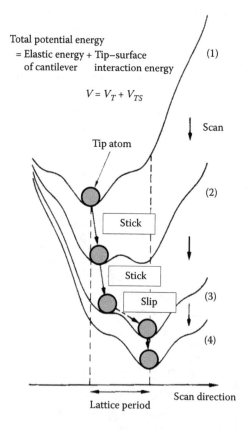

FIGURE 1.4 Schematic illustration of the total energy V obtained by the sum of the elastic energy of the cantilever spring V_T and the tip–surface interaction V_{TS}. (1)–(4) denote the time evolutions of the potential by the cantilever scan. Several metastable points corresponding to local minima appear, and the tip atom jumps to the neighboring local minima at these points, when the barrier between two local minima disappears.

between the two local minima disappears. This continuous and discrete motion is periodically repeated, which is called stick–slip motion. The simulation has been performed under constant-height mode. Therefore, (x_s, y_s) is varied with z_s fixed, and the total energy V is minimized for each (x_s, y_s), based on the Polak–Ribiere-type conjugate gradient method [34]. Then, the optimized position of the tip atom (x, y, z), and the lateral force $F_i (i = x, y)$ acting on the lever basal position are obtained. It should be remarked that the actual cantilever support position can be defined using the natural length of the cantilever spring. If the natural spring lengths along the x, y, and z directions are denoted as l_x, l_y, and l_z, respectively, the actual cantilever support position is represented as $(x_s + l_x, y_s + l_y, z_s + l_z)$ as shown in Figure 1.3. Therefore, in a constant-height mode, the tip is scanned in the condition of $z_s = $ const. under the finite mean loading force $<F_z>$. This static simulation has successfully reproduced the observed two-dimensional frictional force maps at a scan velocity of $v \simeq 100$ nm/s [9,14,19,20].

1.2.2.2 Graphite/C$_{60}$/Graphite Interface

Figure 1.5a and b shows the models of the graphite/C$_{60}$/graphite interface (C$_{60}$ molecular bearing) and the graphite/graphite/graphite interface (graphite system), respectively [22–26]. It is assumed that only the intercalated C$_{60}$ molecule and graphene sheet sandwiched between the upper and lower rigid monolayer graphene sheets can deform in the simulation. The periodic boundary condition is applied to the 1×1 unit cell of the C$_{60}$ molecular bearing system (parallelogram in the

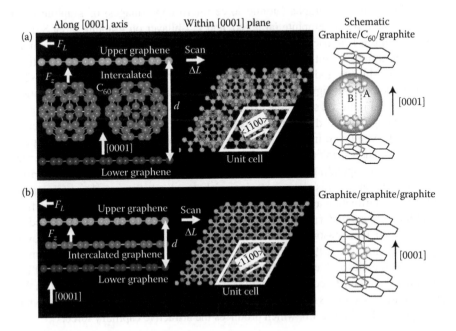

FIGURE 1.5 Schematic illustration of the models of (a) C$_{60}$ bearing system (graphite/C$_{60}$/graphite interface) and (b) graphite system (graphite/graphite/graphite interface). The parallelogram shows a 1×1 unit cell.

middle of Figure 1.5a), within the (0001) plane. The same periodic condition is applied to the graphite system (parallelogram in the middle of Figure 1.5b). Here, the orientation where six-membered rings of the C_{60} molecule face parallel to those of the upper and lower graphene sheets is considered, as illustrated in the right panel of Figure 1.5a. Here, the covalent bonding and nonbonding energies, Tersoff potential function V_{cov} [35], and modified Lennard–Jones potential function V_{vdw} [36,37] are used, respectively. The initial orientation between two graphite sheets is set to be that of AA stacking. First, the upper rigid graphene sheet is moved to change the graphene interlayer distance d along the [0001] direction. Next, for each fixed d, the upper rigid graphene sheet is scanned to change the lateral position L. One example of the scan along the $<1\bar{1}00>$ direction is shown in Figure 1.5a and b. Finally, for each d and L, the metastable structure of the graphite/C_{60}/graphite interface is calculated by minimizing the total energy $V_{total} = V_{cov} + V_{vdw}$, using the Polak–Rebiere-type conjugate gradient (CG) method [34]. Here, the convergence criterion is set so that the maximum absolute value of all the forces acting on movable atoms is lower than 1.6×10^{-4} nN. Thus, the vertical loading force acting on the upper graphene F_z and the lateral force opposite to the scan direction F_L are obtained per 1×1 unit cell.

1.3   RESULTS

1.3.1   Tip/Graphite Interface

Figure 1.6 shows topographs (atomic force microscope images) of graphite (area S_A), C_{60} monolayer on graphite (area S_B), and C_{60} bilayer on graphite (area S_C). As shown in Figure 1.7a, frictional force maps for graphite (a) show that the tip exhibits one-dimensional straight stick–slip and two-dimensional zigzag stick–slip motions [38,39] for $<11\bar{2}0>$ and $<1\bar{1}00>$ scanning directions, respectively [9,14,18–20], although they exhibit significant load dependence at small loads [9,14,20] as shown in Figure 1.8. Figure 1.8 shows the experimental and simulated frictional force maps scanned along $<11\bar{2}0>$ directions for two different loading forces. It can be clearly seen that the simulated image patterns are in very good agreement with the experimental ones, and exhibit marked load dependence. At a lower loading force, the zigzag pattern corresponding to the C–C bond of the graphite lattice appears. However, as the load increases, this zigzag pattern vanishes and only the straight pattern parallel to the scan direction appears. The experimental loading force for measuring frictional force image patterns in our experiment is much smaller than that reported by others [38,39]. One reason for this difference is ascribed to the effect of water covering the graphite surface [9], which can be explained as follows: Since our experiment is performed under ambient conditions, where the relative humidity is around 50%, a long-range capillary force, F_{cap}, is dominant as a tip–surface interaction force. In this case, the water meniscus is formed between the tip and graphite surface as shown in the right panel of Figure 1.8, which produces the attractive capillary force,

$$F_{cap} = -\frac{4\pi\gamma R\cos\theta}{1 + z/R(1 - \cos\phi)}, \tag{1.3}$$

Topographic image of
C_{60} adsorbed graphite

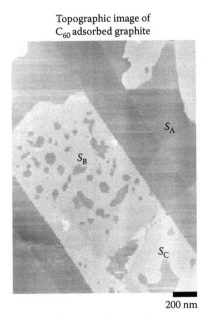

FIGURE 1.6 Measurement of superlubricity of C_{60} monolayer sandwiched system. Topographs (atomic force microscope images) of graphite (area S_A), C_{60} monolayers on graphite (area S_B), and C_{60} bilayers on graphite (area S_C).

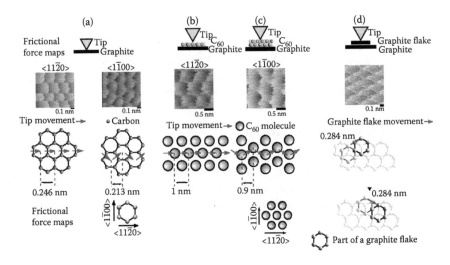

FIGURE 1.7 Frictional force maps and tip movements for representative scan directions for the tip/graphite (a), tip/C_{60} monolayer/graphite (b), tip/C_{60} bilayer/graphite (c), and flake movement for graphite flake/graphite (d).

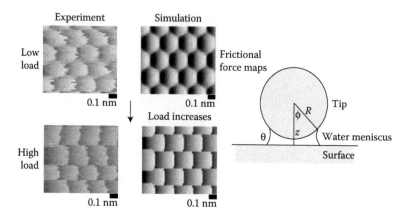

FIGURE 1.8 Experimental and simulated frictional force maps of graphite at two differ-ent loading forces for the tip/graphite interface. The inset shows the water meniscus formed between the tip and graphite surface.

where z is the tip–surface distance, R is the radius of curvature of the tip of the atomic force microscope, and θ and ϕ are the contact and meniscus angles, respectively [40], and $\gamma = 0.07$ N/m is the surface tension of water. Using estimated tip radius $R = 20$ nm, $\theta = 0$ deg, and $\phi = 10$ deg, the calculated F_{cap} for a tip height $z = 0.2$ nm is about −11 nN, which corresponds well with the measured pull-off force of $F_{pull-off} \cong -8$ nN in our current experiment. Thus, F_{cap} significantly contrib-utes to the total loading force in our experiment, which is a physical origin for the difference between the previously reported loading forces and ours.

1.3.2 Tip/C$_{60}$/Graphite Interface

As shown in Figure 1.7b and c, frictional force maps show that the tip exhibits straight and zigzag stick–slip motions for $<11\bar{2}0>$ and $<1\bar{1}00>$ scan directions of a C$_{60}$(111) surface, respectively [28,29]. The frictional force map of Figure 1.7b shows that the C$_{60}$ monolayer consists of close-packed C$_{60}$ molecules.

1.3.3 Graphite/Graphite Interface

When a graphite flake is placed on area S_A of Figure 1.6, the graphite flake on the graphite (which we call graphite/graphite) is obtained. The frictional force map of Figure 1.7d shows that the graphite flake moves on the graphite such that AB stack-ing of the graphite is maintained [9]. The motion of the flake on the graphite surface can be converted to that of a center of the flake mass in the following effective potential:

$$V = V_0\left\{2\cos\left(\frac{2\pi}{a}x\right)\cos\left(\frac{2\pi}{\sqrt{3}a}y\right) + \cos\left(\frac{4\pi}{\sqrt{3}a}y\right)\right\}, \tag{1.4}$$

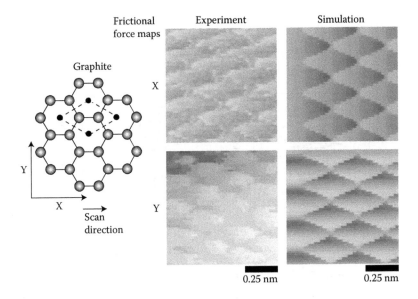

FIGURE 1.9 Experimental and simulated frictional force maps of graphite/graphite interface.

where $a = 0.142$ nm $\times 2 = 0.284$ nm indicates a distance between the neighboring stable positions corresponding to the AB stacking positions of the graphite lattice. This effective potential is determined so that it takes the minimum value at the natural stacking position of the graphite. This simple preliminary simulation can explain the experimental frictional force maps quite well as shown in Figure 1.9.

1.3.4 C_{60} SANDWICHED GRAPHITE

When a graphite flake is placed on area S_B of Figure 1.6, graphite/C_{60}/graphite interface is obtained [13,14]. Figure 1.10 shows transitions of the frictional force image patterns of the graphite/C_{60}/graphite interface depending on the load, and eventually exhibit a chain-like structure. Here, it should be noted that the maps illustrated in Figure 1.10 are composed of friction at the tip/graphite and the graphite/C_{60} interfaces. Considering that the frictional-image patterns of Figure 1.10 are clearly different from those at the tip/graphite interface of Figure 1.7a, effect of the friction at the graphite/C_{60} interface appears more clearly than that at the tip/graphite interface. As a result, the graphite flake moves together with the tip during a scan. In this case, the C_{60} molecule can slide or roll at the graphite/C_{60}/graphite interface as illustrated in Figure 1.11. As shown in Figure 1.10, the frictional force map at a loading force of 9 nN has a periodicity of 1 nm along the scan direction, which reflects the close-packed C_{60} molecular arrangement (shown in Figure 1.7b and c). On the other hand, as shown in Figure 1.10, it has a periodicity of 1.0 nm along the scan direction and 2.6 nm along the direction vertical to the scan direction, which shows a supercell structure.

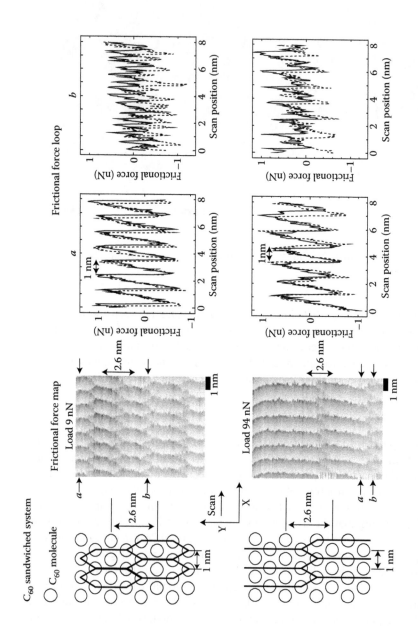

FIGURE 1.10 Lateral force loops of graphite/C$_{60}$/graphite interface (C$_{60}$ molecular bearing) for forward (solid) and backward (broken) scans under the loading force of 9 and 94 nN.

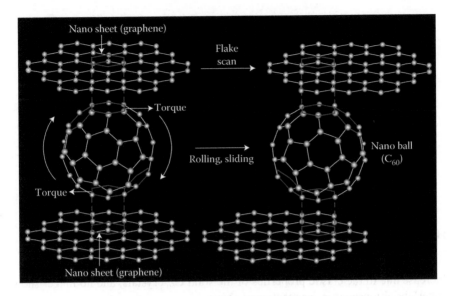

FIGURE 1.11 Original concept of "single molecular bearing." For example, the C_{60} molecule rolls by means of torque, which is produced by the nano-gears of six-membered rings between C_{60} molecules and the upper and lower graphite sheets.

The frictional loops obtained from the line profiles (solid and dotted sawtooth lines shown in Figure 1.10) indicated by arrow a in the frictional force maps exhibit little hysteresis, which indicates that the dynamic frictional or mean lateral forces are nearly zero and thus there is little energy dissipation.[*] As shown in the frictional force loop a in Figure 1.10, the mean lateral force at the position a is nearly zero even under the high-loading condition. On the other hand, the mean lateral force at the position b shows some hysteresis with a slightly larger energy dissipation, in which two-dimensional zigzag motion of a graphite flake occurs. Furthermore, Figure 1.10 shows that unit cells of frictional force maps become enlarged along the y direction, as the loading force increases from 9 to 94 nN, which shows that the region where the mean lateral force is nearly zero, as indicated by arrow a, becomes larger as the load increases. Therefore, the superlubricity region increases as the load increases, which is useful for industrial application as a lubricant. The load dependence of dynamic frictional force is shown in Figure 1.12. The maximum lateral force in this experiment is estimated to be smaller than 1 nN, which is comparable with the shear force (0.4 nN) between a single C_{60} molecule and graphite [29]. However, when there are more than two layers of C_{60} molecules inserted between graphite sheets, the observed frictional forces rapidly

---

[*] The hysteresis of the piezoactuator is extremely small within a nanometer scan. A trivial change in C_{60} molecular arrangements between one direction and the opposite direction scannings may occur, which gives the possibility that some hysteresis is produced. However, the magnitude of the hysteresis is less than a few percents in this experiment. Hence, it can be concluded that within experimental errors, no hysteresis exists and mean frictional forces are zero.

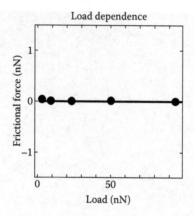

FIGURE 1.12 Load dependence of frictional force of graphite/C_{60}/graphite interface.

increase due to the elastic properties of the solid C_{60} crystals, and no longer show superlubricity properties as mentioned above.

1.3.5 C_{60} INTERCALATED GRAPHITE

1.3.5.1 Structure

HRTEM images of the C_{60} intercalated graphite thin film (mean thickness: 500 μm) are shown in Figure 1.13a and b within the (0001) plane of the graphite and along the <0001> axis, c-axis of graphite, respectively, where the indices used are the same as those of graphite [15,16]. These images show the close-packed C_{60} monolayers with the nearest-neighbor distance of 1 nm between C_{60} molecules within the (0001) plane of graphite, and the periodic spacing of 1.3 nm normal to the (0001) plane of graphite. This interlayer spacing of 1.3 nm can be reproduced by simulation as discussed in Section 1.4.

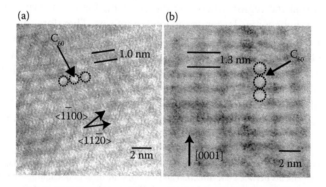

FIGURE 1.13 TEM images of the C_{60} intercalated graphite (a) parallel and (b) perpendicular to the (0001) surface.

However, it should be noted again that the HRTEM images were taken in very thin regions, and may not be a representative of the entire sample. Therefore, in the newly developed sample, close-packed C_{60} monolayers and graphite layers are not necessarily periodically repeated. Furthermore, HRTEM images of the fullerene intercalated graphite film quite often have Moiré patterns, which exhibit that the fullerene close-packed monolayers rotate by a finite angle slightly different from each other around the c-axis of the graphite. If the close-packed C_{60} monolayers distribute randomly around the c-axis of the film, ultralow friction is expected to be observed in all the scan directions. Essentially, the same mechanism, that is, friction-induced reorientation of the (0001) basal planes of the MoS_2 grains has already been pointed out by Martin et al. [5].

1.3.5.2 Superlubricity

Figure 1.14 shows the lateral force maps and the lateral force loops for the C_{60} intercalated graphite film (2.3 mm × 2.3 mm × 0.2 mm) using a frictional force microscope [15,16]. When the loading force is lower than 100 nN, the friction force becomes ultralow with less than 0.1 nN. Furthermore, this feature of the ultralow friction force was observed in all the scan directions, which is confirmed by rotating the scanner underneath the C_{60} intercalated graphite film. The load dependence of the mean lateral force $<F_L>$ exhibits the friction coefficient $\mu < 0.001$, which is smaller than $\mu < 0.002$ for MoS_2 observed by Martin et al. [5] and $\mu \cong 0.001$ for graphite previously observed by our group [9].

However, when the loading force increases to nearly 100 nN, which we call the critical loading force, a stripe pattern with a period of 1 nm appears, which corresponds to the nearest-neighbor spacing between C_{60} molecules within a C_{60} close-packed monolayer surface. This critical loading force ranges from 80 to 120 nN on the entire surface of the film. This result indicates the possibility that the motion of C_{60} molecules is inhibited by the squeezing action of the graphite walls and/or by the formation of chemical bonds between C_{60} molecules and those between C_{60} molecules and graphite [41]. This speculation also indicates a possibility that an existence of fluid layers confined by a solid surface would be important for smooth sliding of the graphite sheet. Simulations by other groups [42,43] have shown that the dynamics of C_{60} molecule exhibit complicated features derived from the effect of lattice stacking. Therefore, in the following sections, simulated superlubricity of graphite/C_{60}/graphite interface will be discussed.

1.3.6 GRAPHITE/C_{60}/GRAPHITE INTERFACE

1.3.6.1 Simulated Structure

If the structures of graphite/C_{60}/graphite interface with various initial orientations of C_{60} molecules for $d = 1.3$ nm are optimized assuming that the upper and lower graphite sheets are located as AA stacking, three different types of metastable orientations of C_{60} molecule are obtained [22,23]. As illustrated in Figure 1.15a, orientations where six-membered rings, single carbon bonds, and single carbon atoms of C_{60} molecules face the upper and lower graphite sheets, are called "AB stacking type

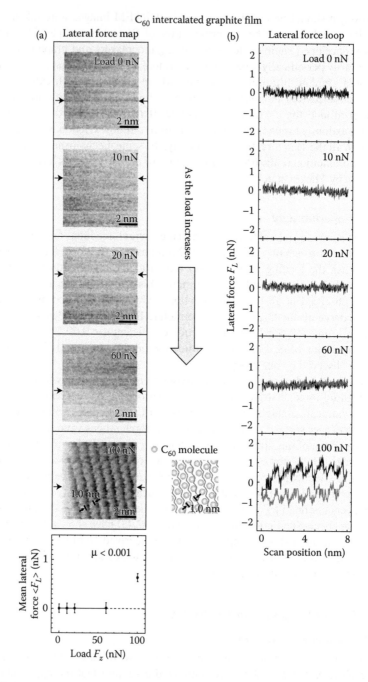

FIGURE 1.14 Measurement of superlubricity of C_{60} intercalated graphite films. Load dependence of two-dimensional frictional maps (a) and lateral force loops (b). Load dependence of friction exhibits a friction coefficient of less than 0.001. For less than 100 nN, both static and dynamic frictional forces are nearly zero within the measurement accuracy of FFM.

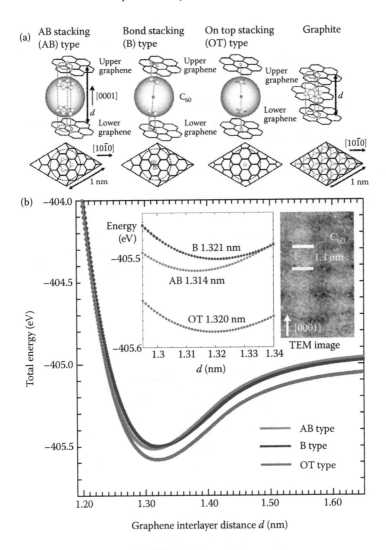

FIGURE 1.15 (a) Three different kinds of metastable orientations of C_{60} molecules: "AB stacking (AB)," "Bond stacking (B)," and "On top stacking (OT)" types. (b) Total energy as a function of the graphite interlayer distance d for AB, B, and OT types. The observed TEM image indicates an interlayer distance of 1.3 nm.

(AB)," "Bond stacking type (B)," and "On top stacking type (OT)," respectively. For B type, C_{60} molecule tilts by several degrees from [0001] axis. OT type is the same as "frustrated AB stacking" obtained by Legoas et al. [42] using structural optimization. Although the most stable interface is OT type shown in Figure 1.15b, AB type is also a local metastable structure. AB, B, and OT types give the stable interlayer distances $d = 1.314$, 1.321, and 1.320 nm, respectively, all of which reproduce well $d \cong 1.3$ nm observed in our previous TEM measurements [15,16]. This result confirms the validity of both our simulation and experiment itself. Our simulation shows that the graphite/C_{60}/graphite interface is stable only for $d \leq 1.4$ nm. It is noted that the stable interlayer distance of graphite/graphite/graphite interface is calculated to be $d = 0.68$ nm.

1.3.6.2   Simulated Anisotropic Superlubricity

In this section, frictional properties are discussed by scanning the graphite sheet. Anisotropy of superlubricity of the C_{60} bearing system (graphite/C_{60}/graphite interface) (Figure 1.16a) is simulated and is compared with that of a graphite system (graphite/graphite/graphite interface) (Figure 1.16b) [24]. Marked anisotropy and its load dependence are observed.

In the simulation, the lateral position L of the upper rigid graphene sheet is varied along the scan direction, $-30° \leq \theta \leq 90°$, and the direction opposite to it, with a fixed distance d maintained between the upper and lower rigid graphene sheets. Here, the initial structure is set so that the AB-stacking registry between the upper graphene sheet and the intermediate C_{60} molecule or graphene sheet is conserved. The mean

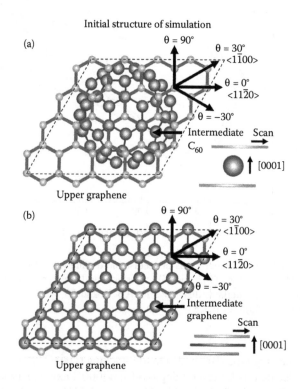

FIGURE 1.16   Model of the (a) C_{60} bearing system (graphite/C_{60}/graphite interface) and (b) graphite system (graphite/graphite/graphite interface). The upper graphene sheet and the intermediate C_{60} molecule or graphene sheet are illustrated. The broken parallelograms show the 1×1 unit cells of the C_{60} bearing system. The initial orientation between the upper and lower graphene sheets is set as the AA-stacking registry.

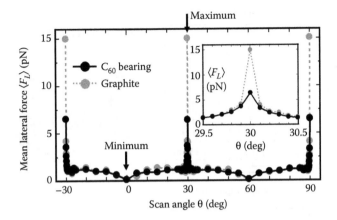

FIGURE 1.17 Mean lateral force $<F_L>$ as a function of the scan angle θ ($-30° \le \theta \le 90°$) of the upper graphene sheet under the condition of $<F_z> = 0.27$ nN for the C_{60} bearing (solid) and graphite (dotted) systems. A magnified view of the region $29.5° \le \theta \le 30.5°$ is shown in the inset.

loading force $<F_z>$ and the lateral force $<F_L>$ are calculated for $0 \le L \le 5$ nm for the forward and backward scans. Here, the interlayer distance between the upper and lower graphene sheets is set so that $<F_z> = 0.27$ nN is obtained.

Figure 1.17 shows the mean lateral force $<F_L>$ plotted as a function of the scan angle θ of the upper graphene sheet for the mean loading force $<F_z> = 0.27$ nN. Here, $\theta = -30°$, $30°$, and $90°$ are equivalent to the commensurate $<1\bar{1}00>$ direction among the intercalated C_{60} molecule (graphene sheet) and the upper and lower graphene sheets. Similarly, $\theta = 0°$ and $60°$ are equivalent to the commensurate $<11\bar{2}0>$ direction. Figure 1.17 clearly shows a periodicity of $60°$. $<F_L>$ of both the C_{60} bearing and graphite systems has a minimum value of nearly zero at $\theta = 0°$ and $60°$, and a maximum value at $\theta = -30°$, $30°$, and $90°$. $<F_L>$ has an almost constant value of about 1 pN except for a narrow region of $\theta \simeq 30° \pm 0.5°$, as shown in the inset of Figure 1.17. The maximum peak value at $\theta = 30°$ of the C_{60} bearing system, 6.4 pN, is about 40% of that of the graphite system, 15 pN. This difference in peak value can be explained by the effects of the rolling and elastic contact of the C_{60} molecule at the C_{60}/graphene interface, which will be discussed in Section 1.3.6.3. Within the region of $\theta \simeq 30° \pm 0.1°$, $<F_L>$ is sensitive to θ, and the peak rapidly decreases to about 60% and 30% for the C_{60} bearing and graphite systems, respectively (inset of Figure 1.17).

The scan-directional dependence shown in Figure 1.17 reflects the difference in the atomic-scale motion of the intermediate C_{60} molecule and the graphene sheet. Since the intermediate graphene sheet exhibits a similar behavior to the C_{60} molecule, only the case of the C_{60} bearing system is discussed. The frictional loops of F_L and the types of C_{60} motion are shown in Figure 1.18a–c. First, for $\theta = 30°$, the frictional loop exhibits sawtooth behavior with a lattice periodicity of $a_1 = 3a_0 = 0.44$ nm along the $<1\bar{1}00>$ direction (Figure 1.18a). Here,

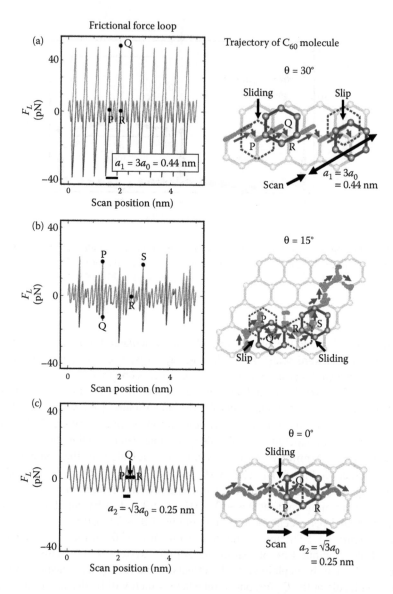

FIGURE 1.18 Frictional loops of the C_{60} bearing system for $<F_z> = 0.27$ nN. The scan angles correspond to $\theta =$ (a) 30°, (b) 15°, and (c) 0°. Frictional loops are comprised of the forward and backward scans. The right panel shows the trajectories of the six-membered ring of the C_{60} molecule on the lower graphene lattice for the forward scan. Some periodic trajectories [(a) and (c)] and all the nonperiodic trajectories (b) are shown.

$a_0 = 0.146$ nm is the carbon–carbon bond length. In this case, the C_{60} molecule slides above the carbon bond along the $<1\bar{1}00>$ direction (P → Q) and discretely slips to the neighboring AB stacking position (Q → R), as illustrated in the right panel of Figure 1.18a. Since this stick–slip motion is irreversible, F_L has a large

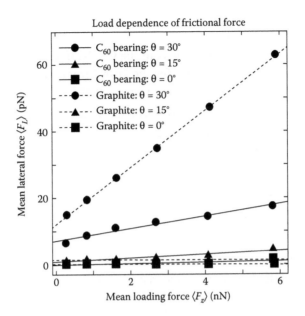

FIGURE 1.19 Mean lateral force $<F_L>$ as a function of the mean loading force $<F_z>$ for the C_{60} bearing (solid) and graphite (dotted) systems for $\theta = 0°$, 15°, and 30°.

hysteresis, which produces the peak of $<F_L>$ in Figure 1.17. Next, for $\theta = 15°$, the frictional loop shows a nonperiodic shape with a small hysteresis due to the incommensurate scan (Figure 1.18b). As shown in the right panel of Figure 1.18b, the C_{60} molecule exhibits essentially the same motion as that for $\theta = 30°$, with sliding along the carbon bond (not necessarily along the $<1\bar{1}00>$ direction) $(R \rightarrow S)$ and a discrete slip to the neighboring AB-stacking position $(P \rightarrow Q)$. Lastly, for $\theta = 0°$, the frictional loop exhibits a sinusoidal behavior with a lattice periodicity of $a_2 = \sqrt{3}a_0 = 0.25$ nm along the $<11\bar{2}0>$ direction. As shown in the right panel of Figure 1.18c, the C_{60} molecule slides almost exclusively along the carbon bonds and exhibits zigzag motion $(P \rightarrow Q \rightarrow R)$. The important feature of this zigzag motion is that it is nearly continuous and reversible. Therefore, the hysteresis loop nearly disappears, which leads to a mean frictional force of nearly zero, $<F_L> \approx 0$.

The scan-directional dependence has a characteristic load dependence, as shown in Figure 1.19. The friction coefficients obtained by linear fitting for all the cases are presented in Table 1.1. Here, $\mu < 10^{-4}$ is assumed to be $\mu = 0$. For $\theta = 30°$, $<F_L>$ shows a clear load dependence for both C_{60} bearing and graphite systems, and is proportional to $<F_z>$, where the relation $\mu_G(\theta = 30°) > \mu_{C_{60}}(\theta = 30°)$ is satisfied. For $\theta = 0$ and 15°, $<F_L>$ is nearly zero for the graphite system. On the other hand, for the C_{60} bearing system, $<F_L>$ slightly increases as $<F_z>$ increases. Thus, the relations $\mu_{C_{60}}(\theta = 0°) > \mu_G(\theta = 0°) \approx 0$ and $\mu_{C_{60}}(\theta = 15°) > \mu_G(\theta = 15°) \approx 0$ are satisfied.

TABLE 1.1
Friction Coefficients of the C_{60} Bearing and
Graphite Systems for the Scan Directions
$\theta = 0°$, $15°$, and $30°$ for $<F_z> = 0.27$ nN

	Friction Coefficient, μ	
θ (deg)	C_{60} Bearing System: $\mu_{C_{60}}$	Graphite System: μ_G
0	0.2×10^{-3}	0
15	0.6×10^{-3}	0
30	1.8×10^{-3}	8.5×10^{-3}

1.3.6.3 Peak along <1$\bar{1}$00> Direction ($\theta = 30°$)

The mean lateral force $<F_L>$ takes the maximum peak value at $\theta = 30°$ as shown in Figure 1.17. In this section, lateral stiffness of the C_{60} molecular bearing system along the <1$\bar{1}$00> direction is evaluated relative to that of the graphite system [25]. In simulation, the lateral position ΔL of the upper rigid graphene sheet is moved along the <1$\bar{1}$00> direction with a fixed distance d between the upper and lower rigid graphene sheets. Here, ΔL is defined as the displacement from the starting position of the sticking part of the C_{60} molecule. For each ΔL and d, the metastable structures of the C_{60} bearing and graphite systems, the vertical loading force acting on the upper graphene sheet F_z, and the lateral force applied opposite the scan direction F_L are obtained per 1×1 unit cell by structural optimization. The mean loading force $<F_z> = 1/\Delta L \int F_z dL$ is calculated in the sticking region ($0 \leq \Delta L \leq 0.11$ nm). Here, the d between the upper and lower graphene sheets of the C_{60} bearing system is varied between $d = 1.3$ and 1.2 nm, corresponding to $<F_z> = 0.26$ and 5.8 nN, respectively.

Figure 1.20a shows effective lateral spring constants of the C_{60} bearing and graphite systems, $k_{C_{60}}$ and k_G, respectively, plotted as a function of the mean loading force $<F_z>$. For the loading region of $<F_z>$ shown in Figure 1.20a, $k_{C_{60}} < k_G$ clearly holds along the commensurate scan direction of $\theta = 30°$, which means that the superlubricity of the C_{60} bearing system is much higher than that of the graphite system. If the tilting (rotation) of the C_{60} molecule derived from the sliding of the graphene sheet is prohibited in the simulation, the lateral spring constant without the tilting of the C_{60} molecule, $k'_{C_{60}}$, can be evaluated. Comparison of $k'_{C_{60}}$ with $k_{C_{60}}$ gives us information about the effect of rotation of C_{60} molecule on superlubricity.

For the lower loading region, $k_{C_{60}} \leq k'_{C_{60}} < k_G$ holds, where the effect of rotation of C_{60} molecule is relatively small. However, as the loading force $<F_z>$ increases, $k'_{C_{60}}$ increases more rapidly than $k_{C_{60}}$, which results in $k_{C_{60}} < k_G < k'_{C_{60}}$, and the effect of rotation of C_{60} molecule clearly appears. The rotation angle of the C_{60} molecule along the scan direction of the graphene sheet is nearly proportional to the loading

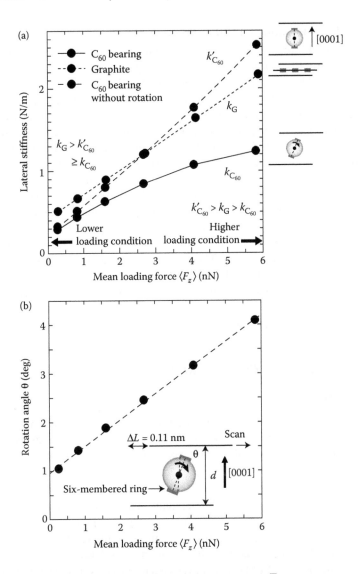

FIGURE 1.20 Simulation of superlubricity along $\theta = 0°$ <$1\bar{1}00$> direction of the C_{60} molecular bearings. (a) Effective lateral stiffness $k'_{C_{60}}$, k_G, and $k'_{C_{60}}$, and (b) rotational angle of C_{60} molecule, plotted as a function of the mean loading force <F_z>. $k'_{C_{60}}$ denotes the lateral stiffness without rotation of C_{60} molecule.

force <F_z> as shown in Figure 1.20b. Thus, it is clarified that the effect of C_{60} tilting mainly contributes to the decrease of the lateral spring constant for the higher loading region. On the other hand, the difference between $k'_{C_{60}}$ and $k_{C_{60}}$ can be explained by considering the difference in atomic-scale elasticity between C_{60}/graphene and graphene/graphene interfaces.

1.3.6.4   Minimum along <11$\bar{2}$0> Direction ($\theta = 0°$)

The mean lateral force $<F_L>$ takes the minimum value of nearly zero at $\theta = 0°$ as shown in Figure 1.17. This tendency clearly appears, particularly, for the lower loading region. This can be explained as follows: the C_{60} molecule does not tilt (rotate) along the scan direction and takes a translational motion along the carbon bond on the graphite substrate, facing one of its six-membered rings nearly parallel to the upper and lower graphene sheets. For the forward and backward scans of the graphene sheet, the C_{60} molecule moves on the same sinusoidal trajectory. Therefore, the C_{60} molecule moves continuously and the lateral force curve has no hysteresis. As a result, the mean lateral force becomes nearly zero, $<F_L> \cong 0$ [26]. The mechanism of this conservative motion of the C_{60} molecule is clarified by the analysis of the total potential energy surface, V_{total}, as shown in the upper panel of Figure 1.21. The two-dimensional map of V_{total} markedly changes during the scan process of the graphene as illustrated in Figure 1.21a–e. However, there exists one feature that does not change. There always exists an energy barrier between the minimum positions where the C_{60} molecule is located now and the nearest-neighboring minimum position. Therefore, for nearly absolute zero temperature, $T \cong 0$ K, the C_{60} molecule does not take stick–slip motion, and continuously moves trapped at the minimum position.

1.4   SUMMARY AND CONCLUSION

1.4.1   SUMMARY

In this chapter, we discussed the superlubricity of the following carbon hybrid interfaces: AFM tip/graphite, AFM tip/C_{60}/graphite, graphite/graphite, and graphite/C_{60}/graphite. The following experimental and simulated results are obtained.

1. For the tip/graphite and graphite/graphite interfaces, simulated frictional force maps are in good agreement with the experimental observations.
2. For the graphite/C_{60}/graphite interface, the magnitude of the (mean) lateral force becomes smaller than the atomic resolution of the frictional force microscopy (FFM). This novel system provides us a way to develop more practical and effective superlubricants using intercalated graphite. Controlling superlubricity of graphite/C_{60}/graphite interface leads to the reduction of energy loss, the increase of durability, and a solution to current energy and environmental problems.
3. Simulated graphite interlayer distance included in the graphite/C_{60}/graphite interface reproduces well $d \cong 1.3$ nm observed in our TEM measurements. This result gives us a proof for the validity of both our simulation and experiment.
4. The simulated superlubricity of graphite/C_{60}/graphite interface shows a marked anisotropy. This reflects the lattice periodicity of the graphite/C_{60}/graphite interface.

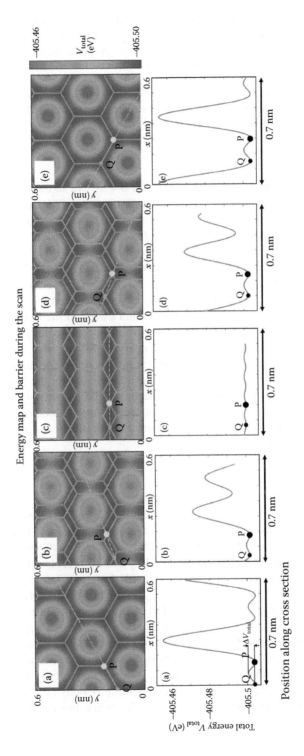

FIGURE 1.21 Simulation of superlubricity along $\theta = 0°$ ($<1\bar{1}20>$) direction of the C_{60} molecular bearings. The total potential energy surface V_{total} (upper half) and the cross section along the broken line connecting neighboring minima P and Q (lower half) for the scan position of the graphene sheet $L =$ (a) 0, (b) 0.067, (c) 0.127, (d) 0.186, and (e) 0.253 nm. C_{60} molecule is always located at the minimum position P.

5. For $\theta = 30°$, along the $<1\bar{1}00>$ direction, the simulated $<F_L>$ has a maximum peak. It was clarified that the C_{60} bearing system has a lower friction coefficient and a lower lateral stiffness than the graphite system, which is attributed to the C_{60} rotation and the elastic contact at the C_{60}/graphene interface.

6. For $\theta = 0°$, along the $<11\bar{2}0>$ direction, $<F_L>$ has a minimum value of near zero. This was attributed to the small hysteresis derived from the nearly reversible sliding above the carbon bond of the C_{60} molecule.

1.4.2 DISCUSSION

The simulated $<F_L>$ exhibits a spike-like peak for $\theta = 30°$. In actual experiments, the peak width observed by the frictional force microscopy (FFM) tip is expected to broaden, because the upper graphene sheet tends to slide along the commensurate $<1\bar{1}00>$ direction even if the sliding direction of the FFM tip deviates slightly from the $<1\bar{1}00>$ direction. This peak width broadening is also expected to be induced by thermal effect. The difference between the C_{60} bearing and graphite systems clearly appears in this region, because the effect of the intercalated C_{60} rotation and the difference in elastic contact between the intercalated C_{60} and graphene are most enhanced in this region owing to the commensurate (AB-stacking) contact. However, outside of this narrow peak region, $<F_L>$ is not sensitive to the scan angle θ and has a small nearly constant value of less than 1 pN. Here, the C_{60} bearing and graphite systems exhibit nearly the same superlubricity owing to the incommensurate contact at the interface for the low-loading condition, as illustrated in Figure 1.17. Furthermore, the anisotropy of the C_{60} bearing system can be explained by the contact of its six-membered ring with the upper and lower graphene sheets, which causes a qualitatively similar effect to that of the graphite system.

1.4.3 CONCLUSION

The load dependence of the superlubricity also exhibits a marked scan-directional dependence. From previous experiments with the graphite and C_{60} bearing systems, we showed that $\mu_G \simeq 10^{-3}$ and $\mu_{C_{60}} < 10^{-3}$, respectively. Considering the average of the simulated μ values for $\theta = 0°$, 15°, and 30° (Table 1.1), the order of magnitude of the simulated μ is comparable with those of experimental μ. For the actual multilayered system (graphite)$_n$/C_{60}/(graphite)$_n$ ($n \geq 2$), the deformation of the graphene sheets, the slight rotation of the graphene sheet around the [0001] axis, and the effect of the tip [12] can markedly affect the superlubricity. The deformed graphene and C_{60} molecules are expected to stick to each other like "nano-gears," which can increase friction. Superlubricity strongly depends on the alone factors. On the other hand, C_{60} rotation that occurs below room temperature will decrease friction. The effect of sliding velocity on superlubricity is also important. Detailed studies of the above effects are under investigation in our group.

ACKNOWLEDGMENTS

We thank Dr. N. Itamura, Seikei University, and Mr. Kamiya and Mr. D. Tsuda, Aichi University of Education, for the simulation and experiment, respectively. This work was supported by the Grant-in-Aid for Scientific Research (B) (Nos. 16340089, 17760030, 18340087, 20360022, and 23360023), Challenging Exploratory Research (No. 21654041), and Specially Promoted Research (No. 21000008) from the Japan Society for the Promotion of Science. Kouji Miura acknowledges the financial support provided by the Practical Application Research, Science and Technology Incubation Program in Advanced Regions from JST. Naruo Sasaki is thankful for the financial support provided by Organization and Functions, PRESTO-JST, and a Grant-in-Aid for Building Strategic Research Infrastructures from the Japan Ministry of Education, Culture, Sports, Science, and Technology.

REFERENCES

1. C. M. Mate, G. McClelland, R. Erlandsson, and S. Chiang, Atomic-scale friction of a tungsten tip on a graphite surface, *Phys. Rev. Lett.* **59**, 1942–1945, 1987.
2. G. McClelland, Friction at weakly interacting interfaces, in: *Adhesion and Friction, Springer Series in Surface Sciences,* Vol. 17, M. Grunze and H. Kreuzer (eds.), pp. 1–16, Springer-Verlag, Berlin, 1989.
3. M. Hirano, K. Shinjo, R. Kaneko, and Y. Murata, Anisotropy of frictional forces in muscovite mica, *Phys. Rev. Lett.*, **67**, 2642–2645, 1991.
4. M. Hirano, K. Shinjo, R. Kaneko, and Y. Murata, Observation of superlubricity by scanning tunneling microscopy, *Phys. Rev. Lett.*, **78**, 1448–1451, 1997.
5. J. M. Martin, C. Donnet, Th. Le Mogne, and T. Epicier, Superlubricity of molybdenum disulphide, *Phys. Rev., B.* **48**, 10583–10586, 1993.
6. K. Miura and S. Kamiya, Observation of the amontons-Coulomb law on the nanoscale: Frictional forces between MoS_2 flakes and MoS_2 surfaces, *Europhys. Lett.*, **58**, 610–615, 2002.
7. P. E. Sheehan and C. M. Lieber, Nanotribology and nanofabrication of MoO_3 structures by force microscopy, *Science*, **272**, 1158–1161, 1996.
8. S. Kamiya, D. Tsuda, K. Miura, and N. Sasaki, $MoS_2(0001)/MoO_3(010)/MoS_2(0001)$ friction-reducing system, *Wear*, **257**, 1133–1136, 2004.
9. K. Miura, N. Sasaki, and S. Kamiya, Friction mechanisms of graphite from a single-atomic tip to a large-area flake tip, *Phys. Rev.*, B**69**, 075420–075428, 2004.
10. M. Dienwiebel, G. S. Verhoeven, N. Pradeep, J. W. M. Frenken, J. A. Heimberg, and H. W. Zandbergen, Superlubricity of graphite, *Phys. Rev. Lett.*, **92**, 1261011–1261014, 2004.
11. K. Matsushita, H. Matsukawa, and N. Sasaki, Atomic scale friction between clean graphite surfaces, *Solid State Commun.*, **136**, 51–55, 2005.
12. N. Sasaki, H. Saitoh, K. Terada, N. Itamura, and K. Miura, Simulation of atomic-scale wear of graphite—nanotip induced graphene formation, *e-J. Surf. Sci. Nanotech.*, **7**, 173–180, 2009.
13. K. Miura S. Kamiya, and N. Sasaki, C_{60} molecular bearings, *Phys. Rev. Lett.*, **90**, 0555091–0555094, 2003.
14. N. Sasaki and K. Miura, Key issues of nanotribology for successful nanofabrication—From basis to C_{60} molecular bearings, *Jpn. J. Appl. Phys.*, **43**, 4486–4491, 2004.
15. K. Miura, D. Tsuda, and N. Sasaki, Superlubricity of C_{60} intercalated graphite films, *e-J. Surf. Sci. Nanotech.*, **3**, 21–23, 2005.

16. K. Miura, D. Tsuda, N. Itamura, and N. Sasaki, Superlubricity of fullerene intercalated graphite composite, *Jpn. J. Appl. Phys.*, **46**, 5269–5274, 2007.
17. J. Colchero, O. Marti, and J. Mlynek, Friction on an atomic scale, in: *Forces in Scanning Probe Methods, NATO ASI Series E Vol. E:286*, H.-J. Guntherodt, D. Anselmetti, and E. Meyer (Eds.), pp. 345–352, Kluwer Academic Publishers, Dordrecht, 1995.
18. N. Sasaki, K. Kobayashi, and M. Tsukada, Atomic-scale friction image of graphite in atomic-force microscopy, *Phys. Rev.*, **B54**, 2138–2149, 1996.
19. N. Sasaki, M. Tsukada, S. Fujisawa, Y. Sugawara, S. Morita, and K. Kobayashi, Analysis of frictional-force image patterns of graphite surface, *J. Vac. Sci. Technol.*, **B15**, 1479–1482, 1997.
20. N. Sasaki, M. Tsukada, S. Fujisawa, Y. Sugawara, S. Morita, and K. Kobayashi, Load dependence of frictional-force microscopy image pattern of graphite surface, *Phys. Rev.*, **B57**, 3785–3786, 1998.
21. A. Socoliuc. R. Bennewitz, E. Gnecco, and E. Meyer, Transition from stick – slip to continuous sliding in atomic friction: Entering a new regime of ultralow friction, *Phys. Rev. Lett.*, **92**, 1343011–1343014, 2004.
22. N. Sasaki, N. Itamura, and K. Miura, Atomic-scale ultralow friction—Simulation of superlubricity of C_{60} molecular bearing, *J. Phys.: Conf. Ser.*, **89**, 01200101–01200110, 2007.
23. N. Sasaki, N. Itamura, and K. Miura, Simulation of atomic-scale ultralow friction of graphite/C_{60}/graphite interface along [10$\overline{1}$0] direction, *Jpn. J. Appl. Phys.*, **46**, L1237–L1239, 2007.
24. N. Itamura, K. Miura, and N. Sasaki, Simulation of scan directional dependence of superlubricity of C_{60} molecular bearings and graphite, *Jpn. J. Appl. Phys.*, **48**, 0602071–0602073, 2009.
25. N. Itamura, K. Miura, and N. Sasaki, Analysis of mechanism of low lateral stiffness of superlubric C_{60} bearing system, *Jpn. J. Appl. Phys.*, **48**, 0302141–0302143, 2009.
26. N. Itamura, H. Asawa, K. Miura, and N. Sasaki, Unique near-zero friction regime of C_{60} molecular bearings along [12$\overline{3}$0] direction, *J. Phys.: Conference Series*, **258**, 0120131–0120139, 2010.
27. K. Miura, T. Takagi, S. Kamiya, T. Sahashi, and M. Yamauchi, Natural rolling of zigzag multiwalled carbon nanotubes on graphite, *Nano Lett.*, **1**, 161–163, 2001.
28. S. Okita, M. Ishikawa, and K. Miura, Nanotribological behavior of C_{60} films at an extremely low load, *Surf. Sci.*, **442**, L959–L963, 1999.
29. S. Okita and K. Miura, Molecular arrangement in C_{60} and C_{70} films on graphite and their nanotribological behavior, *Nano Lett.*, **1**, 101–103, 2001.
30. T. Nakajima and Y. Matsuo, Formation process and structure of graphite oxide, *Carbon*, **32**, 469–475, 1994.
31. C. Guohua, W. Dajun, W. Wehgui, and W. Cuiling, Synthesis of carbon nanostructures on nanocrystalline Ni-Ni$_3$P catalyst supported by SiC whiskers, *Carbon*, **41**, 579–625, 2003.
32. V. Gupta, P. Scarf, K. Rich, H. Romans, and R. Müller, Synthesis of C_{60} intercalated graphite, *Solid State Commun.*, **131**, 153–155, 2004.
33. N. Sasaki and M. Tsukada, Effect of the tip structure on atomic-force microscopy, *Phys. Rev. B*, 52, 8471–8482, 1995.
34. W. H. Press, S. A. Teukolsky, W. T. Vetterling, and B. P. Flannery, *Numerical Recipes: The Art of Scientific Computing*. 2nd ed. Cambridge University Press, New York, 1992.
35. J. Tersoff, Empirical interatomic potential for carbon, with applications to amorphous carbon, *Phys. Rev. Lett.*, **61**, 2879–2882, 1988.
36. S. D. Stoddard and J. Ford, Numerical experiments on the stochastic behavior of a Lennard – Jones gas system, *Phys. Rev.*, **A8**, 1504–1512, 1973.

37. J. P. Lu, X.-P. Li, and R. M. Martin, Ground state and phase transitions in solid C_{60}, *Phys. Rev. Lett.*, **68**, 1551–1554, 1992.
38. S. Morita, S. Fujisawa, and Y. Sugawara, Spatially quantized friction with a lattice periodicity, *Surf. Sci. Rep.*, **23**, 1–41, 1996.
39. S. Fujisawa, K. Yokoyama, Y. Sugawara, and S. Morita, Analysis of experimental load dependence of two-dimensional atomic-scale friction, *Phys. Rev.*, B**58**, 4909–4916, 1998.
40. J. N. Israelachvili, *Intermolecular and Surface Forces*, Third Edition, Academic Press, London, 2010.
41. Wanlin Guo, C. Z. Zhu, T. X. Yu, C. H. Woo, B. Zhang, and Y. T. Dai, Formation of sp^3 bonding in nanoindented carbon nanotubes and graphite, *Phys. Rev. Lett.*, **93**, 2455021–2455024, 2004.
42. S. B. Legoas, R. Giro, and D. S. Galvao, Molecular dynamics simulations of C_{60} nanobearings, *Chem. Phys. Lett.*, **386**, 425–429, 2004.
43. J. Kang and H. Hwang, Fullerene nano ball bearings: An atomistic study, *Nanotechnology*, **15**, 614–621, 2004.

2 Synthesis and Tribological Properties of Surface-Modified Lead Nanoparticles

Lei Sun, Yanbao Zhao, Zhishen Wu, and Zhijun Zhang

CONTENTS

Surface-modified lead nanoparticles were synthesized in water medium through a mild chemical reduction method, using sodium stearate as the modification agent. The morphology and structure of modified Pb nanoparticles were investigated by means of transmission electron microscopy, x-ray powder diffraction and Fourier transform infrared spectroscopy. The tribological properties of the as-synthesized surface-modified Pb nanoparticles as an additive in liquid paraffin were evaluated on a four-ball test machine. The results show that the surface-modified Pb nanoparticles have particle size in the range 10–20 nm and exhibit good dispersion without aggregation, compared to nonmodified Pb nanoparticles. The surface-modified Pb nanoparticles as

33

lubrication oil additive exhibit excellent antiwear and friction reduction properties; it also improved the load-bearing capacity of the base oil.

2.1 INTRODUCTION

In recent years, metal nanoparticles have been widely exploited for use, for example, in catalysis, magnetic recording media, biological labeling, and formulation of magnetic ferrofluids [1–3]. Many methods have been developed to produce metal nanoparticles. Examples include vapor deposition [4], electrochemical reduction [5], radiolytic reduction [6], chemical reduction [7,8], and thermal decomposition [9]. The chemical reduction in solution has been proven to be the most versatile and simple approach for preparation of metal nanoparticles. The reduction agent usually used in this method may be hydrazine [10], sodium boron hydride [11], or formaldehyde [12].

During the last decade, a number of attempts have been made to synthesize various inorganic nanoparticles as lubricant additives. The main obstacle in using nanoparticles as oil additives is their incompatibility. This problem can be solved by using dispersing agents or by employing surface modification techniques. If the surface modification agents are high-molecular-weight hydrocarbons, the nanoparticles thus prepared will have good dispersibility in organic solvents. In this tribology field, a remarkable progress has been achieved by Chinese research teams represented by Henan University and Lanzhou Institute of Chemical Physics, Chinese Academy of Sciences. Zhang in Henan University first developed a novel technique that incorporated the growth of inorganic nanoparticles and *in situ* surface modification into one chemical procedure so that the size and dispersibility of the particles could be simultaneously controlled [13]. After that, a great upsurge in the research of nano-lubricant additives has begun. Nanoparticles of various materials, soft metals, metal sulfides, oxides, borates, polymers, rare earth compounds, polyoxometalates compounds, and so on are tested, and the improvements of tribology performance brought about by applying the nanoparticles additives have been confirmed [14–19]. As for the antiwear and friction reduction mechanism of nanoparticles in lubricant oil, it can be interpreted as microbearings formed by inserted particles, the formation of chemical reaction boundary films, and deposition layers from melted metal particles. In all of the above research results, copper nanoparticles were demonstrated as the most effective tribologically active lubricant additives due to the formation of deposition film on the worn surface and the self-repairing effect. Based on their original study [20,21], Zhang's group successfully enlarged the dialkyldithiophosphates (DDP)-coated Cu nanoparticles from laboratory synthesis to industry production. The commercial DDP-coated Cu oil-soluble additives are now widely used in the market of automobile engine oil and wind turbines in China. This outstanding transfer of scientific and technological achievements is valuated as a representative application of nanoscience and technology by the Chinese government and scientists.

Lead is one of the metals with a low melting temperature, liable to melt and deposit in the high-temperature tribological region [22]. In this chapter, we report on the synthesis of Pb nanoparticles modified by an aliphatic acid salt. This mild and scalable method is based on the reduction of lead salt and *in situ* surface modification. The characterization of the as-prepared products was performed with a variety of methods, including x-ray diffraction (XRD), transmission electron microscopy (TEM), and Fourier transform infrared (FT-IR) spectroscopy. Their tribological properties were investigated on a four-ball test machine.

2.2 EXPERIMENTAL SECTION

2.2.1 Materials

Lead nitrate $Pb(NO_3)_2$, sodium boron hydride ($NaBH_4$), and absolute alcohol (EtOH) were all analytical reagent (A.R.) grade, and sodium stearate ($CH_3(CH_2)_{16}COONa$) was chemically pure (C.P.) reagent. These were purchased from the Chemical Reagent Corporation of the Chinese National Medical Group and used without further purification. Distilled water was used as the solvent.

2.2.2 Instruments

TEM images were obtained using a JEM-100CX transmission electron microscope (JEOL Ltd., Tokyo, Japan) at an acceleration voltage of 200 kV. The samples were prepared by placing a drop of primary sample on a copper grid. The TEM samples were allowed to dry completely at ambient temperature. The resulting specimen was then subjected to TEM analysis.

XRD patterns were collected on an X'pert Pro X-ray powder diffractometer (Philips, Amsterdam, the Netherlands) using Cu Kα radiation ($\lambda = 1.5418$ Å). The operating voltage and current were 40 kV and 40 mA, respectively. The scan rate was 10°/min.

FT-IR spectra were taken on a Nicolet AVATAR 360 Fourier transform infrared spectrometer (Thermo Electron Corporation, Madison, WI, USA), which covered the range from 4000 to 400 cm^{-1}, to characterize the surface structure of Pb nanoparticles. The as-synthesized Pb nanoparticles were mixed with KBr powder and pressed into a pellet for measurement. Background correction was made using a reference blank KBr pellet.

2.2.3 Synthesis of Surface-Modified Pb Nanoparticles

A series of nanoparticle samples were synthesized. The preparation methods for all nanoparticles are essentially the same. They differ only in the mole ratio of $Pb(NO_3)_2$ to sodium stearate used. A typical procedure for $[Pb^{2+}]:[CH_3(CH_2)_{16}COO^-] = 4:1$ was as follows: 0.15 g (0.50 mmol) of sodium stearate dissolved in 50 mL distilled water by heating was placed in a 250 mL flask, and then, 0.66 g (5.0 mmol) of $Pb(NO_3)_2$ dissolved previously in 30 mL distilled water was added to the flask with stirring. The transparent solution was colorless. Subsequently, 0.76 g (20 mmol) $NaBH_4$ dissolved in distilled water was added to the flask dropwise. The color of the reaction solution changed to black, and the solution was opaque. A large number of bubbles could be seen on the solution surface. The reaction was allowed to continue for 4 h at 70°C under magnetic stirring. At the end of the reaction, the solution was allowed to stand for 12 h at ambient temperature, and then it was vacuum filtered. The precipitate was rinsed several times with distilled water and EtOH, successively. It was then dried in a degassed desiccator at ambient temperature for 2 days. Finally, the target product, a gray powder of sodium stearate-modified Pb (Pb–St) nanoparticles, was obtained. Nonmodified Pb nanoparticles were also prepared using the same procedure without adding sodium stearate.

2.2.4 FRICTION AND WEAR EVALUATION

The tribological properties of the as-synthesized surface-modified Pb nanoparticles were determined on an MRS-10A four-ball long duration antiwear testing machine (Ji'nan testing machine factory, Ji'nan, China) at 1450 rpm in ambient conditions. The 12.7 mm diameter balls used in the test were made of bearing steel (composition: 0.95–1.05% C, 0.15–0.35% Si, 0.24–0.40% Mn, <0.027% P, <0.020% S, 1.30–1.67% Cr, <0.30% Ni, and <0.025% Cu) with a Rockweel hardness (HRc) of 61–64. The base oil was chemically pure liquid paraffin (Tianjin Kermel Chemical Reagent Co. Ltd., Tianjin, China), which has a distillation range of 180–250°C, density of 0.835–0.855 g/cm^3, and viscosity of 14.2–17.2 cps. Before each test, the balls and specimen holders were ultrasonically cleaned in petroleum ether (normal alkane with a boiling point of 60–90°C), and then dried in hot air. At the end of each test, the wear scar diameters (WSD) of the three lower balls were measured on a 15J digital microscope (Shanghai Optical Instruments Co. Ltd., Shanghai, China) to an accuracy of 0.01 mm. Then, the average WSD from the three balls was calculated.

2.3 RESULTS AND DISCUSSION

2.3.1 PARTICLE DISPERSIBILITY

The as-synthesized Pb–St nanoparticles can disperse in various organic solvents, such as chloroform, benzene, methylbenzene, and liquid paraffin with the assistance of ultrasonic agitation. The homogeneous suspensions formed were stable for several hours without particles settling. As a comparison, nonmodified Pb particles cannot disperse in organic solvents even after ultrasonication. It showed that after the surface modification process, the dispersibility of Pb–St nanoparticles in oil was significantly improved. It can be concluded that the surface of Pb nanoparticles was coated with stearate anions, which have the hydrophobic aliphatic chain extending into the solvent. Therefore, the compatibility of modified Pb nanoparticles and aliphatic solvents was improved. Surfactants such as sodium stearate play an important role in the dispersion of nanoparticles, this effect is the so-called steric stabilization, which are often used by scientists to prepare size-controlled nanoparticles and endow them compatibility with different medium [23].

2.3.2 XRD PATTERNS

Figure 2.1 shows the XRD pattern of the as-synthesized surface-modified Pb nanoparticles. The peaks seen at 2θ = 31.2°, 36.1°, 52.2°, 62.1°, 65.3°, 85.5°, and

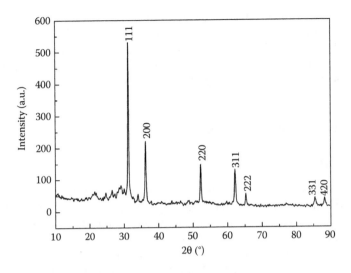

FIGURE 2.1 XRD pattern of surface-modified Pb nanoparticles.

83.3° are assigned to diffractions from the (111), (200), (220), (311), (222), (331), and (420) lattice planes of Pb, respectively, which are in agreement with face-centered cubic (fcc) lead phase (Joint Committee on Powder Diffraction Standards File No. 04-0686). It can also be seen from Figure 2.1 that there are some relatively weak peaks before $2\theta = 30°$, which arise from the diffraction of the modification layer on the coated nanoparticles. Thus, it is concluded that the Pb nanoparticles with fcc structure were successfully prepared, and the existence of a modification layer can prevent oxidation of Pb nanocrystals.

2.3.3 TEM IMAGES

Figure 2.2 shows the TEM images of surface-modified Pb nanoparticles (a) and nonmodified Pb nanoparticles (b). It is seen from Figure 2.2b that nonmodified Pb

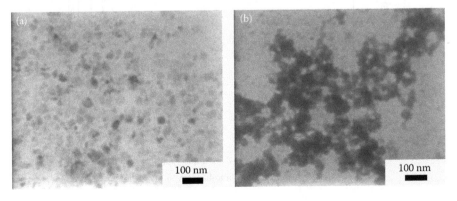

FIGURE 2.2 TEM images of surface-modified (a) and unmodified (b) Pb nanoparticles.

nanoparticles tend to agglomerate owing to their high surface energy, and have a mean particle size of about 50–80 nm. On the contrary, the TEM image of surface-modified Pb nanoparticles shows a relatively small size, with an average size of about 15 nm, and no obvious aggregation. This is because the surfactant stearate is adsorbed on the surface of the nanoparticles, which reduces the surface energy of the nanoparticles, and prevents their aggregation through steric stabilization. The results show that the existence of a surfactant layer can reduce the size of particles, and also improve the dispersion ability of the nanoparticles.

2.3.4   FT-IR Analysis

Figure 2.3 shows the FT-IR spectrum of surface-modified Pb nanoparticles. The band at 1631 cm^{-1} is attributed to the bending vibration of –CH$_3$. The bands at 2916 and 2849 cm^{-1} correspond to the asymmetric and symmetric stretching vibrations of –CH$_2$–, and the vibration in the 703 cm^{-1} region is a characteristic of a minimum of four methyl groups, (CH$_2$)$_4$, in a row and assigned to the methylene rocking vibration. The bands at 1562 and 1402 cm^{-1} correspond to the asymmetric and symmetric stretching vibrations of COO$^-$, respectively. Hence, it can be concluded that the long alkyl acid radical does exist in the surface-modified Pb nanoparticles, which improves the compatibility of nanoparticles and aliphatic solvents. The broad band at 3430 cm^{-1} is attributed to the stretching vibration of –OH, which arises from the adsorbed water on/in the coated nanoparticles.

2.3.5   Tribological Properties

Figure 2.4 shows the tribological properties of liquid paraffin with different concentrations of surface-modified Pb nanoparticles additive. The test conditions were

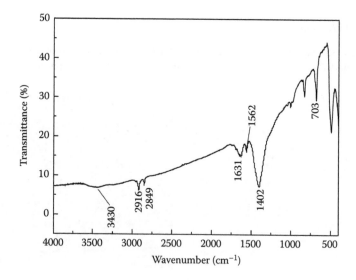

FIGURE 2.3   FT-IR spectrum of surface-modified Pb nanoparticles.

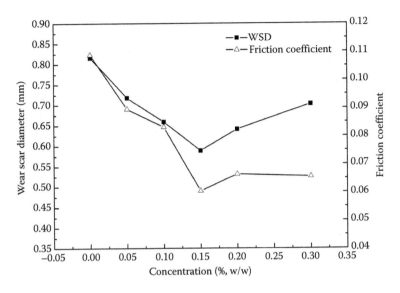

FIGURE 2.4   Effect of concentration (%, w/w) of surface-modified Pb nanoparticles in liquid paraffin on tribological properties (four-ball test machine; applied load: 300 N; speed: 1450 rev/min; test duration: 30 min; room temperature).

as follows: rotation speed 1450 rev/min, load 300 N, test duration 30 min, ambient temperature. The results show that surface-modified Pb nanoparticles can improve the antiwear properties (reduce the values of WSD) of the base oil appreciably, even at relatively low concentration (0.05%). With increase in Pb nanoparticles additive concentration, the values of WSD first decreased, and then, increased slightly. There is a minimum point in the curve. The optimum antiwear property was observed at the additive concentration of 0.15% w/w. At this concentration, the WSD of liquid paraffin was reduced to about three fourth. The friction reducing property of the surface-modified Pb nanoparticles showed a similar behavior. The results indicate that the modified Pb nanoparticles exhibit good tribological properties.

Figure 2.5 compares the tribological properties of liquid paraffin with and without 0.15% w/w surface-modified Pb nanoparticles as a function of applied load. (The conditions were ambient temperature; rotation speed 1450 rev/min; and test duration 30 min.) It can be seen that paraffin oil with surface-modified Pb nanoparticles resulted in extending the load-carrying capacity of paraffin oil from 300 to 500 N. In addition, under identical load, the WSD with Pb nanoparticles was smaller than that of pure paraffin oil. Hence, it can be concluded that, at low loads, the surface-modified Pb nanoparticles additives provide good antiwear properties and can improve the load carrying capacity of base oil.

2.4   CONCLUSIONS

Surface-modified Pb nanoparticles with sodium stearate were synthesized by the chemical reduction method. Their morphology and structure were characterized by XRD, TEM, and FT-IR. The modified Pb nanoparticles can be dispersed in organic

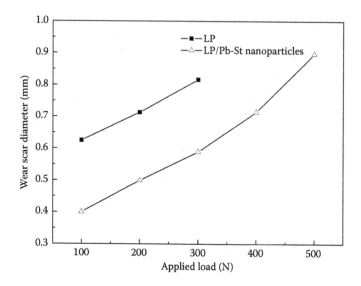

FIGURE 2.5 Effect of load on wear scar diameter of paraffin oil with and without 0.15% (w/w) surface-modified Pb nanoparticles (four-ball test machine; room temperature; speed: 1450 rev/min; test duration: 30 min).

solvents well, and had an average particle size of 15 nm. The tribological properties of the nanoparticles as lubricant additive in paraffin oil were investigated using a four-ball tribometer. The results show that the surface-modified Pb nanoparticles exhibit good antiwear and friction reduction ability.

ACKNOWLEDGMENTS

The authors are grateful for the financial support provided by the National Nature Science Foundation of China (50701016), the China Postdoctoral Science Foundation (2011M500787), and Scientific Research Foundation for the Returned Overseas Chinese Scholars of Henan Province (2001-28).

REFERENCES

1. P. V. Kamat, Photophysical, photochemical and photocatalytic aspects of metal nanoparticles, *J. Phys. Chem. B*, 106, 7729–7744, 2002.
2. C. T. Black, C. B. Murray, and R. L. Standstorm, Spin-dependent tunneling in self-assembled cobalt-nanocrystal superlattices, *Science*, 290, 1131–1134, 2000.
3. M.-P. Pileni, Magnetic fluids: Fabrication, magnetic properties, and organization of nanocrystals, *Adv. Funct. Mater.*, 11, 323–336, 2001.
4. R. P. Andres, J. D. Bielefeld, and J. I. Henderson, Self-assembly of a two-dimensional superlattice of molecularly linked metal clusters, *Science*, 273, 1690–1693, 1996.
5. L. Rodríguez-Sánchez, M. C. Blanco, and M. A. López-Quintela, Electrochemical synthesis of silver nanoparticles, *J. Phys. Chem. B*, 104, 9683–9688, 2000.
6. A. D. Belapurkar, S. Kapoor, and S. K. Kulshreshtha, Radiolytic preparation and catalytic properties of platinum nanoparticles, *Mater. Res. Bull.*, 36, 145–151, 2001.

7. Z. Gui, R. Fan, and W. Mo, Synthesis and characterization of reduced transition metal oxides and nanophase metals with hydrazine in aqueous solution, *Mater. Res. Bull.*, 38, 169–176, 2003.
8. M. L. Wu, D. H. Chen, and T. C. Huang, Synthesis of Au/Pd bimetallic nanoparticles in reverse micelles *Langmuir*, 17, 3877–3993, 2001.
9. C. Nayral, T. Ould-Ely, and A. Maisonnat, A novel mechanism for the synthesis of tin/tin oxide nanoparticles of low size dispersion and of nanostructured SnO_2 for the sensitive layers of gas sensors, *Adv. Mater.*, 11, 61–63, 1999.
10. H. Zheng, J. Liang, and J. Zeng, Preparation of nickel nanopowders in ethanol-water system, *Mater. Res. Bull.*, 36, 947–952, 2001.
11. N. R. Jana, L. Gearheart, and C. J. Murphy, Evidence for seed-mediated nucleation in the chemical reduction of gold salts to gold nanoparticles, *Chem. Mater.*, 13, 2313–2322, 2001.
12. K. S. Chou and C. Y. Ren, Synthesis of nanosized silver particles by chemical reduction method, *Mater. Chem. Phys.*, 64, 241–246, 2000.
13. Z. J. Zhang, J. Zhang, and Q. J. Xue, Synthesis and characterization of a molybdenum disulfide nanoclusters, *J. Phys. Chem.*, 98, 12973–12977, 1994.
14. L. Sun, Z. J. Zhang, and Z. S. Wu, Synthesis and characterization of DDP coated Ag nanoparticles, *Mater. Sci. Eng. A*, 379, 378–383, 2004.
15. Z. J. Zhang, Q. J. Xue, and J. Zhang, Synthesis, structure and lubricating properties of dialkyldithiophosphate-modified Mo-S compound nanoclusters, *Wear*, 209, 8–12, 1997.
16. S. Chen, W. M. Liu, and L. G. Yu, Preparation of DDP-coated PbS nanoparticles and investigation of the antiwear ability of the prepared nanoparticles as additive in liquid paraffin, *Wear*, 218, 153–158, 1998.
17. Q. J. Xue, W. M. Liu, and Z. J. Zhang, Friction and wear properties of a surface-modified TiO_2 nanoparticle as an additive in liquid paraffin, *Wear*, 213, 29–32, 1997.
18. J. F. Zhou, Z. S. Wu, and Z. J. Zhang, Study on an antiwear and extreme pressure additive of surface coated LaF_3 nanoparticles in liquid paraffin, *Wear*, 249, 333–337, 2001.
19. L. Sun, J. F. Zhou, and Z. J. Zhang, Synthesis and tribological behavior of surface modified $(NH_4)_3PMo_{12}O_{40}$ nanoparticles, *Wear*, 256, 176–181, 2004.
20. J. F. Zhou, Z. S. Wu, and Z. J. Zhang, Tribological behavior and lubricating mechanism of Cu nanoparticles in oil, *Tribol. Lett.*, 8, 213–218, 2000.
21. L. Sun, Z. S. Wu, and Z. J. Zhang, Synthesis, characterization and tribological properties of sodium stearate–coated copper nanoparticles, In: *Surfactants in Tribology*, G. Biresaw and K. L. Mittal (Eds), pp. 213–220. CRC Press, Boca Raton, FL, 2008.
22. Y. Zhao, Z. Zhang, and H. Dang, Fabrication and tribological properties of Pb nanoparticles, *J. Nanopart. Res.*, 6, 47–51, 2004.
23. D. H. Napper and A. Netschey, Studies of the steric stabilization of colloidal particles, *J. Coll. Inter. Sci.*, 37, 528–535, 1971.

3 Molecular Organization and Friction Dynamics of Confined Semidilute Polymer Solutions

J. Cayer-Barrioz, D. Mazuyer, and A. Tonck

CONTENTS

The viscosity of lubricating oils is generally dependent on temperature. To reduce this dependence, additives commonly known as viscosity index improvers have been developed. Their efficiency is strongly controlled by the nature of their affinity with the base oil and their interactions with surfaces in the state of confinement. A molecular tribometer derived from a surface force apparatus has been used to characterize the mechanical and tribological behavior of two semidilute polymer solutions. At large distances, two regimes can occur:

a. The polymer can be repelled from the wall and a depletion layer is built up.
b. The polymer sticks to the wall and an adsorption layer is depicted.

The structure of these layers results in a gradient of viscosity from the wall to the bulk that can be mechanically and physically modeled. The confinement at short distances governs the tribological behavior of the polymer layer formed close to the surface. The frictional response of the interface has been observed for various normal loads up to 1 mN and sliding velocities ranging from 0.1 to 500 nm/s. In the case of the adsorbed polymer layer, the sliding stress is strongly dependent on the velocity. The behavior of this system is theoretically accounted for using a model based on the kinetics of formation and rupture of adhesion bonds between the two shearing surfaces. This approach allows the correlation of frictional properties to the molecular organization on the surfaces.

3.1 INTRODUCTION

Soluble polymers, known as viscosity modifiers, are currently added to almost all multigrade engine oils where their main role is to modify the bulk rheological properties of the fluids in which they are blended. These additives typically reduce the extent of the decrease and/or the increase in viscosity as the temperature is raised and/or lowered. Viscosity modifiers are usually high-molecular-weight polymers that change viscosity–temperature response of a base oil to increase the viscosity index of the oil [1]. On the other hand, dispersants are employed in lubricants because of their ability to interact with surfaces, other polar substances, and to keep impurities in suspension. Mixtures of conventional dispersants with polymeric viscosity modifiers are often used but such combinations are costly and may adversely affect low-temperature viscometric performance due to competitive or synergetic interactions. Therefore, multifunctional additives that provide both viscosity-modifying properties and dispersant properties, that is, dispersant-viscosity modifier (DVM), are generally prepared by functionalizing high-molecular-weight hydrocarbon polymers. The question of interactions between polymer additives and surfaces remains crucial in several technological areas since it clearly controls the structure and dynamic properties of the polymer solutions located in the vicinity of solid surfaces.

In fluid mechanics, one usually relies on the assumption that when a liquid flows over a solid surface, the liquid molecules adjacent to the solid are stationary relative to the solid and that the viscosity is equal to the bulk viscosity [2]. This no-slip boundary condition (BC) has been demonstrated in numerous macroscopic experiments [3]. Navier first proposed that a liquid may slip on a solid surface. He introduced the idea of "slip length," which currently is the most commonly used concept to quantify the slip of a liquid on a solid surface [3]. The slip length, b, is the distance beyond the liquid/solid interface at which the liquid velocity extrapolates to zero. In order to take into account possible slip at the wall, Vinogradova [4] has presented a derivation of the hydrodynamic force acting between two approaching surfaces showing that the rate of drainage of a confined Newtonian fluid is increased [3] in the case of a slip at the wall. In the case of complex fluids, Sanchez-Reyes et al. state that the occurrence of slip can arise [5] from

- The breakage of bonds between the fluid and the solid surface by modifying the substrate surface energy

- The interruption of momentum transport between the fluid adsorbed on the solid surface and that present in the bulk at stresses and/or shear rates above critical values
- An accumulation of a low-viscosity fluid at the fluid/solid interface

On the other hand, experimental evidence for adsorption on the surfaces has also been reported [6–9], and it clearly modifies the boundary condition. For example, for various DVM, soluble in oil, in rolling concentrated contacts, Smeeth et al. [10] proposed the following mechanisms of boundary film formation: at high speeds, thick-film conditions persist and the bulk polymer solution is entrained into the contact and thus forms an elastohydrodynamic film consistent with the high shear rate viscosity of the blend; at low speed where film thickness is significantly reduced, the contact inlet—or convergent zone—is filled with the adsorbed polymer layer. The localized concentration of the adsorbed polymer layer creates a higher surface viscosity relative to the bulk, and this results in a thicker-than-expected film. These adsorbed layers result in a decrease in friction. It is, therefore, clear from the above that the ability of polymers to be entrained into the contact and form boundary films depends strongly on their affinity for the surface, relative to the base oil interaction with the surface. Consequently, the performance of a lubricant depends on its dynamic properties and interaction forces with the shearing surfaces [11,12].

Extensive previous experimental [13–18] and theoretical [19–20] works have focused on the organization of various molecules onto the surfaces and their response under confinement. The friction response of a simple liquid or liquid-like lubricant squeezed in a low-pressure contact (<0.1 GPa) under steady-state sliding has also been extensively studied over the past few years [6–7,12,21–24]. The friction has been measured between mica, cobalt, or carbon surfaces separated by a nanometer-thick film. The behavior of these fluids in thin films clearly differs from their respective bulk behavior. Moreover, the topography of the surfaces and the contact area seem to determine the friction force rather than the load or contact pressure [12]. Theoretical work has also been carried out in order to model and predict the viscous friction between two polymer layers grafted on parallel solid surfaces under rubbing conditions [25]. However, this focused on polymer in the brush regime: the specific feature of the brush configuration is the high stretching of the polymer chains due to steric repulsions between monomers and this leads to a different behavior from that of the usual polymer solutions [25]. Another molecular theory of friction in confined polymer melts between atomically smooth surfaces has been developed by Subbotin et al. [26] taking polymer bridges between the surfaces into account. In addition, extensive molecular dynamics simulations have set the framework for a much better understanding of the behavior of fluids under shear [27–30]. This approach reveals a sequence of dramatic changes in the static and dynamic properties of fluid films [30] as their thickness reduces to the molecular level for spherical or short-chain molecules. The entire film may undergo ordering and a phase transition to a crystalline or glassy state under confinement and shear. However, only a few studies have focused on the correlation between friction and molecular organization of DVM molecules used in engine oils. In addition, the originality of this work lies in the combined rheological/tribological approach, that is, on the association of dynamic measurements (linear viscoelastic properties of the confined layer) with the frictional response of the interface to long sliding distances.

In this chapter, a molecular tribometer derived from a surface force apparatus has been used to investigate the drainage behavior and the sliding dynamics of two DVM layers confined between metallic surfaces [31–32]. First, the BC and the molecular organization onto the surfaces are discussed in terms of a mechanical model, composed of dashpots in series, to describe the gradient of monomers from the wall to the bulk. The validity of this approach seems to be confirmed by the theoretical model of de Gennes [19] on adsorption and depletion of semidilute polymer solutions. The effect of confinement and sliding velocity on frictional rheology is also dealt with by a general model based on the kinetics of formation and rupture of adhesion bonds within the contact zone [32–33]. This approach allows to interpret the correlation between the confinement-induced molecular organization of the polymer layer and its frictional behavior.

3.2 EXPERIMENTAL

3.2.1 POLYMER SOLUTIONS

Two blends containing 3%w/w of polymers added to 150 Neutral Solvent base oil were studied. At room temperature, the bulk viscosity of the base oil measured with a capillary viscometer n°501-13 from Schott Instruments GmbH, Mainz, with a constant of 0.02858 mm^2/s^2 at an estimated shear rate of 0.7 s^{-1} is 0.051 Pa.s and it drastically increased with the addition of the polymer molecules. The two polymer molecules used were functionalized and nonfunctionalized olefinic copolymers whose properties are summarized in Table 3.1. The nonfunctionalized polymer was a random olefin ethylene/propylene copolymer, referred to as OCP. The bulk viscosity of this blend is 0.58 Pa.s at room temperature. The molecular weight, M_n, of the copolymer is about 45 kg/mol and its molecular radius of gyration, R_g, is about 13 nm at 100°C. The intrinsic viscosity [η] of the copolymer is 0.085 L/g. Knowledge of the intrinsic viscosity [η] also allows us to calculate the overlap threshold c* [34]. c* represents

TABLE 3.1
Properties of Fluids Tested in This Work

Fluid	Polymer (%w/w) in 150 NS	c (g/L)	Molecular Weight (kg/mol)	R_g (nm)	c* (g/L)	η_b (Pa.s) from Capillary Viscometer
OCP	3	24	45	10	11	0.58
DOCP	3	24	82	6	13	0.27

Note: The concentration of polymer diluted in 150 Neutral Solvent is indicated. The radius of gyration R_g at room temperature, the critical concentration, c*, above which the polymer coils begin to overlap, the concentration of polymer, c, and the viscosity of the bulk fluid measured from capillary viscometer, η_b, are also reported.

the critical concentration above which the polymer coils begin to be densely packed and entangled. It is defined as a region of crossover between dilute ($c < c^*$) and semi-dilute ($c > c^*$) regimes [34–36]. Since $c^* = 1/[\eta] = 11$ g/L and $c = 24$ g/L, the studied blend is a semidilute solution.

The functionalized polymer was a dispersant-substituted olefin copolymer, termed DOCP. The solution has a bulk viscosity of 0.27 Pa.s at room temperature. The molecular weight, M_w, of the DOCP samples is about 82 kg/mol. Its radius of gyration, given by the Wall relation [37] $R_g = 0.0198 M_w^{0.5}$, is estimated to be 6 nm at room temperature. The intrinsic viscosity [η] of the molecule is 0.076 L/g. Since $c^* = 1/[\eta] = 13$ g/L and $c = 24$ g/L, the studied blend is also a semidilute solution.

3.2.2 MOLECULAR TRIBOMETER

The molecular tribometer used in this study has been widely described in the literature [38]. The picture of the apparatus is shown in Figure 3.1. It derives from a surface force apparatus in which a sphere can be moved toward and away from a plane in three directions (X, Y, and Z) using expansion and vibration of a piezoelectric crystal. The measurement of the surface relative displacements is performed with a resolution better than 0.01 nm in each direction using three specifically designed capacitive transducers (see [38] for detailed procedure). Three closed feedback loops are used to control the high-voltage amplifiers associated with the piezoelectric actuators. Double cantilever sensors measuring normal and tangential forces support the sample holder. Each cantilever is equipped with high-resolution capacitive transducers in spite of their low compliance (up to 2.5×10^{-6} m/N). These transducers measure the quasistatic normal and tangential forces (F_z and F_x, respectively) with a resolution up to 10^{-8} N. Two displacement closed feedback loops allow controlling the tangential displacements x and y, while the operations in the normal direction z can be carried out either in displacement or normal force control. Thermal drift is taken into account by comparing the variation of the capacitance with distance D during loading and unloading. Regarding the force measurement, only a little thermal drift is deduced by measuring the force while at very large distances.

An oscillating normal and/or tangential motion can be superimposed on the linear normal (and/or tangential) displacement. For this purpose, appropriate amplitudes and frequencies are chosen in order to avoid microslip and disturbance during

FIGURE 3.1 Picture of the molecular tribometer derived from a surface force apparatus. The inset shows the sphere/plane contact with a droplet of oil solution.

the squeeze and/or sliding process. Moreover, the modulation frequencies differ in the normal and the tangential directions in order to avoid system resonance. The resulting dynamic displacements and forces are measured with double-phase lock-in amplifiers resulting in the normal and tangential mechanical transfer functions of the interface between the sphere and the plane. From these signals, the elastic and viscous properties of the interface can be derived in both normal and tangential directions [38]. The mechanical impedance of the interface in normal and tangential directions is divided into two additive components: the first is the conservative part coming from the in-phase response of the interface, which gives its elastic stiffness $K(\omega)$, and the second is the dissipative part coming from the out-of-phase response of the interface, which gives its viscous damping $A(\omega)$.

3.2.3   PROCEDURE

A pair of solid surfaces (a millimeter diameter sphere and a plane), consisting of a fused silicate glass and a silicon wafer, respectively, was prepared. The sphere was manufactured from a drop of melted glass to have a perfect control of the radius over the whole surface of the probe. The radii of the spheres R were, respectively, 2.30 mm for the OCP experiments and 1.68 mm for the DOCP experiments. Both surfaces were coated with a 50-nm-thick metallic cobalt layer using a sputtering technique. Despite the presence of a very thin oxide layer of about 0.3 nm thick on each surface, both surfaces were highly conductive. Therefore, they permitted measurement of the electrical capacitance of the contact that was used to establish the zero of the displacement scale according to Mazuyer et al. [38] as follows. When the surfaces are far from each other (at a distance longer than 10 times the size of the adsorbed molecules), their elastic deformation can be neglected and the normal displacement, Z, is the sphere/plane distance, D. The electrical capacitance of the contact C follows the relation:

$$\frac{dZ}{dC(Z)} = \frac{dD}{dC(D)} = \frac{D}{2\pi\varepsilon\varepsilon_0 R} \tag{3.1}$$

where ε is the dielectric constant of the liquid, ε_0 is the permittivity of vacuum, and R is the sphere radius. The straight line that extrapolates the $dZ/dC(Z)$ curve to a short distance intercepts the displacement, Z axis at a point corresponding to the location where the metal surfaces should touch. This point, which is never reached because of the confinement effect, is referred to as the origin of the contact.

For the cobalt-coated plane, over a scan length of 1 µm, the surface cumulative peak to valley roughness, measured with an atomic force microscope (AFM), did not exceed 0.5 nm. In comparison with the thickness of the adsorbed layers, the solid surfaces can be considered as very smooth.

When the distance between the sphere and the plane reached 10 µm, a droplet of liquid was deposited between the two surfaces. Before the actual experiments, an adsorption time of 20 h was allowed at this distance.

With this technique, the contact area is not directly measured. Thus, the contact dimensions that give the size of the mechanical probe should be estimated through an appropriate modeling of the contact mechanics (see [38] for a detailed description). As far as the short sphere/plane distances are concerned, the confinement of the fluid molecules is responsible for the elastic deformation of the solids. Nonadhesion or adhesion contact theory can be used to estimate the contact radius, assuming that the confined interface behaves as a rigid wall. As a consequence, the sphere/plane distance D, taking into account the elastic deformation of the solids, can be calculated [38].

From an initial operating distance of 1 μm, quasistatic force was measured as a function of D by making inward and outward motions at a constant speed of 1 nm/s. Small reloading processes showed that the normal speed was slow enough to avoid any disturbance from the viscous force. This quasistatic squeeze of the lubricant within the contact allowed to determine the thickness of the surface layers [38]. The loading–reloading experiments also served as a systematic control for examining the adsorbed layer integrity following shear measurements.

In addition, frictional experiments were performed at constant normal load, in the range 0.1–1 mN, and for various sliding distances and velocities (from 0.1 to 500 nm/s).

Dynamic measurements were superimposed in both normal and sliding directions at respective frequencies of 38 and 70 Hz with respective vibration amplitudes of 0.1 and 0.03 nm. The latter permit to link the evolution of the interface stiffness to the friction dissipation and velocity accommodation processes.

All the experiments presented in this work were performed at room temperature (23°C).

3.3 RESULTS AND DISCUSSION

3.3.1 INTERFACE CHARACTERIZATION

In the dynamic mode, via superimposition of an oscillatory motion of given amplitude and pulsation ω, the viscous damping $A_z(\omega)$ of the interface was measured for a squeeze velocity of 1 nm/s. The plot of $1/(\omega_z \cdot A_z)$ as a function of D, at a frequency of 38 Hz and oscillatory amplitude of 0.1 nm, is presented in Figure 3.2a for both polymer solutions. For large sphere/plane distances and with a homogeneous Newtonian liquid, the Stokes' law describes the hydrodynamic flow under the no-slip BC, that is to say that the velocity of the fluid near the solid surface is zero. The associated damping function A_z is given by

$$A_z = \frac{6\pi\eta R^2}{D} \tag{3.2}$$

where D is the sphere/plane distance, η is the bulk viscosity of the liquid, and R is the sphere radius.

The presence of adsorbed layers on the two solid surfaces drifts the wall where the flow velocity vanishes toward a distance, $2L_H$, over the solid surface. This defines

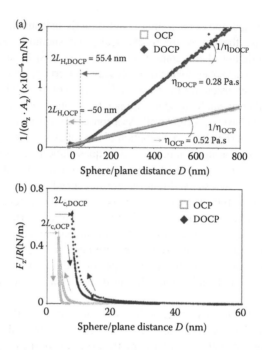

FIGURE 3.2 Variation of the (a) reciprocal of the damping function $\omega_z \cdot A_z$ and (b) the normal force F_z/R versus the sphere–plane distance D during a cycle of quasistatic compression and decompression of the interface at a normal squeeze velocity of 1 nm/s for OCP ($\square$) and DOCP ($\blacklozenge$). For large distances, that is, $D > 300$ nm, the bulk viscosity is measured according to Equation 3.3 or 3.4 for both polymer solutions and the values are consistent with those measured with a capillary viscometer. The boundary conditions and the molecular organization in the vicinity of the surfaces can be discussed from the evolution of the reciprocal of the damping function. Marked hysteresis is observed between compression and decompression of both interfaces. Characteristic layer thicknesses are determined and reported in Table 3.2.

the infinite viscosity layer within which the molecules are not perturbed by the flow [38,39]. Thus, considering this new boundary condition, the damping function becomes

$$A_z = \frac{6\pi\eta R^2}{D - 2L_H} \tag{3.3}$$

In the case of slip at both walls, the analysis of hydrodynamics shows that the wall of no-slip is moved from a distance $2b$ below the solid surfaces. Thus, considering the new boundary condition, Equation 3.2 becomes

$$A_z = \frac{6\pi\eta R^2}{D + 2b} \tag{3.4}$$

The plot of $1/(\omega_z \cdot A_z)$ as a function of D, presented in Figure 3.2a, allows to determine the bulk viscosity of the lubricant and the hydrodynamic thickness L_H or the possible slip length b. At large distances, that is, $D > 300$ nm, for OCP, the linear

TABLE 3.2

Characteristic Layer Thicknesses at First Repulsion (2L), Hydrodynamic (2L_H), and Under Various Confinements (2L_c), Surface Coverage Ratio (L_H/L), and Bulk Viscosity Obtained from Dynamic Measurements (η_d)

Polymer	2L (nm)	2L_H (nm)	2L_c (nm at 0.1 mN)	2L_c (nm at 1 mN)	L_H/L	η_d (Pa.s) from Figure 3.2a
OCP	22		6	4		0.52
DOCP	80	55	11	9	0.68	0.28

slope of the curve gives a viscosity of 0.52 Pa.s at 38 Hz corresponding to a shear rate of 0.025 s^{-1}, which is consistent with the value of the bulk viscosity (0.58 Pa.s) determined with the capillary viscometer. Besides, the linear slope intersects the horizontal axis at a negative value of −50 nm. According to Equation 3.4, this value can be directly interpreted [40,41] as twice the opposite of the slip length b on each surface. However, the value of the slip is rather large.

The slope value in the case of DOCP gives a viscosity of 0.28 Pa.s, also consistent with the value determined with the capillary viscometer of 0.27 Pa.s. In addition, the best fit of the curve, at large distances, intercepts the distance axis at 2L_H, which defines the hydrodynamic thickness of about 55 ± 0.4 nm. This would tend to demonstrate the occurrence of adsorption of DOCP molecules on the surfaces despite a large value of L_H compared to the radius of gyration R_g of the molecules. Values of viscosity and L_H are reported in Table 3.2 for both polymers. For shorter distances, the evolution of the normal force as a function of the sphere/plane distance D is presented in Figure 3.2b and discussed in the following sections.

3.3.1.1 Effect of OCP on Interface Slip

In the case of OCP, no attraction was detected during loading. Marked hysteresis was observed between the loading and unloading curves. During unloading, an attractive force was measured. Its origin may be related to bridging [42]. After some time of contact and confinement, the chains might establish bridges between the two surfaces. When pulling apart the two surfaces, these bridges should retract and act as elastic springs since the bridging kinetics is commonly slow compared to the unloading velocity. Thus, the corresponding adhesion force of about −2.8 × 10^{-6} N was taken into account in the contact model to determine the parameters of the elastic contact versus normal force F_z, that is, contact radius, mean contact pressure, and solid elastic deformation within the framework of the Johnson–Kendall–Roberts theory. The thickness of the adsorbed layer on each surface before and after confinement, as defined in [38], is reported in Table 3.2. 2L represents the distance of onset of repulsion and can be considered as twice the radius of gyration, R_g, of the copolymer molecules; therefore, R_g is about 10 nm at room temperature. This value at 23°C is in good agreement with the value at 100°C (13 nm). 2L_c, the distance under confinement, is measured at a normal load of 10^{-3} N corresponding to a mean contact pressure of 30 × 10^6 Pa and is about 4 ± 0.6 nm. $L \gg L_c$ shows that the confinement is progressive and slow.

Consequently, this leads to a contradictory result as it can be concluded that polymer chains remain confined within the contact. In order to confirm the occurrence of an effective slip at the wall in the experiments presented, the most accredited slip mechanisms have been looked at in detail. It should be noted that the slip of concern in this chapter is that of Newtonian liquids and differs from that of sheared polymer melt systems. The latter is supposed to be due to disentanglement of polymer chains attached to the surface from those in the bulk liquid [5,43,44] and it occurs at stresses and/or flow rates above critical values. This assumption relies on the fact that the OCP solution behaves as a Newtonian fluid for the shear rates considered as discussed in the following. To explain "slip at the wall," molecules of liquid are considered to slip directly over the solid surface [3–4] because of (a) stronger viscous friction between liquid molecules at the interface than between liquid molecules and solid surfaces or (b) the ratio between dimensions of the liquid molecules and surface roughness (if the dimensions of the liquid molecules are of comparable size to the roughness periodicity, the molecules can be trapped in the surface pits [45]; if their dimensions are much smaller or much larger, they may slide over the surface [3]).

However, neither model is believed to be universally correct [3] and many questions regarding slip mechanisms remain unanswered.

Several parameters are known to affect slip occurrence, including surface wettability, surface roughness, presence of gaseous layers, nanobubbles or impurities, shear rate, and so on. The influence of each of these parameters is discussed in the next section.

A 3 μL droplet of the studied blend was deposited on a cobalt-coated silicon wafer. A contact angle of 19° was measured using a PG-X measuring head from The Pocket Goniometer Company. This demonstrates that the lubricant wets the solid surface.

Boundary slip is clearly linked to surface roughness. On an atomic scale, roughness will influence the number of neighboring surface atoms with which an interfacial lubricant molecule can interact. On a larger scale, roughness will alter the pattern of liquid flow [46,47]. However, there is no general agreement on the influence of roughness on slip. In this work, over a scan length of 1 μm, the cumulative peak to valley surface roughness did not exceed 0.5 nm. Relative to the thickness of the adsorbed confined layers ($2L_c = 4$ nm), the solid surfaces can be considered as very smooth.

Moreover, before the droplet of lubricant was introduced into the contact zone, a preliminary approach of the sphere toward the plane (only few nanometers under nitrogen-controlled atmosphere) was performed to check the cleanliness of the substrate and to assure that no long-range force was measured between the prepared surfaces. A number of features also clearly oppose an interpretation involving nanobubbles in the contact area [41]:

• First, there was no normal force signal for distances larger than 30 nm
• Second, no discontinuities in the normal forces were observed within the experimental resolution
• Third, the experiments were highly reproducible

Regarding the shear rate influence on slip, one may consider shear rates induced by the squeeze experiments. It is possible to estimate the squeeze-induced shear rate corresponding to a squeeze velocity of 1 nm/s: it remains less than 1 s^{-1}. For this range of shear rates, bulk rheological measurements show that the OCP solution behaves as a Newtonian fluid. After this detailed examination of the most accredited slip mechanisms, the occurrence of effective slip at the wall seems doubtful for OCP solution.

3.3.1.2   Effect of DOCP on Interface Organization

In the case of DOCP, no adhesion was observed during loading as shown in Figure 3.2b. It shows little hysteresis between the loading and unloading parts of the curve and reveals that the interactions between the compressed layers are purely repulsive. This behavior is in agreement with the work of Klein et al. [48] and with the theoretical prediction of de Gennes [42] when equilibrium is reached. The distance of onset of repulsive interaction corresponds to $2L$ of about 80 ± 0.6 nm, where L can be seen as the thickness of the adsorbed layer of polymer on each surface before confinement. As the confinement increases, the normal force increases and the thickness of each adsorbed layer decreases from L to L_c, the thickness of the confined layer on each surface. L_c measured at a normal load of 0.1 mN, corresponding to a mean contact pressure of about 13×10^6 Pa, is about 5.5 ± 0.4 nm while it is equal to 4.5 ± 0.4 nm for a normal load of 1 mN, that is, a mean contact pressure of 34×10^6 Pa. From the values in Table 3.2, it can be deduced for the DOCP solution that

- $L \gg L_c$ indicates that the confinement is progressive and slow.
- The thickness of each adsorbed layer before confinement $L = 40$ nm is much larger than the radius of gyration R_g of the DOCP molecules that has been estimated at 23°C around 6 nm. It can be assumed that at low confinement, the solvent molecules remain within the contact.
- As confinement is increased, the distance L reaches the value of $L_c = 5.5 \pm 0.4$ nm at $F_z = 0.1$ mN and then the value of $L_c = 4.5 \pm 0.4$ nm at $F_z = 1$ mN. One may suppose that the solvent molecules are finally squeezed out and that the DOCP polymer chains remain confined within the contact at severe confinement.
- The hydrodynamic layer L_H on each surface is measured at 27.5 nm. It appears to be much larger than the diameter of the free coil.
- The surface coverage ratio L_H/L is rather low, at about 0.68.

This preliminary description of the DOCP interface seems insufficient; in particular, the high value of hydrodynamic thickness L_H is inconsistent with the diameter of the coil.

3.3.2   Modeling of Molecular Organization

In order to account for these results, a simple mechanical model of the interface has been proposed [49]. In this model, the confined interface is composed of a bulk viscosity fluid film squeezed between two thin viscous layers close to the immobile hydrodynamic layer L'_H (when it exists) adsorbed on each solid. The total thickness of this viscous layer is $2h_0$ and its viscosity, η_0, is lower or higher than that of the bulk lubricant, η. This allows to simulate the "apparent slip" phenomenon [3–4,47] when $\eta_0 < \eta$ (in the absence of hydrodynamic layer on the surface $L'_H = 0$), that is to

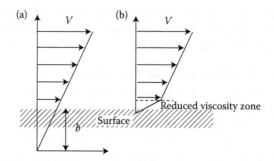

FIGURE 3.3 Schematic representations of (a) partial slip boundary condition and (b) apparent slip. The velocity of the liquid (V) extrapolates to 0 at a distance b inside the wall due to a reduced viscosity zone in the vicinity of the solid surface depicted using the shaded area.

say that the velocity profile of the liquid extrapolates to 0 at a distance inside the wall because of the existence of a reduced viscosity zone near the surface (see Figure 3.3). On the other hand, this permits to describe the gradient in adsorption from the wall to the bulk when $\eta_0 > \eta$ and L'_H exists.

The interface is modeled using two dashpots in series as depicted in Figure 3.4a for the case of OCP and Figure 3.4b for the case of DOCP. The damping function, a_1 and a_2, of the dashpot 1 and 2, respectively, is given by [49]

$$a_1 = \frac{6\pi \, \eta \, R}{(z - 2h_0)^2} \cdot r \, dr \tag{3.5}$$

$$a_2 = \frac{6\pi \, \eta_0 \, R}{2h_0(2z - 2h_0)} \cdot r \, dr \tag{3.6}$$

where h_0 is the thickness of the layer with the modified viscosity on each solid surface. The equivalent dashpot is

$$a_z = \frac{6\pi \, \eta \, R \, r \, dr}{z^2 - z \cdot 4h_0 \cdot (1 - \eta/\eta_0) + 4h_0^2 \cdot (1 - \eta/\eta_0)} \tag{3.7}$$

Taking into account the possible existence of the hydrodynamic layer of thickness L'_H and for $D \gg h_0$, the damping function A_z, which results from the integration of the elementary dashpot, a_z, over the whole contact becomes [49]

$$A_z = 6\pi\eta R^2 \int_{D-2L'_H}^{+\infty} \frac{dz}{\left[z - 2h_0 \cdot (1 - \eta/\eta_0)\right]^2} = \frac{6\pi \, \eta \, R^2}{D - 2L'_H - 2h_0 \cdot (1 - \eta/\eta_0)} \tag{3.8}$$

Equation 3.8 is used to fit data in Figure 3.2a. This allows to estimate η, η_0, h_0, and L'_H.

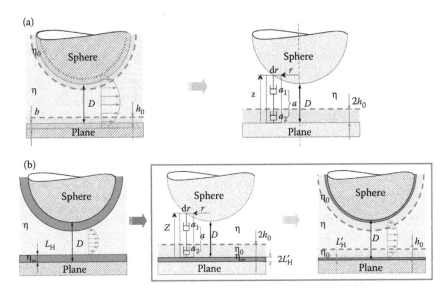

FIGURE 3.4 Modeling of the multilayer interface between a sphere and a plane. Each layer and viscosity are represented by a dashpot of damping function a_1 and a_2 in a ring of thickness dr. The integration of the overall damping function a_z over the contact allows to simulate the existence of a layer of a (a) reduced and (b) higher viscosity near the surface. D is the distance between the sphere and the plane in the z direction. h_0 represents the thickness of the near-surface layer of viscosity η_0 while the bulk layer has a viscosity η. b is the slip length while the hydrodynamic layer has a thickness L'_H and a viscosity η_∞. ((a) From J. Cayer-Barrioz et al. *Tribol Lett* 32, 81–90, 2008 and (b) from J. Cayer-Barrioz et al. *Langmuir* 25, 10802–10810, 2009. With permission.)

3.3.3 DESCRIPTION OF THE DEPLETION LAYER: OCP SOLUTION

In the case of the OCP solution, the model links the apparent slippage effect with a decrease in the viscosity of the boundary layer close to the surface. Consequently, there is no hydrodynamic layer: $L'_H = 0$.

At large distances, far from the surfaces, the viscosity η of the lubricant is equal to that of the bulk, that is, $\eta = 0.52$ Pa.s. In the vicinity of the surfaces, a layer of reduced viscosity, $\eta_0 = 0.28$ Pa.s can be estimated from the slope of the curve $1/(\omega_z \cdot A_z) = f(D)$ at small distances. According to Equation 3.8, the thickness, h_0, of this reduced viscosity layer is approximately 20 nm. Considering the interface as a biviscous layer with reduced viscosity from the surface to the bulk permits to model the whole experimental curve $1/(\omega_z \cdot A_z) = f(D)$ as shown in Figure 3.5.

The measured signal of the tangential viscous response is too small and does not allow to determine the slip length. The measured signal corresponds to the whole contact and it is more sensitive to the less viscous part if one describes the interface with dashpots in series. It can, therefore, be concluded that the viscosity of the contact in the tangential direction is very low. This also confirms the reduced viscosity layer interpretation.

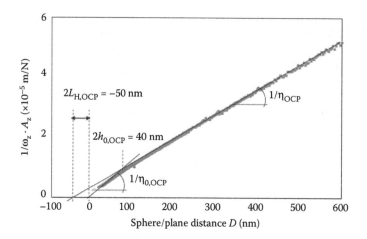

FIGURE 3.5 Application of the multilayer interface model (Equation 3.8) to the OCP solution. The bulk viscosity (η_{OCP}), the surface layer viscosity ($\eta_{0,OCP}$), and its thickness ($2h_{0,OCP}$) are determined. This approach allows to fit the whole experimental curve and to estimate $2L_{H,OCP}$.

Nevertheless, even if the former model allows to accurately simulate the variation of the damping function with the sphere/plane distance, questions regarding the origin of the molecular segregation in the vicinity of the solid surfaces and the way the molecular organization of the OCP controls the properties of the reduced viscosity surface layer remain unanswered.

It is well known that when a fluid separating two solids contains high-molecular-weight polymer molecules, these large entities may be excluded from a thin region of fluid between the surfaces [50–52]. This phenomenon is referred to as depletion. The surface energy, γ_0, of the solvent is 25 mJ/m². The surface energy, γ_p, of the ethylene/polypropylene copolymer is 31 mJ/m² [35]. Thus, depletion can be assumed to take place at the wall/polymer interface and OCP molecules are considered to be repelled from the surface over a distance ξ. A physical model has been proposed by de Gennes [19,42] to describe the depletion layer for a semidilute polymer solution near a solid wall by using a free sticking energy, γ_1, which is negative in the case of a repulsive wall. This approach is based upon the construction of concentration profiles, which divides the adsorbed polymer layer into three regions (see Figure 3.6).

The proximal region close to the surface has a thickness, e, given by

$$e = \frac{1}{a^2}\left(\frac{kT}{|\gamma_1|}\right)^{3/2} \tag{3.9}$$

where a is the size of one monomer estimated as 0.2 nm.

The central region, which has a thickness equal to that of the depletion layer, extends to a distance, ξ_b from the wall. ξ_b is the correlation length of the polymer solution given by

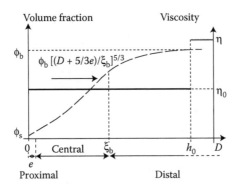

FIGURE 3.6 Qualitative profile of the monomer concentration ϕ (in broken line) for a repulsive wall as a function of the distance from the wall, D. Three distinct regions are depicted: the proximal region for $D < e$, the central region for $e < D < \xi_b$, and the distal region, for $D > \xi_b$ and $\phi = \phi_b$. The corresponding schematic evolution of the modeled viscosity is plotted in solid line as a function of D with η, the bulk viscosity and η_0, the surface layer viscosity such as $\eta > \eta_0$. (From J. Cayer-Barrioz et al. *Tribol Lett* 32, 81–90, 2008. With permission.)

$$\xi_b = a \cdot \phi_b^{-3/4} \tag{3.10}$$

where ϕ_b is the volume fraction of the monomer in the bulk solution. The depletion length, δ, is then

$$\delta = \xi_b - e \tag{3.11}$$

The distal region where the concentration tends to that of the bulk solution extends over a distance from the wall is more than ξ_b. Equations 3.9 through 3.11 are valid if $a \ll e \ll \xi_b$. The bulk monomer volume fraction, ϕ_b, is 6.10^{-3}, which gives $\xi_b = 9.4$ nm. Assuming that γ_i is due to short-range Van der Waals interaction with a Hamaker constant of about 5×10^{-19} J, we find that the size of the proximal zone is $e = 3.5$ nm and that the sticking energy, $|\gamma_i|$ is 15 mJ/m². It can be verified that $|\gamma_i|$ is less than kTa^{-2}, which ensures that the surface concentration is much lower than 1. According to these calculations, the value of the depletion length is 5.9 nm.

This theoretical approach makes valid the hypothesis of depletion at the surface, with the depletion length of the same order of magnitude as R_g. The gradient of concentration of monomers results in a gradient of viscosity from the surface to the bulk over a distance corresponding to the thickness of the depletion layer, δ, which is of the same order of magnitude as the thickness of the reduced viscosity layer that has been observed experimentally ($h_0 = 20$ nm). This gradient of viscosity can, therefore, be modeled as a two-viscosity layer organization near the solid surface. Therefore, at large sphere/plane distances (i.e., $D \gg 2h_0$), this confirms that the interface can be viewed as a biviscous film (see Figure 3.7). Moreover, the very high value of concentration $c > c^*$ indicates that the nondilute regime is reached in the bulk layer [34–36]. The polymer coils interpenetrate each other enough so that the molecular

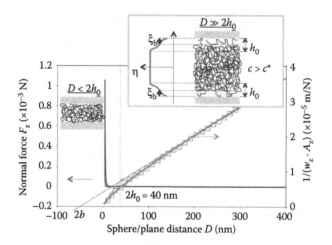

FIGURE 3.7 Schematic of the OCP interface versus D compared to the thickness of lower viscosity layer $2h_0 = 40$ nm. The variation of the viscosity η with the distance D is schematically indicated where the bulk concentration, c, is higher than the semidilute critical concentration, c^*. The correlation length of the polymer solution is noted ξ_b. This is also associated with the evolution of the normal force F_z/R in solid line and the reciprocal of the damping function, $w_z \cdot A_z$, ($\square$). For large distances, a depletion length is estimated at $\delta = 9$ nm. At short distances, the weak depletion effect is suppressed and OCP molecules remain confined at the interface. (From J. Cayer-Barrioz et al. *Tribol Lett* 32, 81–90, 2008. With permission.)

motions of one chain are greatly slowed by the interfering effects of other chains. These interferences are attributed to molecular entanglements. As a consequence, at large distances D, the interface can be schematically described as in Figure 3.7.

At small distances ($D < 20$ nm), the contact pressure increases. Owing to the slight difference between the surface energy of the solvent and that of the polymer, the depletion effect is weak and it can be assumed that the solvent molecules are squeezed out of the contact and that the polymer chains are compressed onto the surfaces due to the increased confinement (as depicted in Figure 3.7). Moreover, from the measurements of normal and tangential stiffnesses plotted in Figure 3.8, compressive and shear elastic moduli (E and G, respectively) can be estimated as described in detail elsewhere [38]:

$$E = \frac{K_z}{2a_c} \tag{3.12}$$

$$G = \frac{K_x \cdot D}{\pi a_c^2} \tag{3.13}$$

where K_z and K_x are, respectively, the normal and tangential stiffnesses, a_c is the contact radius calculated from the JKR theory. The compressive elastic modulus, E, is about 2 GPa while the shear elastic modulus, G, is approximately 80 MPa. As a consequence, the confined interface can be seen as a single layer of glassy polymer [31].

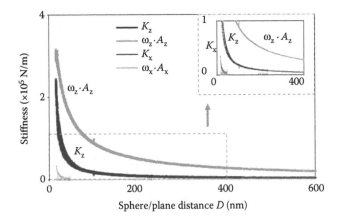

FIGURE 3.8 Variation of the normal (respectively tangential) in-phase K_z (respectively K_x) and out-of-phase, $\omega_z \cdot A_z$ (respectively $\omega_x \cdot A_x$) parts of the OCP interface mechanical properties with D at 38 Hz (respectively 70 Hz). The insert highlights the much lower tangential properties compared to the normal ones. (From J. Cayer-Barrioz et al. *Tribol Lett* 32, 81–90, 2008. With permission.)

3.3.4 Description of the Adsorption Layer: DOCP Solution

In the case of the DOCP solution, Equation 3.8 is used to fit data from Figure 3.2a. This leads to estimate η, η_0, and L'_H. At large distances, far from the surfaces, the viscosity η of the lubricant is equal to that of the bulk, that is, $\eta = 0.28$ Pa.s. In the vicinity of the surfaces, a homogeneous layer of enhanced viscosity, $\eta_0 = 0.50$ Pa.s can be calculated from the slope of the curve $1/(\omega_z \cdot A_z) = f(D)$ at small distances by using the best linear fit for the viscosity. The thickness of the immobile hydrodynamic layer L'_H is deduced from the intercept of this slope with the axis, as reported in Figure 3.9. Derived from Equation 3.8, the thickness of the higher viscosity layer, h_0, can be calculated as

$$h_0 = \frac{L_H - L'_H}{1 - \eta/\eta_0} \tag{3.14}$$

with $L_H = 27.5$ nm, $L'_H = 8$ nm, $\eta = 0.28$ Pa.s, and $\eta_0 = 0.50$ Pa.s. It is noteworthy that L'_H is of the same order of magnitude as the diameter of the coil [42]. This gives $h_0 = 44$ nm. Considering the interface as a multilayer medium from the surface to the bulk results in modeling the whole experimental curve $1/(\omega_z \cdot A_z) = f(D)$ (see Figure 3.9).

Once again, the former model allows to simulate accurately the variation of the damping function with the sphere/plane distance. Regarding the physical origin of the molecular segregation, de Gennes [19] describes the adsorption layer for a semidilute polymer solution near a solid wall using a free sticking energy, γ_1, which is positive in the case of an attractive wall. Concentration profiles that divide the adsorbed polymer layer into three regions [19] can be built.

The proximal region close to the surface and where the effects of short-range forces between a monomer and the wall are important has a thickness, e, still given

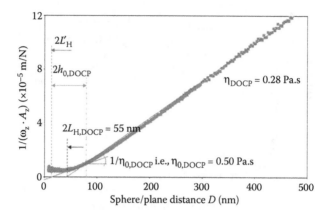

FIGURE 3.9 Application of the multilayer interface model to the DOCP solution (see Equation 3.8). The bulk viscosity (η_{DOCP}), the surface layer viscosity ($\eta_{0,DOCP}$), and its thickness ($2h_{0,DOCP}$) as well as the thickness of the hydrodynamic layer ($2L'_H$) are determined that is of the same order of magnitude than the radius of gyration of the molecule.

by Equation 3.9 with a the size of one monomer estimated at 0.2 nm. In this region, the concentration is higher than the bulk and the monomer volume fraction near the wall, ϕ_s, is given by

$$\phi_s = \left(\frac{a}{e}\right)^{\frac{4}{3}} \gg \phi_b \tag{3.15}$$

The central region, which has a thickness equal to that of the adsorption layer, extends to a distance, ξ_b from the wall. ξ_b is the correlation length of the polymer solution such as defined in Equation 3.10 with ϕ_b the volume fraction of the monomer in the bulk solution. In this region, the concentration profile becomes independent of the bulk concentration and varies with the distance D from the wall as

$$\phi = \left(\frac{a}{D + \frac{4}{3} \cdot e}\right)^{\frac{4}{3}} \tag{3.16}$$

The distal region where the concentration tends to that of the bulk solution is larger than ξ_b. In this region, the concentration relaxes exponentially toward the bulk value [19]. Equations 3.9, 3.10, 3.15, and 3.16 are valid if $a \ll e \ll \xi_b$.

Assuming that the sticking energy $|\gamma_i|$ is due to short-range Van der Waals interaction [19], a value for $|\gamma_i|$ can be estimated as 20 mJ/m². Moreover, assuming that the size of the monomer a is 0.2 nm, the proximal zone is $e = 2$ nm, and the bulk monomer volume fraction, ϕ_b is 4.8×10^{-3}, we obtain $\xi_b = 10$ nm. This value is consistent with L'_H and R_g as predicted theoretically by de Gennes [19].

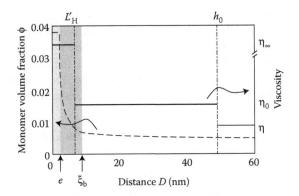

FIGURE 3.10 Quantitative plot of the monomer volume fraction ϕ (in broken line) as a function of D for the DOCP solution. The corresponding modeled viscosity (in solid line) as well as the hydrodynamic layer of thickness $L'_H \sim \xi_b$ the correlation length $\sim R_g$ the radius of gyration is also presented. η is the bulk viscosity while η_0 is that of the increased viscosity layer and η_∞ corresponds to the infinite viscosity of the hydrodynamic layer of thickness L'_H. e defines the proximal zone. (From J. Cayer-Barrioz et al., *Langmuir* 25, 10802–10810, 2009. With permission.)

According to these calculations, the value of the surface polymer volume fraction, ϕ_s, is $0.04 \gg \phi_b$. The quantitative profile of the monomer volume fraction is represented in Figure 3.10 as a function of the distance D from the surface and the three zones are depicted. The corresponding modeled evolution of the viscosity is also plotted.

The number of monomers in contact with the wall per unit area Γ is given by [19]

$$\Gamma = a^{-2} \cdot \phi_s = s^{-2} \tag{3.17}$$

where s is the mean interanchor spacing. Equation 3.17 gives $\Gamma = 10^{18}$ monomers/m^2 and $s = 1$ nm. These values are consistent with those found in the literature for various polymers [24,48,53]. In addition, assuming a good solvent condition, it is also possible to estimate the number of polymer chains per unit area, $\Gamma_{polymer}$, as follows:

$$\Gamma_{polymer} \approx \frac{\Gamma}{N} = \Gamma \cdot \left(\frac{a}{R_g}\right)^{5/3} = d^{-2} \tag{3.18}$$

where N is the number of monomers per chain, and d is the mean distance between polymer chains. Equation 3.18 gives $\Gamma_{polymer} \sim 4 \times 10^{15}$ polymers/m^2 and $d \sim 16$ nm. This allows to calculate the surface coverage ratio S as

$$S = \frac{\pi \cdot R_g^2}{d^2} = 0.4 \tag{3.19}$$

This value is lower than the previous calculated coverage ratio ($L_H/L = 0.68$). The discrepancy might be an indication that the initial measurement of L might be erroneous. It can be assumed that solvent molecules remained in the contact at low confinement and L_H represents an overestimate of the real thickness of the hydrodynamic layer as demonstrated by the two-viscosity-layer model. Moreover, the assumed good solvent condition might not hold here at the high monomer concentration at the interface.

3.3.5 NANORHEOLOGY OF THE ADSORBED LAYER UNDER CONFINEMENT

In the dynamic mode, using an oscillating normal and tangential motion, superimposed on the linear normal displacement during the squeeze of the interface, allows to measure the interface viscoelastic properties in both normal and tangential directions. This assumes that amplitudes and frequencies of oscillation are appropriately chosen (0.1 nm and 38 Hz in the normal direction and 0.03 nm and 70 Hz in the tangential direction) not to disturb the friction process. Figure 3.11 presents the variation of the elastic stiffness and the viscous damping in the normal and tangential directions versus the sphere/plane distance D. It can be seen in the insert that the solution is clearly viscoelastic at large distances since the ratio of viscous damping to elastic stiffness is about 2.

After some time of contact, one could expect some bridging process between the polymer chains from one surface with the other surface. However, no attractive normal force was detected while unloading at the slow squeeze velocity (1 nm/s) [18,42]. No hysteresis was observed in the normal stiffness. Another way to detect bridging is to observe the evolution of tangential stiffness K_x. It slowly increases from $D = 80$ nm during the loading process but it reaches zero at D = 20 nm after unloading. No hysteresis was observed in the normal stiffness. Under these conditions, one may conclude absence of bridging after normal loading.

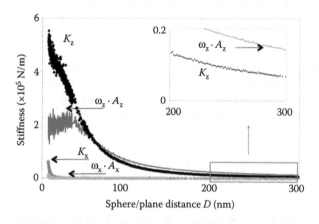

FIGURE 3.11 Evolution of the normal (respectively tangential) in-phase K_z (respectively K_x) and normal (respectively tangential) out-of-phase, $\omega_z \cdot A_z$ (respectively $\omega_x \cdot A_x$) parts of the DOCP interface mechanical properties at 38 Hz (respectively 70 Hz). The mechanical impedance is much higher in the normal direction than in the tangential direction. As shown in the inset, this polymer solution is clearly viscoelastic since the ratio $\omega_z \cdot A_z/K_z$ is 2. (From J. Cayer-Barrioz et al. *Langmuir* 25, 10802–10810, 2009. With permission.)

The model of adsorption developed by de Gennes [19] describes the interface as a self-similar structure. At distances $D > \xi_b$ from the wall, in the distal zone, the solution can be pictured as a random network of mesh ξ_b; for $D < \xi_b$, the mesh size gets smaller but remains of the order D. Figure 3.12 presents the evolution of the tangential stiffness K_x versus the distance D within the different zones (proximal, central, and distal) taking into account the fact that in this experiment, the two surfaces were covered with adsorbed polymer chains and both obeyed the de Gennes' model. A fit of the variation of the stiffness with distance shows that K_x varies as D^{-3} for an interface thickness smaller than 40 nm. Moreover, assuming that the interface is homogeneously sheared over all its thickness in its elastic domain, the ratio of the shear elastic modulus of the interface G over the contact pressure P can be deduced from the relation [38] and experimental measurements as

$$\frac{G}{P} = \frac{K_x \cdot D}{F_z} \tag{3.20}$$

Thus, the influence of confinement on mechanical properties can be quantified. The shear elastic modulus G of the interface is about 12×10^6 Pa under a normal load F_z of 0.1 mN. Its value increases to 15×10^6 Pa at a higher confinement under a normal load of 1 mN.

To conclude, confinement has a profound effect on molecular organization and mechanical properties of adsorbed layers. Molecule adsorption on the surfaces leads to a gradient of concentration that results in a gradient of properties in terms of viscosity, as depicted by the two-viscosity-layer model, and also in a gradient of elastic properties. Confinement induces a reduction of the mesh size, which explains the increase in mechanical properties [32].

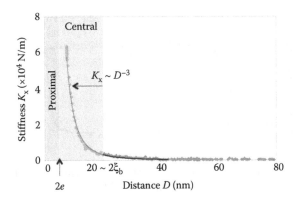

FIGURE 3.12 Plot of the tangential stiffness K_x versus D for the DOCP confined interface. The three zones defined by the de Gennes' adsorption model [19] are also indicated. Moreover, the fact that the two surfaces brought into contact, are covered with adsorbed polymer chains is taken into account. In the central zone, K_x varies with D^{-3}. $2e$ corresponds to the proximal zone. (From J. Cayer-Barrioz et al., *Langmuir* 25, 10802–10810, 2009. With permission.)

3.3.6 FRICTIONAL RESPONSE

Sliding friction experiments were carried out on the OCP solution at a constant normal force varying from 0.01 to 1 mN, for a sliding distance of 100 nm and sliding velocities ranging from 0.2 to 10 nm/s. In these conditions, the steady-state friction coefficient reaches a high value of 0.3 independently of the normal force or sliding velocity as shown in Figure 3.13 for $F_z = 0.3$ and 1 mN. This friction value is consistent with the description of the confined OCP interface as a glassy layer [31].

A completely different behavior was observed in the case of the adsorbed layer of DOCP at constant normal force F_z of 0.1 and 1 mN and varying sliding velocities from 0.1 to 500 nm/s. Figure 3.14 shows the evolution of the friction force versus sliding distance X at a constant normal load of 1 mN and the kinetics of its variations when a sequence of increasing and decreasing sliding velocities from 2 to 120 nm/s is applied. The corresponding evolution of the tangential stiffness K_x and the variation of the interface thickness are also plotted. The sliding distance is less than 1 μm, that is, much less than one contact diameter of about 6 μm. The absence of static peak in the friction force can be due to the repulsion between polymer chains that reduces creep and therefore an increase in contact area or strengthening of the nanojunctions [54].

However, the main results regarding the highest confinement ($F_z = 1$ mN, i.e., contact pressure of 34 MPa) are detailed as follows:

- The friction force increases sublinearly with increasing velocity.
- The velocity effect is reversible.
- The confined layer exhibits rather good lubricating properties since the friction coefficient remains lower than 0.04 in the studied range of velocities.

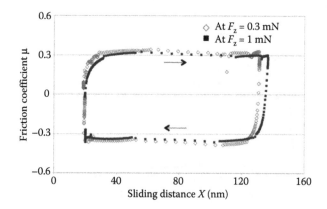

FIGURE 3.13   Friction coefficient μ versus the sliding distance X at a constant normal force F_z of 0.3 and 1 mN, and varying sliding velocities for OCP in a confined interface. The velocity first increases before it decreases in three successive steps of 0.2, 1, and 5 nm/s. The arrows indicate the direction of sliding. It can be seen that the friction coefficient is independent of sliding velocity and the applied normal force in the range of the experiment. (From J. Cayer-Barrioz et al. *Tribol Lett* 32, 81–90, 2008. With permission.)

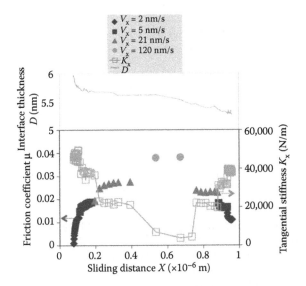

FIGURE 3.14 Friction coefficient μ versus sliding distance, X, for DOCP friction trace induced by a series of successive increasing and decreasing sliding velocities V_x from 2 to 120 nm/s for a constant normal force $F_z = 1$ mN. It can be seen that the interface accommodates shear through a progressive and reversible increase or decrease in friction coefficient (♦ for $V_x = 2$ nm/s, ■ for $V_x = 5$ nm/s, ▲ for $V_x = 21$ nm/s, ● for $V_x = 120$ nm/s) and a simultaneous variation of the tangential stiffness (o). The DOCP interface thickness is also plotted (solid line on top) and it shows continuous variation during shear. (From J. Cayer-Barrioz et al. *Langmuir* 25, 10802–10810, 2009. With permission.)

- The film thickness, initially equal to 6 nm, decreases to 5.6 nm and then remains constant. After 457 nm, it starts decreasing again to finally reach 5.2 nm. No dilatancy is associated with the change in velocity.
- The tangential stiffness decreases reversibly with increasing velocity while the tangential dissipative part remains low. Therefore, it is expected that the viscous contribution to the friction remains low. However, the ratio between the dissipative and the conservative parts (respectively $\omega_x \cdot A_x$ and K_x) increases with the velocity.
- No static friction force is observed when the sliding motion is initiated.

These results are confirmed at lower confinement ($F_z = 0.1$ mN and contact pressure of 13 MPa). However, slight differences appear in the low confined interface behavior:

- The confinement seems to enhance the lubricating properties with decreasing friction coefficient with increasing confinement.
- The film thickness first decreases to reach 8 nm after a sliding distance of 19 nm. Then, it does not vary with the sliding velocity.

Figure 3.15 presents the evolution of the friction force with sliding velocity for DOCP at both studied confinement loads. It shows that friction increases as logarithmic

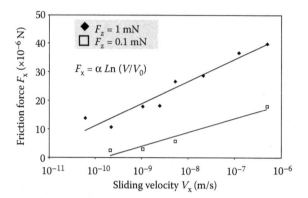

FIGURE 3.15 Friction force F_x versus sliding velocity V_x for the DOCP confined interface at a constant normal force 0.1 mN (□) and 1 mN (♦). A logarithmic velocity dependence of the friction force is observed. The variation of the slope with the normal force indicates that the velocity effect depends on the confinement load. (From J. Cayer-Barrioz et al. *Langmuir* 25, 10802–10810, 2009. With permission.)

of sliding velocity. The velocity dependence also varies with the confinement or the applied normal force. The logarithmic velocity dependence of friction can be theoretically accounted for by a general model based on the kinetics of formation and rupture of links between two shearing surfaces [21,32,33,55]. Indeed, in this model, at any time of shearing, the total contact area, A, is assumed to consist of N independent bonds or adhesion nanodomains, called "nanojunctions," each of average area δA. During motion, the whole contact area A does not slide as a single unit. Individual junctions are continually formed and broken incoherently. The dissipation arises from the elastic energy stored during the deformation of pinned nanojunctions and then irreversibly lost when they are depinned at the critical deformation l* [21,32,33]. The model predicts the existence of various friction regimes, provided that the mean time to (re)activate a junction thermally is negligible compared to the mean time, τ_0, to break a junction due to thermal fluctuations under zero shear force. At lower velocities, the main contribution to the dissipation comes from the elastic component while the viscous contribution dominates at higher velocities. One of the predicted regimes, corresponding to sliding velocity comparable to l*/τ_0, describes the logarithmic velocity dependence of the friction force. In this regime, the forced rupture process competes with the thermal mechanism with lowered activation energy barriers [21,32,33], giving in the first approximation the friction force F_x as

$$F_x = \alpha \, Ln\left(\frac{V}{V_0}\right) \tag{3.21}$$

with

$$\alpha = \frac{\pi a_c \cdot k_B T}{\gamma_{\text{interf}} \cdot \delta A} \tag{3.22}$$

$$V_0 = \frac{k_B T}{\gamma_{\text{interf}} \cdot \delta A} \cdot \frac{D}{G \cdot \tau_0} \qquad (3.23)$$

where k_B is the Boltzmann constant, a_c is the contact radius, δA is the mean size of a nanojunction, τ_0 is the mean time to break a junction, G is the shear elastic modulus as defined in Equation 3.20, and γ_{interf} is the thickness of the interpenetration zone between the two shearing surfaces. The interpenetration zone is expected to be rather thin due to limited mutual interactions between the polymer chains [25,28,53,56]. Moreover, for small interface thickness D, it is expected to decrease with increasing sliding velocity [24,28] as a result of polymer chain stretching and disentanglement from the two surfaces. It may also increase with confinement [24,48,53]. Beyond this logarithmic velocity-dependent regime, from a velocity V_1, the junctions are mainly elastically depinned and the shear stress is almost velocity independent [21,32]. According to the model

$$1^* = V_0 \cdot \tau_0 \cdot Ln\left(\frac{V_1}{V_0}\right) \qquad (3.24)$$

During sliding, the interface thickness D irreversibly decreases as the sliding distance X increases, independent of the velocity variation for both confinements. The thickness variation is usually considered as the sum of two contributions: the first is the creep of the viscoelastic interface itself and depends mainly on the contact pressure and on the sliding distance, while the second is correlated to the variation of the interpenetration zone thickness and is strongly dependent on the sliding velocity [37]. It is shown in Figure 3.14 that the interface thickness varies independently of the sliding velocity. Therefore, one can assume that the creep effect is preponderant over the interpenetration effect in the interface thickness evolution.

An accommodation of tangential stiffness with sliding velocity is experimentally observed and the tangential stiffness reversibly varies with the sliding velocity. In this framework, the evolution of tangential stiffness presented in Figure 3.14 for a constant normal force of 1 mN can be considered as the result of polymer chain disentanglement within the interpenetration zone. As the velocity increases, the number of interactions between chains diminishes leading to a decrease in tangential stiffness. Consequently, the variation of the tangential stiffness can be correlated to that of the thickness of the interpenetration zone γ_{interf}. The interpenetration zone can be depicted with n elementary springs of stiffness k_{el} in parallel and in this description, $K_x = n\, k_{\text{el}}$ and $\gamma_{\text{interf}} = n\, \gamma_{\text{el}}$ where γ_{el} is the thickness of an element of stiffness k_{el}. Therefore, one can compute that

$$\frac{\Delta K_x}{K_x} = \frac{\Delta n \cdot k_{\text{el}}}{n \cdot k_{\text{el}}} = \frac{\Delta \gamma_{\text{interf}}}{\gamma_{\text{interf}}} \qquad (3.25)$$

One may consider the initial interpenetration zone thickness as the thickness of the confined layer at rest, that is, $\gamma_{\text{interf}}\,(V_x = 0) = 6$ nm. This results in the variation of the interpenetration zone thickness γ_{interf} with the sliding velocity

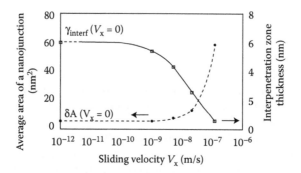

FIGURE 3.16 Evolution of the average area of a nanojunction (δA) and of the interpenetration zone thickness γ_{interf} with the sliding velocity (V_x) for the DOCP interface in broken and solid lines, respectively. Values of δA and γ_{interf}, corresponding to a static contact (i.e., $V_x = 0$), are also indicated. The interpenetration zone becomes thinner and the nanojunction larger as the sliding velocity increases in order to maintain the activation volume constant.

as presented in Figure 3.16. It is shown that the interpenetration zone thickness decreases with increasing velocity as predicted by Grest [28]. Since $\gamma_{interf} < D$, this confirms the assumption that only a thin layer undergoes and accommodates shear in a sliding contact.

From the logarithmic slope of the friction force-sliding velocity curves (see Figure 3.15), according to Equation 3.22, one may compute an activation volume, $\gamma_{interf} \cdot \delta A$, that can be considered as a measure of the average size of the clusters involved in the elementary events. One thus obtains for a contact pressure of 13 MPa, $(\gamma_{interf} \cdot \delta A)_{0.1mN} = 14$ nm^3, and for a contact pressure of 34 MPa, $(\gamma_{interf} \cdot \delta A)_{1mN} = 34$ nm^3. The elementary cluster size retains its order of magnitude, though showing a trend toward increasing with increasing pressure. This activation volume remains constant in the range of studied velocity. From a prior estimate of the thickness γ_{interf}, it is, therefore, possible to calculate the evolution of the average size of the nanojunction with sliding velocity, for a constant normal force $F_z = 1$ mN, as shown in Figure 3.16.

In Figure 3.15, the graphical determination of V_0 gives $V_0 = 1.4 \times 10^{-10}$ m/s at 0.1 mN and $V_0 = 3.7 \times 10^{-12}$ m/s at 1 mN and it can be seen that V_0 decreases with increasing load [21,32]. From Equation 3.23, one may calculate the mean time τ_0 to break a junction: it is equal to 4 s (respectively 10 s) for a contact pressure of 13 MPa (respectively 34 MPa). This seems physically reasonable because the larger the load, the longer will be the time to break a junction [21,32].

Previous results [33] show that the threshold sliding distance X^* measured by the ratio of the steady-state friction force to the tangential stiffness of the interface gives the critical deformation l* at the yield point beyond which all junctions are depinned following equation:

$$l^* = 2X^* \tag{3.26}$$

However, according to Figure 3.17, presenting the evolution of X^* with the normal force F_z, the critical deformation l* can be estimated for both loads: l*$_{0.1\,mN} = 2$ nm

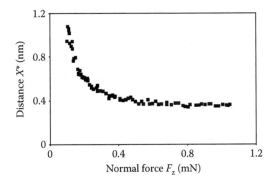

FIGURE 3.17 Plot of the critical distance X^*, defined as F_x/K_x, versus the normal force F_z for the DOCP interface. X^* represents the threshold beyond which the interface is no longer elastically stretched and starts sliding with energy loss. This length reaches a stable value of 0.4 nm for a normal force above 0.5 mN. (From J. Cayer-Barrioz et al. *Langmuir* 25, 10802–10810, 2009. With permission.)

while $l^*_{1\ mN} = 0.8$ nm. On the other hand, Figure 3.15 shows that $V_1 \geq 0.5$ μm/s because the logarithmic dependence is well obeyed for sliding velocity below 0.5 μm/s. It is, therefore, also possible to estimate l^* following Equation 3.24, it results: $l^*_{0.1\ mN} \geq 4$ nm and $l^*_{1\ mN} \geq 0.4$ nm. Although a good agreement is found at large load, a discrepancy is observed at low load even if the calculated values of l^* are of the same order of magnitude. We may invoke that solvent molecules remain within the contact and the adsorbed layers are not completed at 0.1 mN.

The experimental input data used to model the friction behavior shown in Figure 3.15 as well as the output data are reported in Table 3.3.

Using the self-similar model of de Gennes to describe the interface of adsorbed DOCP allows to relate the confinement effect to the friction behavior by means of the molecular organization and the mesh size ξ, of the critical deformation l^* and of the mechanical properties. The sliding layers at 0.1 and 1 mN are both in the central zone. However, at 0.1 mN (respectively at 1 mN), the interface thickness is very close to that corresponding to the distal (respectively proximal) region. The gradient of monomer volume fraction leads to various friction behaviors in terms of activation volume $\gamma_{interf} \cdot \delta A$ and critical deformation l^* of the nanojunctions. The theoretical friction modeling confirms the preponderance of the elasticity over the viscous contribution during friction in this experimental range of low sliding velocities, in particular, at high confinement. At a larger load, the layers sliding past each other consist of a concentrated monomer with reduced monomer mobility [32].

3.4 SUMMARY AND CONCLUSIONS

A molecular tribometer based on a surface force apparatus has been used to investigate the drainage and molecular organization of semidilute polymer solutions onto metallic surfaces. An attempt to correlate friction dynamics to the structure of confined layers was tentatively made.

TABLE 3.3
Input Data Obtained from Experiments and Output Data Obtained from the Fit of the Logarithmic Velocity Dependence for Normal Confinement Loads (F_z) of 0.1 and 1 mN in the Case of DOCP

	$F_z = 0.1$ mN	$F_z = 1$ mN
Contact radius a_c ($\times 10^{-6}$ m)	1.5	3
Shear elastic modulus G ($\times 10^6$ Pa)	5	14
Interface thickness D ($\times 10^{-9}$ m)	11	6
Activation volume $\gamma_{interf} \cdot \delta A$ ($\times 10^{-27}$ m³)	14	34
V_0 (m/s)	1.4×10^{-10}	3.7×10^{-12}
τ_0 (s)	4	10
l* ($\times 10^{-9}$ m) from Equation 3.26	2	0.8
V_1 (m/s)	$\geq 5 \times 10^{-7}$	$\geq 5 \times 10^{-7}$
l* ($\times 10^{-9}$ m) from Equation 3.24	≥ 4	≥ 0.4

Note: Values of contact radius (a_c) and interface thickness (D) are indicated. It can be seen that the activation volume ($\gamma_{interf} \cdot \delta A$), and the time ($\tau_0$) to break a junction are velocity independent for both loads. Both values of critical velocities (V_0 and V_1) are also noted. A comparative estimate of l* from Equations 3.24 and 3.26 gives the same order of magnitude. It is noteworthy that the values of shear elastic modulus G, corresponding to another contact location on the plane are slightly lower than those computed from Equation 3.20: this can be related to the heterogeneity of the adsorbed layers.

Semidilute solutions of nonfunctionalized (OCP) and functionalized (DOCP) olefinic copolymers in 150 Neutral Solvent base oil were investigated. Confinement loads varied in the range 0.1–1 mN. At low confinement, for large distances between surfaces, two regimes can occur depending on the polymer/surface interactions:

a. Possible existence of slip at the wall was observed and discussed, which led to another interpretation of the behavior of OCP. The polymer was repelled from the wall and a depletion layer was formed.
b. DOCP stuck to the wall due to its functionalized groups and an adsorption layer was observed.

The structure of these layers results in a gradient of monomer concentration, leading to a gradient of viscosity from the wall to the bulk. This observation can be well described using either a mechanical model that consists of multilayer interface or a physical model based on the work of de Gennes, which depicts the interface as a self-similar structure of mesh size ξ_b in the case of adsorption.

As confinement increases, both kinds of molecules are squeezed and finally adsorb onto the surfaces. Dynamic measurements gave the viscoelastic properties of both interfaces in the normal and tangential directions. In the case of OCP, confinement results in a glassy polymer layer within the contact. For DOCP, the increase in elastic properties is linked to a diminution of the network mesh size as predicted by the scaling theory of polymers.

The frictional response of both confined interfaces was observed for various normal forces up to 1 mN and sliding velocities ranging from 0.1 to 500 nm/s. In the case of OCP, the layers gather to form a glassy film. The homogeneous elastic shearing of the latter leads to Amontons-like friction with negligible adhesion contribution. For DOCP, Amontons' proportionality does not hold. The more severe the confinement, the higher the normal force, and the lower the friction. Besides, the friction force was strongly dependent on the sliding velocity and increased logarithmically with velocity. Physical modeling of the shearing of these confined layers based on the kinetics of formation and rupture of bonds between two shearing surfaces was used to interpret this frictional behavior from the dynamics of shear-activated pinned and free junctions. It is noteworthy that the input parameters of the model were derived from experimental data. Moreover, the properties of the interpenetration zone between the polymer chains and its dynamics can be deduced. This approach allowed correlation of the frictional response of the confined interface to the molecular organization on the surfaces.

REFERENCES

1. G.W. Stachowiak and A.W. Batchelor, *Engineering Tribology.* 2nd ed., Butterworth Heinemann, Boston, 2001.
2. E. Bonaccurso, M. Kappl, and H.J. Butt, Hydrodynamic force measurements: Boundary slip of water on hydrophilic surfaces and electrokinetic effects, *Phys Rev Lett* 88, 076103(4), 2002.
3. C. Neto, D.R. Evans, E. Bonaccurso, H.J. Butt, and V.S.J. Craig, Boundary slip in Newtonian liquids: A review of experimental studies, *Rep Prog Phys* 68, 2859–2897, 2005.
4. O.I. Vinogradova, Slippage of water over hydrophobic surfaces, *Int J Miner Process* 56, 31–60, 1999.
5. J. Sanchez-Reyes and L.A. Archer, Interfacial slip violations in polymer solutions: Role of microscale surface roughness, *Langmuir* 19, 3304–3312, 2003.
6. J.M. Georges, S. Millot, A. Tonck, R.C. Coy, A.G. Schlijper, and B.P. Williamson, Nanorheology of polyisoprene solutions confined between two solid surfaces, In: *Tribology for Energy Conservation*, D. Dowson (Ed.), pp. 51–62, Tribology Series 32, Elsevier, Amsterdam, 1998.
7. E. Georges, J.M. Georges, and S. Hollinger, Contact of two carbon surfaces covered with a dispersant polymer, *Langmuir* 13, 3454–3463, 1997.
8. D. Mazuyer, E. Varenne, A.A. Lubrecht, J.M. Georges, and B. Constans, Shearing of adsorbed polymer layers in an elastohydrodynamic contact in pure sliding, In: *Lubrication at the Frontier*, D. Dowson (Ed.), pp. 493–504, Elsevier, Amsterdam, 1999.
9. H.A. Spikes, Direct observation of boundary layers, *Langmuir* 12, 4567–4573, 1996.
10. M. Smeeth, S. Gunsel, and H.A. Spikes, Boundary film formation by viscosity index improvers, *Tribol Trans* 39, 726–734, 1996.
11. H. Mitsui, H.A. Spikes, and Y. Suita, Boundary film formation by low molecular weight polymers In: *Elastohydrodynamics' 96*, D. Dowson (Ed.), pp. 487–500, Elsevier, Amsterdam, 1997.

12. P.M. McGuiggan, M.L. Gee, H. Yoshizawa, S.J. Hirz, and J.N. Israelachvili, Friction studies of polymer lubricant surfaces, *Macromolecules* 40, 2126–2133, 2007.

13. S.C. Bae, Z. Lin, and S. Granick, Conjugated polymers confined and sheared: Photoluminescence and absorption dichroism in a surface forces apparatus, *Macromolecules* 38, 9275–9279, 2005.

14. H.K. Christenson, D.W.R. Gruen, R.G. Horn, and J.N. Israelachvili, Structuring in liquid alkanes between solid surfaces: Force measurements and mean-field theory, *J Chem Phys* 87, 1834–1841, 1987.

15. R.G. Horn and J.N. Israelachvili, Direct measurement of structural forces between two surfaces in a nonpolar liquid, *J Chem Phys* 75, 1400–1411, 1981.

16. J.N. Israelachvili, *Intermolecular and Surface Forces*. 2nd ed., Academic Press, New York, 1991.

17. J.N. Israelachvili and S.J. Kott, Liquid structuring at solid interfaces as probed by direct force measurements: The transition from simple to complex liquids and polymer fluids, *J Chem Phys* 88, 7162–7166, 1988.

18. J. Klein, Forces between polymer-bearing surfaces: The question of constrained equilibrium, *J Phys: Condens Matter* 2, 323–328, 1990.

19. P.G. de Gennes, Polymer solutions near an interface. 1. Adsorption and depletion layers, *Macromolecules*, 14, 1637–1644, 1981.

20. P.G. de Gennes, Polymer solutions near an interface. 2. Interaction between two plates carrying adsorbed polymer layers, *Macromolecules* 15, 492–500, 1982.

21. C. Drummond, J.N. Israelachvili, and P. Richetti, Friction between two weakly adhering boundary lubricated surfaces in water, *Phys Rev E* 67, 066110(16), 2003.

22. D. Gourdon and J.N. Israelachvili, Transitions between smooth and complex stick-slip sliding of surfaces, *Phys Rev E* 68, 021602(10), 2003.

23. L. Léger, H. Hervet, and P.G. de Gennes, Role of surface-anchored polymer chains in polymer friction, In: *Lubrication at the Frontier*, D. Dowson (Ed.), pp. 3–9, Elsevier, Amsterdam, 1999.

24. R. Tadmor, J. Janick, J. Klein, and L.J. Fetters, Sliding friction with polymer brushes, *Phys Rev Lett* 91, 115503(4), 2003.

25. J.F. Joanny, Lubrication by molten polymer brushes, *Langmuir* 8, 989–995, 1992.

26. A. Subbotin, A. Semenov, and M. Doi, Friction in strongly confined polymer melts: Effect of polymer bridges, *Phys Rev E* 56, 623–630, 1997.

27. J. Gao, W.D. Luedtke, and U. Landman, Friction control in thin-film lubrication, *J Phys Chem B* 102, 5033–5037, 1998.

28. G.S. Grest, Interfacial sliding of polymer brushes: A molecular dynamics simulation, *Phys Rev Lett* 76, 4979–4982, 1996.

29. P.A. Thompson and M.O. Robbins, Origin of stick-slip motion in boundary lubrication, *Science* 250, 792–794, 1990.

30. P.A. Thompson, M.O. Robbins, and G.S. Grest, Simulations of lubricant behavior at the interface with bearing solids, In: *Thin Film in Tribology*, D. Dowson (Ed.), pp. 347–360, Elsevier, Amsterdam, 1993.

31. J. Cayer-Barrioz, D. Mazuyer, A. Tonck, and E. Yamaguchi, Drainage of a wetting liquid: Effective slippage or polymer depletion? *Tribol Lett* 32, 81–90, 2008.

32. J. Cayer-Barrioz, D. Mazuyer, A. Tonck, and E. Yamaguchi, Frictional rheology of a confined adsorbed polymer layer, *Langmuir* 25, 10802–10810, 2009.

33. D. Mazuyer, J. Cayer-Barrioz, A. Tonck, and F. Jarnias, Friction dynamics of confined weakly adhering boundary layers, *Langmuir* 24, 3857–3866, 2008.

34. P.G. de Gennes, *Scaling Concepts in Polymer Physics*. Cornell University Press, Ithaca, NY, 1979.

35. J.E. Mark, *Physical Properties of Polymers Handbook*. American Institute of Physics Press, New York, 1996.

36. R.G. Larson, *The Structure and Rheology of Complex Fluids*. Oxford University Press, New York, 1999.
37. J.M. Georges, A. Tonck, J.L. Loubet, M. Mazuyer, E. Georges, and F. Sidoroff, Rheology and friction of compressed polymer layers adsorbed on solid surfaces, *J Phys II France*, 6, 57–76, 1996.
38. D. Mazuyer, A. Tonck, and J. Cayer-Barrioz, Friction control at the molecular level: From superlubricity to stick-slip, In: *Superlubricity*, A. Erdemir and J.M. Martin (Eds.), pp. 397–426, Elsevier, Amsterdam, 2007.
39. J.M. Georges, S. Millot, J.L. Loubet, and A. Tonck, Drainage of thin films between relatively smooth surfaces, *J Chem Phys* 98, 7345–7360, 1993.
40. J. Baudry, E. Charlaix, A. Tonck, and D. Mazuyer, Experimental evidence for a large slip effect on a non-wetting fluid-solid interface, *Langmuir* 17, 5232–5236, 2001.
41. C. Cottin-Bizonne, B. Cross, A. Steinberger, and E. Charlaix, Boundary slip on smooth hydrophilic surfaces: Intrinsic effects and possible artefacts, *Phys Rev Lett* 94, 056102(4), 2005.
42. P.G. de Gennes, Polymers at an interface: A simplified view, *Adv Colloid Interface Sci* 27, 189–209, 1987.
43. H.A. Spikes, Slip at the wall—Evidence and tribological implications, In: *Tribological Research and Design for Engineering Systems*, D. Dowson (Ed.), pp. 525–535, Elsevier, Amsterdam, 2003.
44. B.V. Derjaguin, and N.V. Churaev, Structure of water in thin layers, *Langmuir* 3, 607–612, 1987.
45. R. Pit, H. Hervet, and L. Léger, Direct experimental evidence of slip in hexadecane: Solid interfaces, *Phys Rev Lett* 85, 980(4), 2000.
46. E. Bonaccurso, H.J. Butt, and V.S.J. Craig, Surface roughness and hydrodynamic boundary slip of a Newtonian fluid in a completely wetting system, *Phys Rev Lett* 90, 144501(4), 2003.
47. S. Granick, Y. Zhu, and H. Lee, Slippery questions about complex fluids flowing past solids, *Nat Mater* 2, 221–227, 2003.
48. J. Klein, E. Kumacheva, E. Perahia, and L.J. Fetters, Shear forces between sliding surfaces coated with polymer brushes: The high friction regime, *Acta Polym* 49, 617–625, 1998.
49. A. Tonck, Développement d'un appareil de mesure de forces de surface et de nanorhéologie. PhD dissertation, *Ecole Centrale de Lyon*, N°89-12, 1989.
50. R. Horn, O.I. Vinogradova, M.E. Mackay, and N. Phan-Thien, Hydrodynamic slippage inferred from thin film drainage measurement in a solution of non-adsorbing polymer, *J Chem Phys* 112, 6424–6433, 2000.
51. E. Donath, A. Krabi, G. Allan, and B. Vincent, A study of polymer depletion layers by electrophoresis: The influence of viscosity profiles and the nonlinearity of the Poisson-Boltzmann equation, *Langmuir* 12, 3425–3430, 1996.
52. T.L. Kuhl, A.D. Berman, J.N. Sek Wen Hui, and J.N. Israelachvili, Part 1. Direct measurement of depletion attraction and thin film viscosity between lipid bilayers in aqueous polyethylene glycol solutions, *Macromolecules* 31, 8250–8257, 1998.
53. J. Klein, E. Kumacheva, D. Mahalu, D. Perahia, and L.J. Fetters, Reduction of frictional forces between solid surfaces bearing polymer brushes, *Nature* 370, 634–636, 1994.
54. F.P. Bowden and D. Tabor, *Friction, An Introduction to Tribology*. Heinemann, London, Science Study Series N°41, 1973.
55. Y.B. Chernyak and A.I. Leonov, On the theory of the adhesive friction of elastomers, *Wear* 108, 105–138, 1986.
56. T. Witten, L. Leibler, and P. Pincus, Stress relaxation in the lamellar copolymer mesophase, *Macromolecules* 23, 824–829, 1990.

4 Tribological Behavior of Ionic Polymer Brushes in Aqueous Environment

Motoyasu Kobayashi, Masami Terada,
Tatsuya Ishikawa, and Atsushi Takahara

CONTENTS

Surface wettability and tribological properties of ion-containing polymer brushes prepared by surface-initiated controlled radical polymerization were investigated. Polymer brushes consisting of (2-methacryloyloxyethyl)trimethylammonium chloride (MTAC), 3-sulfopropyl methacrylate potassium salt (SPMK), and 2-methacryloyloxyethyl phosphorylcholine (MPC) gave superhydrophilic surfaces, which showed extremely low water contact angle in air. In addition, air bubbles in water cannot adsorb on the hydrated brush surface. Macroscopic frictional properties of brush surfaces were characterized by sliding a glass ball probe in water using a ball-on-plate-type tribotester under a load of 0.49 N at a sliding velocity of 10^{-5}–10^{-1} m s^{-1} at 25°C. Poly(SPMK) and poly(MPC) showed significantly low friction coefficients below 0.02 in water and in humid air condition. A drastic reduction in friction coefficient of polyelectrolyte brushes in aqueous solution was observed at a sliding velocity of 10^{-3}–10^{-2} m s^{-1} due to the hydrodynamic lubrication effect; however, an increase in salt concentration in the aqueous solution led to an increase in friction coefficients of poly(MTAC) and poly(SPMK) brushes. The repulsive interaction between like charges along the polyelectrolyte brushes led to the reduction in friction force between the opposite brush surfaces in water. Poly(SPMK) brush in water demonstrated a stable and extremely low friction coefficient even after 450 friction cycles.

4.1 INTRODUCTION

The surface-tethering of polymers has become a widely used method for improving the surface physicochemical properties, such as wettability, adhesion, and friction, of solid surfaces. An assembly of polymer chains end-grafted to a solid surface at a sufficiently high graft density in a good solvent is generally referred to as a "polymer brush" [1,2]. The graft density is the number of tethered chains at the surfaces per unit area, which largely depends on the preparation process, such as "grafting-to" or "grafting-from." Over the last decade, various types of well-defined, high-density polymer brushes have been prepared via surface-initiated controlled radical polymerization (CRP). Such brushes are grown from surface initiating sites that are immobilized on solid surfaces or substrates through covalent bonding. As a result, the brush chains are strongly anchored and are hardly detached from the substrate, even in a good solvent as well as under large shear deformation. Therefore, the polymer brushes with nanometer-scale thickness can act as an efficient lubricant in friction.

In particular, water-soluble polymer or polyelectrolyte brushes are expected to be used as a water lubrication system, which is closely related to biological friction in aqueous environment, such as synovial joints, which have extremely low friction coefficients in the range of 0.001–0.03 [3]. For example, glycoproteins are ion-containing graft polymers with many branching chain, like a molecular bottle-brush [4], which work as a biological lubricant by immobilizing large amounts of water molecules to aid lubrication in the cartilage of mammalian joints and cornea tissue. The water lubrication systems using polyelectrolyte brushes can contribute to innovations in novel environmentally friendly technologies.

In principle, densely grafted polymer chains in a good solvent stretch from the surface to reduce their interaction with other chains and avoid overlapping of polymer chains. Also, the state of chain stretching is determined by the balance between osmotic pressure due to high polymer concentration and elastic restoring force of polymer chain. When two opposing, polymer brush-covered surfaces are brought into contact in a good solvent, they normally repel each other because of the excluded volume effect among polymer segments and this can suppress the mutual interpenetration of two compressed brushes. This is the classical lubrication mechanism for efficient lubrication of solvated polymer brushes based on repulsive steric forces [5].

The reduction of frictional forces between solid surfaces bearing polymer brushes was first reported by Klein and coworkers [6] using a surface force balance, and they also reported that the polyelectrolyte brushes could act as efficient lubricants between mica surfaces in an aqueous solution [7–9].

In the case of polyelectrolyte brushes in aqueous media, the osmotic pressure of free counterions within the charged brush also contributes to the extremely low friction property [9–12]. The hydration sheaths bound to the charges, and the fluidity of the hydrating water also play important roles in boundary lubrication [13,14]. In addition, the lubrication properties of polyelectrolyte brushes are affected by many factors, such as graft density [15–17], ionic strength [18], solvent quality [19–22], and the repulsive and attractive interactions of polar functional groups [23,24], which are directly measured by surface force balance [25,26] and atomic force microscopy (AFM) [27].

Macroscopic friction properties [28–30] are also important for the application of polymer brushes in practical use. However, the number of published papers related to high-density polyelectrolyte brushes is still limited. We have prepared anionic, cationic, and zwitterionic polymer brushes on silicon substrates by surface-initiated CRP to investigate the friction and wear properties in an aqueous environment using a reciprocating ball-on-plate-type tribometer [31–34]. This chapter discusses macroscopic tribological characteristics of ion-containing high-density polymer brushes in air, water, and aqueous salt solutions and brings out the dependence of friction coefficient on the sliding velocity of friction probe and salt concentration. When two surfaces covered with ionic polymer brushes having like charges are brought into contact, the coulombic repulsion is expected to reduce the interfacial frictional forces under wet conditions.

4.2　EXPERIMENTAL

4.2.1　Preparation of Ion-Containing Polymer Brushes

The chemical structures of polymer brushes and the polymerization scheme are illustrated in Figure 4.1. All brush samples were prepared by the surface-initiated atom transfer radical polymerization (ATRP) [35] from a silicon wafer with an immobilized surface initiator as follows [36]. Silicon wafers are washed with a mixture solution of sulfuric acid and hydrogen peroxide at 100°C for 1 h and washed with deionized water before use. The surface initiator, (2-bromo-2-methyl)propionyloxyhexyl triethoxysilane (BHE) [34,37], was immobilized on a silicon wafer by chemical vapor adsorption method [38]. The BHE-immobilized silicon wafers, 4.0 mL of 2-(methacryloyloxy)ethyltrimethylammonium chloride (MTAC, Aldrich) diluted with 2,2,2-trifluoroethanol (TFE, Acros) solution (2.0 M), and 0.30 mL of 2-propanol were charged in a well-dried glass tube with a stopcock and then degassed using a freeze–thaw process that was repeated three times. A catalyst solution containing CuBr (0.020 mmol, Wako Pure Chemical, Osaka), 2,2-bipyridyl (Aldrich, St. Louis, MO, 98%) (0.040 mmol), and ethyl 2-bromoisobutylate (EB, Tokyo Chemical Industry, Tokyo, 99%) (0.020 mmol) diluted with TFE was injected into the monomer solution. The resulting reaction mixture was degassed again by repeated freeze–thaw cycles to remove the oxygen and then stirred in an oil bath at 363 K for 16 h under

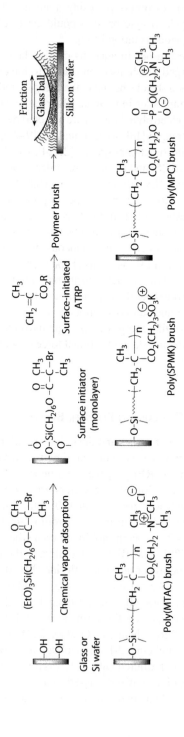

FIGURE 4.1 Preparation scheme of polymer brushes by surface-initiated atom transfer radial polymerization and the chemical structures of the polyelectrolyte brushes.

argon, which simultaneously generated poly(MTAC) brushes from the substrate and free (unbound) poly(MTAC) from EB. The reaction was stopped by opening the glass vessel to air at 273 K. The reaction mixture was poured into tetrahydrofuran (THF) to precipitate the free polymer and unreacted MTAC. The silicon wafers were washed with methanol using a Soxhlet apparatus for 12 h to remove the free polymer adsorbed on their surfaces and dried under nitrogen at room temperature.

The poly(2-methacryloyloxyethyl phosphorylcholine) (MPC) brush was synthesized in a similar manner using CuBr and 4,4'-dimethyl-2,2'-bipyridyl (Aldrich, St. Louis, MO, 98%) in methanol at 303 K for 12 h [39]. Surface-initiated ATRP of 3-sulfopropyl methacrylate potassium salt (SPMK, Tokyo Chemical Industry, Tokyo) was carried out in methanol/water (5/2, v/v) (1.7 M of SPMK) at 298 K for 15 h using a reaction system consisting of SPMK/CuBr/CuBr$_2$/4,4'-dimethyl-2,2'-bipyridyl (200/2/0.4/4, molar ratio) [40].

4.2.2 MOLECULAR WEIGHT MEASUREMENT

The number-average molecular weight (M_n) and molecular weight dispersity of the unbound polymers were determined by size exclusion chromatography (SEC) using a Shimadzu HPLC system equipped with a multiangle light scattering detector (MALS; Wyatt Technology DAWN-EOS) and refractive index detector (Shimadzu RID-10A, tungsten lamp with a wavelength of 470–950 nm). The Rayleigh ratio at a scattering angle of 90° was based on that of pure toluene at a wavelength of 632.8 nm at 298 K. SEC analysis on poly(MPC) was performed with three polystyrene gel columns connecting with two super AW3000 and super AW4000 (Tosoh Bioscience, Tokyo) using water containing 0.01 M LiBr as an eluent at a rate of 0.5 ml min^{-1} at 40°C. In the case of poly(MTAC), three polystyrene gel columns of G3000PW$_{XL}$-CP and two G5000PW$_{XL}$-CP (Tosoh Bioscience, Tokyo) were used with an acetic acid aqueous solution (500 mM) containing sodium nitrate (200 mM) as an eluent at a rate of 0.6 mL min^{-1}.

The M_n of the surface-grafted poly(MTAC) was also measured by SEC after isolation of grafted polymer cleaved from the substrate surface by acidic hydrolysis. Both surface-grafted and unbound polymers had almost the same M_n [36].

4.2.3 THICKNESS OF POLYMER BRUSHES

The thickness of the polymer brushes immobilized on a silicon wafer in air (the relative humidity was ca. 45%) was determined by a spectroscopic ellipsometer MASS-102 (Five Lab Co., Kanagawa, Japan) with a xenon arc lamp (wavelength 380–890 nm) at a fixed incident angle of 70°. The brush thickness L (nm) and the M_n gave the graft density σ of polymer brush using the following equation:

$$\sigma = \frac{dLN_A 10^{-21}}{M_n} \quad (4.1)$$

where d and N_A are the assumed density of bulk polymer at 293 K and the Avogadro's number, respectively. The graft density of poly(MTAC) and poly(MPC) were determined to be 0.20 and 0.23 chains nm^{-2}, respectively. In this study, the polymer brushes with approximately 100 nm of thickness were prepared for the friction tests.

4.2.4 CONTACT ANGLE MEASUREMENT

The water contact angles on polymer brush surfaces were recorded with a drop shape analysis system, DSA10 Mk2 (KRÜSS Optronic GmbH, Hamburg, Germany), equipped with a video camera. A 2.0 µL droplet of water was placed on the surface using a micropipette to measure the static contact angle. All these evaluations were conducted in ambient air at room temperature (approximately 298 K). The relative humidity was approximately 40%. The contact angle of a captive air bubble in water was measured in a square-shaped transparent glass vessel filled with deionized water. The upside-down substrate was immersed in water, and then the air bubble (10 µL) was released from beneath the brush substrate using a microsyringe. After the bubble floats upwards and touches the brush surface, the contact angle was measured.

4.2.5 FRICTION MEASUREMENT USING A TRIBOMETER

Macroscopic friction tests on polymer brushes were carried out on a conventional ball-on-plate-type reciprocating tribotester, Tribostation Type32 (Shinto Scientific Co. Ltd., Tokyo, Japan), by sliding a glass ball on the substrates at a reciprocating distance of 20 mm and a sliding velocity of 10^{-5}–10^{-1} m s^{-1} in a dry nitrogen atmosphere and in water under a normal load of 0.49 N at room temperature. The friction force was measured by a strain gauge attached to the arm of the tester and was recorded as a function of time. The friction coefficient was given by the friction force divided by the normal load. Every friction test used a fresh surface area on the brush substrate to measure the friction coefficient of the first reciprocating scan of the sliding probe. The brush substrate was pinned in a stainless-steel trough filled with water, which was fixed on a moving stage. In the case of a non-modified silicon wafer under a normal load of 50 g (0.49 N), the theoretical contact area between the glass probe and substrate was estimated to be 3.51×10^{-9} m^2 by Hertz's contact mechanics theory* and the average pressure on the contact area was estimated to be 139 MPa.

---

* If a circle with radius a (m) is regarded as the contact area between a glass ball and substrate under a normal load P (0.49 N), Hertz's theory gives the following relationship using Young's modulus of glass and silicon wafer, E_A (7.16×10^{10} Pa), E_B (1.30×10^{11} Pa), and Poisson's ratio υ_A (0.23), υ_B (0.28), respectively:

$$\frac{1}{E} = \frac{1 - \upsilon_A^2}{E_A} + \frac{1 - \upsilon_B^2}{E_B}$$

$$a = \left(\frac{3PR_A}{4E} \right)^{1/3}$$

where R_A is curvature radius (5.00×10^{-3} m) of glass ball and E is the elastic modulus. The contact area can be calculated by πa^2.

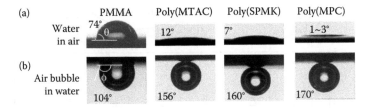

FIGURE 4.2 Photographs (side view) of water droplets on polymer brushes in air, and air bubbles in contact with the brush surfaces in water. (a) Angle θ is static contact angle of liquid on a solid surface in air; (b) φ is the contact angle of an air bubble in water.

4.3 RESULTS AND DISCUSSION

4.3.1 WETTABILITY OF POLYELECTROLYTE BRUSHES

Figure 4.2 shows photographs (side view) of water droplets on polyelectrolyte brushes in air, and air bubbles in contact with the brush surfaces in water. A water droplet on the polyelectrolyte brush surface showed a lower contact angle θ compared with poly(methyl methacrylate) (PMMA) brush. Poly(MTAC), poly(SPMK), and poly(MPC) are water-soluble polymers, whereas PMMA is insoluble in water. The polyanion, polycation, and zwitterion-type polyelectrolyte brushes showed lower contact angles. In particular, a very low contact angle of water (3°) was observed on the poly(MPC) brush surface because poly(MPC) absorbs moisture from the atmosphere like a deliquescent material. The surface free energies of polyelectrolyte brushes evaluated by the Owens–Wendt method [41]* were higher than 70 mJ/m², which were much higher than those of the PMMA brush (43.8 mJ m⁻²).

The contact angles φ of the air bubble in water were measured by Hamilton's method [42,43]. Photographs of air bubbles attached to the brush surfaces in water are also shown in Figure 4.2. The contact angle φ of the air bubble on the poly(MPC) was 170°. The two contact angles φ and θ were almost supplementary angles. Most of the air bubbles in contact with the polyelectrolyte brush surfaces formed sphere-like shapes. The surfaces of the polyelectrolyte brushes repelled air bubbles in water. When the air bubble in water rises from the lower position and touches the polyelectrolyte brush surface, the air bubble bounces several times, indicating high efficiency of polyelectrolyte brushes with water.

4.3.2 FRICTION OF POLYELECTROLYTE BRUSHES IN AIR AND WATER

Figure 4.3 shows the friction coefficients of silicon wafer and ion-containing polymer brushes measured by a conventional ball-on-plate-type reciprocating tribotester using a glass ball probe (diameter = 10 mm) sliding on the substrates along a distance

---

* Surface free energies of polymer brush surfaces were determined with contact angles of water and diiodomethane droplets using the following parameters: $\gamma_{Water} = 72.80$ mJ m⁻², $\gamma_{Water}^{d} = 21.80$ mJ m⁻², $\gamma_{Water}^{p} = 51.00$ mJ m⁻², $\gamma_{CH2I2} = 50.80$ mJ m⁻², $\gamma_{CH2I2}^{d} = 49.50$ mJ m⁻², $\gamma_{CH2I2}^{p} = 1.30$ mJ m⁻².

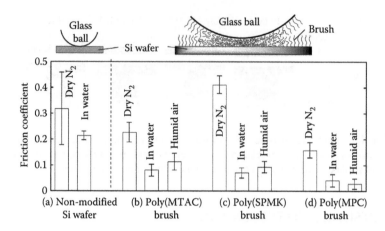

FIGURE 4.3 Friction coefficients between glass ball and (a) unmodified silicon wafer, (b) poly(MTAC) brush, (c) poly(SPMK) brush, and (d) poly(MPC) brush in dry nitrogen atmosphere, water, and humid air (relative humidity >75%); glass ball (10 mm diameter) with corresponding immobilized polymer brush over a distance of 20 mm at a sliding velocity of 1.5×10^{-3} m s^{-1} under a load of 0.49 N at 298 K.

of 20 mm at a sliding velocity of 1.5×10^{-3} m s^{-1} in air and in water under a normal load of 0.49 N at 298 K.

Friction coefficient of nonmodified glass ball sliding on the silicon wafer was 0.32 under dry N_2 condition, and was reduced to 0.22 in water due to the fluid lubrication effect. Friction coefficients larger than 0.2 were also observed under the dry N_2 condition for the glass ball covered with polyelectrolyte brushes sliding on the identical brush substrates. In contrast, the polyelectrolyte brushes in water showed much lower friction coefficients than that of nonmodified glass ball and silicon wafer. For example, the friction coefficient of poly(MPC) brush on silicon wafer in water was 0.04 by sliding a glass ball with immobilized poly(MPC) brush. Klein and coworkers also reported that the poly(MPC) brush surface in an aqueous solution showed friction coefficient as low as 0.0004 at pressure as high as 7.5 MPa, which was measured by the surface force balance technique [44,45]. The lowest friction coefficient in water was observed on the poly(SPMK) brush. The opposing swollen brushes in water would form a thicker boundary layer to restrict the direct contact of the glass probe with the silicon substrate. Interestingly, the friction coefficient of the poly(MPC) brush in the highly humidified air condition was significantly reduced to 0.02, which is lower than that in water. Since poly(MPC) is a superhydrophilic polymer, it is supposed that water molecules in humid air adsorbed on the brush surface and worked as a lubricant to reduce the friction force.

Interestingly, the reduction in the friction coefficient in humid atmosphere was not observed in the other brushes, except for poly(MPC) brush. The friction coefficient of poly(MTAC) and poly(SPMK) brushes in water was lower than that in humid air condition. Although the mechanism for the unique frictional property of poly(MPC) brush in humid air condition is still unclear, the authors suppose that swollen and extended poly(MPC) brush chains in water should have larger actual contact area between brush and probe to result in a higher friction coefficient, compared with that observed in humid air.

The thickness of the polyelectrolyte brush in dry state in air and swollen state in deuterium oxide (D_2O) were measured by neutron reflectivity [46]. The thickness of the swollen brush was twice that of dry state thickness because the hydrated brushes form a relatively extended chain conformation due to large osmotic pressure and high graft density. The polyelectrolyte brush in water formed a hydrated layer containing a high concentration of polymer to assist lubrication by water.

The water contact angles on poly(MTAC), poly(SPMK), and poly(MPC) are lower than 10°. Lower friction coefficients in water were observed on the brush surfaces with lower water contact angles. Thus, low friction coefficients in wet condition are related to the affinity of the polymer brush to water, that is, water is a good solvent for these polymers. The higher hydrophilic brush revealed better aqueous lubrication. It is reported that the surface-attached hydration layers produce strong repulsive forces that can support the normal load and prevent the contact of opposing surfaces [15]. In addition, water molecules bound to ions or ionized surface or polymers have sufficient fluidity even at high compressions to give a very low friction coefficient. Furthermore, swollen brush layers with aqueous phases overcome the drawback of the low viscosity of pure water, and can contribute to hydrodynamic lubrication over a large range of sliding velocity [47].

4.3.3 SLIDING VELOCITY DEPENDENCE OF FRICTION COEFFICIENT OF POLYELECTROLYTE BRUSHES

Figure 4.4 shows the friction coefficients of poly(MTAC), poly(SPMK), and poly(MPC) brushes in water at a sliding velocity of 10^{-5}–10^{-1} m s^{-1}. The polymer brushes were immobilized on surfaces of both silicon wafer as well as sliding glass ball. The friction coefficients of these brushes were 0.1–0.2 at slower sliding velocity

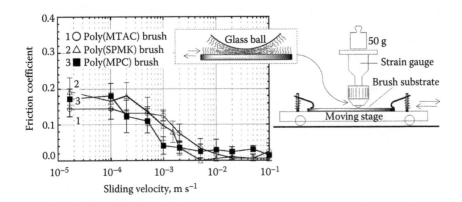

FIGURE 4.4 Illustration of the tribometer setup (right) and sliding velocity dependence of the friction coefficient (left) of poly(MTAC), poly(SPMK), and poly(MPC) brushes in water by sliding a glass ball (10 mm diameter) covered with the corresponding polyelectrolyte brushes over a distance of 20 mm under a load of 0.49 N at 298 K.

of 10^{-5}–10^{-3} m s^{-1}, whereas the friction coefficient dropped to 0.01–0.03 at higher sliding velocity 10^{-2}–10^{-3} m s^{-1}. Drastic reduction in the friction coefficient at a certain sliding velocity could be caused by the change in friction mechanism. At a low sliding velocity, the interaction between the opposing brushes and the interpenetration of brushes under the pressure of 138 MPa dominated friction to give a large friction coefficient, similar to that in the boundary friction regime. One recent study showed that water molecules bound to ionic groups or charged segments maintained fluidity even at a highly compressed situation leading to low friction [48]. In contrast, our experiment showed relatively large friction coefficients in water at a low sliding velocity of 10^{-5}–10^{-3} m s^{-1}. With an increase in the sliding velocity, a thicker liquid layer would be formed between the sliding surfaces by the hydrodynamic lubrication effect to reduce the actual contact area and thus the friction force. By analogy to the Stribeck curve [49], decreasing friction with increasing sliding velocity suggests that this system is in the mixed lubrication region. At higher sliding rates, hydrodynamic lubrication partially occurs at the interface between the oppositely sliding swollen brushes, to reduce friction.

Poly(SPMK) showed significantly lower friction coefficients at the sliding velocity of 10^{-3}–10^{-1} m s^{-1} probably because of water lubrication, the high osmotic pressure of brush, and the electrostatic repulsive interactions between the anionic groups.

4.3.4 EFFECT OF SALT CONCENTRATION ON FRICTION COEFFICIENT OF THE POLYELECTROLYTE BRUSHES

Lubrication in water is of interest in the biolubrication of catheter guide-wire, artificial joints, and environmentally friendly lubrication systems for rotating parts of wind turbines and tidal turbines. The study of friction in aqueous salt solutions is important in biolubrication application. The friction coefficient of polyelectrolyte brushes in aqueous solution is affected by salt concentration. Figures 4.5 and 4.6a display the friction coefficients of poly(MTAC) and poly(SPMK) brushes in 0–5000 mM NaCl aqueous solution at a sliding velocity of 10^{-5}–10^{-1} m s^{-1}. The poly(MTAC) brush in a 10–100 mM aqueous NaCl solution showed the same friction coefficient as in pure water. The reduction in the friction coefficient was observed at a sliding velocity of 10^{-2}–10^{-3} m s^{-1}. The friction coefficient in a 1000–5000 mM NaCl aqueous solution increased from 0.02 to 0.1 at a higher sliding velocity. The low friction coefficient, below 0.02, was observed in poly(SPMK) brush in salt-free solution at a sliding velocity of 10^{-3}–10^{-1} m s^{-1}; however, the friction coefficient increased to 0.1 in a 5000 mM NaCl aqueous solution.

In general, the dimension and intermolecular interactions of unbound polyelectrolytes bearing only positive or negative charges strongly depend on the ionic strength of the solution [50,51]. Polyanion or polycation chains in aqueous solution with low ionic strength form a relatively uniaxially stretched chain shape due to intermolecular repulsive interaction. On the other hand, the polyelectrolyte dissolved in solution with higher salt concentration behaves like an electrically neutral polymer to give small dimension, hydrodynamic radius, and small second virial coefficient because the electrostatic interactions are screened by hydrated salt ions

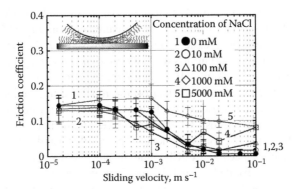

FIGURE 4.5 Sliding velocity dependence of the friction coefficient of poly(MTAC) brush in aqueous NaCl solution by sliding a glass ball (10 mm diameter) covered with poly(MTAC) brush over a distance of 20 mm under a load of 0.49 N at 298 K.

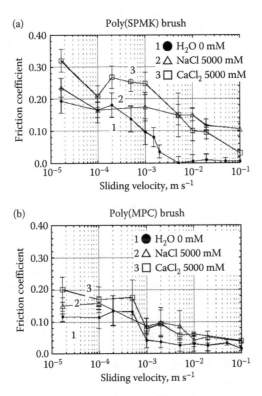

FIGURE 4.6 Sliding velocity dependence of the friction coefficient of (a) poly(SPMK) and (b) poly(MPC) brushes in water, 5000 mM NaCl, and 5000 mM CaCl$_2$ aqueous solutions by sliding a glass ball (10 mm diameter) with corresponding immobilized polymer brushes over a distance of 20 mm under a load of 0.49 N at 298 K.

[52,53]. A similar phenomenon takes place in surface-grafted polyelectrolyte in an aqueous salt solution. Neutron reflectivity measurement on poly(MTAC) brush in D_2O demonstrated that the thickness of a swollen brush layer drastically decreased at a concentration above 1000 mM $NaCl/D_2O$ solution, indicating the collapse state of brush in aqueous salt solution [46]. Therefore, an increase in salt concentration would lead to the reduction of the electrostatic repulsive interaction between the brushes to result in a higher friction coefficient. In addition, the collapse of polymer chains may increase the viscosity of the brush layer and increase its friction [54].

Figure 4.6 shows the friction coefficients of poly(SPMK) and poly(MPC) brushes in water, and aqueous NaCl and $CaCl_2$ solutions at room temperature. Poly(SMPK) brush, in particular, revealed a much higher friction coefficient in aqueous $CaCl_2$ solution probably due to the strong coordination of Ca^{2+} with sulfonate groups, forming the bridge between the contacting polymer brushes. On the other hand, the friction coefficient of poly(MPC) also increased, but to a lesser degree compared with the poly(SPMK) case. It is not well understood why the friction coefficient of poly(MPC) brush is not affected so much by salt concentration. As mentioned above, chain conformation of a typical polyelectrolyte, such as poly(MTAC) and poly(SPMK), in aqueous solution is usually affected by ionic strength. In contrast, the chain dimension and the second virial coefficient of poly(MPC) in NaCl aqueous solution are independent of salt concentration, which has been experimentally demonstrated by dynamic light scattering [55], small angle x-ray scattering [56], and neutron reflectivity measurements [46], although the detailed reason is still unclear. Poly(MPC) brush maintained a relatively low friction coefficient even in an aqueous $CaCl_2$ solution.

4.3.5 Stability of the Friction Coefficient of Polymer Brushes

Wear resistance is the most important tribological property for a high-density polymer brush. Since the tethered polymer chain end is bound to the substrate by covalent bonds and multiple hydrogen bonds, the brush layer can hardly be scratched off by a sliding probe. Figure 4.7 shows the evolution of friction coefficient of the poly(SPMK) brush during continuous reciprocating sliding of a brush-immobilized glass ball probe in water. The polymer thin films prepared by conventional coating, such as spin coating method, are easily scratched away by a sliding friction probe. In contrast, friction test on poly(SPMK) brush under a load of 0.49 N in water showed a significantly low friction coefficient of around 0.01 continuously, even after 450 friction cycles. Although a wear track was formed on the brush surface by the friction test, the components of the brush remained in the wear tracks, which were confirmed by x-ray photoelectron spectroscopy (XPS). The abrasion of the brush was supposed to be prevented owing to a good affinity of poly(SPMK) brush for water forming a water lubrication layer, and electrostatic repulsive interaction among the brushes bearing sulfonic acid groups.

One of the attractive applications of the polyelectrolyte brushes is surface modification of artificial hip joints. The hip joint is the spherical joint between the femoral head and the acetabulum covered with elastic hydrogels and filled with lubricant fluid. Some lubrication simulations of human hip joint reported that the hip joint receives

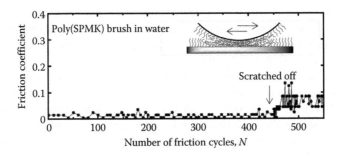

FIGURE 4.7 Friction coefficient versus the number of friction cycles N for the surface of poly(SPMK) brush in water by sliding a glass ball (10 mm diameter) with immobilized poly(SPMK) brush over a distance of 20 mm at a sliding velocity of 1.5×10^{-3} m s^{-1} under a load of 0.49 N at 298 K.

2–3 times of body weight as vertical load during the standard daily walking activity [57,58]. Although the load changes with the hip angles, angular velocity, and the position in the share head, in the case of a standard man with a body weight of 750 N, the corresponding pressure is calculated to 2–8 MPa, which is much lower than the normal pressure in our ball-on-plate-type friction test under 130 MPa estimated by Hertz's contact theory. A much lower friction coefficient and better wear resistance could be expected to the polyelectrolyte brushes under a low normal load condition.

Moro and his coworkers have already started the application of poly(MPC) brush to the artificial hip joint [57–59] and knee joint [60] consisting of a polyethylene (PE) acetabular liner and a cobalt–chromium alloy femoral head as a bearing couple. Friction torque of poly(MPC)-grafted PE liner against the femoral head was 1/10 of the nongrafted PE liner. In addition, the wear particles from the poly(MPC)-grafted PE liner surface were hardly observed due to the extremely small amount even after 5×10^6 cycles of loading against femoral head in a hip-joint wear simulator, which performs a physiological walking simulation with continuous cyclic motion (frequency = 1 Hz) under the load (maximum force = 2744 N) in aqueous bovine calf serum solution. Wear resistance is very important not only for duration elongation of the artificial joint but also the prevention of osteolysis around the implant because osteolysis is known to be triggered by the host inflammatory responses to the PE wear particles. Therefore, a surface-grafted poly(MPC) is expected to improve the performance of the artificial joint.

4.4 CONCLUSIONS

This chapter has described macroscopic friction properties between mutually compressed surfaces bearing high-density ion-containing polymer brushes prepared by surface-initiated CRP. Significant reduction in friction coefficient was observed on ion-containing polymer brush surfaces in water using a brush-tethered glass ball probe. Hydrated brushes would form a sufficient boundary layer to restrict direct contact of the friction probe with the substrate. Friction coefficient of polycation and polyanion brushes increased in aqueous salt solution involving conformational transition of the brush from the extended chains to a collapsed state with added salt. On the other hand, in contrast to poly(MTAC) and poly(SPMK) brushes, the zwitter-ionic poly(MPC) brush exceptionally maintained relatively low friction coefficient even in aqueous NaCl or CaCl$_2$ solution.

These tribological behaviors of polyelectrolyte brushes on flat substrate surface in an aqueous environment are closely related to the fundamental solution properties of unbound polyelectrolyte, such as salt concentration dependency of the dimension of hydrated polymer chains, hydrodynamic radius, radius of gyration, Kuhn segment length, cross-sectional diameter of a polymer chain, the second virial coefficient, which can be determined by light scattering and small angle x-ray scattering of unbound poly-electrolyte as well as colloidal particle bearing brushes with well-regulated chemical structure and narrow molecular weight dispersity. These parameters are useful for pre-diction of the brush chain conformation in aqueous solution, osmotic pressure, electro-static interaction between the brush and hydrated ions, and stiffness of brush. Therefore, it would be important to collaborate with a solution characterization study to improve understanding of the friction behavior of polyelectrolyte brushes in aqueous media.

ACKNOWLEDGMENT

The authors thank Professor K. Ishihara of the University of Tokyo for donating the MPC monomer and for helpful discussions.

REFERENCES

1. A. M. Granville and W. J. Brittain, Recent advances in polymer brush synthesis, in *Polymer Brushes: Synthesis, Characterization, Applications*, R. C. Advincula, W. J. Brittain, K. C. Caster, and J. Rühe (Eds), Wiley-VCH, Weinheim, Germany, pp. 35–50, 2004.
2. Y. Tsujii, K. Ohno, S. Yamamoto, A. Goto, and T. Fukuda, Structure and properties of high-density polymer brushes prepared by surface-initiated living radical polymeriza-tion, *Adv. Polym. Sci.*, 197, 1–46, 2006.
3. C. W. McCutchen, The frictional properties of animal Joints, *Wear*, 5, 1–17, 1962.
4. L. Han, D. Dean, C. Ortiz, and A. J. Grodzinsky, Lateral nanomechanics of cartilate aggrecan macromolecules, *Biophys. J.*, 92, 1384–1398, 2007.
5. J. Klein, Interactions, friction and lubrication between polymer-bearing surfaces, in *Fundamentals of Tribology and Bridging the Gap between the Macro- and Micro/ Nanoscales,* B. Bhushan (Ed.), Kluwer Academic Publishers, Dordrecht, pp. 177–198, 2001.
6. J. Klein, D. Perahia, and S. Warburg, Forces between polymer-bearing surfaces under-going shear, *Nature*, 352, 143–145, 1991.
7. J. Klein, E. Kumacheva, D. Mahalu, D. Perahia, and L. J. Fetters, Reduction of frictinal forces between solid surfaces bearing polymer brushes, *Nature*, 370, 634–636, 1994.
8. M. Kampf, J.-F. Gohy, R. Jérôme, and J. Klein, Normal and shear forces between a poly-electrolyte brush and a solid surface, *J. Polym. Sci. Part B Polym. Phys.*, 43, 193–204, 2005.
9. U. Raviv, S. Giasson, N. Kamph, J.-F. Gohy, R. Jérôme, and J. Klein, Lubrication by charged polymers, *Nature*, 425, 163–165, 2003.
10. J. Ruhe, M. Ballauff, M. Biesalski, P. Dziezok, F. Gröhn, D. Johannsmann, N. Houbenov et al. Polyelectrolyte brushes, *Adv. Polym. Sci.*, 165–166, 79–150, 2004.
11. I. Luzinov, S. Minko, and V. V. Tsukruk, Adaptive and responsive surfaces through con-trolled reorganization of interfacial polymer layers, *Prog. Polym. Sci.*, 29, 635–698, 2004.

12. S. J. Miklavic and S. Marčelja, Interaction of surfaces carrying grafted polyelectrolytes, *J. Phys. Chem.*, 92, 6718–6722, 1988.

13. J. Klein, U. Raviv, S. Perkin, N. Kampf, L. Chai, and S. Giasson, Fluidity of water and of hydrated ions confined between solid surfaces to molecularly thin films, *J Phys: Condens Matter*, 16, S5437–S5448, 2004.

14. U. Raviv and J. Klein, Fluidity of bound hydration layers, *Science*, 297, 1540–1543, 2002.

15. R. Israels, F. A. M. Leermakers, G. J. Fleer, and E. B. Zhulina, Charged polymeric brushes: Structure and scaling relations, *Macromolecules*, 27, 3249–3261, 1994.

16. Y. V. Lyatskaya, F. A. M. Leermakers, G. J. Fleer, E. B. Zhulina, and T. M. Birshtein, Analytical self-consistent-field model of weak polyacid brushes, *Macromolecules*, 28, 3562–3569, 1995.

17. E. B. Zhulina, J. K. Wolterink, and O. V. Borisov, Screening effects in a polyelectrolyte brush: Self-consistent-field theory, *Macromolecules*, 33, 4945–4953, 2000.

18. P. Pincus, Colloid stabilization with grafted polyelectrolytes, *Macromolecules*, 24, 2912–2919, 1991.

19. R. S. Ross and P. Pincus, The polyelectrolyte brush: Poor solvent, *Macromolecules*, 25, 2177–2183, 1992.

20. V. A. Pryamitsyn, F. A. M. Leermakers, G. J. Fleer, and E. B. Zhulina, Theory of the collapse of the polyelectrolyte brush, *Macromolecules*, 29, 8260–8270, 1996.

21. A. Nomura, K. Okayasu, K., Ohono, T. Fukuda, and Y. Tsujii, Lubrication mechanism of concentrated polymer brushes in solvents: Effect of solvent quality and thereby swelling state, *Macromolecules*, 44, 5013–5019, 2011.

22. A. Nomura, A. Goto, K. Ohno, E. Kayahara, S. Yamago, and Y. Tsujii, *J. Polym. Sci. Part A Polym. Chem.*, 49, 5284–5292, 2011.

23. H. J. Taunton, C. Toprakcioglu, L. J. Fetters, and J. Klein, Forces between surfaces bearing terminally anchored polymer chains in good solvent, *Nature*, 332, 712–714, 1988.

24. E. Eiser, J. Klein, T. A. Witten, and L. J. Fetters, Shear of telechelic brushes, *Phys. Rev. Lett.*, 82, 5076–5079, 1999.

25. S. Hayashi, T. Abe, N. Higashi, M. Niwa, and K. Kurihara, Polyelectrolyte brush layers studied by surface forces measurement: Dependence on pH and salt concentrations and scaling, *Langmuir*, 18, 3932–3944, 2002.

26. N. Kampf, D. Ben-Yaakov, D. Andelman, S. A. Safran, and J. Klein, Direct measurement of sub-Debye-length attraction between oppositely charged surface, *Phys. Rev. Lett.*, 103, 118304, 2009.

27. T. W. Kelley, P. A. Shorr, D. J. Kristin, M. Tirrell, and C. D. Frisbie, Direct force measurements at polymer brush surfaces by atomic force microscopy, *Macromolecules*, 31, 4297–4300, 1998.

28. M. Kobayashi, H. Ishida, M. Kaido, A. Suzuki, and A. Takahara, Frictional properties of organosilane monolayers and high-density polymer brushes, in *Surfactants in Tribology*, G. Biresaw and K. L. Mittal (Eds.), CRC Press, Boca Raton, FL, pp. 89–110, 2008.

29. M. Kobayashi, Z. Wang, Y. Matsuda, M. Kaido, A. Suzuki, and A. Takahara, Tribological behavior of polymer brush prepared by the "grafting-from" method, in *Polymer Tribology*, S. S. Kumar (Ed.), Imperial College Press, London, UK, pp. 582–602, 2009.

30. Y. Ohsedo, R. Takashina, J. P. Gong, and Y. Osada, Surface friction of hydrogels with well-defined polyelectrolyte brushes, *Langmuir*, 20, 6549–6555, 2004.

31. H. Sakata, M. Kobayashi, H. Otsuka, and A. Takahara, Tribological properties of poly(methyl methacrylate) brushes prepared by surface-initiated atom transfer radical polymerization, *Polym. J.*, 37, 767–775, 2005.

32. T. Ishikawa, M. Kobayashi, and A. Takahara, Macroscopic frictional properties of poly(1-(2-methacryloyloxy)ethyl-3-butyl imidazolium bis(trifluoromethane sulfonyl) imide) brush surfaces in an ionic liquid, *Appl. Mater. Interfaces*, 2, 1120–1128, 2010.

33. M. Kobayashi and A. Takahara, Tribological properties of hydrophilic polymer brushes under wet condition, *Chem. Record*, 10, 208–216, 2010.

34. M. Kobayashi and A. Takahara, Synthesis and frictional properties of poly(2,3-dihydroxypropyl methacrylate) brush prepared by surface-initiated atom transfer radical polymerization, *Chem. Lett.*, 34, 1582–1583, 2005.

35. K. Matyjaszewski and J. Xia, Atom transfer radical polymerization, *Chem. Rev.*, 101, 2921–2990, 2001.

36. M. Kobayashi, M. Terada, Y. Terayama, M. Kikuchi, and A. Takahara, Direct synthesis of well-defined poly[{2-(methacryloyloxy)ethyl}trimethyl ammonium chloride] brush via surface-initiated ATRP in fluoroalcohol, *Macromolecules*, 43, 8408–8415, 2010.

37. K. Ohno, T. Morinaga, K. Koh, Y. Tsujii, and T. Fukuda, Synthesis of monodisperse silica particles coated with well-defined, high-density polymer brushes by surface-initiated atom transfer radical polymerization, *Macromolecules*, 38, 2137–2142, 2005.

38. T. Koga, M. Morita, H. Ishida, H. Yakabe, S. Sasaki, O. Sakata, H. Otsuka, and A. Takahara, Dependence of the molecular aggregation state of octadecylsiloxane monolayers on preparation methods, *Langmuir*, 21, 905–910, 2005.

39. M. Kobayashi, Y. Terayama, N. Hosaka, M. Kaido, A. Suzuki, N. Yamada, N. Torikai, K. Ishihara, and A. Takahara, Friction behavior of high-density poly(2-methacryloyloxyethyl phosphorylcholine) brush in aqueous media, *Soft Matter*, 3, 740–746, 2007.

40. M. Ramstedt, N. Cheng, O. Azzaroni, D. Mossialos, H. J. Mathieu, and W. T. S. Huck, Synthesis and characterization of poly(3-sulfopropyl methacrylate) brushes for potential antibacterial applications, *Langmuir*, 23, 3314–3321, 2007.

41. D. K. Owens and R. C. Wendt, Estimation of the surface free energy of polymers, *J. Appl. Polym. Sci.*, 13, 1741–1747, 1969.

42. W. C. Hamilton, A technique for the characterization of hydrophilic solid surfaces, *J. Colloid Interface Sci.*, 40, 219–222, 1972.

43. M. Kobayashi, Y. Terayama, H. Yamaguchi, M. Terada, D. Murakami, K. Ishihara, and A. Takahara, Wettability and antifouling behavior on the surfaces of superhydrophilic polymer brushes, *Langmuir*, 28, 7212–7222, 2012.

44. M. Chen, W. H. Briscoe, S. P. Armes, and J. Klein, Lubrication at physiological pressures by polyzwitterionic brushes, *Science*, 323, 1698–1701, 2009.

45. M. Chen, W. H. Briscoe, S. P. Armes, H. Cohen, and J. Klein, Polyzwitterionic brushes: Extreme lubrication by design, *Eur. Polym. J.*, 47, 511–523, 2011.

46. M. Kobayashi, K. Mitamura, M. Terada, N. L. Yamada, and A. Takahara, Characterization of swollen states of polyelectrolyte brushes in salt solution by neutron reflectivity, *J. Phys. Conf. Ser.*, 272, 012019, 2011.

47. P. C. Nalam, J. N. Clasohm, A. Mashaghi, and N. D. Spencer, Macrotribological studies of poly(L-lysine)-graft-poly(ethylene glycol) in aqueous glycerol mixtures, *Tribol. Lett.*, 37, 541–552, 2010.

48. Y. Leng and P. T. Cummings, Fluidity of hydration layers nanoconfined between mica surfaces, *Phys. Rev. Lett.*, 94, 026101, 2005.

49. H. A. Spikes, Boundary lubrication and boundary films, in *Thin Films in Tribology*, D. Dowson, C. M. Talor, T. H. C. Childs, M. Godet, and G. Dalmz (Eds.), Elsevier, Amsterdam, Netherlands, pp. 331–346, 1993.

50. A. Takahashi, N. Kato, and M. Nagasawa, The osmotic pressure of polyelectrolyte in neutral salt solutions, *J. Phys. Chem.*, 74, 944–946, 1970.

51. M. Nagasawa and Y. Eguchi, The charge effect in sedimentation. I. Polyelectrolytes, *J. Phys. Chem.*, 71, 880–888, 1967.

52. M. Beer, M. Schmidt, and M. Muthukumar, The electrostatic expansion of linear polyelectrolytes: Effects of gegenions, co-ions, and hydrophobicity, *Macromolecules*, 30, 8375–8385, 1997.

53. J. Yashiro and T. Norisuye, Excluded-volume effects on the chain dimensions and transport coefficients of sodium poly(styrene sulfonate) in aqueous sodium chloride, *J. Polym. Sci, Part B Polym. Phys.*, 40, 2728–2735, 2002.

54. M. Gelbert, M. Biesalski, J. Rühe, and D. Johannsmann, Collapse of polyelectrolyte brushes probed by noise analysis of a scanning force microscope cantilever, *Langmuir*, 16, 5774–5784, 2000.

55. Y. Matsuda, M. Kobayashi, M. Annaka, K. Ishihara, and A. Takahara, Dimensions of a free linear polymer and polymer immobilized on silica nanoparticles of a zwitterionic polymer in aqueous solutions with various ionic strengths, *Langmuir*, 24, 8772–8778, 2008.

56. M. Kikuchi, Y. Terayama, T. Ishikawa, T. Hoshino, M. Kobayashi, H. Ogawa, H. Masunaga et al., Chain dimension of polyampholytes in solution and immobilized brush states, *Polym. J.*, 44, 121–130, 2012.

57. L. Mattei, F. Di Puccio, B. Piccigallo, and E. Ciulli, Lubrication and wear modelling of artificial hip joints: A review, *Tribol. Int.*, 44, 532–549, 2011.

58. T. Moro, Y. Takatori, K. Ishihara, T. Konno, Y. Takigawa, T. Matsushita, U. Chung, K. Nakamura, and H. Kawaguchi, Surface grafting of artificial joints with a biocompatible polymer for preventing periprosthetic osteolysis, *Nat. Mater.*, 3, 829–836, 2004.

59. T. Moro, H. Kawaguchi, K. Ishihara, M. Kyomoto, T. Karita, H. Ito, K. Nakamura, and Y. Takatori, Wear resistance of artificial hip joints with poly(2-methacryloyloxyethyl phosphorylcholine) grafted polyethylene: Comparisons with the effect of polyethylene cross-linking and ceramic femoral heads, *Biomaterials*, 30, 2995–3001, 2009.

60. T. Moro, Y. Takatori, M. Kyomoto, K. Ishihara, K. Saiga, K. Nakamura, and H. Kawaguchi, Surface grafting of biocompatible phospholipid polymer MPC provides wear resistance of tibial polyethylene insert in artificial knee joints, *Osteoarthritis Cartilage*, 18, 1174–1182, 2010.

5 Modification of Biopolymers Friction by Surfactant Molecules

Maurice Brogly, Ahmad Fahs, and Sophie Bistac

CONTENTS

Green materials such as hydroxypropyl methylcellulose (HPMC) appear to be a successful biomaterial alternative for films and coating formulations to replace gelatin-based products for pharmaceutical dosage forms (capsules, tablets, etc.). Friction and surface properties of such biopolymer films play an important role during the capsule opening and closing processes. This chapter demonstrates that incorporation of low concentration of stearic acid additive (0.1% w/w) in film formulations allows to tune the friction properties of HPMC film. The surface energy of HPMC film is strongly reduced by the incorporation of fatty acids. Friction measurements at nanoscale were performed using friction force microscopy. At macroscale, a pin-on-disk friction test was conducted to access tribological properties. The correlation of the results between nano- and macro-tests was evidenced and shows that surface properties are governed by phase separation between hydrophobic additive and HPMC matrix films, and that accumulation of stearic acid molecules at the surface favors surface sliding and thus reduces friction.

5.1 INTRODUCTION

In the pharmaceutical and packaging industries, films are defined as a fine layer to protect drugs or food materials. Films can be formed into pouches, capsules, bags, or casings through further fabrication processes. However, coatings are a particular form of films directly applied to the surface of materials. They usually consist of biopolymers and food-grade additives able to provide mechanical strength and other important functions [1]. Films mainly limit diffusion of water vapor, oxygen, carbon dioxide, and so on, and enhance the quality of drugs and food products, protecting them from physical, chemical, and biological deterioration. Film-forming biopolymers can be proteins (gelatin, gluten, etc.), polysaccharides (starch, cellulose, chitosan, etc.), or lipids (waxes, fatty acids, etc.). They can also be hydrophilic or hydrophobic and the solvents used are restricted to water and ethanol. Surfactants, plasticizers, and other additives are combined with the film-forming biopolymers to modify the physical properties or functionality of films [2–5]. In the last few years, research has focused on the development of films and coating from environmentally friendly materials. Among these, polysaccharide polymers such as cellulose ethers (carboxymethyl cellulose (CMC), methylcellulose (MC), hydroxypropyl methylcellulose (HPMC)), and chitosan have been particularly studied [6–9]. The use of these materials presents many advantages including widespread availability, low cost, and biodegradability.

In this work, HPMC is used as a matrix for film formation. During the formulation process, an additive that plays the role of a surfactant molecule is incorporated in order to modify the film surface characteristics. Fatty acids frequently incorporated into HPMC films lead to decrease water affinity and moisture diffusion due to their hydrophobic properties [7,10–12]. Surface-active agents or surfactants are added to coating solutions and films to reduce the surface tension and modify the wetting properties by increasing the spreadability of the coating material [13]. Moreover, coatings intended for pharmaceutical dosage forms (capsules and tablets) must exhibit specific friction properties, which are considered important depending on their applications. Indeed, friction properties of film surface are the key factors for controlling hard capsule opening and closing processes, especially when the surface possesses a high roughness [14], or to reduce the friction arising at the interface of tablet coatings and dry walls during compression and ejection processes.

5.2　EXPERIMENTAL

5.2.1　MATERIALS

Cellulose ethers are a class of semisynthetic polymers obtained by chemical reaction of the hydroxyl groups at positions two, three, and/or six of the anhydroglucose residues of cellulose, which are made of D-glucopyranose units with chair conformation, bonded through $\beta(1 \rightarrow 4)$ glycosidic linkages. HPMC, one of the cellulose ethers, contains two types of substituents: the methoxy group (OCH_3) and the hydroxypropyl group (OC_3H_6OH). The chemical structure of HPMC is shown in Figure 5.1a. The physicochemical properties of HPMC polymers are strongly affected by (i) the methoxy group content, (ii) the hydroxypropoxy group content, and (iii) the molecular weight [15]. HPMCs are described by the degree of substitution (DS) and the molar substitution (MS). Each anhydroglucose unit in the cellulose chain has three hydroxyl groups available for modification. Thus, if all three available positions on each unit are substituted, the DS is designated as three, and if two available positions on each unit are substituted, the DS would be two (Figure 5.1b). The term DS is reserved for substituents that block reactive hydroxyl groups (methoxy groups). Derivatization of a reactive hydroxyl group with propylene oxide generates a replacement hydroxyl site for further reaction, followed by the formation of internal and external hydroxypropoxy groups. Thus, the substitution is described by the MS, that is, the number of moles of hydroxypropyl groups per mole of anhydroglucose in the chain (Figure 5.1c). The United States Pharmacopeia (USP) distinguishes different types of HPMCs, classified according to their relative methoxy and hydroxypropoxy contents: HPMC 1828, HPMC 2208, HPMC 2906, and HPMC 2910. The first two numbers indicate the weight percent of methoxy groups and the last two numbers the weight percent of hydroxypropoxy groups, determined after drying at 105°C for 2 h. Among the HPMC that have been commercially available worldwide for many years, HPMC 2910 has the best solubility in organic solvents, and so it has often been used for aqueous and organic-solvent-based films. In the present study, HPMC 2910 was used, which was kindly supplied by Colorcon, France. The quantification of methoxyl and hydroxypropoxyl contents were carried out [16] by infrared and ^{13}C-NMR spectroscopy. The HPMC 2910 has a DS of 1.91 (methoxyl substituents) and an MS of 0.25 (hydroxypropyl substituents). DS can also be expressed in terms of weight percent. Thus, for HPMC 2910, the amount of OCH_3 groups is 28.8% (w/w) and hydroxypropyl groups is 9.1% (w/w). HPMC polymers are available in a number of viscosity grades, defined as the nominal viscosity of a 2% w/w aqueous solution at 20°C. In the present study, we used the grade that has a nominal viscosity of 3 mPa · s, which corresponds to a molecular weight of 10,000 g · mol^{-1} as determined by size exclusion chromatography (SEC). The glass transition temperature was determined by differential scanning calorimetry (DSC) and was found to be 138°C.

　　Stearic acid (SA) ($C_{18}H_{36}O_2$) was purchased from Sigma-Aldrich and was used as received as an additive for the film-forming process. The purity of SA is higher than 99%, and the melting point of the white crystalline powder was in the range 69–72°C.

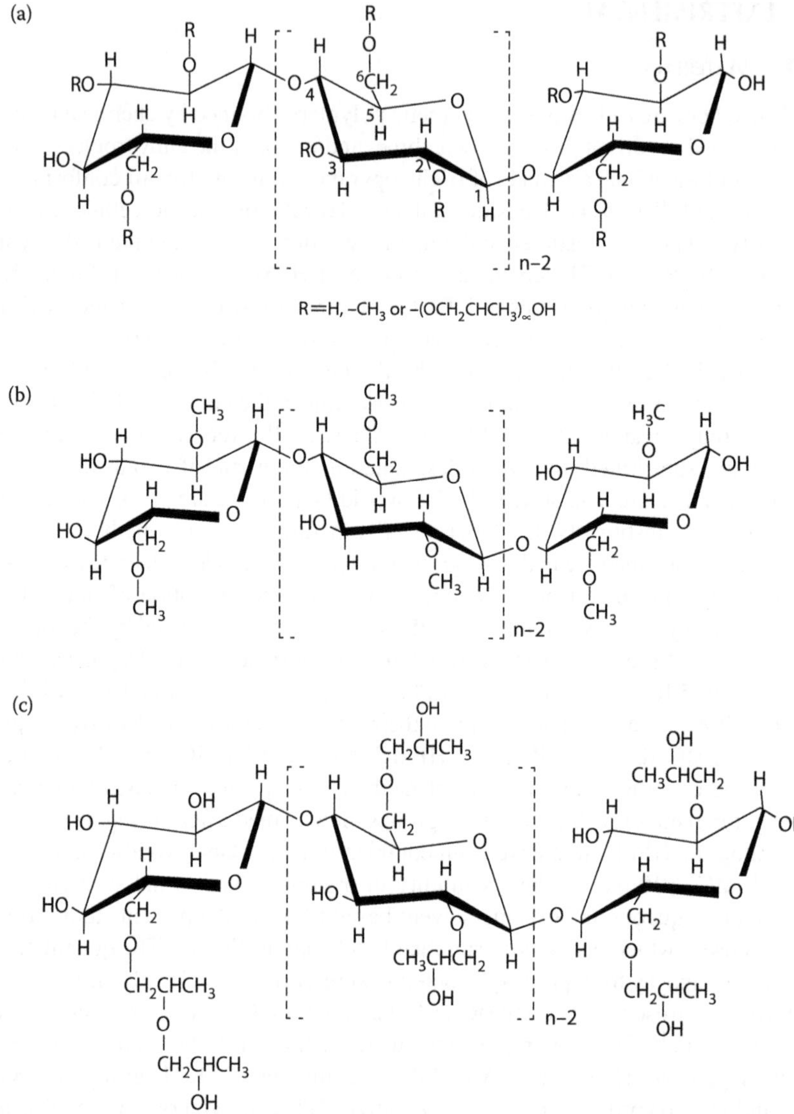

(a)

R=H, –CH$_3$ or –(OCH$_2$CHCH$_3$)$_\infty$OH

(b)

(c)

FIGURE 5.1 (a) Chemical structure of hydroxypropyl methylcellulose (HPMC). (b) Chemical structure of methylcellulose (MC) with two hydroxyl groups of each anhydroglucose unit that were substituted (i.e., DS of two). (c) Chemical structure of hydroxypropylcellulose (HPC) that has 2 moles of hydroxypropyl groups per mole of anhydroglucose in the chain (i.e., MS of two).

5.2.2 METHODS

5.2.2.1 Film Preparation

Biopolymer films were prepared by the "solvent process" or "casting," which is based on the drying of the film-forming solution. It is used to form edible films or coatings.

This process is generally adapted for coating seeds and foods, for making cosmetic masks or varnishes, and for making pharmaceutical capsules. The polymer solution was cast onto a glass plate substrate and the solvent was evaporated. The film was then easily removed off the substrate. The solvents used were restricted to water and ethanol.

The formation of HPMC films requires preparing HPMC solutions (6% w/w) and casting them onto a Petri dish, and then drying at ambient conditions. An HPMC solution was prepared by dispersing HPMC powder under moderate agitation in deionized water at 80–90°C to prevent lumping. After HPMC was dissolved by hydration, the solution was cooled down to $25 \pm 2°C$. A clear solution was obtained after cooling. To achieve the maximum hydration of HPMC polymer, the solution was preserved at 5°C for 24 h.

HPMC-formulated films were prepared by incorporation of SA additive into the HPMC solution. The amount of additive was expressed as a weight percentage of HPMC powder. For HPMC-SA films, the fatty acid was first dissolved in 10 mL of absolute ethanol, and then added to the HPMC solution under magnetic stirring. Four different solutions with various concentrations of SA were prepared: 0.05%, 0.1%, 0.5%, and 1% w/w. All solutions were homogenized for 2 h and preserved for 24 h at 5°C before use.

Film-forming solutions were spread onto a Petri dish to form uniform layers (1.6 mm thick), and dried under natural conditions. The films obtained were 100 μm thick. In order to compare sample properties, all formulated films were preserved at $25 \pm 2°C$ and $30 \pm 5\%$ RH before experiments.

5.2.2.2 Surface Energy and Wettability

The surface energy of a solid represents the discontinuity existing between the bulk of the solid and its surface [16]. It quantifies the disruption of intermolecular bonds that occurs when a surface is created. The surface energy may, therefore, be defined as the excess energy at the surface of a material compared with the bulk [17]. The presence of surfactant molecules incorporated into HPMC may strongly affect the surface energy, γ_s, of the HPMC-SA films. The most common way used to determine the surface energy consists of measuring the contact angles of reference liquids onto the solid surface of interest [18]. The drop was imaged and the resulting contact angle was determined by drop shape analysis software. The instrument used was a Krüss G2 Goniometer (Krüss GmbH, Hamburg, Germany). Contact angles were measured in open air, relative humidity of $30 \pm 5\%$ RH, and room temperature ($22 \pm 2°C$). The three reference liquids were water, diiodomethane, and α-bromonaphthalene. The volume of droplets was in the range 2–3 μL. Ten droplets were imaged at different regions of the same film. The contact angle was averaged over these 10 measurements. The average contact angle was used to calculate surface energy of formulated films using the Owens and Wendt [19] method.

5.2.2.3 Friction Force Microscopy

Friction force microscopy (FFM) is a powerful tool for surface characterization of materials and helps explain friction phenomena at the nanoscale. Under a given applied load, which corresponds to a given vertical deflection of the cantilever (expressed in volt on the microscope photodetector), the sample is scanned back and forth (producing a friction loop) in a direction perpendicular to the long axis of an atomic force microscope (AFM) cantilever. The friction force between the film and the tip causes twisting of the AFM cantilever. This twisting is given by the output signal of the two horizontal quadrants of the microscope photodetector (Figure 5.2a).

The differential signal between the left (L) and right (R) quadrants of the photo-detectors is denoted as FFM signal [(L − R)/(L + R)]. This signal can be related to the degree of twisting, and hence to the magnitude of friction force, which can also be influenced by adhesion force at the tip–sample interface. Nanoscale friction is quantified in terms of trace minus retrace (TMR) (in volts) equal to the difference between friction forces measured for left-to-right scanning and right-to-left scanning. In this case, to calculate the coefficient of friction (μ_{nano}), friction force (F_{TMR}) should be divided by the sum of applied load (F_N) and adhesion force (F_{ADH}). The coefficient of friction is thus given by [20]

$$\mu_{nano} = \frac{F_{TMR}}{F_N + F_{ADH}} \tag{5.1}$$

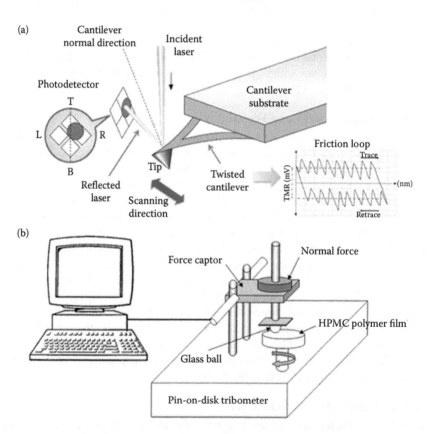

FIGURE 5.2 (a) Friction force microscope (FFM) principle showing the deformation of tip and cantilever during sample sliding. (b) Schematics of the pin-on-disk tribometer.

A higher TMR value corresponds to a larger friction force. Friction forces (in volts) can be converted into force units (in newtons) and absolute friction forces can be obtained using complex calibration methods [21,22]. The lateral spring constant depends critically on the tip length and the tip should be centered at the free end. In our study, we did not use the calibration method in FFM; thus, the study will provide a comparison between samples using the same cantilever for all HPMC-SA films, and nano-friction measurements were made using square-pyramidal-shaped tips. The Si_3N_4 probe has a tip radius of 45 ± 3 nm and nominal spring constant of $0.3 \ N \cdot m^{-1}$. TMR values were measured for different applied normal loads (expressed in volts and ranging from 0.1 to 3 V). The friction properties of different HPMC films were compared using TMR values. The nano-tribological tests were carried out in an ambient atmosphere. For a given applied load, the TMR value during the sliding of the AFM probe on the film surface was constant. The friction force (TMR value) increases when the applied load increases. Friction force is the average of TMR values at different locations on the sample surface.

5.2.2.4 Pin-On-Disk Tribological Experiments

Tribological investigations of HPMC-formulated films at macroscale were carried out on a conventional pin-on-disk tribometer (CSM Instruments SA, Peseux, Switzerland) (Figure 5.2b) by sliding a glass ball (diameter = 6 mm) on substrates [23]. The HPMC films were deposited on a glass disk and a glass ball was brought into contact with HPMC films, under a selected normal load. The glass disk was then rotated at a given speed and the force opposing the motion of the ball (tangential force), which corresponds to the friction force, was recorded. The coefficient of friction (μ) was calculated by dividing the measured friction force (F_T) by the specified applied normal load (F_N).

$$\mu = \frac{F_T}{F_N} \tag{5.2}$$

The experiment was performed to a sliding distance of 100 mm, at a sliding velocity of $5 \ mm \cdot s^{-1}$, under a normal load of 2, 5, and 10 N. Data were collected at 25°C and at an acquisition rate of 100 Hz.

5.3 RESULTS AND DISCUSSION

5.3.1 Effect of Stearic Acid on Wettability and Surface Energy

The Owens and Wendt approach [19] was used to determine the surface energy of the different HPMC and HPMC-SA films. According to Owens and Wendt, the surface energy of a solid has two components: first one due to dispersive interactions and second one due to nondispersive (mainly polar) interactions. This theory is derived from the combination of the well-known Young's equation, which relates the contact angle (θ) to the surface energies of the solid (γ_S) and liquid (γ_L) and to the interface tension (γ_{SL}) (Figure 5.3), and Good's equation, which relates the interface tension to

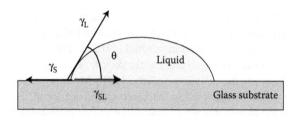

FIGURE 5.3 Schematics of a liquid drop on a glass substrate.

the nondispersive (γ^{ND}) and dispersive (γ^D) components of the surface energy. The resulting principle equation is

$$\frac{\gamma_L (\cos\theta + 1)}{2 \cdot \sqrt{\gamma_L^D}} = \frac{\sqrt{\gamma_S^{ND}} \cdot \sqrt{\gamma_L^{ND}}}{\sqrt{\gamma_L^D}} + \sqrt{\gamma_S^D} \qquad (5.3)$$

This equation has the form of $y = ax + b$, with

$$y = \frac{\gamma_L (\cos\theta + 1)}{2 \cdot \sqrt{\gamma_L^D}} \quad a = \sqrt{\gamma_S^{ND}} \quad x = \frac{\sqrt{\gamma_L^{ND}}}{\sqrt{\gamma_L^D}} \quad b = \sqrt{\gamma_S^D}$$

As such, the dispersive and nondispersive components of the solid's surface energy are determined by the slope and intercept of the resulting $y = f(x)$ graph. Of course, in order to make that graph, knowing the surface energy (γ_L) of the probe liquid is not enough, as it is necessary to know specifically its dispersive and nondispersive components such as $\gamma_L = \gamma_L^D + \gamma_L^{ND}$. The probe liquids used in this study were water ($\gamma_L^D = 21.8$ mJ·m^{-2}; $\gamma_L^{ND} = 51$ mJ·m^{-2}), diiodomethane ($\gamma_L^D = 48.5$ mJ·m^{-2}; $\gamma_L^{ND} = 2.3$ mJ·m^{-2}), and α-bromonaphthalene ($\gamma_L^D = 44.4$ mJ · m^{-2}; $\gamma_L^{ND} = 0$ mJ·m^{-2}). The accuracy and precision of this method is supported largely by the confidence level of the results for appropriate liquid/solid combinations. The Owens and Wendt approach is typically applicable to surfaces with low charge and moderate polarity, which is the case for HPMC and HPMC-SA films.

The evolution of the surface hydrophobic character with incorporation of SA was first investigated using the contact angle of water (Figure 5.3). As mentioned above, the contact angle is defined as the angle between the substrate surface and the tangent line at the point of contact of the liquid droplet with the substrate. Table 5.1A shows the contact angle values of water droplets obtained for HPMC-SA films. It is known that the water contact angle increases with increasing surface hydrophobic character. The results confirmed that the addition of SA increases the water contact angle value of HPMC-formulated films. The contact angles of diiodomethane and α-bromonaphthalene were also measured on the different HPMC-SA films in order to determine the surface energy of the films using the Owens and Wendt [19] method. The results show that the surface energy decreases with increasing SA concentration in the film, that is, the hydrophobic character of the film surface (Table 5.1B). Even for low SA concentration (0.05% and 0.1% w/w), a strong decrease of surface energy

TABLE 5.1A

Water (H₂O), Diiodomethane (CH₂I₂), and α-Bromonaphthalene (α-Br) Contact Angle Values of HPMC-Stearic Acid Films

Liquid	Pure HPMC Film	+0.05% SA	+0.1% SA	+0.5% SA	+1% SA
H_2O	69° ± 2°	76° ± 2°	84° ± 2°	91° ± 2°	94° ± 2°
CH_2I_2	55° ± 2°	55° ± 2°	56° ± 2°	57° ± 2°	58° ± 2°
α-Br	43° ± 2°	44° ± 2°	45° ± 2°	46° ± 2°	47° ± 2°

TABLE 5.1B

Surface Energy Values of HPMC-Stearic Acid Films

Surface Energy	Pure HPMC Film	+0.05% SA	+0.1% SA	+0.5% SA	+1% SA
γ_s (mJ·m^{-2})	43 ± 1	38 ± 1	36 ± 1	34 ± 1	31 ± 1

was observed. The water contact angle increases from 69° to 94° with the incorporation of 1% w/w SA (Table 5.1A). As a result, the addition of 1% w/w of SA decreased the surface energy from 43 to 31 mJ·m^{-2}. The addition of SA significantly affects the surface energy of HPMC-formulated films, reflecting therefore the reduction in hydrophilicity of films and shows the presence of nonpolar aliphatic chains at the top of the film surface. This result indicates that the SA chains have the possibility of diffusing to and accumulating at the film surface.

In order to quantify the migration of SA to the film surface, one must determine the surface fraction of SA on the film surface. We used the Cassie–Baxter approach [24,25], which describes the apparent liquid contact angle, θ^*, measured on a heterogeneous solid surface composed of two different materials (from the chemical point of view). The Cassie equation can be expressed by

$$\cos \theta^* = f_1 \cos \theta_1 + f_2 \cos \theta_2 \tag{5.4}$$

where, for a water droplet deposited on the film surface, θ_1 is the contact angle of water on pure HPMC film ($\theta_{H_2O/HPMC} = 69°$), θ_2 is the contact angle of water experimentally determined by deposing a droplet of water on a pure SA film ($\theta_{H_2O/SA} = 89°$), and θ^* is the experimental contact angle on HPMC-SA film surface. The factors f_1 and f_2 represent the corresponding surface fractions of HPMC and SA, respectively, and $f_1 = 1 - f_2$.

Thus, Equation 5.4 can be rearranged to

$$\cos \theta^* = \cos \theta_1 + f_2 (\cos \theta_2 - \cos \theta_1) \tag{5.5}$$

Using the measured values of θ^* (Table 5.1A), θ_1 and θ_2, one can calculate the surface fraction of SA (f_2) on the film surface. Figure 5.4 shows the calculated values

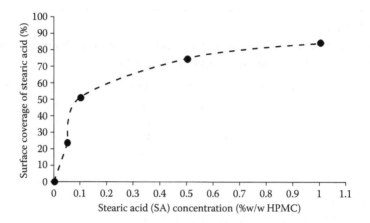

FIGURE 5.4 Variation of area fraction (%) of stearic acid (SA) as a function of its concentration in HPMC films.

of f_2 (in percent) as a function of SA concentration in the formulation (% w/w). The surface coverage of SA sharply increases with its concentration. Addition of 1% of SA to the film formulation is sufficient to induce a coverage of 80% of the HPMC film surface with SA molecules. This result indicates that a phase separation could occur that causes the migration of the hydrophobic additive (SA) to the top of the film surface, resulting in a decrease of surface energy of HPMC-formulated films (Table 5.1B).

5.3.2 FRICTION AT NANOSCALE USING FFM

5.3.2.1 Effect of Stearic Acid and Applied Load on Friction Force

During FFM friction measurements, friction force data were recorded during both forward (trace T) and backward (retrace R) scans (Figure 5.5). The TMR values for all HPMC-SA film samples, at different normal loads (expressed in volts), are summarized in Table 5.2. Each TMR value is the average of data at different locations

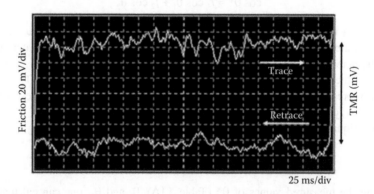

FIGURE 5.5 Friction curve (TMR) of pure HPMC films obtained by FFM (applied load 0.1 V).

TABLE 5.2

Friction Force at the Nanoscale (TMR Value Recorded in Volts) of HPMC-Stearic Acid Films for Different Applied Loads (0.1–3 V ± 0.01 V)

	Applied Load			
Sample	0.1 V	1 V	2 V	3 V
HPMC film	132	183	247	309
+0.05% SA	111	157	213	269
+0.01% SA	93	128	183	233
+0.5% SA	82	118	169	210
+1% SA	79	107	156	188

on the sample surface. For all samples, the data showed that the increase of SA concentration causes a decrease of the TMR value and, thus, of the friction force regardless of the value of the applied normal load. Pure HPMC film presents the highest friction forces at all applied normal loads. In order to study the effect of the hydrophobic additive on friction properties of HPMC films, we estimated the coefficient of friction. Figure 5.6 shows a plot of the friction force (TMR value) as a function of applied normal load from Table 5.2. As shown in Figure 5.6, the friction force has a linear dependence on applied load, which can be described by a modified form of Amonton's law, given by [26]

$$F_{TMR} = F_T^0 + \mu_{nano} F_N \qquad (5.6)$$

μ_{nano} is the apparent coefficient of friction, F_N is the normal load, F_T^0 is the friction force when the external load is zero, and F_{TMR} is the experimental friction force.

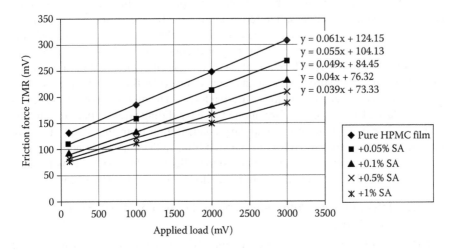

FIGURE 5.6 Friction forces (TMR values) versus applied normal load for different composition of HPMC-SA films.

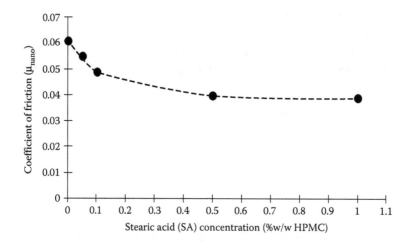

FIGURE 5.7 Apparent coefficient of friction of HPMC films versus SA concentration.

The slope of the friction-versus-load curve can be used as the apparent coefficient of friction because it is directly proportional to the actual coefficient of friction [27]. Figure 5.7 shows the variation of apparent coefficient of friction for various formulations of HPMC films. It can be seen that the pure HPMC film has the highest coefficient of friction (0.061). Addition of 1% SA to HPMC film decreased the nanofriction coefficient by 36%. Indeed, friction force decreased when SA concentration increased. Taking into account the effect of SA surface hydrophobicity (Table 5.1A and B, Figure 5.4), one can suspect that addition of SA in HPMC films decreases the magnitude of polar interactions inside the AFM tip–film surface contact zone. Indeed and relative to HPMC, SA molecules have a lower surface energy equal to 22.8 mJ·m^{-2} as determined by wettability on pure SA film. Moreover, a strong decrease of capillary condensation forces around the tip apex also occurs when the amount of SA increases due to the accumulation by phase separation of hydrophobic SA molecules at the film surface. Lowering of friction force with increasing concentration of SA shows that a decrease of capillary contribution is a dominant factor. In other words, the hydrophobic chains reduce the AFM tip–HPMC films surface interactions and thereby lead to lower friction forces. The migration of SA molecules to the surface also induces the formation of an easy sliding layer on the film surface. These observations indicate that capillary condensation is an important factor for the frictional properties of HPMC-formulated films under ambient conditions. This observation agrees with the results obtained for cellulosic films [28], polymer-coated silicon surface [29], and peptide-containing alkylsiloxane monolayers [26].

5.3.2.2 Effect of Nano-Adhesion on Nano-Friction of HPMC-Stearic Acid Films

The increase of applied load induces a higher friction force, in agreement with the modified Amonton's law (Equation 5.6). The data of TMR versus applied load are linear for both HPMC and HPMC-formulated films (Figure 5.6). The intercept of

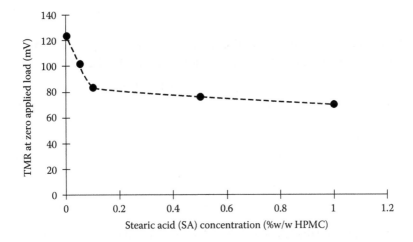

FIGURE 5.8   Variation of TMR (at zero applied load) as a function of SA concentration (% w/w).

the friction force versus normal load line at zero applied load (0 V) corresponds to the nano-adhesion force. TMR values obtained at zero applied load are plotted in Figure 5.8 as a function of SA concentration. A constant decrease of TMR (at 0 V) is observed. The change of TMR at zero load as a function of SA concentration shows that adhesion and friction mechanisms, at a local scale, are similar (compare Figures 5.7 and 5.8). The study of the friction properties of HPMC films at nanoscale underlines clearly that nano-adhesion plays an important role in friction measurement of HPMC films. SA causes the formation of a weak boundary layer (WBL) [30] at the film surface. WBL is formed by migration and surface segregation of SA molecules. This WBL has low cohesion and leads to easy sliding and low adhesion properties during tip–film surface contact (hydrophobic and low molecular weight compared with pure HPMC).

5.3.3   TRIBOLOGICAL PROPERTIES OF HPMC-STEARIC ACID FILMS AT MACROSCALE

5.3.3.1   Coefficient of Friction

The macro tribological properties of HPMC-SA films were investigated from friction of a glass ball against the film deposit on a glass disk. The coefficients of friction were measured as a function of sliding time and sliding distance. Before each test, the glass ball was chemically cleaned in a 3:1 mixture of concentrated sulfuric acid and hydrogen peroxide (piranha solution). Each data point is the average of three measurements. The average coefficient of friction over all sliding distances was automatically computed. No change of coefficient of friction with sliding distances was observed at the macroscale. The mean coefficient of friction of the film was determined from the slope of the plot of friction force (tangential force) versus applied normal load. Table 5.3 is a summary of the coefficients of friction obtained for

TABLE 5.3

Values of Coefficients of Friction at the Macroscale, μ_{macro}, of HPMC-Stearic Acid Films for Different Applied Loads (2–10 N ± 0.01 N)

	Applied Load		
Sample	2 N	5 N	10 N
HPMC film	0.48	0.43	0.4
+0.05% SA	0.19	0.16	0.18
+0.01% SA	0.14	0.11	0.12
+0.5% SA	0.1	0.11	0.1
+1% SA	0.12	0.1	0.09

different samples of HPMC and HPMC-SA films. The values of coefficients of friction show little changes due to increased load for pure HPMC film. It also remained constant for HPMC-SA films, regardless of the applied load in the range 2–10 N. The addition of a very small amount of SA (0.05% w/w) sharply reduced the coefficient of friction, which remained constant above 0.1% w/w of SA in the HPMC films.

5.3.3.2 Effect of Stearic Acid on the Friction Properties of HPMC-SA Films

To study the effect of applied load on the macro tribological properties of HPMC films, the friction forces were evaluated at different concentrations of SA (Figure 5.9) with increasing applied loads ranging from 2 to 10 N. The friction forces were calculated, using the data in Table 5.3, as follows:

$$F_T = \mu_{macro} \times F_N \qquad (5.7)$$

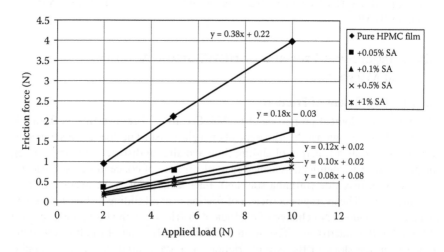

FIGURE 5.9 Friction force versus applied normal load for HPMC-SA films.

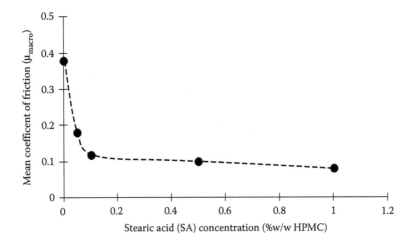

FIGURE 5.10   Evolution of the mean friction coefficient versus SA concentration (% w/w) in HPMC-SA films.

where F_N is the applied normal load (N) and μ_{macro} is the mean coefficient of friction.

For all samples, the mean coefficient of friction μ_{macro} is the slope in the friction force (N) versus the applied load (N) plot (Figure 5.9). A high dependence of friction force on applied load was observed for pure HPMC films. The high coefficient of friction of pure films may be due to the hydrophilic contact of the glass ball (which is considered to have a rather polar surface) with polar HPMC made of long polymer chains having low flexibility. As a consequence, adhesion between the glass and the pure HPMC is strong and prevents easy sliding. Results of the mean coefficient of friction as a function of SA concentration are shown in Figure 5.10. For a pure HPMC film, the mean coefficient of friction is 0.38. On the other hand, for HPMC-SA-formulated films, addition of even small amounts of SA (0.05% and 0.1% w/w) sharply decreased the mean coefficient of friction. Moreover, extrapolation of friction force curve to zero normal load shows very low intercept values. These results also support the hypothesis of the formation of a WBL of SA, which plays the role of an easy sliding agent. As a consequence, the coefficient of friction decreases both with applied normal load and with SA concentration.

5.3.4   NANO-FRICTION/MACRO-FRICTION CORRELATION

Generally speaking, the results of nano- and macro-friction data show the same trend with variation of the chemical composition of the films. On a small scale, the high sensitivity of AFM and FFM modes provides a precise evaluation of any modification of the surface properties. On a larger scale, since the adhesion force is not significant even for pure films, the changes induced due to SA concentration (in the range 0.05–1%) remain highly relevant in terms of friction force, and reveal the high impact of SA on the mean coefficient of friction, which drops sharply even for very low amounts of SA (0.05 and 0.1% w/w).

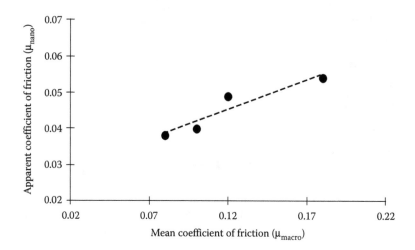

FIGURE 5.11   Correlation of apparent coefficient of friction (μ_{nano}) versus mean coefficient of friction (μ_{macro}).

SA has a great influence on the frictional properties at macroscale. Incorporation of SA allows HPMC films to exhibit easy sliding characteristics. Its strong effect on HPMC films is believed to be caused by the polar nature of hydroxyl groups on cellulosic chains that induces phase separation and surface migration of SA. In addition, the low molecular weight of the fatty acid favors surface sliding and thus reduces friction. These results are in good agreement with the friction properties of HPMC-SA at the nanoscale that show the same tendency as those obtained at the macroscale. Large decrease of coefficient of friction was observed by the addition of 0.1% w/w SA to HPMC, and a good nano/macro correlation is observed (Figure 5.11). All these results support the formation of a WBL at the film surface. The driving force is the thermodynamic phase separation, migration, and accumulation of SA at the film surface.

5.4   CONCLUSION

The frictional properties of HPMC and HPMC-formulated films at two different scales were investigated by AFM/FFM and a ball-on-disk friction test. SA, which was incorporated as a hydrophobic additive to HPMC films, had a strong effect on surface properties. The lowering of the surface energy of HPMC by SA incorporation was confirmed by wettability measurements. Surface coverage of the film surface by SA molecules was also estimated from wettability data. The results show that even for a very low concentration of SA, the phase separation process leads to accumulation of SA at the film surface. A hypothesis of surface migration of a hydrophobic additive is proposed and confirmed with the help of the Cassie–Baxter equation. A strong correlation between surface energy and friction force at the nanoscale was observed.

Nano- and macroscale techniques show that surface properties are governed by phase separation between a hydrophobic additive and HPMC matrix, and surface accumulation of SA. The friction force between the AFM tip and the film surface was highly influenced by the surface energy and adhesion at the nanoscale. The experiments revealed that water capillary force acts as the driving force that has a large influence on friction at the nanoscale. A good correlation was observed between nano-friction and nano-adhesion (i.e., friction force at zero load).

In addition, the macroscopic results showed a good agreement with those at the nanoscale. The macro-friction between a glass ball and the HPMC-formulated films surface was also lowered with SA concentration. These results support the surface migration and surface phase separation phenomena between polar HPMC matrix and hydrophobic fatty acid. The low-molecular-weight SA molecules favor surface sliding and thus reduce friction. The nano and macro properties of HPMC-formulated films were strongly dependent on SA concentration. This study provided evidence for a relationship between nano- and macro-friction properties for systems where the additive plays an important role as a surfactant and a lubricant. The results clearly underline the strong dependence of film surface properties on additive property, concentration, and additive–biopolymer matrix compatibility.

ACKNOWLEDGMENT

The authors are grateful to Région Alsace for funding this work.

REFERENCES

1. J. H. Han, *Innovations in Food Packaging*, Elsevier Academic Press, San Diego, CA, 2005.
2. J.A.Q. Gallo, F. Debeaufort, and A. Voilley, Interactions between aroma and edible films. 1. Permeability of methylcellulose and low-density polyethylene films to methyl ketones, *J. Agr. Food Chem.*, 47, 108–113, 1999.
3. J. J. Kester and O. Fennema, Edible films and coatings, *Food Technol.*, 40, 47–59, 1986.
4. H. Möller, S. Grelier, P. Pardon, and V. Coma, Antimicrobial and physicochemical properties of chitosan—HPMC-based films, *J. Agr. Food Chem.*, 52, 6585–6591, 2004.
5. I. Sebti, E. Chollet, P. Degraeve, C. Noel, and E. Peyrol, Water sensitivity, antimicrobial, and physicochemical analyses of edible films based on HPMC and/or chitosan, *J. Agr. Food Chem.*, 55, 693–699, 2007.
6. R. Villalobos, P. Hernández-Munõz, and A. Chiralt, Effect of surfactants on water sorption and barrier properties of hydroxypropyl methylcellulose films, *Food Hydrocolloids*, 20, 502–509, 2006.
7. I. Sebti, F. Ham-Pichavant, and V. Coma, Edible bioactive fatty acid-cellulosic derivative composites used in food-packaging applications, *J. Agr. Food Chem.*, 50, 15, 4290–4294, 2002.
8. N. E. Suyatma, L. Tighzert, A. Copinet, and V. Coma, Effects of hydrophilic plasticizers on mechanical, thermal, and surface properties of chitosan films, *J. Agr. Food Chem.*, 53, 3950–3957, 2005.
9. F. Debeaufort and A. Voilley, Methylcellulose-based edible films and coatings: 2. Mechanical and thermal properties as a function of plasticizer content, *J. Agr. Food Chem.* 45, 685–689, 1997.
10. R. D. Hagenmaier and P. E. Shaw, Moisture permeability of edible films made with fatty acid and (hydroxy propyl) methylcellulose, *J. Agr. Food Chem.*, 38, 1799–1803, 1990.

11. E. Ayranci and S. Tunc, The effect of fatty acid content on water vapor and carbon diox-ide transmission of cellulose-based edible films, *Food Chem.*, 72, 231–236, 2001.
12. V. Coma, I. Sebti, P. Pardon, A. Deschamps, and H. Pichavant, Anti-microbial edible packaging based on cellulosic ethers, fatty acids and nisin incorporation to inhibit *Listeria innocua* and *Staphylococcus aureus*, *J. Food Protect.*, 64, 470–475, 2001.
13. R. Villalobos, P. Hernandez-Munoz, A. Albors, and A. Chiralt, Barrier and optical prop-erties of edible hydroxypropyl methylcellulose coatings containing surfactants applied to fresh cut carrot slices, *Food Hydrocolloids*, 23, 526–535, 2009.
14. J. W. McGinity and L. A. Felton, *Aqueous Polymeric Coatings for Pharmaceutical Dosage*, Informa Healthcare Inc., New York, 2008.
15. J. Siepmann and N.A. Peppas, Modeling of drug release from delivery systems based on hydroxypropyl methylcellulose (HPMC), *Adv. Drug Deliv. Rev.*, 48, 139–157, 2001.
16. A. Fahs, PhD Thesis, Université de Haute Alsace, France, 2010.
17. J. N. Israelachvili, *Intermolecular and Surface Forces*, Academic Press Ltd, London, 1991.
18. F. M. Etzler, *Contact Angle, Wettability and Adhesion*, Vol. 3, K. L. Mittal (Ed.), pp. 219–264, VSP, Utrecht, 2003.
19. D. K. Owens and R. C. Wendt, Estimation of the surface free energy of polymers, *J. Appl. Polym. Sci.*, 13, 1741–1747, 1969.
20. B. Bhushan, *Nanotribology and Nanomechanics: An Introduction*, Springer, New York, 2005.
21. K. Jradi, PhD Thesis, Université de Haute Alsace, France, 2007.
22. S. Bistac and A. Galliano, Nano and macro tribology of elastomers, *Tribol. Lett.*, 18, 21–25, 2005.
23. A. Galliano, S. Bistac, and J. Schultz, The role of free chains in adhesion and friction of poly(dimethylsiloxane) (PDMS) networks, *J. Adhes.*, 79, 973–991, 2003.
24. A. B. D. Cassie and S. Baxter, Wettability of porous surfaces, *Trans. Faraday Soc.*, 40, 546–551, 1944.
25. A. Marmur, Soft contact: Measurement and interpretation of contact angles, *Soft Matter*, 2, 12–17, 2006.
26. S. Song, S. Ren, J. Wang, S. Yang, and J. Zhang, Preparation and tribological study of a peptide-containing alkylsiloxane monolayer on silicon, *Langmuir*, 22, 6010–6015, 2006.
27. N. J. Brewer, B. D. Beake, and G. Leggett, Friction force microscopy of self-assembled monolayers: Influence of adsorbate alkyl chain length, terminal group chemistry, and scan velocity, *Langmuir*, 17, 1970–1974, 2001.
28. N. Garoff and S. Zauscher, The influence of fatty acids and humidity on friction and adhesion of hydrophilic polymer surfaces, *Langmuir*, 18, 6921–6927, 2002.
29. S. L. Ren, S. R. Yang, J. Q. Wang, W. M. Liu, and Y. P. Zhao, Preparation and tribologi-cal studies of stearic acid self-assembled monolayers on polymer coated silicon surface, *Chem. Mater.*, 16, 428–434, 2004.
30. S. J. O'Shea, M. E. Welland, and T. Rayment, Atomic force microscope study of bound-ary layer lubrication, *Appl. Phys. Lett.*, 61, 2240–2242, 1992.

6 Characterization and Tribological Behavior of Polymer Brush Functionalized with Ionic Liquid Moiety

Tatsuya Ishikawa, Motoyasu Kobayashi, and Atsushi Takahara

CONTENTS

Ionic poly[1-(2-methacryloyloxy)ethyl-3-butylimidazolium bis(trifluorometh-anesulfonyl)imide] (PMIS) and nonionic poly(n-hexyl methacrylate) (PHMA) brushes were prepared on silicon wafers with immobilized initiator by sur-face-initiated atom transfer radical polymerization. Macroscopic frictional properties of the brushes were characterized by a ball-on-plate-type tribom-eter under reciprocating motion in a dry nitrogen atmosphere, water, metha-nol, toluene, and 1-ethyl-3-methylimidazolium bis(trifluoromethanesulfonyl) imide (EMImTFSI). EMImTFSI on PMIS brush showed much lower fric-tion coefficients than on nonmodified silicon wafer in the speed range of 10^{-4}–10^{-1} m s^{-1}. It is attributed that the high affinity of the PMIS brush to EMImTFSI led to a reduction in the interaction between the brush and the friction probe, which resulted in a low friction coefficient. The friction coef-ficient gradually decreased to 0.01 with increasing sliding velocity from 10^{-4} to 10^{-1} m s^{-1}. The friction coefficient of the PMIS brush remained low until 800 friction cycles in the dry nitrogen atmosphere, while the PHMA brush was abraded away within 150 friction cycles. The x-ray photoelectron spec-troscopy measurements of the worn surfaces on the PMIS brush indicated that the brush was gradually abraded away by friction but the brush structure still remained after 400 friction cycles.

6.1   INTRODUCTION

Since the use of room-temperature ionic liquids as novel lubricants was first reported by Liu and coworkers in 2001 [1–3], ionic liquids have attracted consid-erable attention as a new class of lubricants [4,5]. This is because of their unique properties such as negligible volatility, low flammability, high thermal stability [6], a low melting point, and a wide liquid range [7]. These characteristics are derived from ionic interactions between large organic cations and inorganic or organic noncoordinating anions, which together form salts with melting points below room temperature [8]. Ionic liquids are expected to be ideal candidates for new lubricants under severe conditions such as ultrahigh vacuum and extreme tem-peratures [9,10]. The tribological properties of some ionic liquids have been evalu-ated extensively. Ionic liquids exhibited excellent friction reduction, better wear resistance, and high load capacity than conventional lubricants such as synthetic hydrocarbons and fluoroether polymers [1]. The tribological properties of ionic liquids largely depend on the chemical structure of the organic cations and anions [11]. Liu and coworkers proposed that ionic liquids could be easily adsorbed on the sliding surfaces of frictional pairs because of their polar structure. These liquids can form an effective boundary film, which would reduce friction and wear [1–3]. Xiao and coworkers measured the film thicknesses of ionic liquids at high pres-sures up to 3 GPa in real time employing the relative optical interference intensity measurement method [12]. The film thicknesses of ionic liquids were greater than those of silicone oils of similar viscosities, indicating better film-forming ability of ionic liquids. Liu and coworkers further suggested that a tribochemical reaction between an ionic liquid and a friction surface under severe contact conditions forms surface-protective films, which was confirmed by Mori and his coworkers [13–15]. However, some tribochemical reactions involving the decomposition of the ionic liquid would cause corrosive wear. In the case of an N-alkyl imidazolium-derived ionic liquid, it has been reported that imidazolium with a short alkyl chain and a reactive anion, such as tetrafluoroborate (BF_4^-) or hexafluorophosphate (PF_6^-), increased wear through tribocorrosive attack on steel and aluminum surfaces [16]. One of the reasons for the increased wear is that BF_4^- and PF_6^- produce corrosive hydrogen fluoride upon hydrolysis, which can damage frictional systems [17]. Jiménez et al. reported that trifluoromethanesulfonate or 4-methylbenzene-sulfonate anions can reduce tribocorrosion and, consequently, friction and wear, despite the presence of the short alkyl chains of the imidazolium cation [18]. In contrast, BF_4^- of imidazolium cations with longer chains results in lower friction and wear compared with the corresponding PF_6^- salt [18,19]. The pressure viscosity coefficient of ionic liquids is also an important factor in the formation of a lubrication film [20]. Therefore, the lubrication film for hydrodynamic lubrication and the boundary layer on the friction surface are important for reducing friction and wear using ionic liquids as well as conventional oil-based lubricants. In this study, we propose another type of boundary film prepared by grafting a polymer film consisting of an ionic liquid moiety on the friction surface to improve the tribological properties of ionic liquids. It is expected that the grafted polymer bearing ionic liquid moiety would assist the retention of ionic liquids between the sliding pairs and work as an effective lubrication layer in combination with an ionic liquid.

Grafting polymer or alkyl chains on a solid surface is a promising surface modification method for controlling the surface wettability, adhesion, and tribological properties [21–23]. Generally, polymer chains tethered to a surface with sufficiently high graft density are named "polymer brushes" [24]. These structures can be prepared by chemical grafting of the end monomers of the chains or by adsorption of a diblock copolymer on the surface: one of the blocks serves as an anchor. The densely grafted brush chains assume a highly extended conformation in a good solvent due to osmotic pressure. The reduction in the friction forces between solid surfaces bearing polymer brushes, especially in a good solvent, has been studied well, both theoretically [25,26] and experimentally [27–30]. The efficient lubrication of the solvated brushes is attributed to the osmotic pressure of the compressed chains, together with the weak interpenetration of the two brushes [30]. Klein et al. have studied the normal and shear forces between opposing polystyrene (PS) brushes immersed in toluene using a surface force apparatus (SFA). In this study, the brushes were prepared by the "grafting-to" method. The brushes were found to function as an extremely efficient lubricant only up to a moderate pressure because the interpenetration between opposing brushes was suppressed by configurational entropy effects [28,31]. In a theta solvent, large shear forces were detected even at milder levels of compression [32]. This was attributed to changes in the strength of the frictional interaction [32], interpenetrating depth [33], and vitrification of the compressed polymer layers [34,35].

Tsujii et al. reported the tribological properties of high-density poly(methyl methacrylate) (PMMA) brushes in toluene. The brush was prepared by surface-initiated atom transfer radical polymerization (SI-ATRP) [36]. These authors measured the interaction force between the grafted layer and a silica probe attached to an atomic force microscope (AFM) cantilever. The result revealed that a large osmotic pressure led to the extension of the grafted polymer chains in the vertical direction, giving rise to an extremely strong resistance to normal loading force [37]. The steric repulsion between polymer brushes supporting high normal loads results in extremely low frictional forces between the brush-bearing surfaces. Takahara and coworkers also investigated the tribological properties of high-density PMMA brushes from a macrotribological viewpoint [38]. They showed that PMMA brushes could be used as a very effective lubricant between sliding surfaces immersed in a good solvent for PMMA.

Therefore, a high-density polymer brush bearing ionic liquid moieties such as imidazolium salts can be expected to act as a good boundary layer. In fact, several types of poly(ionic liquid) brushes have already been synthesized. Yu and coworkers prepared a poly(1-ethyl 3-(2-methacryloyloxyethyl) imidazolium chloride) brush on a gold surface and investigated its swelling/collapsing behavior in different electrolyte solutions and also its electrochemical properties [39]. Yang and coworkers prepared a polystyrene brush with an imidazolium hexafluorophosphate unit in the side chain by SI-ATRP [40]. They found that the surface wettability of the poly(ionic liquid) brushes could be controlled by exchanging their counter anions. Hao and coworkers prepared brush-like poly(ionic liquids)-grafted multiwalled carbon nanotubes (MWCNTs-g-PILs). They evaluated the tribological property of MWCNTs-g-PILs as additives in the base lubricant 1-methyl-3-butylimidazolium hexafluorophosphate, confirming that MWCNTs-g-PILs were excellent antiwear and friction-reducing additives [41].

In this study, a poly(methacrylate) brush bearing an N-alkylimidazolium-type ionic liquid unit in the side chain was prepared on a silicon wafer by SI-ATRP. In order to achieve low friction surface and inhibit corrosive wear, a combination of an n-butyl group and bis(trifluoromethanesulfonyl)imide (TFSI) was selected as the N-alkyl chain for the imidazolium and counter anions. Subsequently, the macrotribological properties of poly(1-(2-methacryloyloxy)ethyl-3-butylimidazolium bis(trifluoromethanesulfonyl)imide) (PMIS) brushes in a dry nitrogen atmosphere, water, methanol, and an ionic liquid were investigated by ball-on-plate-type tribometer. A tribotest on a nonionic poly(n-hexyl methacrylate) (PHMA) brush surface was also carried out as a control experiment.

6.2 EXPERIMENTAL

6.2.1 MATERIALS

Copper(I) bromide (CuBr, Wako Pure Chemical Industries, Ltd. (Wako), Osaka, Japan, 99.9%) was purified by successive washing with acetic acid and ethanol and then dried in vacuum. Acetonitrile (Wako, 99.5%), ethyl 2-bromoisobutylate (EB, Tokyo Chemical Inc. (TCI), Tokyo, Japan, 98%), n-hexyl methacrylate (TCI, 98%), and (−)-sparteine (TCI, 98%) were dried and distilled over CaH$_2$ before use. Anisole was stirred with small pieces of sodium at 100°C for 6 h, followed by distillation from sodium under reduced pressure. Commercially available 2,2′-bipyridyl (Wako, 99.5%), methanol (Wako, 99.5%), toluene (Wako, 99.8%), and 1-ethyl-3-methyl-imidazolium bis(trifluoromethanesulfonyl)imide (EMImTFSI, Kanto Chemical Co., Tokyo, Japan, 98%) were used without further purification. 1-(2-Methacryloyloxy)ethyl-3-butylimidazolium bis(trifluoromethanesulfonyl)imide (MIS) monomer was prepared using a previously reported procedure [42]. The procedure used for the synthesis of (2-bromo-2-methyl)propionyloxyhexyltriethoxysilane (BHE) has been described in previous papers [43,44]. Silicon (111) wafers (SUMCO Corp., Tokyo, Japan) were cleaned by washing with freshly prepared piranha solution (H_2SO_4/H_2O_2 = 7/3, v/v) at 373 K for 1 h followed by exposure to vacuum ultraviolet ray (VUV) generated from an excimer lamp (UER20-172V, USHIO Inc., Tokyo, Japan, $\lambda = 172$ nm) for 10 min under reduced pressure (30 Pa). The BHE monolayer was immobilized on the silicon wafer by chemical vapor adsorption technique [45,46]. Deionized water used for contact angle measurement was purified by the Simpli Lab (Nihon Millipore Ltd., Tokyo, Japan).

6.2.2 Polymer Brush Preparation

A few sheets of the silicon wafers with immobilized BHE, CuBr (0.045 mmol), and 2,2′-bipyridyl (0.091 mmol) were introduced into a well-dried glass tube with a stop-cock. The glass tube was then degassed by several cycles of vacuum pumping and flushing with argon. Degassed acetonitrile (4.4 mL), MIS (8.54 g, 17 mmol), and EB (unbound initiator, 0.025 mmol) in a separable Schlenk flask were added to the catalyst in the glass tube. The polymerization solution was degassed by repeatedly carrying out freeze–pump–thaw cycles to remove dissolved oxygen. The polymerization reaction was conducted at 318 K for 73 h under argon atmosphere to simultaneously form the PMIS brush from the initiator on silicon substrate and unbound PMIS from EB (Scheme 6.1). The reaction was stopped by opening the glass vessel to the air, and the reaction mixture was poured into chloroform to precipitate the unbound polymer. The resultant polymer solution was passed through an alumina column using tetrahydrofuran to remove the catalyst. The silicon wafers were washed with acetonitrile using a Soxhlet apparatus for 12 h to remove the unbound polymer adsorbed on their surfaces, and then dried under reduced pressure for 2 h. SI-ATRP of HMA was also carried out in a similar manner in anisole as a solvent using CuBr and (−)-sparteine at 358 K for 17 h to produce a PHMA brush on the silicon wafer.

6.2.3 Methods

6.2.3.1 Size Exclusion Chromatography

Size exclusion chromatography (SEC) on the unbound PMIS was performed to determine the molecular weight and molecular weight distribution with a Shimadzu liquid chromatography system (Shimadzu Corp., Kyoto, Japan) connected to three PS gel columns of Shodex GF-310 HQ × 2 + GF-510 HQ and equipped with a multiangle light scattering (MALS) detector (DAWN-EOS, Wyatt Technology Corp., Santa Barbara, CA, wavelength: $\lambda = 690$ nm) and a refractive index (RI) detector

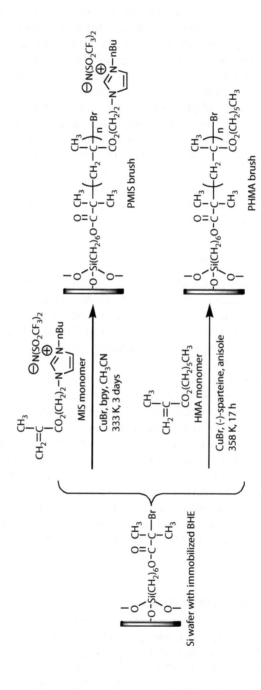

SCHEME 6.1 Surface-initiated ATRP of MIS and HMA from silicon wafer with immobilized initiator.

(RID-10A, Shimadzu) using methanol/water (9/1, v/v) containing acetic acid (0.5 M) and sodium nitrate (0.2 M) as an eluent at a flow rate of 0.6 mL min^{-1}. The Rayleigh ratio at a scattering angle of 90° was based on that of pure toluene at a wavelength of 632.8 nm at 25°C. The corrections for sensitivity of 17 detectors at angles of other than 90° and volume delay between the MALS and RI detectors were performed using the scattering intensities of 0.20 wt% aqueous solution of poly(ethylene oxide) standard with $M_w = 2.22 \times 10^4$ g mol^{-1} and $M_w/M_n = 1.08$.

6.2.3.2   Differential Scanning Calorimetry

The differential scanning calorimetry (DSC) measurement was conducted using an EXSTAR6000 (Seiko Instruments Inc., Chiba, Japan). Samples of ca. 5 mg were used in the tests. They were sealed in aluminum pans and heated at a heating rate of 10 K min^{-1} in a temperature range of 173–473 K. The glass transition temperature (T_g) was determined from a third scan.

6.2.3.3   X-Ray Photoelectron Spectroscopy

X-ray photoelectron spectroscopy (XPS) measurements were carried out on an APEX (ULVAC-PHI, Inc., Kanagawa, Japan) at 10^{-6} Pa using a monochromatic Al-Kα x-ray source of 150 W. All XPS data were collected at a take-off angle of 45°, and a low-energy (25 eV) electron flood gun was used to minimize sample charging. The survey spectra (0–1000 eV) and high-resolution spectra of C_{1s}, O_{1s}, F_{1s}, N_{1s}, S_{2p}, and Si_{2p} were acquired at pass energies for the analyzer of 100.0 eV and 25.0 eV, respectively. The x-ray beam was irradiated onto an area with a diameter of ca. 0.2 mm.

6.2.3.4   Contact Angle Measurement

The static contact angles of water and EMImTFSI on silicon wafers were recorded with a drop shape analysis system, DSA10 Mk2 (Krüss Inc., Hamburg, Germany), equipped with a video camera. A droplet of 2.0 mL was used for contact angle measurement. Angles from different spots were measured 10 times and statistically compiled.

6.2.3.5   Surface Analysis and Thickness of Polymer Brush

AFM observations were carried out with an SPI4000 (SII NanoTechnology Inc., Chiba, Japan) using a Si_3N_4 tip on a cantilever with a spring constant of 0.09 N m^{-1} under vacuum. The thicknesses of the polymer brushes on the silicon substrates were evaluated by an imaging ellipsometer (Nippon Laser and Electronics Lab., Aichi, Japan) equipped with a YAG laser ($\lambda = 532.8$ nm). Optical micrographs were obtained using an Eclipse E400 (Nikon Corporation, Tokyo, Japan).

6.2.3.6   Macroscopic Friction Test Using a Ball-on-Plate Tribometer

Macroscopic friction tests on polymer brushes were carried out on a conventional ball-on-plate-type reciprocating tribometer (Tribostation Type32, Shinto Scientific Co. Ltd., Tokyo, Japan). The test involves sliding a glass ball probe on the substrates at a reciprocating distance of 20 mm at a rate of 1.5×10^{-3} m s^{-1}. Tests were conducted in a dry nitrogen atmosphere, water, methanol, and EMImTFSI under a normal load of 0.49 N at room temperature. The friction force was measured by a strain gauge attached to the arm of the tester and was recorded as a function of time.

The friction coefficient was given by the friction force divided by the normal load. Every friction test used a virgin surface area on the brush substrate to measure the friction coefficient for the first reciprocating scan of the sliding probe. The friction coefficients for at least five trials were measured and averaged. In the case of a bare silicon wafer, the friction coefficients of the fourth scan were stored to avoid the influence of the oxide layer on the silicon surface. The test piece substrate was pinned by a Teflon belt and screws in a Teflon trough filled with an ionic liquid. The Teflon trough was fixed on the moving stage. In the case of a nonmodified silicon wafer under a normal load of 50 g (0.49 N), the theoretical contact area between the glass probe and substrate could be estimated to be 3.51×10^{-9} m^2 by Hertz's contact mechanics theory* and the average pressure on the contact area was estimated to be 139 MPa. Although the actual contact area and pressure on the brush surface cannot be estimated by Hertz's contact mechanics theory, the normal pressure would have been over 10^2 MPa in our experiments.

6.3 RESULTS AND DISCUSSION

6.3.1 PREPARATION OF POLY[1-(2-METHACRYLOYLOXY)ETHYL-3-ETHYLIMIDAZOLIUM CHLORIDE] BRUSH

The SI-ATRP of MIS at 318 K from the silicon wafer with immobilized initiator proceeded slowly for 3 days to yield a PMIS brush with a thickness of 54 nm, as shown in Scheme 6.1. The formation of the PMIS brush was confirmed by XPS. The surface atomic concentrations of carbon, oxygen, fluorine, nitrogen, and sulfur measured by XPS revealed good agreement with the theoretical values calculated from the atomic composition of MIS monomer. The AFM observation revealed that a homogeneous polymer layer was formed on the substrate, and the surface roughness was less than 0.5 nm *in vacuo* in a 5 μm × 5 μm scan area. A 60-nm-thick PHMA brush was also prepared by a similar SI-ATRP protocol.

SI-ATRP in the presence of the unbound initiator simultaneously produced a polymer brush and an unbound polymer. The number-average molecular weight (M_n) value of unbound PMIS estimated by MALS-SEC was 316,000, and its polydispersity index (M_w/M_n) was 1.32. The M_n of free PHMA and M_w/M_n measured by SEC using THF as an eluent were 65,400 and 1.74, respectively. The M_n of the surface-grafted polymer on the silicon wafer was not directly measured, but it is widely known that a polymer brush should have the same molecular weight as the corresponding free

---

* If a circle with radius a (m) is regarded as the contact area between a glass ball and substrate under a normal load P (0.49 N), Hertz's theory affords the following relationship using Young's modulus of glass and silicon wafer, E_A (7.16×10^{10} Pa), EB (1.30×10^{11} Pa), and Poisson's ratio υ_A (0.23), υ_B (0.28), respectively:

$$1/E = (1 - \upsilon_A^2)/E_A + (1 - \upsilon_B^2)/E_B$$
$$a = (3 \times P \times R_A /4E)^{1/3}$$

where R_A is curvature radius (5.00×10^{-3} m) of glass ball. The contact area can be calculated by πa^2.

polymer [47–50]. The unbound PMIS was a highly viscous polymer at room temperature because the glass transition temperature (T_g) obtained by DSC was 268.5 K, which is very close to the T_g of PHMA (268.0 K) [51]. The unbound PMIS dissolved well in acetone, acetonitrile, methanol, and EMImTFSI, but was insoluble in water, diiodomethane, and toluene.

As will be described below, the relationship between the solubility of PMIS and the frictional properties of the brush is important. The graft density, σ, of the PMIS brush was estimated to be 0.14 chains nm^{-2} based on the relationship between the thickness L (nm) and M_n as follows [50]:

$$\sigma = d \times L \times N_A \times 10^{-21}/M_n, \tag{6.1}$$

where d and N_A are the assumed density of bulk polymer (g cm^{-3}) at 293 K and the Avogadro's number, respectively. This value is lower than the typical graft density of a PMMA brush (0.60–0.70 chains nm^{-2}) or a PHMA brush (0.56 chains nm^{-2}) prepared by the "grafting-from" method. The low value is attributed to the fact that PMIS has a bulky imidazolium group on the side chain, which occupies a larger cross-sectional area on the surface compared with PMMA and PHMA.

6.3.2   FRICTIONAL PROPERTIES OF THE PMIS BRUSH

6.3.2.1   PMIS Brush versus Bare Silicon Wafer: Effect of Sliding Velocity

Figure 6.1 shows the effect of sliding velocity of a glass ball probe on friction coefficients against PMIS brush surface and nonmodified silicon wafer surface in EMImTFSI, over a distance of 20 mm under a load of 0.49 N at 298 K. The chemical structures of the brush and the ionic liquid are given in Table 6.1. The friction coefficients at each sliding velocity were measured on a virgin surface area of the brush

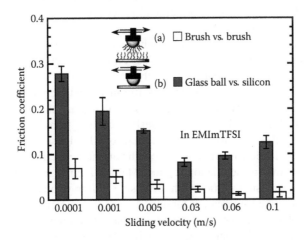

FIGURE 6.1   Sliding velocity dependence of the friction coefficient of (a) glass ball probe with immobilized PMIS brush versus PMIS brush and (b) bare glass ball probe versus silicon wafer in EMImTFSI under a load of 0.49 N at 298 K.

TABLE 6.1
Polymers and Solvents Used in This Study

	Abbreviation	Structure
Poly[1-(2-methacryloyloxy)ethyl-3-ethylimidazolium bis(trifluoromethanesulfonyl)imide]	PMIS	$\left[CH_2-C\right]_n$ with CH_3 and $CO_2(CH_2)_2-N \curvearrowright N-nBu$; $^{\ominus}N(SO_2CF_3)_2$
Poly(hexyl methacrylate)	PHMA	$\left[CH_2-C\right]_n$ with CH_3 and $CO_2(CH_2)_5CH_3$
1-Ethyl-3-methylimidazolium bis(trifluoromethanesulfonyl)imide	EMImTFSI	$^{\ominus}N(SO_2CF_3)_2$; $C_2H_5-N \curvearrowright N-CH_3$

and silicon substrate. It can be seen that the friction coefficients of the PMIS brush were much lower than the bare silicon wafer surface in the sliding velocity range of 10^{-4}–10^{-1} m s^{-1}.

These results indicated that the combination of PMIS brush and EMImTFSI showed an excellent friction reduction. One reason for reduction in the friction is that the PMIS brush has a particular affinity for EMImTFSI. Therefore, EMImTFSI easily permeated into the PMIS brush and the PMIS brush would be highly solvated with EMImTFSI to form an effective boundary layer. It is thought that the swollen brush layer retained EMImTFSI under friction condition and the swollen brush with an ionic liquid can support the normal load to reduce the interaction between the substrate and glass probe.

The decrease in the friction coefficient with an increase in the sliding velocity was observed in both the brush-on-brush and glass-on-silicon friction systems. The friction coefficient of the silicon wafer gradually decreased from 0.28 to 0.08 when the sliding velocity increased from 10^{-4} to 3×10^{-2} m s^{-1} and increased with a further increase in sliding velocity after 3×10^{-2} m s^{-1}. In contrast, the friction coefficient of the PMIS brush gradually and continuously decreased from 0.07 to 0.01 with an increase in the sliding velocity over a wide range of 10^{-4}–10^{-1} m s^{-1}.

The drastic reduction in the friction coefficient at a certain velocity could be caused by transition in the friction regime. Making a comparison to the Stribeck curve [52], the decreasing friction with increasing sliding velocity suggests that this system is in the mixed lubrication regime. At low sliding velocity, the interaction between the polymer brushes and their interpenetration dominated the friction to give a large friction coefficient (boundary or interfacial friction). With an increase in the sliding velocity, a thicker liquid layer would be formed between the sliding surfaces by the hydrodynamic lubrication effect to reduce the effective contact area and the friction force (mixed lubrication region). At higher sliding rates (or viscosities or lower pressure), the Stribeck curve will move to the elastohydrodynamic and hydrodynamic regimes, and an increase in the friction force with the sliding velocity is expected due to the shear resistance of the fluids. Hydrodynamic lubrication by poly(styrene) brush in a solvent has been already confirmed by microtribological test on an AFM, which was well described by the relationship between shear velocity and degree of swelling of a poly(styrene) brush [53]. Unfortunately, in our experiments, the friction coefficients could not be measured at sliding rates faster than 10^{-1} m s^{-1} due to the limitation of the tribometer.

6.3.2.2   PMIS versus PHMA Brush against Glass Ball: Effect of Solvent

Figure 6.2 shows the friction coefficients of the PMIS and PHMA brushes measured by sliding a glass ball probe over a distance of 20 mm at a sliding velocity of 1.5×10^{-3} m s^{-1} in a dry nitrogen atmosphere, water, methanol, toluene, and EMImTFSI under a normal load of 0.49 N at 298 K. A high friction coefficient was observed under the dry N_2 condition for both brushes, whereas lower friction coefficients were observed in the water and organic solvent due to fluid lubrication effect. The friction coefficient of the nonmodified silicon wafer was larger than 0.2 in the dry N_2 atmosphere, although the corresponding value is not shown in Figure 6.2. It is notable that the friction coefficient appears to be dependent on the solvent quality. A remarkable reduction in the friction coefficient was observed for the PMIS brush in methanol and EMImTFSI, which are good solvents for PMIS, while a higher friction coefficient was observed in water and toluene, which are poor solvents for PMIS.

The effect of the solvent quality on the frictional properties of polymer brushes in various solvents has been previously reported. When the solvent was changed from a good solvent to a theta solvent, Kilbey et al. detected larger shear forces between sliding surfaces with immobilized PS brushes using SFA [32]. Spencer and coworkers observed that the friction coefficient of a poly(ethylene glycol) (PEG) brush surface in water, measured by colloidal-probe lateral force microscopy, increased from 0.2 to 0.6 when the volume fraction of 2-propanol exceeded 85% [54]. They found that a hydrated PEG brush in water adopted an extended chain conformation to

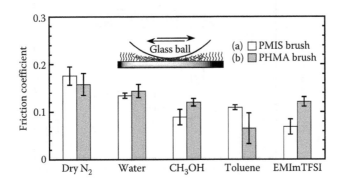

FIGURE 6.2   Friction coefficient of (a) PMIS and (b) PHMA brushes in dry N_2 atmosphere, water, methanol, toluene, and in EMImTFSI by sliding a glass ball over a distance of 20 mm at a sliding velocity of 1.5×10^{-3} m s^{-1} under a load of 0.49 N at 298 K.

afford effective boundary lubricants, whereas an increase in the 2-propanol fraction resulted in the collapse of the brush-like structure to a more random-coil-like polymer conformation. High-density PMMA brushes in toluene and acetone exhibited lower friction coefficients than in hexane and cyclohexane when a stainless-steel ball was used as the sliding probe [39]. Therefore, the magnitude of the polymer/solvent interaction must have played an important role in the solvent quality effect. In this study, EMImTFSI and methanol were regarded as good solvents whereas water and toluene were inferior in quality for PMIS. A PMIS brush would be highly solvated with EMImTFSI to form a swollen boundary layer. EMImTFSI would moderate the interaction between the brush surface and the glass ball probe to give a lower friction coefficient. In contrast, the PMIS chains in water or toluene would be unwilling to be in contact with solvent molecules, but prefer to interact with the polymers or friction probe rather than solvent molecules, thus giving a higher friction coefficient. A similar trend was observed with the PHMA brush. The PHMA brush showed a much lower friction coefficient in toluene, which is a good solvent for PHMA. On the other hand, water, methanol, and EMImTFSI are poor solvents for PHMA. Therefore, the friction coefficient of the PHMA brush was not reduced, even in EMImTFSI. These results indicated that the imidazolium moiety in the polymer brush contributed to the reduction in the friction coefficient, especially when combined with an ionic liquid.

When a glass ball probe with immobilized PMIS brush was used as the sliding probe, as shown in Figure 6.3, the friction coefficient of the PMIS brush was further reduced. For instance, the brush-versus-brush friction coefficient in an ionic liquid was 0.048, which was lower than that of the brush-versus-glass (0.078). We proposed that this reduction in the friction coefficient was caused by the presence of a thicker boundary film produced by the swollen brush and high osmotic pressure from densely grafted polymer chains with the approach of brush-bearing surfaces.

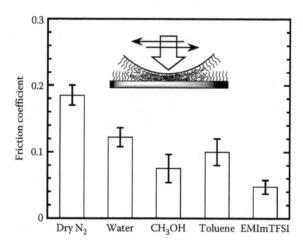

FIGURE 6.3 Friction coefficient of PMIS brush in dry N_2 atmosphere, water, methanol, toluene, and in EMImTFSI by sliding a glass ball with immobilized PMIS brush over a distance of 20 mm at a sliding velocity of 1.5×10^{-3} m s^{-1} under a load of 0.49 N at 298 K.

(a)   EMImTFSI on PHMA brush

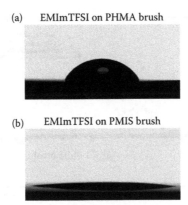

(b)   EMImTFSI on PMIS brush

FIGURE 6.4   Contact angles of EMImTFSI droplet on (a) PHMA brush surface and (b) PMIS brush surface at 298 K.

6.3.2.3   Contact Angle Measurement

Contact angles of EMImTFSI droplet on the PMIS brush and the PHMA brush clearly showed the difference in affinity of EMImTFSI for each surface (Figure 6.4). The contact angle of EMImTFSI droplet on the PMIS brush surface was 14° just within 1 min after the EMImTFSI was dropped. However, the ionic liquid spread continuously and swelled the brush [55], and the contact angle reached nearly zero degrees within 5 min. On the other hand, the contact angle of the EMImTFSI droplet on the PHMA brush was 61°. These results showed the excellent affinity of the PMIS brush to EMImTFSI.

6.3.2.4   PMIS versus PHMA Brush against Bare Glass: Effect of Friction Cycles

Figure 6.5 shows the variations in the friction coefficients of the PMIS brush and PHMA brush in a dry nitrogen atmosphere with the number of friction cycles. The friction coefficient of the PHMA brush film began to increase in the early stage of the friction test, attaining a magnitude of 0.12 within 150 tracking cycles. This result indicates that the PHMA brush was abraded away by the sliding glass probe. In contrast, the relatively low friction coefficient of the high-density PMIS brush was continuously observed, as shown in Figure 6.5 (curve b), implying a better wear resistance compared with the PHMA brush. Actually, the high-density PMIS brush maintained a friction coefficient of around 0.16 even after 800 friction cycles.

6.3.3   Optical and XPS Surface Analyses

Figure 6.6 displays an optical microscope image of the wear track after 400 friction cycles by the glass probe sliding at a sliding velocity of 1.0×10^{-2} m s^{-1} under a load of 1.96 N in a dry N$_2$ atmosphere at 298 K and high-resolution XPS spectra of C$_{1s}$ and Si$_{2p}$ regions for the virgin surface and the worn surface inside the wear track after 400 friction cycles. Even with the wear track formed on the brush surface by the friction test, the components of the brush would remain in the wear tracks. With an increase in the number of friction cycles from 0 to 400, the atomic ratio of carbon

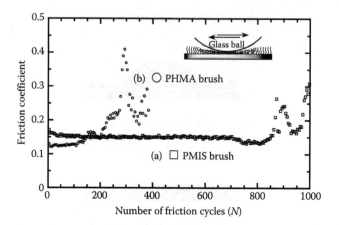

FIGURE 6.5 Evolution of the friction coefficient versus the number of friction cycles N for the surface of (a) PMIS brush and (b) PHMA brush at a sliding velocity of 1.0×10^{-2} m s^{-1} under a load of 1.96 N in dry N$_2$ atmosphere.

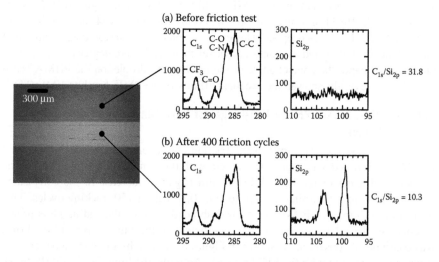

FIGURE 6.6 Optical microscope image of the wear track of the PMIS brush surface after 400 friction cycles by glass ball probe sliding at a sliding velocity of 1.0×10^{-2} m s^{-1} under a load of 1.96 N in dry N$_2$ atmosphere at 298 K and high-resolution XPS spectra of C$_{1s}$ and Si$_{2p}$ peak regions for (a) the original surface and (b) the worn surface inside the wear track after 400 friction cycles.

and silicon (C$_{1s}$/Si$_{2p}$) decreased from 31.8 to 10.3, probably due to the reduction of the brush thickness. The C$_{1s}$ spectra of brush surfaces before and after the friction tests showed similar peak patterns, indicating that the chemical structure of PMIS still remained. We also confirmed that the atomic ratio of carbon and fluorine decreased drastically after 1000 friction cycles, whereas the sulfur component still remained on the worn surface. These results reveal that the chemical decomposition of the PMIS and counter anions took place after the brush layer peeled off [43].

As mentioned in Section 6.1, the tribological properties of an ionic liquid largely depend on counter anions. Minami [56] and Itoh et al. [57] reported excellent tribological properties when using hydrophobic anions, such as trifluorotris(pentafluoroethyl) phosphate and perfluoroalkyl sulfate, probably because the low moisture content of ionic liquids retards unfavorable chemical reactions. In this work, only TFSI was investigated as a counter anion for an imidazolium-type poly(ionic liquid). Further improvement in the tribological properties of poly(ionic liquids) can be expected by optimizing the combination of organic anions and cations.

6.4 CONCLUSIONS

In this chapter, we have reported the tribological properties of a poly(ionic liquid) brush, PMIS, and have compared this brush with a nonionic PHMA brush. The PMIS brush showed a friction coefficient of around 0.16 even after 800 cycles of reciprocating sliding against a glass ball probe under a dry nitrogen atmosphere. On the other hand, under identical condition the PHMA brush showed a friction coefficient much higher than 0.3 and was completely worn out after 150 friction cycles. These results indicate that the poly(ionic liquid) brush contributed to the formation of a good lubricant layer. In addition, we showed that the combination of a poly(ionic liquid) brush and an ionic liquid showed a low friction coefficient.

The influences of solvent quality on the tribological properties have been well studied, and in general, a polymer brush immersed in a good solvent would afford extremely low friction coefficient. A similar effect can be considered for the combination of a PMIS brush and the ionic liquid EMImTFSI. There are several reasons for the reduction in the friction coefficient achieved by a swollen PMIS brush and EMImTFSI. First, the PMIS brush was strongly tethered to the substrate by covalent bonds. In addition, the PMIS brush surface had good wettability and high affinity for the ionic liquid, which enabled the EMImTFSI to spread and permeate into the PMIS brush, forming a good lubrication film and reducing the interaction between opposite sliding brushes. The high viscosity of the swollen brush film in EMImTFSI also contributed to maintaining the lubrication film, even under severe friction conditions. In contrast, the PMIS brush immersed in a poor solvent such as water showed a larger friction coefficient. The affinity of the polymer brush for the solvent (or lubricant) plays an important role in controlling the interaction between the brush and the friction probe, as well as the frictional properties. XPS analysis of the wear tracks revealed that the PMIS brush layer was gradually abraded away by sliding the glass ball probe in a dry N_2 atmosphere under a load of 1.96 N, and the layer was peeled off after 1000 friction cycles, which resulted in a much higher friction coefficient. Further improvement in the wear resistance is necessary before a poly(ionic liquid) brush can be put to practical use.

ACKNOWLEDGMENTS

The present work was supported by a grant-in-aid for the Global COE Program, "Science for Future Molecular Systems," and partially supported by a grant-in-aid for young scientists (B) (19750098) from the Ministry of Education, Culture, Science, Sports and Technology of Japan. T.I. acknowledges the financial support of a grant-in-aid for JSPS Fellows. We gratefully acknowledge Idemitsu Kosan Co. Ltd. for the supply of ionic liquids.

REFERENCES

1. C. Ye, W. Liu, Y. Chen, and L. Yu, Room-temperature ionic liquids: A novel versatile lubricant, *Chem. Commun.*, 37, 2244–2245, 2001.
2. W. Liu, C. Ye, Q. Gong, H. Wang, and P. Wang, Tribological performance of room-temperature ionic liquids as lubricant, *Tribol. Lett.*, 13, 81–85, 2002.
3. W. Liu, C. Ye, Y. Chen, Z. Ou, and D. C. Sun, Tribological behavior of sialon ceramics sliding against steel lubricated by fluorine-containing oils, *Tribol. Int.* 35, 503–509, 2002.
4. A. E. Jiménez, M. D. Bermúdez, F. J. Carrión, and G. Martínez-Nicolás, Room temperature ionic liquids as lubricant additives in steel-aluminium contacts: Influence of sliding velocity, normal load and temperature, *Wear*, 261, 347–359, 2006.
5. B. S. Phillips, G. John, and J. S. Zabinski, Surface chemistry of fluorine containing ionic liquids on steel substrates at elevated temperature using Mössbauer spectroscopy, *Tribol. Lett.*, 26, 85–91, 2007.
6. C.-M. Jin, C. Ye, B. S. Phillips, J. S. Zabinski, X. Liu, W. Liu, and J. M. Shreeve, Polyethylene glycol functionalized dicationic ionic liquids with alkyl or polyfluoroalkyl substituents as high temperature lubricants, *J. Mater. Chem.*, 16, 1529–1535, 2006.
7. A. Suzuki, Y. Shinka, and M. Masuko, Tribological characteristics of imidazolium-based room temperature ionic liquids under high vacuum, *Tribol. Lett.*, 27, 307–313, 2007.
8. J. S. Wilkes and M. J. Zaworotko, Air and water stable 1-ethyl-3-methylimidazolium based ionic liquids, *J. Chem. Soc., Chem. Commun.*, 965–966, 1992.
9. M.-D. Bermúdez, A.-E. Jiménez, J. Sanes, and F. J. Carrión, Ionic liquids as advanced lubricant fluids, *Molecules*, 14, 2888–2908, 2009.
10. F. Zhou, Y. Liang, and W. Liu, Ionic liquid lubricants: designed chemistry for engineering applications, *Chem. Soc. Rev.*, 38, 2590–2599, 2009.
11. I. Minami, Ionic liquids in tribology, *Molecules*, 14, 2286–2305, 2009.
12. H. Xiao, D. Guo, S. Liu, G. Pan, and X. Lu, Film thickness of ionic liquids under high contact pressures as a function of alkyl chain length, *Tribol. Lett.*, 41, 471–477, 2011.
13. H. Kamimura, T. Chiba, N. Watanabe, T. Kubo, H. Nanano, I. Minami, and S. Mori, Effects of carboxylic acids on friction and wear reducing properties for alkylmethyl-imidazolium derived ionic liquids, *Tribol. Online*, 1, 40–43, 2006.
14. H. Kamimura, T. Kubo, I. Minami, and S. Mori, Effect and mechanism of additives for ionic liquids as new lubricants, *Tribol. Int.*, 40, 620–625, 2007.
15. I. Minami, N. Watanabe, H. Nanao, S. Mori, K. Fukumoto, and H. Ohno, Improvement in the tribological properties of imidazolium-derived ionic liquids by additive technology, *J. Synth. Lubrication*, 25, 45–55, 2008.
16. A.-E. Jiménez and M.-D. Bermúdez, Ionic liquids as lubricants for steel–aluminum contacts at low and elevated temperatures, *Tribol. Lett.*, 26, 53–60, 2007.
17. B. S. Phillips and J. S. Zabinski, Ionic liquid lubrication effects on ceramics in a water environment, *Tribol. Lett.*, 17, 533–541, 2004.
18. A. E. Jiménez, M. D. Bermúdez, P. Iglesias, F. J. Carrión, and G. Martínez-Nicolás, 1-N-alkyl-3-methylimidazolium ionic liquids as neat lubricants and lubricant additives in steel–aluminium contacts, *Wear*, 260, 766–782, 2006.
19. J. Sanes, F. J. Carrión, M. D. Bermúdez, and G. Martínez-Nicolás, Ionic liquids as lubricants of polystyrene and polyamide 6-steel contacts. Preparation and properties of new polymer-ionic liquid dispersions, *Tribol. Lett.*, 21, 121–133, 2006.

20. A. S. Pensado, M. J. P. Comuñas, and J. Fernández, The pressure–viscosity coefficient of several ionic liquids, *Tribol. Lett.*, 31, 107–118, 2008.
21. M. Kobayashi, H. Ishida, M. Kaido, A. Suzuki, and A. Takahara, Frictional properties of organosilane monolayers and high-density polymer brushes, in: *Surfactants in Tribology*, G. Biresaw and K. L. Mittal (Eds.), pp. 89–110, CRC Press, Boca Raton, FL, 2008.
22. M. Kobayashi, Z. Wang, Y. Matsuda, M. Kaido, A. Suzuki, and A. Takahara, Tribological behavior of polymer brush prepared by the grafting-from method, in: *Polymer Tribology*, S. S. Kumar (Ed.), pp. 582–602, Imperial College Press, London, 2009.
23. H. Ishida, T. Koga, M. Morita, H. Otsuka, and A. Takahara, Macro- and nanotribological properties of organosilane monolayers prepared by a chemical vapor adsorption method on silicon substrates, *Tribol. Lett.*, 19, 3–8, 2005.
24. R. C. Advincula, W. J. Brittain, K. C. Caster, and J. Rühe, *Polymer Brushes*. Wiley-VCH, Weinheim, 2004.
25. S. J. Miklavic and S. Marčelja, Interaction of surfaces carrying grafted polyelectrolytes, *J. Phys. Chem.*, 92, 6718–6722, 1988.
26. G. H. Fredrickson and P. Pincus, Drainage of compressed polymer layers: Dynamics of a "squeezed sponge", *Langmuir*, 7, 786–795, 1991.
27. A. Halperin, M. Tirrell, and T. P. Lodge, Tethered chains in polymer microstructures, *Adv. Polym. Sci.*, 100, 31–71, 1992.
28. J. Klein, E. Kumacheva, D. Mahalu, D. Perahia, and L. J. Fetters, Reduction of frictional forces between solid surfaces bearing polymer brushes, *Nature*, 370 634–636, 1994.
29. H. J. Taunton, C. Toprakcioglu, L. J. Fetters, and J. Klein, Forces between surfaces bearing terminally anchored polymer chains in good solvents, *Nature*, 332, 712–714, 1988.
30. T. A. Witten, L. Leibler, and P. A. Pincus, Stress relaxation in the lamellar copolymer mesophase, *Macromolecules*, 23, 824–829, 1990.
31. P. A. Schorr, T. C. B. Kwan, S. M. Kilbey II, E. S. G. Shaqfeh, and M. Tirrell, Shear forces between tethered polymer chains as a function of compression, sliding velocity, and solvent quality, *Macromolecules*, 36, 389–398, 2003.
32. G. S. Grest, Normal and shear forces between polymer brushes, *Adv. Polym. Sci.*, 138, 149–183, 1999.
33. T. Kreer, M. H. Müser, K. Binder, and J. Klein, Frictional drag mechanisms between polymer-bearing surfaces, *Langmuir*, 17, 7804–7813, 2001.
34. J. Klein, E. Kumacheva, D. Perahia, and L. J. Fetters, Shear forces between sliding surfaces coated with polymer brushes: The high friction regime, *Acta Polymerica*, 49, 617–625, 1998.
35. A. Dhinojwala, and S. Granick, Surface forces in the tapping mode: Solvent permeability and hydrodynamic thickness of adsorbed polymer brushes, *Macromolecules*, 30, 1079–1085, 1997.
36. M. Ejaz, S. Yamamoto, K. Ohno, Y. Tsujii, and T. Fukuda, Controlled graft polymerization of methyl methacrylate on silicon substrate by the combined use of the Langmuir-Blodgett and atom transfer radical polymerization techniques, *Macromolecules*, 31, 5934–5936, 1998.
37. S. Yamamoto, M. Ejaz, Y. Tsujii, and T. Fukuda, Surface interaction forces of well-defined, high-density polymer brushes studied by atomic force microscopy. 2. Effect of graft density, *Macromolecules*, 33, 5608–5612, 2000.
38. H. Sakata. M. Kobayashi, H. Otsuka, and A. Takahara, Tribological properties of poly(methyl methacrylate) brushes prepared by surface-initiated atom transfer radical polymerization, *Polym. J.*, 37, 767–775, 2005.
39. B. Yu, F. Zhou, H. Hu, C. Wang, and W. Liu, Synthesis and properties of polymer brushes bearing ionic liquid moieties, *Electrochim. Acta*, 53, 487–494, 2007.
40. X. He, W. Yang, and X. Pei, Preparation, characterization, and tunable wettability of poly(ionic liquid) brushes via surface-initiated atom transfer radical polymerization, *Macromolecules*, 41, 4615–4621, 2008.

41. X. Pei, Y. Xia, W. Liu, B. Yu, and J. Hao, Polyelectrolyte-grafted carbon nanotubes: Synthesis, reversible phase-transition behavior, and tribological properties as lubricant additives, *J. Polym. Sci.: Part A: Polym. Chem.*, 46, 7225–7237, 2008.
42. T. Ishikawa, M. Kobayshi, and A. Takahara, Macroscopic frictional properties of poly(1-(2-methacryloyloxy)ethyl-3-butylimidazolium bis(trifluoromethanesulfonyl)imide) brush surfaces in an ionic liquid, *ACS Appl. Mater. Interfaces*, 2, 1120–1128, 2010.
43. K. Ohno, T. Morinaga, K. Koh, Y. Tsujii, and T. Fukuda, Synthesis of monodisperse silica particles coated with well-defined, high-density polymer brushes by surface-initiated atom transfer radical polymerization, *Macromolecules*, 38, 2137–2142, 2005.
44. M. Kobayashi and A. Takahara, Synthesis and frictional properties of poly(2,3-dihydroxypropyl methacrylate) brush prepared by surface-initiated atom transfer radical polymerization, *Chem. Lett.*, 34, 1582–1583 2005.
45. M. Kobayashi, Y. Terayama, N. Hosaka, M. Kaido, A. Suzuki, N. Yamada, N. Torikai, K. Ishihara, and A. Takahara, Friction behavior of high-density poly(2-methacryloyloxyethyl phosphorylcholine) brush in aqueous media, *Soft Matter*, 3, 740–746, 2007.
46. K. Hayashi, N. Saito, H. Sugimura, O. Takai, and N. Nagagiri, Regulation of the surface potential of silicon substrates in micrometer scale with organosilane self-assembled monolayers, *Langmuir,* 18, 7469–7472, 2002.
47. R. Matsuno, K. Yamamoto, H. Otsuka, and A. Takahara, Polystyrene- and poly(3-vinylpyridine)-grafted magnetite nanoparticles prepared through surface-initiated nitroxide-mediated radical polymerization, *Macromolecules*, 37, 2203–2209, 2004.
48. T. v. Werne and T. E. Patten, Preparation of structurally well-defined polymer-nanoparticle hybrids with controlled/living radical polymerizations, *J. Am. Chem. Soc.*, 121, 7409–7410, 1999.
49. M. Husseman, E. E. Malmstrom, M. McNamura, M. Mate, D. Mecerreyes, D. G. Benoit, J. L. Hedrick et al., Controlled synthesis of polymer brushes by "Living" free radical polymerization techniques, *Macromolecules*, 32, 1424–1431, 1999.
50. J. Pyun, S. Jia, T. Kowalewski, G. D. Patterson, and K. Matyjaszewski, Synthesis and characterization of organic/inorganic hybrid nanoparticles: Kinetics of surface-initiated atom transfer radical polymerization and morphology of hybrid nanoparticle ultrathin films, *Macromolecules*, 36, 5094–5104, 2003.
51. G. Meier, F. Kremer, G. Fytas, and A. Rizos, Evidence of dynamic heterogeneity as obtained from dielectric and brillouin light-scattering studies of poly(*n*-hexylmethacrylate), *J. Polym. Sci. Part B: Polym. Phys.*, 34, 1391–1401, 1996.
52. H. A. Spikes, Boundary lubrication and boundary films, in: *The Films in Tribology*, D. Dowson, C. M. Taylor, T. H. C. Childs, M. Godet and G. Dalmaz (Eds.), pp. 331–346, Elsevier, Amsterdam, Netherlands, 1993.
53. A. Nomura, K. Okayasu, K. Ohno, T. Fukuda, and Y. Tsujii, Lubrication mechanism of concentrated polymer brushes in solvents: Effect of solvent quality and thereby swelling state, *Macromolecules*, 44, 5013–5019, 2011.
54. T. M. Miller, X. Yan, S. Lee, S. S. Perry, and N. D. Spencer, Preferential solvation and its effect on the lubrication properties of a surface-bound, brushlike copolymer, *Macromolecules*, 38, 3861–3866, 2005.
55. M. A. C. Stuart, W. M. de Vos, and F. A. M. Leermakers, "Why surfaces modified by flexible polymers often have a finite contact angle for good solvents, *Langmuir*, 22, 1722–1728, 2006.
56. I. Minami, M. Kita, T. Kubo, H. Nanao, and S. Mori, The tribological properties of ionic liquids composed of trifluorotris(pentafluoroethyl) phosphate as a hydrophobic anion, *Tribol. Lett.*, 30, 215–223, 2008.
57. T. Itoh, N. Watanabe, K. Inada, A. Ishioka, S. Hayase, M. Kawatsura, I. Minami, and S. Mori, Design of alkyl sulfate ionic liquids for lubricants, *Chem. Lett.*, 38, 64–65, 2009.

Part II

Biobased and Environmentally Friendly Lubricants and Additives

7 A Biobased Nitrogen-Containing Lubricant Additive Synthesized from Epoxidized Methyl Oleate Using an Ionic Liquid Catalyst*

Kenneth M. Doll and Atanu Biswas

CONTENTS

---

* The mention of trade names or commercial products in this chapter is solely for the purpose of providing specific information and does not imply recommendation or endorsement by the U.S. Department of Agriculture. The USDA is an equal opportunity provider and employer.

An aniline adduct was synthesized from methyl oleate utilizing an epoxidation route. An ionic liquid catalyst, 1-methylimidazolium tetrafluoroborate, was found to be the key for this process. The reaction produces a product with the aniline incorporated into the fatty chain at the 9(10) position, and fatty amide was not produced. The product shows antiwear properties when used as an additive in polyalphaolefin base oil, reducing the measured wear scar diameter of a friction and wear test from 0.59 mm to less than 0.47 mm. The material also reduced the pressurized differential scanning calorimetry oxidation onset temperature of the lubricant by more than 25°C.

7.1 INTRODUCTION

Lubricants are a uniform mixture of interacting base oils and additives [1–4] where trying to track down the function of each is a "slippery" endeavor indeed. Biobased materials, once prevalent in the industry [5], have attracted renewed interest in recent years [6–10]. Vegetable oils make a good starting point [11,12] due to favorable life cycle [13], government policies [14], and renewability considerations. However, they suffer from oxidation instability [15–18], especially the highly unsaturated oils [18], and high pour points [19], especially the highly saturated oils. Improvements in these properties have been the focus of many patents where oligomers [20–23], hydroxy vegetable oils [24,25], and ester-forming [26–28] technologies have been used to improve vegetable oil shortcomings.

Even with these improvements, it is obvious that appropriate additives must be developed in order to actually develop a commercial industrial lubricant. Traditional additives, such as ZDDP [29], can be effective in soybean oil, but others can be incompatible. Especially problematic is the area of oxidation, where effective antioxidants are still limited and data show a lack of synergistic effects between antioxidants of different types [30]. An approach of chemical modification using biobased materials as the starting point to achieve more compatible additives has been undertaken by many groups. In addition to the hydroxy products, chemical methods have been utilized to produce oleochemical ethers [31–34], esters [35–40], estolides [41], acetals [42], and sulfides [43]. These were produced either by addition to the olefinic groups of the natural oil or through an epoxidation reaction followed by further chemistry.

Oleochemical epoxides, which are effective lubricant additives in their own right [44,45], have been the subject of further chemical modification for a range of applications where considerable market advantage can be gained by the use of a biobased product. Urethane foams [46–50], elastomeric amines [51], adhesives [52], and structural composites [53–59] have all been made from epoxidized vegetable oils. However, the use of aromatic amines, likely to have positive effects on both pour point and oxidation properties, was not reported prior to our work [60].

The physical properties of ionic liquids have been studied extensively during the past decade [61–64], including recent application in chemical catalysis [65]. One reaction that has been specifically studied is the ability of ionic liquids to ring-open an epoxide to form the desired products [66]. Of specific interest, the ionic liquid, 1-methylimidazolium tetrafluoroborate, has shown the ideal solubility properties to perform oleochemical catalysis, yet still is easily extractable with ordinary washing steps. Our recent report [60] takes advantage of this property in order to synthesize a new lubricant additive of primarily oleochemical origin. In this chapter, we report for the first time, friction-wear reducing and antioxidation properties found by addition of this biobased material to lubricant feedstocks of both natural and synthetic origins.

7.2 SYNTHESIS OF THE LUBRICANT ADDITIVE

7.2.1 MATERIALS

Methyl oleate (>99%, Nu-Check Prep, Elsyian, MN); 1-methylimidazole (99%, Acros, Fairlawn, NJ); tetrafluoroboric acid (48% min w/w aq. solution, Alfa Aesar, Ward Hill, MA); sodium chloride (A.C.S. Reagent, Thermo-Fisher, Fairlawn, NJ); NaHCO₃ (A.C.S. Reagent, Thermo-Fisher, Fairlawn, NJ); hydrogen peroxide (A.C.S. Reagent, 30% Solution, Sigma-Aldrich, St. Louis, MO); formic acid (96%, A.C.S. reagent, Sigma-Aldrich, St. Louis, MO); hexanes (>95%, HPLC grade); and aniline (99.5% reagent grade, Sigma-Aldrich, St. Louis, MO) were all used as received. The lubricant basestocks, soybean oil (RBD grade, KIC Chemicals, New Paltz, NY), and PAO 8 (Synfluid Polyalphaolefin, Chevron-Phillips, Woodlands, TX) were also used as received.

The 1-methylimidazolium tetrafluoroborate catalyst was synthesized by the addition of tetrafluoroboric acid to 1-methylimidazole. First, 61.5 g of 1-methylimidazole was placed in a large three-neck flask with stirring and cooling to 0°C. Next, 118 mL of 40% aqueous tetrafluoroborate solution was added dropwise over 30 min with continued cooling to maintain the temperature below 5°C. After 2 h of reaction, the water was removed by vacuum to produce the colorless ionic liquid product.

7.2.2 SYNTHESIS OF EPOXIDIZED METHYL OLEATE

Epoxides of oleochemicals have found considerable use, recently, as plasticizers, precursors for the production of oleochemicals [42,67–70], surfactants [67], and polymer components [49,50,54]. Epoxidized methyl oleate (methyl 9,10-epoxy stearate; EMO) was synthesized using a method adapted from La Scala and Wool [71], originally based on Swern epoxidation [72]. In short, the reaction, as shown in Figure 7.1, occurs in a biphasic environment where performic acid is formed, dissolves in the oil layer, and then reacts with the olefin. A more detailed account of this synthesis is available elsewhere [39].

7.2.3 RING OPENING REACTION OF EPOXIDIZED MATERIALS

In the ring opening reaction, as shown in Figure 7.2, EMO, typically ~1.56 g, was placed in a 30 mL glass vial where a catalyst, various wt%, and a stir-bar were present. Aniline was added, 1–4 equivalents relative to EMO. The reaction vials were sealed with a septa and the temperature increased to the reaction temperature, 60–105°C. The reaction was monitored by suspending the stirring and taking aliquots from the top layer.

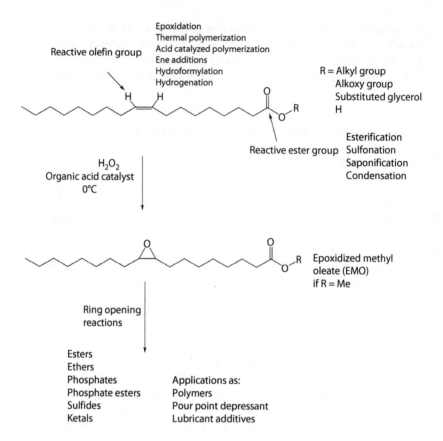

FIGURE 7.1 Schematic for the utilization of epoxidation to synthesize lubricant additives from vegetable oils and other natural oils.

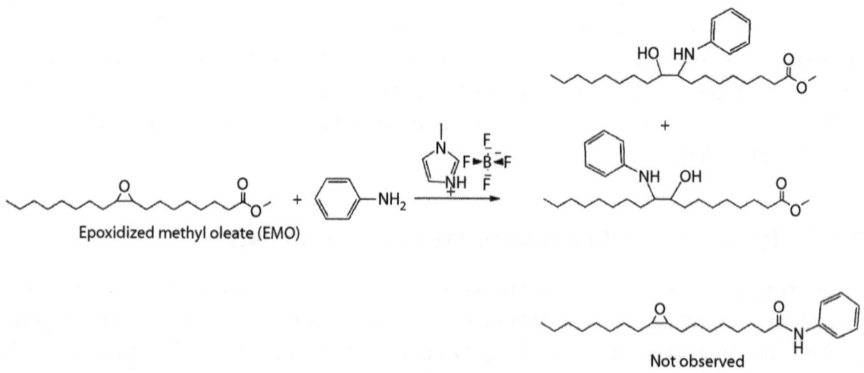

FIGURE 7.2 Ring opening of epoxidized methyl oleate (EMO) with aniline catalyzed by 1-methylimidazolium tetrafluoroborate.

Reactions were typically allowed to run for at least 6 h with aliquots taken at regular time intervals. The product was purified by decanting from the vial, then mixing with 30 mL of ethyl acetate and washing with 50 mL of water three times in a separatory funnel. A final wash was performed with a saturated sodium chloride solution, the ethyl acetate was removed by rotary evaporation and the product dried overnight at 50°C under vacuum.

7.2.4 Reaction Aliquot Characterization

The reaction aliquots were filtered through a 0.4 μm syringe filter and then analyzed by gas chromatography–mass spectrometry (GC-MS). An Agilent (Santa Clara, CA) 7890A gas chromatograph equipped with a 7683B series injector and a 5975C mass detector was used for analysis of the aliquots. The instrument programs and data acquisition were handled by a Windows XP equipped HP-Compaq DC7700 computer with a 3.39 GHz Pentium D processor, using Agilent MSD Enhanced Chemstation Version E01.00.237 software. The GC column was an HP-5MS (Agilent, Santa Clara, CA, 30 m × 0.25 mm, film thickness 0.25 μm). An injection volume of 0.1 μL and a 50:1 split ratio were used with a helium flow rate of ~0.3 mL min^{-1}. The temperature program was Inlet 220°C, detector 220°C, auxiliary transfer line 250°C, MSD 150°C, initial temperature of 150°C for 2 min, ramp to 280°C at 15°C min^{-1} and held for 20 min. The detector was run in the EI mode with scanning set for m/z ratios from 50 to 500 daltons. As a word of caution, the injection of any amount of ionic liquid into the GC column causes a complete loss of separation on future injections. In other words, the sampling must be done carefully in order to obtain only the product layer or else the column can only be used once.

7.2.5 Structural Characterization

The proton and carbon nuclear magnetic resonance (^{1}H NMR and ^{13}C NMR) spectra of the products are shown in Figures 7.3 and 7.4 and confirm the identity of the product. Briefly, the carbonyl area of the ^{13}C NMR definitively shows that the amide structure was not obtained, and the signature from the methyl headgroup was intact, as shown by both ^{13}C NMR and ^{1}H NMR. Patterns from the aniline moiety can be easily assigned as can the methine carbons at the 9–10 position of the fatty chain. These assignments were all confirmed by simulation, as well as by two-dimensional correlations of NMR spectra.

Analysis of the electron impact mass spectrometry data shows the two largest peaks at m/z 218.19 and 262.18 daltons. These two peaks can be assigned to probable fragments, which both contain the aniline moiety. A more complete multinuclear one- and two-dimensional NMR characterization and mass spectrometry identification are available elsewhere [60].

7.2.6 Oxidation Stability Evaluation

Pressurized differential scanning calorimetry (PDSC) was performed on a Q10 differential scanning calorimeter (TA Instruments, New Castle, DE) controlled by Q

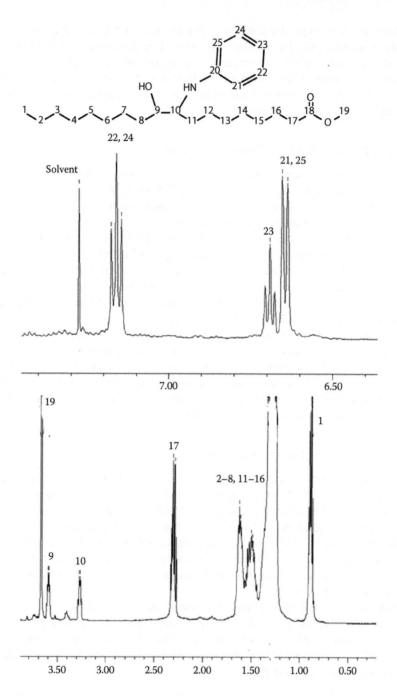

FIGURE 7.3 ¹H NMR spectra of the EMO aniline product, taken in deuterated chloroform solution. Note the isomers with the aniline moiety and the hydroxyl groups at the 9 and 10 positions and the isomer with the groups at the 10 and 9 positions are expected to have similar NMR spectra. Only one set of signals was determined in this work.

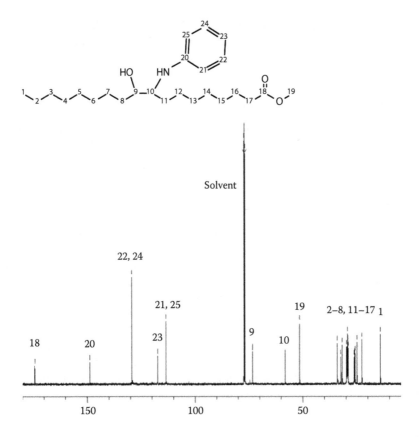

FIGURE 7.4 ^{13}C NMR spectra of the EMO aniline product, taken in deuterated chloroform solution.

Advantage software version 2.8.0 run on an IBM ThinkCentre Pentium 4 processor with a 3.0 GHz processor. Samples of ~2 mg were placed in a hermetically sealed aluminum pan with a pinhole lid. An air pressure of 1379 kPa was kept over the sample throughout the run. The temperature was ramped at 10°C min^{-1} and an oxidation onset temperature was determined by the exothermic reaction putatively accompanying the oxidation of the lubricant.

7.2.7 LUBRICATION EVALUATION

Lubrication testing was performed on a Falex Multi-Specimen Friction and Wear instrument, controlled by Falex 330 Software. Testing was conducted under four-ball configuration according to the American Society for Testing and Materials (ASTM) method D-4172 at a temperature of 75°C, speed of 1200 rpm, and a load of 44 kg. The specimen balls (52100 steel, 12.7 mm diameter, 64–66 Rc hardness and extreme polish) were thoroughly cleaned with dichloromethane and hexane before each experiment. Wear scar diameters (wsd) were measured using a microscope (6SD with L2 fiber-optic light, Leica, Bannockburn, IL) equipped with a digital microscope camera (PAXcam, Vill Park, IL) controlled by Pax-it 6.4 software.

7.2.8 LUBRICATION SAMPLE PREPARATION

Lubrication samples were prepared by mixing the product into lubrication oil base-stocks and stirring overnight. Solutions of 0%, 0.5%, 1%, and 2% product in PAO 8 were prepared and a solution of 2% in soybean oil was also prepared.

7.3 SYNTHESIS TRENDS

The ring opening reaction of EMO with aniline was carried out under varied conditions with aniline:EMO weight ratios from 0 to 4, catalyst loadings from 0 to 2.5 g, and reaction temperatures from 60°C to 105°C. Under all conditions where both catalyst and aniline were present, the reaction proceeded and the major observed product was the compound with aniline incorporated into the fatty acid backbone as shown in Figure 7.2. No significant amount of amide was formed under any of the conditions employed here. However, a reaction where EMO was mixed with a catalyst but no other compound, at 105°C, showed the starting material to be stable for ~4 h. After this initial period, the EMO decomposed into unidentified products. Within the product-forming reactions, several fairly expected trends were uncovered.

7.3.1 CATALYST LOADING

Experiments with varied amounts of catalyst gave the first and most obvious trends. Catalyst amount was varied from 0 to 2.5 g. Analysis of reaction aliquots taken after 1 h of reaction at 105°C show varied reaction progress. In the catalyst-free reaction, no product was formed, whereas with catalyst loading of 0.125 g or greater, the reactions were essentially complete after 1 h. This shows the expected result, that aniline by itself does not possess sufficient reactivity to ring-open EMO through either acidic or basic mechanisms. The effect of catalyst amount on product formation after 1 h of reaction at 105°C is summarized in Figure 7.5.

7.3.2 TEMPERATURE

Ring opening reactions of oxiranes have been shown to exhibit significant temperature dependence under a variety of conditions [73–76]. The ring opening reaction of EMO with aniline was no exception. The reaction progress of the catalyzed reaction at 105°C was more than twice that of a reaction run under the same conditions but at only 60°C. This trend is evident throughout the reaction, as shown in Figure 7.6.

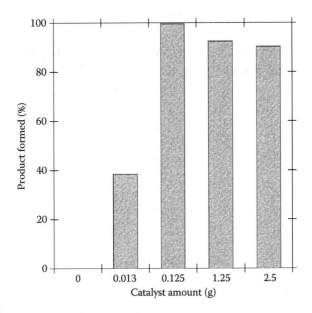

FIGURE 7.5 Effect of catalyst concentration on the ring opening conversion of EMO to product with 1 h reaction time at 105°C.

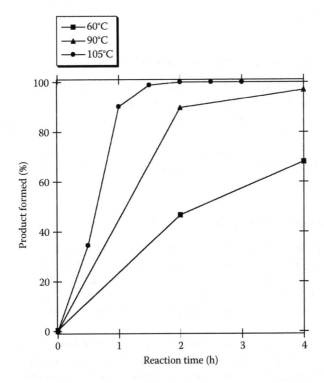

FIGURE 7.6 Effect of temperature on the ring opening reaction of EMO with aniline.

TABLE 7.1

Effect of Changes in the Aniline:EMO Ratio on the Ring Opening Reaction of EMO with Aniline at 105°C with 2.5 g Catalyst Loading

	Reaction Progress by GC (%)		
Wt Ratio (Aniline:EMO)	1 h	2 h	4 h
0:1	0	0	0
1:1	12	44	81
2:1	90	100	100
3:1	88.9	100	100
4:1	100	100	100

7.3.3 REAGENT RATIO

The ratio of aniline to EMO also affects the reaction progress. At a lower amount of aniline, a 1:1 weight ratio, the reaction is slow. It is significantly faster at 2:1 ratio, but further increases in aniline ratio did not cause further increase in reaction completion. This observation points to the active involvement of aniline in the early stages of the reaction where it still has an effect on the kinetics, up to the point where the maximum rate enhancement is generated. Table 7.1 is a summary of the reaction progress after 1–4 h at 105°C for various ratios of aniline to EMO.

7.4 PHYSICAL AND OXIDATION PROPERTIES

The effect of the EMO–aniline product as a lubricant additive was evaluated. Properties tested were friction, wear, and oxidation stability using the four-ball friction test, wear test, and PDSC, respectively. In the friction and wear tests, the selected basestock was the synthetic polyalphaolefin, PAO 8. In the oxidation tests, the additive was evaluated in both PAO 8 and soybean oil. This is the first nonpatent literature report on the lubricant and oxidation properties of this reaction product.

7.4.1 FRICTION AND WEAR INVESTIGATION

This lubricant additive was tested in PAO 8 basestock at concentrations where anti-wear additives are typically used (0.5–2 wt%). The wsd difference was noticeable, reducing from 0.59 mm in the control experiment to below 0.47 mm with additive concentrations of 1% or 2%. Soybean oil already displays a very low wear scar in this test, a value that was not improved by the addition of this additive (data not shown).

It can be illustrative also to examine the coefficient of friction data generated in the wear test, even though it is not part of the ASTM method. Overall, the coefficient of friction remained fairly consistent in the PAO 8 system, as shown in Figure 7.7, as well as in soybean oil (data not shown).

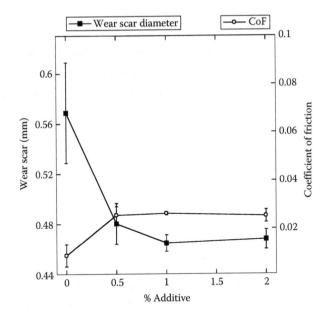

FIGURE 7.7   Coefficient of friction (CoF) and wear scar diameter observed in experiments using PAO 8 basestock with addition of the EMO–aniline product.

7.5   OXIDATION STABILITY

Owing to the aromatic group in the EMO–aniline product, an antioxidation effect is predicted and was evaluated using PDSC. The oxidation stability of lubrication fluids was measured at the same additive levels as in the friction and wear study. The oxidation onset data show a significant improvement of the PAO 8 values, from a value starting below 195°C, to a value above 220°C (Table 7.2). This is about half of the effect which could be achieved utilizing a commercial antioxidant, such as butylated hydroxy toluene (BHT). What is even more pronounced is that a loading of 0.5 wt% of the EMO–aniline adduct corresponds only to ~0.125 wt% of the organic aromatic group, the bulk of the additive being oleo based. The EMO–aniline product also does improve the oxidation shortcomings of soybean oil lubricants at 2% loading.

7.6   CONCLUSION

It is possible to produce a predominantly biobased lubricant additive from chemical synthesis by utilizing EMO and aniline. The key to this product formation is the use of an ionic liquid catalyst, which can be easily removed from the product via a simple washing and extraction procedure. Optimization of the reaction variables leads to straightforward trends. Increased temperature, increased aniline:EMO ratio, and increased catalyst concentration all increase the reaction rate within limits. The aniline product displays some antiwear properties when used as an additive in PAO 8 solution

TABLE 7.2
Oxidation of Lubricant Basestocks with Addition of EMO–Aniline Product

Additive Concentration of (wt%)	Onset Temperature (°C)	
	EMO–Aniline Additive	BHT Antioxidant
PAO 8		
0	191.5 ± 0.6	
0.5	208.2 ± 0.6	240.9 ± 0.7
1	215.9 ± 1.2	246.3 ± 0.2
2	223.8 ± 0.6	251.5 ± 0.04
Soybean Oil		
0	175.3 ± 0.2	
2	180.9 ± 0.3	209.7 ± 1.7

and it also displays impressive antioxidation tendency. Overall, this is a step in the right direction in the search for increased biobased content in industrial lubrication.

ACKNOWLEDGMENTS

We would like to acknowledge Janet Berfield, Cynthia M. Ruder, Jennifer R. Koch, and Brajendra K. Sharma for assistance with the work presented here.

REFERENCES

1. E. R. Booser, Lubrication and lubricants, in: *Lasers to Mass Spectrometry, Kirk-Othmer Encyclopedia of Chemical Technology,* J. I. Kroschwitz and M. Howe-Grant (Eds.) pp. 463–517, John Wiley and Sons, New York, 1995.
2. T. Mang, C. Freiler, and D. Horner, Metalworking fluids, in: *Lubricants and Lubrication,* T. Mang and W. Dresel (Eds.) pp. 384–521, Wiley-VCH, Weinheim, 2007.
3. A. R. Lansdown, *Lubrication and Lubrication Science.* ASME Press, New York, 2004.
4. N. J. Mosey, M. H. Muser, and T. K. Woo, Molecular mechanisms for the functionality of lubricant additives, *Science* 307; 1612–1615, 2005.
5. C. Barrett, Perception vs. reality, *Tribol. Lubr. Technol.* 63; 28–35, 2007.
6. M. P. Schneider, Plant-oil-based lubricants and hydraulic fluids, *J. Sci. Food Agric.* 86; 1769–1780, 2006.
7. S. Z. Erhan, B. K. Sharma, Z. Liu, and A. Adhvaryu, Lubricant base stock potential of chemically modified vegetable oils, *J. Agric. Food Chem.* 56; 8919–8925, 2008.
8. H.-S. Hwang and S. Z. Erhan, Lubricant base stocks from modified soybean oil, in: *Biobased Industrial Fluids and Lubricants,* S. Z. Erhan and J. M. Perez (Eds.) pp. 20–34, AOCS Press, Champaign, IL, 2002.
9. S. Z. Erhan, A. Adhvaryu, and B. K. Sharma, Chemically functionalized vegetable oils, in: *Synthetics, Mineral Oils, and Bio-Based Lubricants Chemistry and Technology,* L. R. Rudnick (Ed.) pp. 14–30, CRC Press, Boca Raton, FL, 2005.
10. K. M. Doll, B. R. Moser, B. K. Sharma, and S. Z. Erhan, Current uses of vegetable oil in the surfactant, fuel, and lubrication industries, *Chemica Oggi/Chemistry Today* 24; 41–44, 2006.

11. G. Biresaw, A. Adhvaryu, and S. Z. Erhan, Friction properties of vegetable oils, *J. Am. Oil Chem. Soc.* 80; 697–704, 2003.

12. N. N. Tupotilov, V. V. Ostrikov, and A. Y. Kornev, Plant oil derivatives as additives for lubricants, *Chem. Technol. Fuels Oils* 42; 192–195, 2006.

13. S. A. Miller, A. E. Landis, T. L. Theis, and R. L. Reich, A comparative life cycle assessment of petroleum and soybean-based lubricants, *Environ. Sci. Technol.* 41; 4143–4149, 2007.

14. Guideline for designating biobased products for federal procurement, *Fed. Register* 70; 1792–1812, 2005.

15. W. E. Artz, P. C. Osidacz, and A. R. Coscione, Acceleration of the thermoxidation of oil by heme iron, *J. Am. Oil Chem. Soc.* 82; 579–584, 2005.

16. A. R. Coscione and W. E. Artz, Vegetable oil stability at elevated temperatures in the presence of ferric stearate and ferrous octanoate, *J. Agric. Food Chem.* 53; 2088–2094, 2005.

17. K. Warner, Effects on the flavor and oxidative stability of stripped soybean and sunflower oils with added pure tocopherols, *J. Agric. Food Chem.* 53; 9906–9910, 2005.

18. A. Kockritz and A. Martin, Oxidation of unsaturated fatty acid derivatives and vegetable oils, *Eur. J. Lipid Sci. Technol.* 110; 812–824, 2008.

19. T. C. Ming, N. Ramli, O. T. Lye, M. Said, and Z. Kasim, Strategies for decreasing the pour point and cloud point of palm oil products, *Eur. J. Lipid Sci. Technol.* 107; 505–512, 2005.

20. S. J. Miller, Conversion of vegetable oils to base oils and transportation fuels, US Patent 20100018109 A1, 2010.

21. S. Z. Erhan and M. O. Bagby, Vegetable oil-based printing ink, US Patent 5122188, 1992.

22. S. Z. Erhan and M. O. Bagby, Vegetable oil-based offset printing inks, US Patent 5713990, 1998.

23. A. L. Cholli, A. Dhawan, and V. Kumar, Post-coupling synthetic approach for polymeric antioxidants, US Patent 20050238979, 2005.

24. H. Benecke, B. Vijayendran, R. and J. Cafmeyer, Lubricant derived from plant and animal oils and fats, WO Patent 2006/020716, 2005.

25. S. Z. Erhan, A. Adhvaryu, and Z. Liu, Chemically modified vegetable oil-based industrial fluid, US Patent 6583302, 2003.

26. S. Z. Erhan, K. M. Doll, and B. K. Sharma, Method of making fatty acid ester derivatives, US Patent 20080154053, 2008.

27. M. M. Zuckerman, Synthesizing and compounding molecules from and with plant oils to improve low temperature behavior of plant oils as fuels, oils, and lubricants, US Patent 20080227993 A1, 2008.

28. M. Kunz, J. Kowalczyk, A. H. Begli, R. Kohlstrung, M. Harperscheid, A. Kesseler, R. Luther, T. Mand, C. Puhl, and H. Wagner, Carbohydrate esters for using as lubricants, US Patent 7220710, 2007.

29. S. J. Asadauskas, G. Biresaw, and T. G. McClure, Effects of chlorinated paraffin and ZDDP concentrations on boundary lubrication properties of mineral and soybean oils, *Tribol. Lett.* 37; 111–121, 2010.

30. B. K. Sharma, J. M. Perez, and S. Z. Erhan, Soybean oil-based lubricants: A search for synergistic antioxidants, *Energy Fuels* 21; 2408–2414, 2007.

31. B. R. Moser and S. Z. Erhan, Preparation and evaluation of a series of α-hydroxyethers from 9,10-epoxystearates, *Eur. J. Lipid Sci. Technol.* 109; 206–213, 2007.

32. B. R. Moser and S. Z. Erhan, Synthesis and evaluation of a series of α-hydroxyethers derived from isopropyl oleate, *J. Am. Oil Chem. Soc.* 83; 959–963, 2006.

33. H.-S. Hwang, A. Adhvaryu, and S. Z. Erhan, Preparation and properties of lubricant basestocks from epoxidized soybean oil and 2-ethylhexanol, *J. Am. Oil Chem. Soc.* 80; 811–815, 2003.

34. H.-S. Hwang and S. Z. Erhan, Modification of epoxidized soybean oil for lubricant formulations with improved oxidative stability and low pour point, *J. Am. Oil Chem. Soc.* 78; 1179–1184, 2001.

35. B. K. Sharma, K. M. Doll, and S. Z. Erhan, Ester hydroxy derivatives of methyl oleate: Tribological, oxidation and low temperature properties, *Bioresour. Technol.* 99; 7333–7340, 2008.

36. K. M. Doll, B. K. Sharma, and S. Z. Erhan, Synthesis of branched methyl hydroxy stearates including an ester from bio-based levulinic acid, *Ind. Eng. Chem. Res.* 46; 3513–3519, 2007.

37. J. Salimon and N. Salih, Preparation and characteristic of 9, 10-epoxyoleic acid-hydroxy ester derivatives as biolubricant base oil, *Eur. J. Sci. Res.* 31; 265–272, 2009.

38. B. K. Sharma, Z. Liu, A. Adhvaryu, and S. Z. Erhan, One-pot synthesis of chemically modified vegetable oils, *J. Agric. Food Chem.* 56; 3049–3056, 2008.

39. B. K. Sharma, K. M. Doll, and S. Z. Erhan, Chemically modified fatty acid methyl esters: Their potential use as lubrication fluids and surfactants, in: *Surfactants in Tribology*, G. Biresaw and K. L. Mittal (Eds.) Vol 2, pp. 387–408, CRC Press, Boca Raton, FL, 2011.

40. K. M. Doll and B. K. Sharma, Surfactant effects on bio-based emulsions used as lubrication fluids, in: *Surfactants in Tribology*, G. Biresaw and K. L. Mittal (Eds.) Vol 2, pp. 173–190, CRC Press, Boca Raton, FL, 2011.

41. S. Cermak, G. Biresaw, and T. Isbell, Comparison of a new estolide oxidative stability package, *J. Am. Oil Chem. Soc.* 85; 879–885, 2008.

42. K. M. Doll and S. Z. Erhan, Synthesis of cyclic acetals (ketals) from oleochemicals using a solvent-free method, *Green Chem.* 10; 712–717, 2008.

43. G. B. Bantchev, J. A. Kenar, G. Biresaw, and M. G. Han, Free radical addition of butanethiol to vegetable oil double bonds, *J. Agric. Food Chem.* 57; 1282–1290, 2009.

44. T. L. Kurth, J. A. Byars, S. C. Cermak, B. K. Sharma, and G. Biresaw, Non-linear adsorption modeling of fatty esters and oleic estolide esters via boundary lubrication coefficient of friction measurements, *Wear* 262; 536–544, 2007.

45. B. K. Sharma, K. M. Doll, and S. Z. Erhan, Oxidation, friction reducing, and low temperature properties of epoxy fatty acid methyl esters, *Green Chem.* 9; 469–474, 2007.

46. I. Javni, W. Zhang, and Z. S. Petrovic, Effect of different isocyanates on the properties of soy-based polyurethanes, *J. Appl. Polym. Sci.* 88; 2912–2916, 2003.

47. I. Javni, Z. S. Petrovic, A. Guo, and R. Fuller, Thermal stability of polyurethanes based on vegetable oils, *J. Appl. Polym. Sci.* 77; 1723–1734, 2000.

48. A. Guo, I. Javni, and Z. Petrovic, Rigid polyurethane foams based on soybean oil, *J. Appl. Polym. Sci.* 77; 467–473, 2000.

49. G. L. Wilkes, S. Sohn, and B. Tamami, Preparations of nonisocyanate polyurethane materials from epoxidized soybean oils and related epoxidized vegetable oils, US Patent 7045577, 2006.

50. B. Tamami, S. Sohn, G. L. Wilkes, and B. Tamami, Incorporation of carbon dioxide into soybean oil and subsequent preparation and studies of nonisocyanate polyurethane networks, *J. Appl. Polym. Sci.* 92; 883–891, 2004.

51. J. Xu, Z. Liu, S. Z. Erhan, and C. J. Carriere, A potential biodegradable rubber—Viscoelastic properties of a soybean oil-based composite, *J. Am. Oil Chem. Soc.* 79; 593–596, 2002.

52. S. P. Bunker and R. P. Wool, Synthesis and characterization of monomers and polymers for adhesives from methyl oleate, *J. Polym. Sci., Part A: Polym. Chem.* 40; 451–458, 2002.

53. Z. Liu, S. Z. Erhan, and J. Xu, Preparation, characterization and mechanical properties of epoxidized soybean oil/clay nanocomposites, *Polymer* 46; 10119–10127, 2005.

54. S. Z. Erhan, Z. Liu, and P. D. Calvert, Extrusion freeform fabrication of soybean oil-based composites by direct deposition, US Patent US 6528571 B1, 2003.

55. F. Li and R. C. Larock, New soybean oil-styrene-divinylbenzene thermosetting copolymers IV: Good damping properties, *Polym. Adv. Technol.* 13; 436, 2002.

56. F. Li, M. V. Hanson, and R. C. Larock, Soybean oil-divinylbenzene thermosetting polymers: Synthesis, structure, properties and their relationships, *Polymer* 42; 1567–1579, 2000.

57. R. C. Larock, X. Dong, S. Chung, C. Kishan Reddy, and L. E. Ehlers, Preparation of conjugated soybean oil and other natural oils and fatty acids by homogeneous transition metal catalysis, *J. Am. Oil Chem. Soc.* 78; 447–453, 2001.

58. F. Li and R. C. Larock, New soybean oil-styrene-divinylbenzene thermosetting copolymers. I. Synthesis and characterization, *J. Appl. Polym. Sci.* 80; 658–670, 2001.

59. F. Li and R. C. Larock, New soybean oil-styrene-divinylbenzene thermosetting copolymers III: Tensile stress-strain behavior, *J. Polym. Sci. B. Polym. Phys.* 39; 60–77, 2001.

60. A. Biswas, B. K. Sharma, K. M. Doll, S. Z. Erhan, J. L. Willett, and H. N. Cheng, Synthesis of an amine oleate derivative using an ionic liquid catalyst, *J. Agric. Food Chem.* 57; 8136–8141, 2009.

61. J. A. Widegren, E. M. Saurer, K. N. Marsh, and J. W. Magee, Electrolytic conductivity of four imidazolium-based room-temperature ionic liquids and the effect of a water impurity, *J. Chem. Thermodyn.* 37; 569–575, 2005.

62. H.-P. Zhu, F. Yang, J. Tang, and M.-Y. He, Bronsted acidic ionic liquid 1-methylimidazolium tetrafluoroborate: A green catalyst and recyclable medium for esterification, *Green Chem.* 5; 38–39, 2003.

63. D. G. Archer, D. R. Kirklin, J. A. Widegren, and J. W. Magee, Enthalpy of solution of 1-octyl-3-methylimidazolium tetrafluoroborate in water and in aqueous sodium fluoride, *J. Chem. Eng. Data* 50; 1484–1491, 2005.

64. J. Dupont, R. F. de Souza, and P. A. Z. Suarez, Ionic liquids (molten salt) phase organometallic catalysis, *Chem. Rev.* 102; 3667–3692, 2002.

65. J. Dupont, P. A. Z. Suarez, A. P. Umpierre, and R. F. De Souza, Organo-zincate molten salts as immobilising agents for organometallic catalysis, *Catal. Lett.* 73; 211–213, 2001.

66. J. Chen, H. Wu, C. Jin, X. Zhang, Y. Xie, and W. Su, Highly regioselective ring-opening of epoxides with thiophenols in ionic liquids without the use of any catalyst, *Green Chem.* 8; 330–332, 2006.

67. K. M. Doll and S. Z. Erhan, Synthesis and performance of surfactants based on epoxidized methyl oleate and glycerol, *J. Surfact. Deterg.* 9; 377–383, 2006.

68. J. Filley, New lubricants from vegetable oil: Cyclic acetals of methyl 9,10-dihydroxystearate, *Bioresour. Technol.* 96; 551–555, 2005.

69. M. Dierker, Oleochemical carbonates—An overview, *Lipid Technol.* 16; 130–134, 2004.

70. J. A. Kenar, Current perspectives on oleochemical carbonates, *INFORM* 15; 580–582, 2004.

71. J. La Scala and R. P. Wool, Rheology of chemically modified triglycerides, *J. Appl. Polym. Sci.* 95; 774–783, 2005.

72. D. Swern, G. N. Billen, T. W. Findley, and J. T. Scanlan, Hydroxylation on monounsaturated fatty materials with hydrogen peroxide, *J. Am. Chem. Soc.* 67; 1786–1789, 1945.

73. A. Campanella and M. A. Baltanás, Degradation of the oxirane ring of epoxidized vegetable oils with solvated acetic acid using cation-exchange resins, *Eur. J. Lipid Sci. Technol.* 106; 524–530, 2004.

74. A. Campanella and M. A. Baltanas, Degradation of the oxirane ring of epoxidized vegetable oils with hydrogen peroxide using an ion exchange resin, *Catal. Today* 107–108; 208–214, 2005.

75. A. Campanella and M. A. Baltanás, Degradation of the oxirane ring of epoxidized vegetable oils in liquid-liquid systems: II. Reactivity with solvated acetic and peracetic acids, *Latin Am. Appl. Res.* 35; 211–216, 2005.

76. A. Campanella and M. A. Baltanas, Degradation of the oxirane ring of epoxidized vegetable oils in liquid-liquid heterogeneous reaction systems, *Chem. Eng. J.* 118; 141–152, 2006.

8 Environmentally Friendly Surface Active Agents and Their Applications

Nabel A. Negm and Salah M. Tawfik

CONTENTS

This chapter discusses the synthetic and natural surface active agents and their potential in several applications. The reviewed data showed that most surface active agents have deep applications in industry, science, and domestic uses. These compounds accumulated in the environment and finally reached the underground water reservoirs and rivers. That causes severe diseases for humans and other living organisms. Surfactants accumulate in liver, kidney, and fat tissues. This chapter classifies the surfactants according to their origin as synthetic from petroleum distillates and natural from animal or plant origins. The presented data showed that the synthetic surfactants are one of the pollutants in the environment. To overcome this problem, replacement of these compounds by other environmentally friendly surface active agents is an urge. The chapter reviews the different types of degradable surfactants or environmentally friendly surfactants, including sugar surfactants, carbonate surfactants, acetal and ketal surfactants, ethoxylated esters, and sugars. Several studies concerning the biodegradation of surfactants are reviewed. The most important conclusion obtained is that surface active agents with natural origin can easily degrade in the open environment through microorganisms and produce most safe products through β- and ω-oxidation pathways.

8.1 DEFINITION AND CLASSIFICATION OF SURFACTANTS

Some compounds are amphiphilic or amphipathic, that is, they have one part that has an ability to dissolve in nonpolar media and another part that has an ability toward polar media. These molecules form oriented monolayers at interfaces and show surface activity. These compounds are termed surfactants, amphiphiles, surface active agents, or tensides.

The polar or ionic head group usually interacts strongly with an aqueous environment, in which case it is solvated via dipole or ion–dipole interactions. In fact, it is the type of the polar head group that is used to classify surfactants into different categories. Compared with the commonly encountered hydrocarbon-based surfactants, incorporating fluorocarbon in the structure creates molecules that are resistant to oxidation (Figures 8.1 and 8.2). This is because of the smaller size of the fluorine atom relative to the hydrogen atom. The fluorinated surfactants are more rigid in structure, and thus have a surface tension-lowering action, water and oil repellency, thermal resistance, chemical resistance, and lubricating ability.

There is also an entire class of surfactants known as microbial surfactants or biosurfactants. These surfactants have some very interesting and complicated structures, and are expensive to produce compared with chemically synthesized surfactants [1].

In the early 1980s, interest in surfactants derived from nonlinear alkyl benzene (non-LAB) sources began to increase, as it became clear that consumer demand for "newer and better" detergents was outpacing the ability of detergent manufacturers to reformulate their products using as the main component the conventional single-head, single-tail amphiphile. Hence, the synthesis of novel surfactants emerged as a

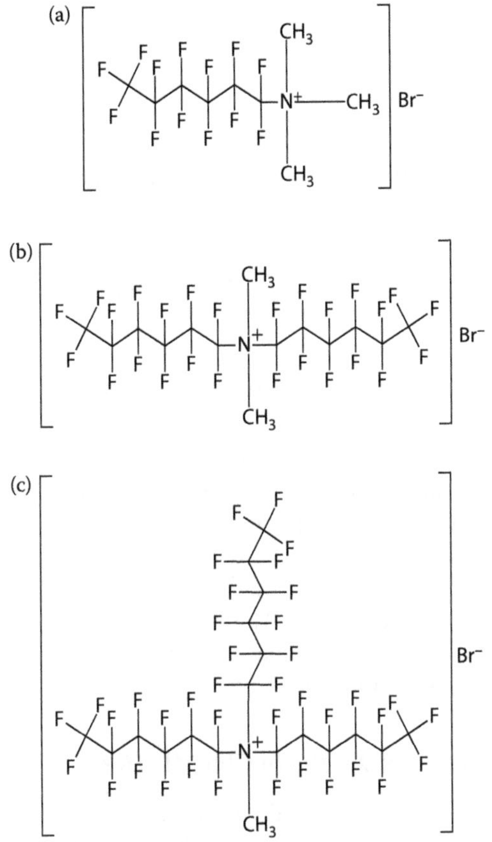

FIGURE 8.1 Examples of fluorocarbon surfactants: (a) 1,1,1,2,2,3,3,4,4,5,5,6,6-florohexyl-trimethyl ammonium bromide, (b) di-(1,1,1,2,2,3,3,4,4,5,5,6,6-florohexyl)-dimethyl ammonium bromide, and (c) tri-(1,1,1,2,2,3,3,4,4,5,5,6,6-florohexyl)-methyl ammonium bromide.

viable and important topic in the literature. In recent surfactant literature, one can find investigation of the properties of vitamin E-based surfactants [2] and sugar-based surfactants, and many others are discussed [3].

One of the most exciting developments in the field of surfactant chemistry was the emergence of the gemini surfactants in the late 1980s and the early 1990s. Examples

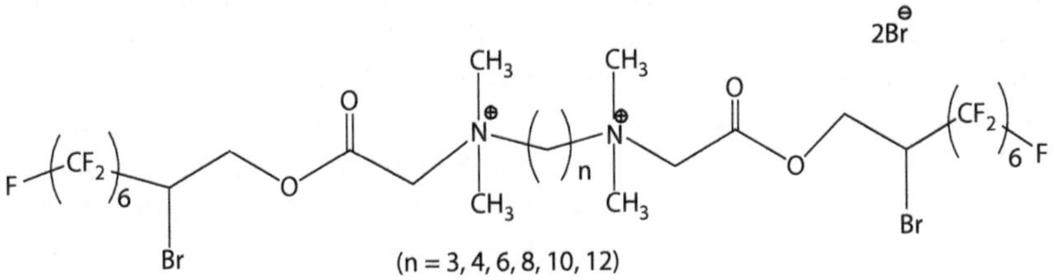

FIGURE 8.2 Example of gemini fluorocarbon surfactants.

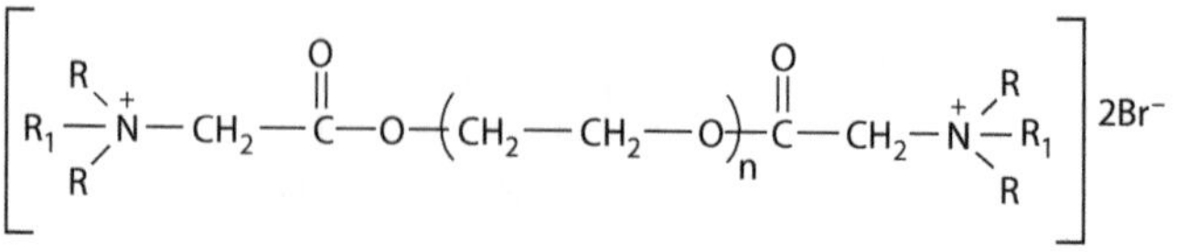

FIGURE 8.3 Example of a trialkyl gemini surfactant containing aliphatic spacer.

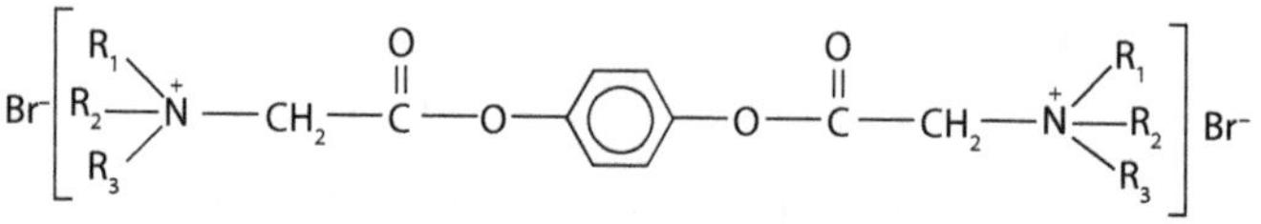

FIGURE 8.4 Example of a trialkyl gemini surfactant containing aromatic spacer.

of a typical gemini surfactant are shown in Figures 8.3 and 8.4. Gemini surfactants possess a number of superior properties [4,5] when compared with conventional single-headed, single-tailed surfactants. The gemini surfactants tend to exhibit lower critical micelle concentration (cmc), increased surface activity and decreased surface tension at the cmc, enhanced solution properties such as hard water tolerance, better wetting times, and lower Krafft points. These properties suggest that the application of gemini surfactants in soil remediation, oil recovery, and commercial detergency will have a favorable cost/performance advantage.

8.2   SURFACTANTS AND GREEN CHEMISTRY

Pollution caused by synthetic chemicals has resulted in an increased demand for environmentally friendly chemical processes. The development of green chemistry has emerged as an important tool to secure the environment for the future. Naturally occurring compounds are those compounds that occur in different organisms, animals, plants, or microorganisms. These compounds are different and widely spread in the plant and animal kingdoms. Fats and oils are typical examples from these resources. The importance of green chemistry is to produce their derivatives in the form of different types of useful surfactants, synthetic lipids, and related structures [6]. Owing to their characteristic surface activity and tendency for molecular self-assembly, they can control and modify both the physical and chemical properties of the phase boundary between different phases of liquids encountered in almost all chemical reactions. A first approach toward environmentally compatible chemical reactions or chemical processes should be to develop systems that do not require organic solvents. Water, the most ubiquitous inorganic solvent, results in heterogeneous reactions requiring tedious reaction control and yielding lower qualities and quantities of reaction products. An area of growth to solve this problem is the application of surfactants or lipids for controlling reactions based on different phases, such as liquid in an inorganic solvent reaction system. As a result, it is possible to achieve several objectives, including inorganic water solvent system, clean reaction conditions, high yields of halide-free products, waste reduction, simplicity of operation, and cost effectiveness. For biphase reactions, the application of phase transfer catalysts (PTCs) to fatty compounds was reported as a useful and superior method related to green chemistry [7]. Another approach to green chemistry is to develop chemical reactions combined with enzymatic action at moderate temperatures, pressures, and pH conditions [8–11]. In the fats and oils area, sugar-based surfactants have been developed from renewable resources. These include sorbitan esters, sucrose esters, alkyl polyglycosides, and fatty acid glucamides [12]. On the other hand, traditional and well-established procedures for the production of surfactants from fats and oils need to be transformed so that they are environmentally friendly.

8.3 CLEAVABLE SURFACTANTS

Surfactants are stable species, including anionic, nonionic, and cationic species. Only alkyl sulfates are chemically unstable under normal conditions. Over the years, the susceptibility of alkyl sulfates to acid-catalyzed hydrolysis has been seen as a serious problem, particularly for sodium dodecyl sulfate. The general opinion has been that weak bonds in a surfactant may cause handling and storage problems. In recent years, however, the attitude toward easily cleavable surfactants has changed. Environmental concerns have become one of the main driving forces for the development of new surfactants, and the biodegradation rate has become a major issue. One of the main approaches taken to produce readily biodegradable surfactants is to build into the structure a bond with only limited stability. For practical reasons, the weak bond is usually the bridging unit between the polar head group and the hydrophobic tail of the surfactant at which the degradation occurs. This is usually called the primary degradation of the surfactant. Biodegradation then proceeds along various routes, depending on the type of primary degradation product. The ultimate decomposition of the surfactant is the quantity of carbon dioxide evolved during a 4-week exposure to appropriate microorganisms, expressed as a percentage of the amount of carbon dioxide that could theoretically be produced. Seemingly, for most surfactants containing easily cleavable bonds, the values for ultimate decomposition are higher than for the corresponding surfactants lacking the weak bond. Thus, the strong trend toward more environmentally friendly products favors the cleavable surfactant approach.

A second motivation for the development of cleavable surfactants is to avoid complications such as foaming or emulsion formation after the use of a surfactant formulation. The weak bond in the surfactant molecule leads to cleavage into water-soluble and water-insoluble products. This approach is of particular interest for surfactants used in natural organic chemistry and in various biochemical applications.

A third application of surfactants with limited stability arises from new functionality of their cleaved product. The surfactant used in personal care formulations may decompose on application to form useful products for the skin. This type of surfactant is referred to as a functional surfactant. Similarly, surfactants that break down into nonsurfactant products in a controlled way find use in specialized applications such as in the biomedical field. In addition, cleavable surfactants that form vesicles or microemulsions can be of interest for drug delivery, provided the metabolites are nontoxic to the host environment.

8.4 HYDROLYZABLE SURFACTANTS

8.4.1 ALKALI HYDROLYZABLE ESTERS

8.4.1.1 Normal Quaternary Esters

The term "ester quat" refers to surface active quaternary ammonium compounds that have the general formula $R_4N^+X^-$, where R is the long-chain alkyl moiety, linked to the charged head group by an ester bond and X^- is the counter ion. Normal ester quats are surfactants based on esters between one or more fatty acids and a quaternized amino alcohol. Figure 8.5 shows examples of three different hydrolyzable ester quats containing two long-chains and short-chain substituents on the nitrogen atom. As can be seen, the ester-containing surfactants contain two carbon atoms between the ester bond and the nitrogen that carries the positive charge. Cleavage of the ester bonds of the hydrolyzable surfactants yields fatty acid soap and a highly water-soluble quaternary ammonium diol or triol. These degradation products exhibit low fish toxicity, and are degraded further by the established metabolic pathways.

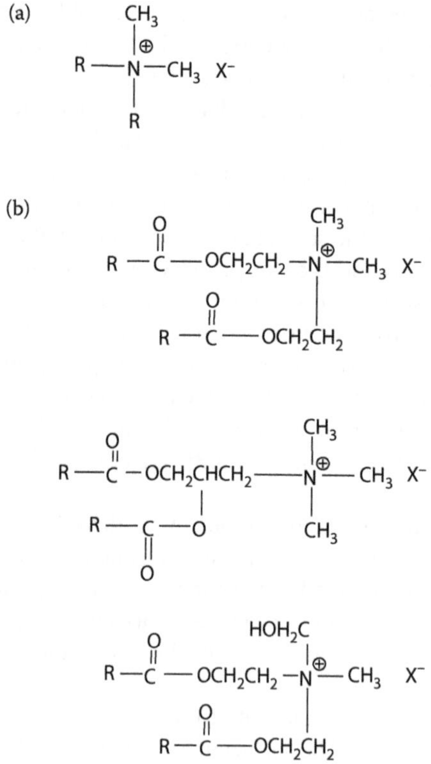

FIGURE 8.5 Structures of cationic surfactants: (a) non-degradable surfactant and (b) three degradable ester quats with two long chain and two short substituents on the nitrogen atom.

The overall ecological characteristics of ester quats are much superior to those of traditional quats represented in Figure 8.5. The switch from stable dialkyl quats to dialkyl ester quats may represent the most dramatic change of product type in the history of surfactants, and it is entirely driven by environmental considerations. Unlike stable quats, ester quats show excellent biodegradability and low aquatic toxicity [13]. Ester quats also have fully or partially replaced traditional quats in other applications of cationics, such as hair care products and various industrial formulations [13].

The positive charge close to the ester bond renders normal ester quats unusually stable to acid but labile to alkali conditions. The strong pH dependence of the hydrolysis can be used as an advantage to induce rapid cleavage of the product. This phenomenon is even more pronounced for betaine esters. Esters of choline have attracted special attention because the primary degradation products, choline and a fatty acid, are both natural metabolites in the body. Thus, choline esters constitute a group of very nontoxic cationic surfactants. Compounds with an alkyl group of 9–13 carbons showed an excellent antimicrobial effect [14]. *In vivo* hydrolysis was rapid due to catalysis by butyryl choline esterase, which is present in human serum and mucosal membranes.

8.4.1.2   Betaine Esters

The rate of alkali-catalyzed ester hydrolysis is influenced by adjacent electron-withdrawing or electron-donating groups. Compounds in Figure 8.5 have two carbon atoms between the ammonium nitrogen and the oxygen of the ester bond. Such esters undergo alkaline hydrolysis at a faster rate than esters lacking the adjacent charge, but the difference is not very large. If the charge is at the other side of the ester bond, the rate enhancement is much more pronounced. Such esters are extremely labile on the alkaline side but very stable under strongly acidic conditions. The large effect of the quaternary ammonium group on the alkaline and acid rates of hydrolysis is due to stabilization/destabilization of the ground state, as illustrated in Figures 8.6 and 8.7. However, the presence of large, polarizable counter ions, such as bromide, can completely outweigh micellar catalysis [15]. The extreme pH dependence of surface active betaine esters

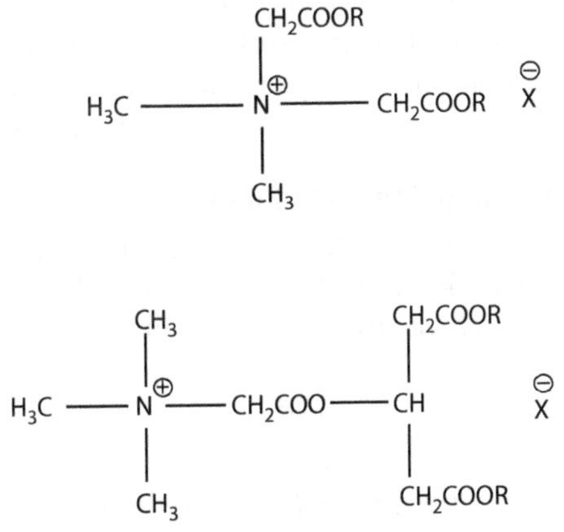

FIGURE 8.6   Examples of betaine esters containing two long alkyl chain groups.

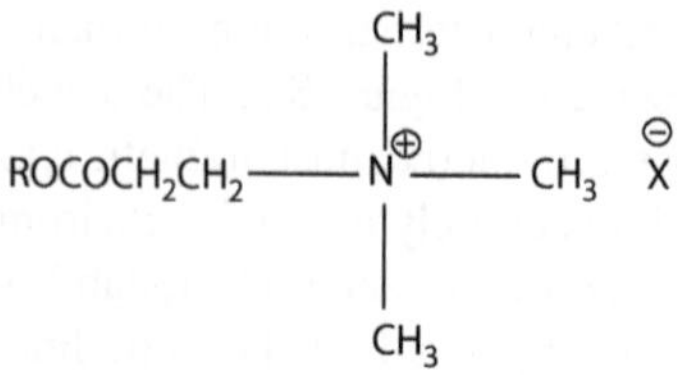

$$\text{ROCOCH}_2\text{CH}_2\!-\!\!-\!\!\overset{\overset{\displaystyle CH_3}{|}}{\underset{\underset{\displaystyle CH_3}{|}}{\overset{\oplus}{N}}}\!\!-\!\!CH_3 \quad X^{\ominus}$$

R = Long-chain alkyl

X = Cl, Br, or CH$_3$SO$_4$

FIGURE 8.7 Examples of betaine esters containing one alkyl group.

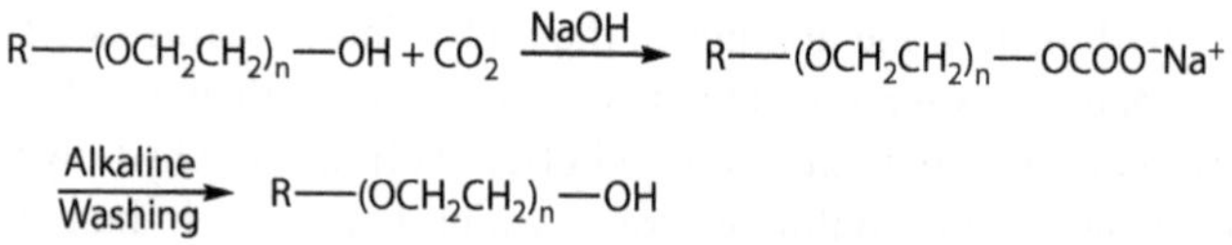

$$R\!-\!(OCH_2CH_2)_n\!-\!OH + CO_2 \xrightarrow{\text{NaOH}} R\!-\!(OCH_2CH_2)_n\!-\!OCOO^-Na^+$$

$$\xrightarrow[\text{Washing}]{\text{Alkaline}} R\!-\!(OCH_2CH_2)_n\!-\!OH$$

FIGURE 8.8 Formation of carbonate salt of nonionic surfactants and subsequent regeneration of the starting surfactant during a washing process. Carbonate salts of non-ionic surfactants are major ingredients in commercial washing detergents.

makes them interesting as cleavable cationic surfactants. The shelf life is long when they are stored under acidic conditions, and the hydrolysis rate will then depend on the pH at which they are used. Single-chain surfactants of this type have been suggested as temporary bactericides for use in hygiene products, and for disinfection in the food industry [15].

8.4.1.3 Monoalkyl Carbonates

Carbonate salts of such surfactants have been used as labile derivatives from which the surfactant can be readily regenerated. Such derivatives are called prosurfactants by analogy to prodrug in medicine. The reaction of alcohol ethoxylate with carbon dioxide gives a solid carbonate salt that decomposes under the alkaline washing conditions to give the starting nonionic surfactant and carbonate, as illustrated in Figure 8.8. Conversion of alcohol ethoxylate into solid carbonate enables the incorporation of high levels of this surfactant into granular detergents of high bulk density.

8.4.1.4 Surfactants Containing Si–O Bond

The silicon–oxygen bond is susceptible to both alkaline and acid hydrolysis. In addition, the bond is specifically cleaved by fluoride ions at a relatively neutral pH. Silicon-based cationic surfactant with the structures shown in Figure 8.9 has been synthesized

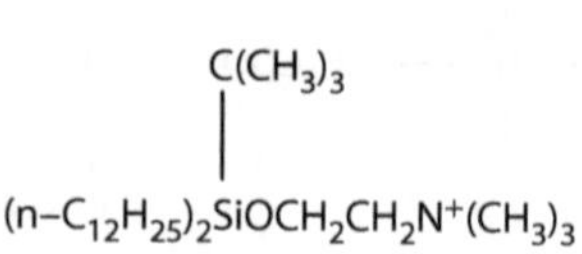

$$\text{(n--C}_{12}\text{H}_{25}\text{)}_2\overset{\overset{\displaystyle C(CH_3)_3}{|}}{Si}OCH_2CH_2N^+(CH_3)_3$$

FIGURE 8.9 Surfactant containing the Si–O bond.

and tested for degradation characteristics. The preparation route of such surfactants is relatively complicated and these surfactants are of limited practical value.

8.4.1.5 Sugar Esters

Sugar esters have received considerable attention mainly because of developments in synthesis procedures. The main advantage of the biochemical route compared with conventional organic synthesis is the much higher regioselectivity obtained in the synthesis. A long reaction time was a typical disadvantage of the enzymatic process. Enzymatic synthesis of sugar esters has been thoroughly reviewed by Vulfson [16]. In a systematic investigation of the effect of the number of condensed hexose units on surfactant properties, mono dodecyl esters of glucose, sucrose (two sugar units), raffinose (three units), and stachyose (four units) were prepared by organic synthesis followed by careful chromatographic purification [17]. As shown in Figure 8.10, all compounds had the acyl substituents at the sixth position of a glucose ring, that is, the ester bond had the same environment in all four surfactants. The surface properties of the compounds were studied. It was concluded

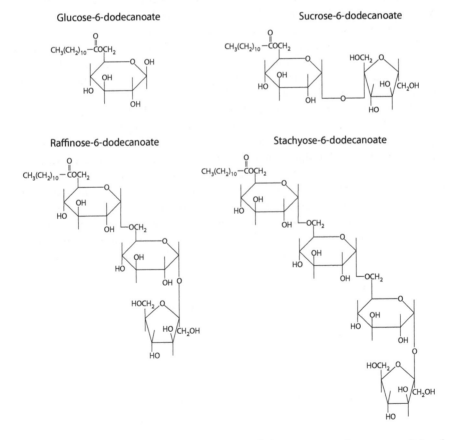

FIGURE 8.10 Examples of monododecyl esters of glucose, sucrose (two sugar units), raffinose (three units), and stachyose (four units).

that the self-assembly of these surfactants was governed primarily by geometric packing constraints, which in turn depend on the size of the polar head group. Enzymatic synthesis of sugar esters can be carried out either in an organic solvent or in solvent-free conditions at reduced pressure. An interesting new development is the use of a microemulsion as a reaction medium. Physicochemical characterization of the sugar esters showed the expected result with the efficiency and effectiveness of the surfactants mainly being dependent on the chain length of the fatty acid. There was little difference in cmc between surfactants based on different sugars and the same fatty acid.

8.4.2 ACID HYDROLYZABLE SURFACTANTS

8.4.2.1 Cyclic Acetals

Cyclic 1,3-dioxolane (five-member ring) and 1,3-dioxane (six-member ring) compounds, illustrated in Figure 8.11, have been studied in depth by Sokolowski et al. [18] as examples of acid-labile surfactants. They are typically synthesized from long-chain aldehydes by reaction with a diol or a higher polyol. The reaction with a vicinal diol gives dioxolane and with 1,3-diol yields dioxane [19]. If the diol contains an extra hydroxyl group, such as in glycerol, a hydroxy acetal is formed and the remaining hydroxyl group can subsequently be derivatized to give anionic or cationic surfactants. It is claimed that glycerol causes ring closure of dioxolane, yielding a free, primary hydroxyl group, but it is likely that some dioxane with a free, secondary hydroxyl group is formed as well. The free hydroxyl group can be treated with SO_3 and then neutralized to give the sulfate. It can also be reacted with propane sultone (sultones are cyclic sulfonate esters of hydroxy sulfonic acids) to give the sulfonate. It can also be substituted by bromide or chloride and then reacted with dimethyl amine to give a tertiary amine as a polar group. An analogous reaction with pentaerythritol as diol yielded a 1,3-dioxane with two unreacted hydroxyl methyl groups that can be reacted further to give a dianionic surfactant. The remaining hydroxyl group may also be ethoxylated, and such acetal surfactants have been commercialized. Hydrolysis splits acetals into aldehydes, which are intermediates in the biochemical oxidation of hydrocarbon chains. Acid-catalyzed hydrolysis of unsubstituted acetals is generally facile and occurs at a reasonable

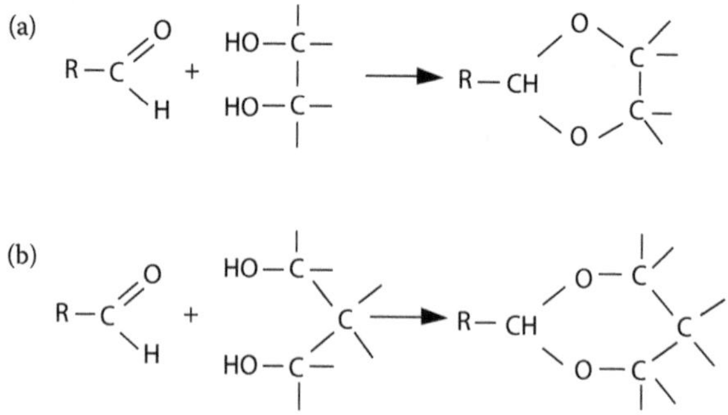

FIGURE 8.11   Five-member ring of cyclic 1,3-dioxolane (a), six-member ring of 1,3-dioxane (b).

rate at pH 4–5. Anionic acetal surfactants are more labile than cationic ones due to the locally high oxonium ion activity around such micelles. Acetal surfactants are stable at neutral and high pH.

8.4.2.2   Acyclic Acetals

Alkyl polyglucosides (APGs) are cyclic compounds, but because the ring does not involve the two geminal hydroxyl groups of the aldehyde hydrate, they are included here in the category of acyclic acetals. Alkyl glucosides are by far the most important type of acetal surfactants. This surfactant class has been the topic of several reviews [20,21]. Alkyl glucosides are made using an acid catalyst either by direct condensation of glucose and a long-chain alcohol or by the trans-acetalization of a short-chain alkyl glucoside, such as ethyl glucoside, with a long-chain alcohol. This procedure leads to some degree of sugar condensation, which governs by the ratio of long-chain alcohol to sugar. The alkyl glucoside surfactants break down into glucose and long-chain alcohol under acidic conditions. On the other hand, they are stable to hydrolysis at very high pH. Their cleavage profile along with their relatively straightforward synthesis route makes these surfactants interesting candidates for various types of cleaning formulations.

Polyoxyethylene-based cleavable surfactants have been synthesized by reacting ethylene oxide (EO) with a long-chain aldehyde. The physicochemical behavior of these surfactants resembles that of normal nonionics. For instance, they have a reversible solubility–temperature relationship and they exhibit a cloud point. Acid hydrolysis of the labile polyoxyethylene-based surfactants yields polyoxyethylene monomethyl ether and a long-chain aldehyde. It was found that the hydrolysis of these noncyclic acetal-linked surfactants was several orders of magnitude faster than that of the cyclic acetal-linked surfactants. This is important because many applications of cleavable surfactants such as textile processing [22], electronics fabrication [23], sample management [24], wastewater processing [25], and cleavable phase transfer reagents [26] receive a rather high rate of breakdown. The hydrolytic reactivity increased as the hydrophobe chain length decreased.

A series of acyclic anionic, nonionic, cationic, and amphoteric acetal surfactants from a common allyl chloride intermediate (Figure 8.12) were investigated [27]. The cmc of these surfactants were lower than those of conventional surfactants of the same alkyl chain length. Furthermore, the efficiency values of the acyclic acetal surfactants (expressed as the concentration of surfactants that required reducing the surface tension of surfactant solution to 51 mN/m) are increased. Evidently, the group connecting the hydrophobic tail and the polar head group gives a hydrophobic contribution to the amphiphilic properties. A systematic study of hydrolysis rates was made with the four classes of surfactant shown in Figure 8.12. The result showed that the micellar surface charge had an effect on the hydrolysis rate. The reaction was very fast with negatively charged micelles, slow with positively charged micelles, and intermediate with uncharged micelles.

8.4.2.3   Ketals

Surfactants containing ketal bonds can be prepared from a long-chain ketone and a diol as shown in the reaction scheme given in Figure 8.13 for the preparation of

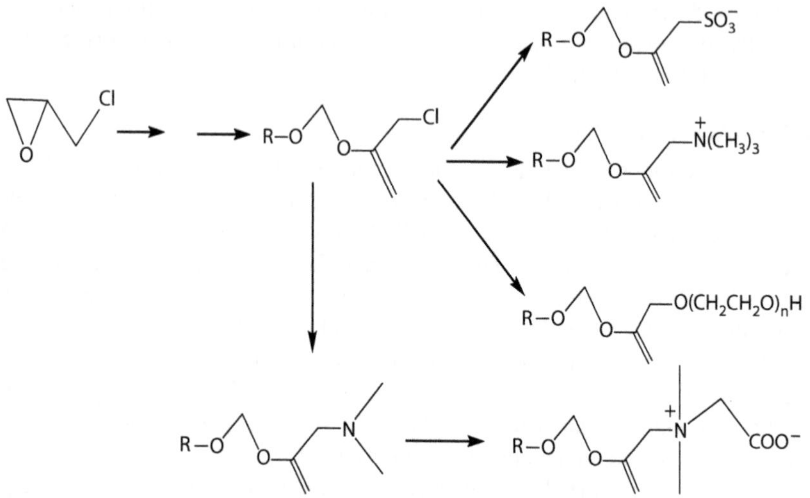

FIGURE 8.12 Schematic route of synthesis for acyclic acetal surfactants.

acetal surfactants. Ketal-based surfactants have also been prepared in good yields from esters of keto acids via two routes.

The degradation rate of ketal surfactants is dependent on the alkyl chain length. The process is markedly faster for the labile surfactants (and particularly for surfactants containing an extra ether oxygen) than for the conventional carboxylate surfactant of the same alkyl chain length. Ketal surfactants are in general more labile than the corresponding acetal surfactants. As an example, a ketal surfactant kept at pH 3.5 was cleaved to the same extent as an acetal surfactant of similar structure kept at pH 3. The relative reactivity of the ketal linkage is due to the greater stability of the carbocation formed during ketal hydrolysis compared with the carbocation formed during acetal hydrolysis. Evidently, there is no strict correlation between ease of biodegradation and rate of chemical hydrolysis. Jaeger et al. [28] have introduced the

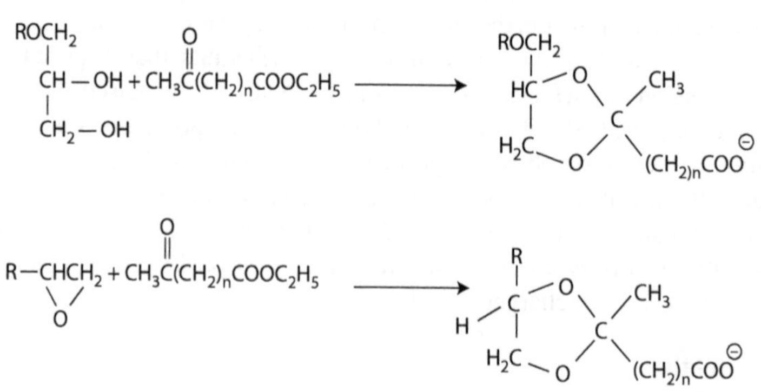

FIGURE 8.13 Ketal-based surfactants.

term "second generation cleavable surfactant" for labile surfactants that on cleavage give another surfactant together with a small water-soluble species upon cleavage. The daughter surfactant generally has a higher cmc value than the parent surfactant. The concept has been applied to a variety of structures, including phospholipid analogs, and several applications of this specific type of cleavable surfactants have been proposed by Jaeger et al. [28]. Double-chain, double-head group second-generation surfactants have also been synthesized. The geometry of the molecules may be varied by the position of the linkage between the hydrocarbon tails and the symmetry of linkage between the head groups [29]. These surfactants can be seen as examples of gemini surfactants. In one approach, labile gemini surfactants were synthesized, which on acid treatment, broke down into single-chain, single-head group surfactants.

8.4.2.4 Ortho Esters

Ortho esters are interesting candidates for acid-labile surfactants and have recently been commercialized [30]. They are easily prepared from homologs of triethyl ortho formate and alcohols, as illustrated in Figure 8.14. They are stable in alkali but decompose in acid by the same general mechanism as acetals and ketals. Hydrolysis gives 1 mol of alkyl formate and 2 mol of alcohol. One or more of the starting alcohols can be an end-capped polyethylene glycol (PEG), in which case a nonionic polyoxyethylene surfactant is obtained. An interesting feature of ortho esters is that they are much more labile in acid than both acetals and ketals. For instance, an ortho ester based on monomethyl PEG decomposes to about 50% at pH 6 and to almost 100% at pH 5 after 1 h at 25°C. The ortho ester concept gives molecules with three similar or different branches. Figure 8.14 shows two

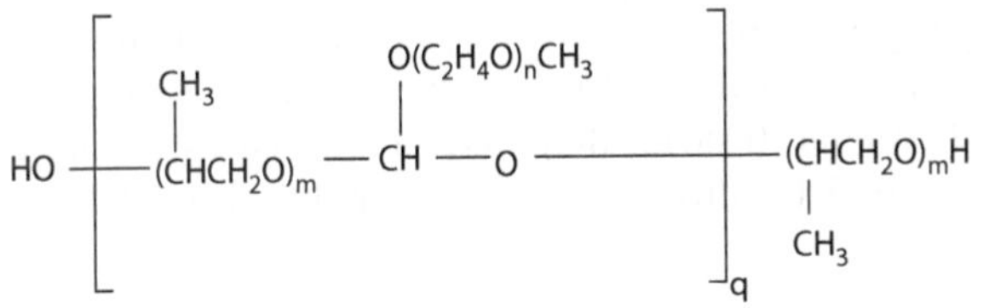

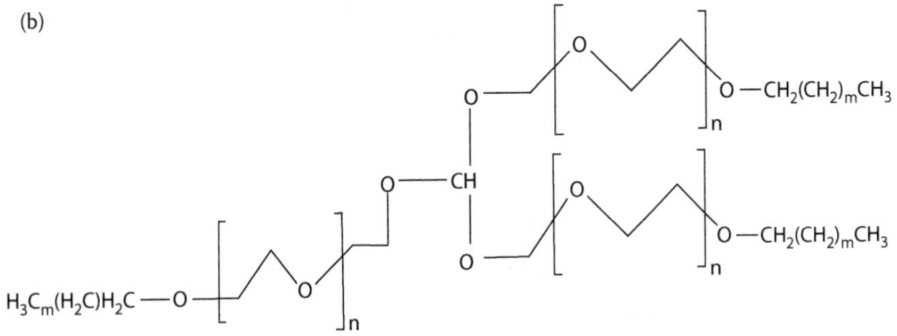

FIGURE 8.14 Examples of ortho-esters surface active agents.

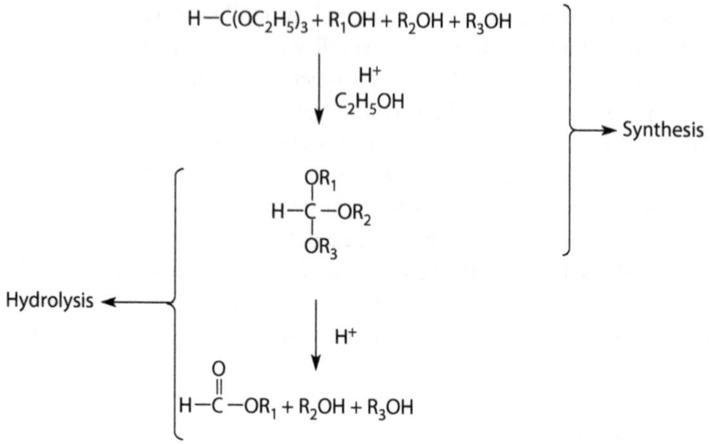

$$H-C(OC_2H_5)_3 + R_1OH + R_2OH + R_3OH$$

$$\downarrow \begin{array}{c} H^+ \\ C_2H_5OH \end{array}$$

$$\begin{array}{c} OR_1 \\ | \\ H-C-OR_2 \\ | \\ OR_3 \end{array}$$

Synthesis

Hydrolysis

$$\downarrow H^+$$

$$\begin{array}{c} O \\ \parallel \\ H-C-OR_1 + R_2OH + R_3OH \end{array}$$

FIGURE 8.15 Route of synthesis and hydrolysis of the ortho-ester surfactants.

examples: (a) a block copolymer with two chains of polyoxypropylene and one chain of polyoxyethylene and (b) a triple-tailed nonionic species connected to the polar head group. The synthesis and hydrolysis of the ortho surfactants are illustrated in Figure 8.15.

8.4.2.5 Surfactants Containing N=C Bond

Jaeger et al. [31] have synthesized surfactants consisting of two parts connected with –N=C– group (azomethine-derived surfactant). Each part is a surfactant on its own with a hydrophobic tail and a polar head group, and the two head groups have different signs (Figure 8.16). The two charges are far apart in the molecule; thus, this type is conceptually different from double-chain zwitterionic surfactants such as phosphatidyl choline. Instead, they may be viewed as a kind of hetero-gemini surfactant. Figure 8.16 illustrates the acid-catalyzed breakdown of these surfactants.

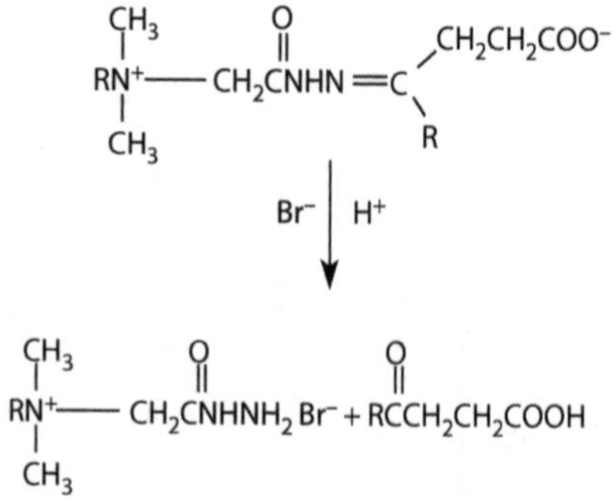

$$\begin{array}{c} CH_3 \\ | \\ RN^+\!\!-\!\!-CH_2CNHN\!=\!\!C \\ | \\ CH_3 \end{array} \begin{array}{c} O \\ \parallel \end{array} \begin{array}{c} CH_2CH_2COO^- \\ \diagup \\ \diagdown \\ R \end{array}$$

$$Br^- \bigg| H^+ \downarrow$$

$$\begin{array}{c} CH_3 \\ | \\ RN^+\!\!-\!\!-CH_2CNHNH_2\ Br^- + RCCH_2CH_2COOH \\ | \\ CH_3 \end{array} \begin{array}{cc} O & O \\ \parallel & \parallel \end{array}$$

FIGURE 8.16 Hydrolysis of the azomethine derived surfactant, R: long alkyl chain.

8.5 CLEAVABLE SURFACTANTS

Apart from the products classes already discussed, which include the most important types of cleavable surfactants, several more or less exotic examples of surfactants with limited half-lives have been reported [32]. For instance, isothionate esters with a very high degree of alkalinity have been developed [32]. The esterification products of alkyl polyoxyethylene carboxylic acid with the sodium salt of isothionic acid have been claimed to partially cleave when applied to the skin.

Cleavable quaternary hydrazinium surfactants have been explored as amphiphiles containing a bond that splits very easily [33]. The surfactants are cleaved by nitrous acid under extremely mild conditions.

Ozone-cleavable surfactants have been developed [34] as examples of environmentally benign amphiphiles. These surfactants, which contain unsaturated bonds, break down easily during ozonization of water, which is a water purification process of growing importance.

Glucose-based surfactants having a disulfide linkage between the anomeric carbon (geometrical isomer such as chair and boot structure of sugar) of the sugar ring and the hydrophobic tail were synthesized [35] and evaluated for use as solubilizing agents for membrane proteins.

Cleavage into nonsurfactant products was performed by addition of dithioerythritol to disulfide linkages under physiological conditions.

Surfactants with thermolabile bonds have been synthesized [36] and evaluated as short-lived surfactants. Amino oxide surfactants with an ether oxygen in the 2-position are examples of such structures. They decompose at elevated temperature to the corresponding vinyl ether. The structures of the cleavable surfactants discussed in this section are shown in Figure 8.17.

8.6 RAW MATERIALS FOR ENVIRONMENTALLY FRIENDLY SURFACTANTS

The development of surfactants based on natural renewable resources is a concept that is gaining recognition in detergents, cosmetics, and green chemistry. This new class of biodegradable and biocompatible products is a response to increasing consumer demand for products that are greener, milder, and more efficient. In order to achieve these objectives, it is necessary to (a) use renewable low-cost biomass that is available in large quantities, and (b) to design molecular structures through green processes that show improved performance, favorable ecotoxicological properties, and reduced environmental impact. In this context, marine algae represent a rich source of complex polysaccharides and oligosaccharides with novel structures and functional properties [37]. These saccharides may find application as starting materials for the development of green surfactants or cosmetic ingredients. Thus, original surfactants based on mannuronates moieties (Figure 8.18) derived from alginates (cell wall polyuronic acids from brown seaweeds) and fatty hydrocarbon chains derived from vegetable sources were developed [38]. Controlled chemical or enzymatic depolymerizations of the algal polysaccharides give saturated and unsaturated functional oligomannuronates. Clean chemical processes allow efficient transformation of the oligomers into

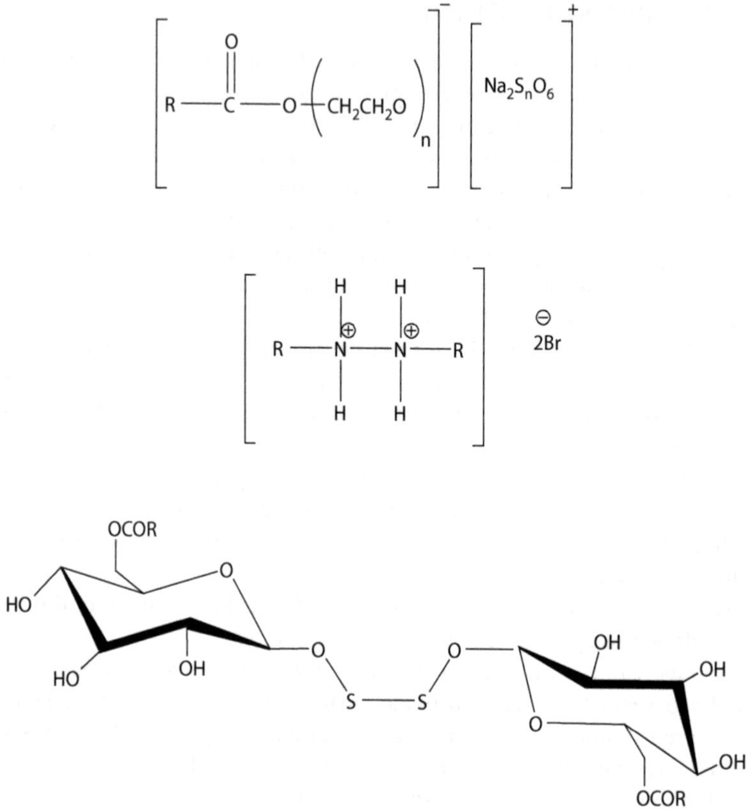

FIGURE 8.17 Cleavable surfactants: Isothionic acid derived surfactant, quaternary hydrazinium surfactant and glucose based surfactants having a disulfide linkage.

neutral or anionic amphiphilic molecules. These materials represent a new class of surface active agents with promising foaming/emulsifying properties. Environmental problems associated with nonbiodegradable surfactants can be solved with the design of biodegradable replacements. A potential problem with biodegradable surfactants is the eutrophication of lakes due to the buildup of organics resulting from the microbial degradation process. However, the principal cause of most lake eutrophication is runoff organic waste and fertilizer from farms. The role of surfactants in, for example, the eutrophication problem is found to be minor [39].

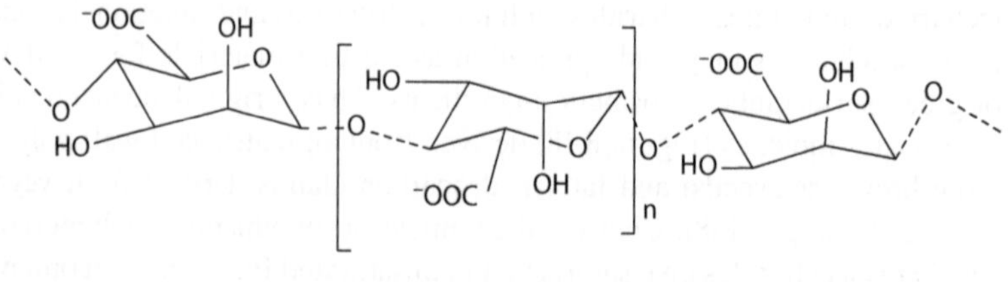

FIGURE 8.18 Structure of mannuronate moiety.

8.7   APPLICATIONS OF ENVIRONMENTALLY FRIENDLY SURFACTANTS

Currently, there is a strong trend at replacing conventional surfactants with more environmentally friendly compounds. One example is the replacement of alkyl aryl by straight-chain alkyl as hydrophobe for nonionic surfactants. Another example is the use of ester-containing cationic surfactants instead of the corresponding hydrolytically stable compounds [40]. Examples of labile bonds inserted between the polar head group and the hydrophobic tail of the surfactant are esters adjacent to a positive charge such as betaine, amides, carbonates, and ortho esters [41]. Novel surfactants based on natural building blocks, such as sugars, are also being synthesized and their solution behavior investigated [42]. Gemini surfactants, or twin surfactants, are another type of surfactant that are currently being studied [43]. The work on gemini surfactants also includes work on cleavable surfactants of the gemini type [44].

For several years, there has been a strong trend toward green surfactants, particularly for the household sector. In this context, the term "natural surfactant" is often used to indicate some natural origin of the compound. However, no surfactants used in any substantial quantities today are truly natural. With few exceptions, they are all manufactured by organic synthesis, usually involving rather severe conditions (high pressure and temperature), which inevitably give unwanted by-products. For instance, monoglycerides are certainly available in nature, but the surfactants sold as monoglycerides are prepared by glycerolysis of triglyceride oils at high temperatures, yielding diglycerol and triglycerol derivatives as by-products. Alkyl glucosides are abundant in living organisms but the surfactants of this class are made in several steps that by no means are natural. A more appropriate approach to the issue of origin is to divide surfactants into oleochemical-based and petrochemical-based surfactants. Surfactants based on oleochemicals are made from renewable raw materials, most commonly vegetable oils [45]. Surfactants from petrochemicals are made from small building blocks, such as ethylene, produced by cracking of naphtha [46].

Quite commonly, a surfactant may be built up with raw materials from both origins. Sometimes the oleochemical and the petrochemical pathways lead essentially to the same products. For instance, linear alcohols in the C_{10}–C_{14} range, which are commonly used as hydrophobes for both nonionics and anionics, are made either by hydrogenation of the corresponding fatty acid methyl esters or via Ziegler–Natta polymerization of ethylene using triethyl aluminum as the catalyst. Both routes yield straight-chain alcohols and their chain lengths are largely governed by the distillation process. Both pathways are used in very large-scale operations. It is not obvious that the oleochemical route will lead to a less toxic and more environmentally friendly surfactant than the petrochemical route. However, from the carbon cycle point of view, chemical production based on renewable raw materials is always preferred.

Linear long-chain alcohols are often referred to as fatty alcohols, regardless of their source. Branched alcohols are also of importance as surfactant raw material. They are invariably produced by synthetic routes, the most common being the so-called oxo process [47], in which an olefin is reacted with carbon monoxide and hydrogen to give an aldehyde, which is subsequently reduced to alcohol by catalytic hydrogenation. A mixture of branched and linear alcohols is obtained and the ratio between the two can be varied, to some extent, by the choice of catalyst and reaction conditions. The commercial oxo alcohols are mixtures of linear and branched alcohols of specific alkyl chain length ranges.

8.7.1 DETERGENTS

The standard surfactant types used in many cleansers and detergents are alkyl phenol ethoxylates (APEs), with the most common members of the family being octyl phenol ethoxylate and nonyl phenol ethoxylate [48]. APEs are versatile, nonionic surfactants that have been used for more than half a century in a variety of applications. They have excellent detergency, wetting, solubilization, and emulsification properties and are one of the lowest-priced bulk surfactants in the market. A drawback of APEs, though, is that they biodegrade very slowly and persist in the environment. They are made from alkyl phenols and oxyethylene polymer chains, both derived from petrochemicals. In addition, their biodegradation products such as phenols and sulfates are more toxic to marine life than the surfactant themselves. Very small amounts can cause harmful effects to the endocrinal systems of fish.

Another example of a highly effective surfactant that persists in the environment is branched tetrapropylene aryl sulfonates [49]. Alkylbenzene sulfonate (ABS) was such an effective surfactant that it flooded the market soon after introduction. Over time, a foaming problem was discovered in sewage treatment and home faucets and ABS was identified as the culprit. The branched nature of the ABS molecule was found to be the cause of the problem. As a result, a linear version of the surfactant, linear alkylbenzene sulfonate (LAS), was developed. LAS is biodegradable, although slowly, and is currently one of the most commonly used surfactants.

Increasing awareness of the effects of surfactants on the environment caused consumers to demand more environmentally friendly cleaning products. In response, chemical manufacturers are producing several new surfactants that are ecofriendly and derived from renewable resources.

Alkyl polyglycosides are a type of environmentally friendly surfactants currently manufactured entirely from natural, renewable resources such as plant oils [50]. They have great environmental compatibility and high biodegradability. Alkyl polyglycosides may be used in a wide range of applications, from home, health, and beauty products to household and industrial cleansers. The lipophile of alkyl polyglycosides is a fatty alcohol, which is derived from natural fats and oils found in plants. The hydrophile of the alkyl polyglycosides is a chain composed of glucose, which is also derived from plants. Several manufacturers of cleaning products have entered into the consumer's desire for more earth-friendly cleaners by introducing products that boast of superior cleaning capabilities.

Several other natural surfactants are manufactured for use in cosmetic formulations. Examples include those produced from lanolin (wool fat), phytosteroids extracted from various plants, and surfactants extracted from bee wax [51]. Unfortunately, these naturally occurring surfactants are not widely used in cosmetics due to their relatively poor physicochemical performance compared with synthetic surfactants.

Another important class of natural surfactants are proteins [52]. As with other synthetic macromolecular surfactants, proteins adsorb strongly and irreversibly at the oil/water interface and can stabilize emulsions effectively. However, the high molecular weight of proteins and their compact structure makes them unsuitable for preparation of emulsions with small droplet sizes. For this reason, many proteins are modified by hydrolysis to produce lower-molecular-weight protein fragments, for example, polypeptides, or by chemical alteration of reactive protein side chains [53]. Protein sugar condensates are sometimes used in skin care formulations. In addition, these proteins impart a lubricous feel to the skin and can be used as moisturizing agents.

8.7.2   GREEN CORROSION AND SCALE INHIBITORS

Chemical inhibitors play an important role in the protection and mitigation strategies for retarding corrosion. Several corrosion inhibitors are nitrogen- or sulfur-based compounds and possess inherent toxic properties. Faced with this challenge, various strategies were investigated to reduce environmental effects while maintaining desired surfactant properties and sources (e.g., amino acid, aspartic acid, and polyaspartates) [54,55]. These chemicals require high dosage rates or have insufficient biodegradation to meet the required environmental regulations. Earlier works showed that agents other than surfactants could provide good corrosion protection under high flow rate conditions. However, the chemistry did not provide complete answers to biodegradation problems. It was found that, under certain conditions, surfactants could provide good protective characteristics and an excellent environmental profile.

The use of an inhibitor is one of the best options for protecting metals against corrosion. This has prompted a search for green corrosion inhibitors. Green corrosion inhibitors are biodegradable, do not contain heavy metals or other toxic compounds, and so are environmentally friendly. Several studies have been carried out on the inhibition of corrosion of metals using plant extracts such as green corrosion and scale inhibitors [56,57].

To protect the environment from the harsh and hazardous effects of chemicals requires the use of low-toxicity and environmentally friendly chemicals and application of green chemistry principles. Stringent regulations for the protection of the environment also limit the number of chemicals allowed for use as inhibitors, which must meet in the three main criteria, that is, their level of biodegradability, bioaccumulation, and toxicity. An ideal green inhibitor, according to the Paris Commission (PARCOM) [58], is nontoxic, readily biodegradable, and shows no bioaccumulation. Thus, the development of green requires (a) knowledge of the pertinent country regulations, (b) evaluation of the performance for the environment to which the inhibitor will be exposed, and (c) the corrosion protection property of the inhibitor [59].

In the study by Negm et al. [60], many ecofriendly cationic surface active agents (Figure 8.19) were synthesized by chemical modification of vanillin. The chemical structures of these surfactants were confirmed using elemental analysis, infrared (IR), and nuclear magnetic resonance (NMR) spectra. Surface activity measurements showed that they have high tendency toward adsorption and micellization, good surface tension reduction, and low interfacial tension. Emulsion stability measurements

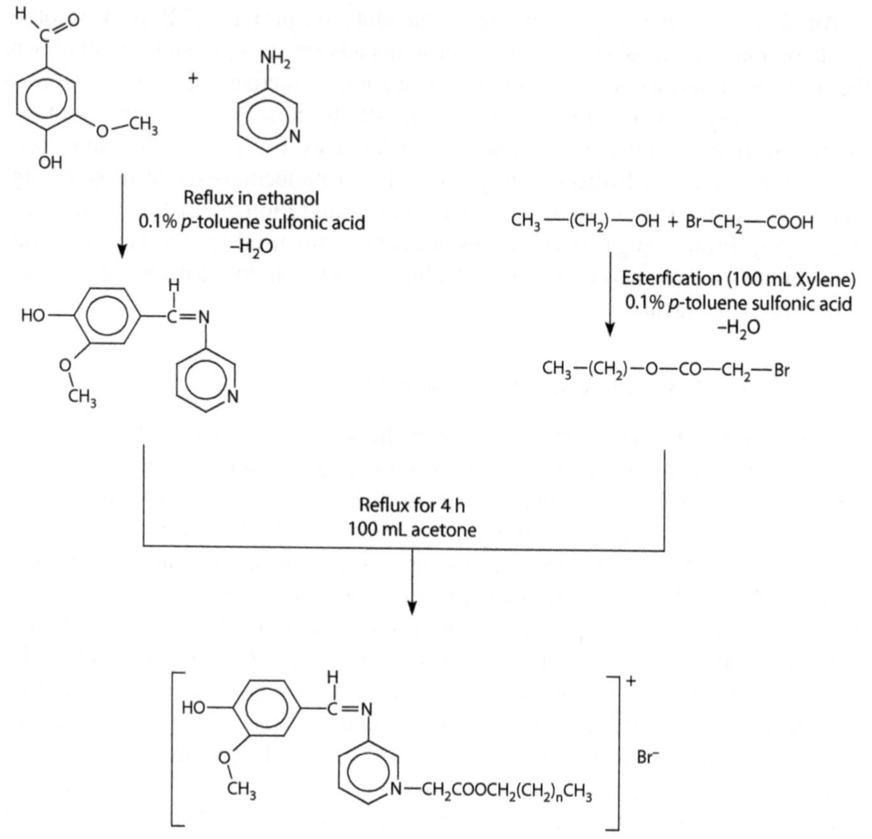

FIGURE 8.19 Synthesis of novel eco-friendly cationic surface active agents.

showed that they also have acceptable efficiency as emulsifying agents for short-term emulsions. Biodegradability tests revealed that these compounds were ecofriendly and had completely degraded in 30 days.

Vanillin is white or lightly yellow solid extracted from the vanilla bean. It is also found in roasted coffee and the Chinese red pine. Several studies were conducted on vanillin and its application as pesticide and green corrosion inhibitor for protection of different metals against the corrosion process.

Oil well green-scale inhibitor is a relatively unexplored area. Large amounts of water containing various types and amounts of residual scale and corrosion inhibitors are produced and discharged into the environment everyday. This has brought the scrutiny of the environmental regulatory bodies who have encouraged researchers to look for greener chemicals.

Tannic acid (Figure 8.20) exists in nature and has an excellent environmental profile. It is highly biodegradable and is also nontoxic. Although it is acidic, it prevents corrosion through passivation of the steel surface through a chelation mechanism. This property encouraged researchers to further investigate its scale inhibition efficiency at oil well bore condition. Static and dynamic studies show

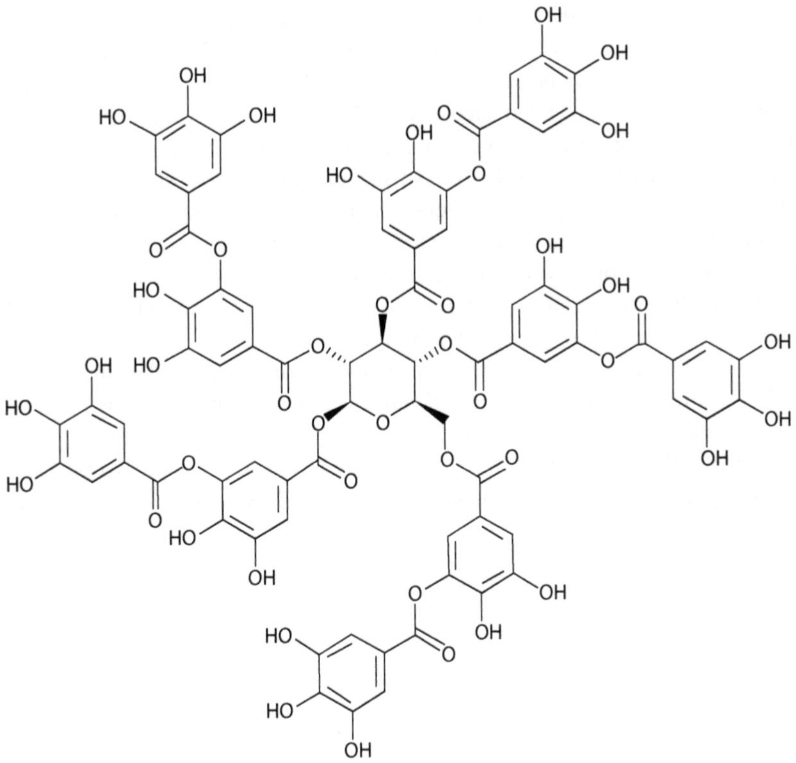

FIGURE 8.20   Structure of tannic acid.

that tannic acid could be a potential scale inhibitor for produced waste water with high scaling tendency. The test was conducted at 110°C on off-shore reservoir fluids. Characterization of inhibited scale crystals through various methods such as Fourier transform infrared (FT-IR), x-ray diffraction (XRD), and scanning electron microscope (SEM) revealed structural deformation of crystals and explained the scale prevention mechanism. High temperature-high pressure core flow studies along with quantitative measurement using ultraviolet (UV) spectroscopy revealed poor adsorption and fast release of tannic acid to form carbonate. The study suggested that tannic acid was an excellent and cheap green chemical for continuous injection into the well bore below the bubble point region for controlling the carbonate scale. Owing to its poor adsorption and retention characteristics, tannic acid cannot be suggested for use as a squeeze inhibitor in its original form [61].

8.7.3   ETHOXYLATED SURFACTANTS AND THEIR APPLICATIONS

Nonionic surfactants have either a polyether or a polyhydroxyl unit as the polar group. In the vast majority of nonionics, the polar group is a polyether consisting of 10 or more oxyethylene units, made by the polymerization of EO dispersants [62]. Ethoxylation is usually carried out under alkaline conditions. Any material containing active hydrogen can be ethoxylated. The most commonly used starting materials are fatty alcohols, alkyl phenols, fatty acids, and fatty amines [63]. Triglyceride oils may be ethoxylated in the one-pot reaction where alcohol is formed followed by partial condensation of the ethoxylated species [64]. Castor oil ethoxylates, used for animal feed applications, constitute an interesting example of triglyceride-based surfactants. Examples of polyhydroxyl polyol-based surfactants are sucrose esters, sorbitan esters, alkyl glucosides, and polyglycerol esters. The polyglycerol ester is actually a combination of polyol and polyether surfactant. Polyol surfactants may also be ethoxylated. A common example of polyol surfactants are fatty acid esters of sorbitan (Span) and the corresponding ethoxylated products (Tween) [65]. Some examples of ethoxylated surfactants discussed above are shown in Figure 8.21.

A five-membered ring structure of sorbitan is formed by the dehydration of sorbitol during manufacture [66]. The sorbitan ester surfactants are edible and, hence, useful for food and drug application.

Acetylenic glycols are surfactants containing a centrally located acetylenic bond and hydroxyl groups at the adjacent carbon atoms [67]. Acetylenic glycol surfactants

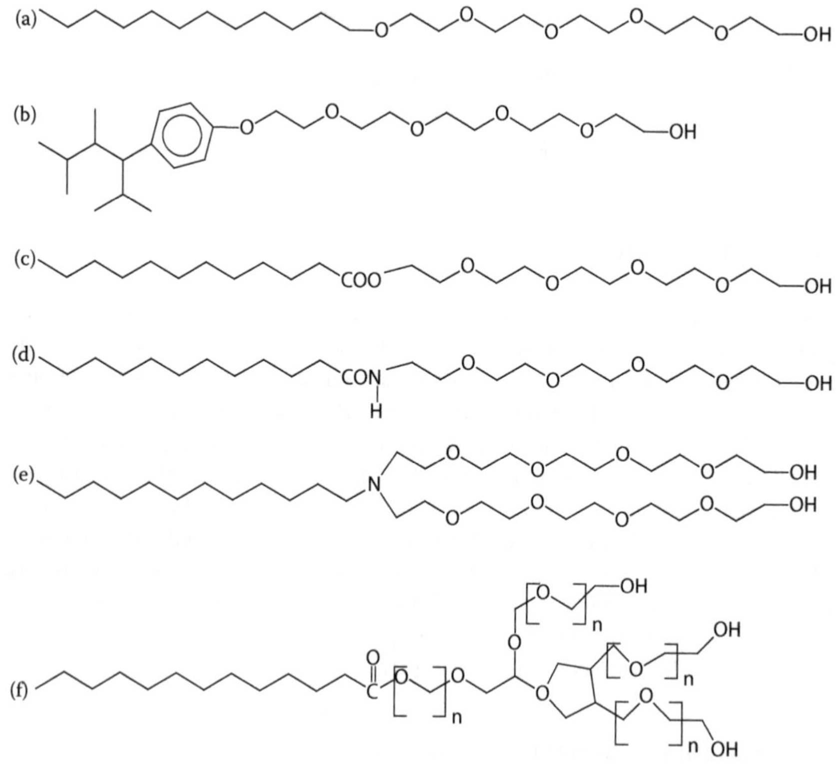

FIGURE 8.21 Examples of some ethoxylated surfactants: (a) fatty alcohol ethoxylated, (b) alkylphenol ethoxylated, (c) fatty acid ethoxylated, (d) fatty amide ethoxylated, (e) fatty amine ethoxylated, and (f) ethoxylated sorbitan alkanoate.

constitute a special type of hydroxyl-based surfactant, and are used as antifoam agents, particularly in coating applications [68].

Commercial oxyethylene-based surfactants consist of a very broad spectrum of structure and molecular weights, broader than for most other surfactant types. Fatty acid ethoxylates are constituted of particularly complex mixtures with high amounts of unreacted poly(ethylene glycol) and fatty acid as by-products [69].

The most important type of nonionic surfactants are fatty alcohol ethoxylates. They are used in liquid and powder detergents as well as in a variety of industrial applications. They are particularly useful as emulsifiers to stabilize oil-in-water emulsions [70]. Fatty alcohol ethoxylates can be regarded as hydrolytically stable in the pH range 3–11. They undergo a slow oxidation in air, and some oxidation products (e.g., aldehydes and hydroperoxides) are more irritating to the skin than the intact surfactant. Ethoxylated surfactants can be synthesized with high precision of the average number of oxyethylene units added to a specific hydrophobe, for example, a fatty alcohol.

However, ethoxylation invariably gives a broad distribution of chain lengths. If all hydroxyl groups had the same reactivity, a Poisson distribution of oligomers would be obtained. Since the starting alcohol is slightly less acidic than the glycol ethers, its deprotonation is unfavored, leading to lower probability for reaction with EO [71].

Nonionic surfactants containing polyoxyethylene chains exhibit reverse solubility in water with temperature. Increasing temperature causes two phases to appear. The temperature at which the solution becomes turbid is referred to as the cloud point [72]. The cloud point depends on both the hydrophobe chain length and the number of oxyethylene units and can be determined with great accuracy. In fact, the cloud point is used to monitor the degree of ethoxylation. The onset of turbidity varies somewhat with surfactant concentration. In the official cloud point test method, the cloud point is determined by heating a 1% aqueous solution to the above clouding and then monitoring the transition from turbid to clear solution on slow cooling [73]. For polyoxyethylene long-chain surfactants, the cloud point may exceed 100°C, and thus the determination is often made in electrolyte solutions since most salts lower the cloud point [74]. Ethoxylated triglycerides, for example, castor oil ethoxylates, have an established position in the market and are often regarded as seminatural surfactants.

In recent years, there has been a growing interest in fatty acid methyl ester ethoxylates. These are made from methyl ester by ethoxylation using a special type of catalyst, such as magnesium aluminum hydroxy carbonate [75]. The methyl ester ethoxylate has an advantage over alcohol ethoxylate in that it is much more soluble in water.

Surfactants that combine high water solubility with appropriate surface activity are needed in various types of surfactant concentrates [76].

Alcohol ethoxylates with the terminal hydroxyl group replaced by a methyl or ethyl ether group constitute a category of natural products. Such end-capped nonionics are made by O-alkylation of the ethoxylate with alkyl chloride or dialkyl sulfate or by hydrogenation of the corresponding acetal [77]. Compared with normal alcohol ethoxylates, the end-capped products are more stable against strong alkali and oxidation, and have unusually low foam.

Microemulsion formation of triglyceride oils at ambient conditions and without the addition of additives is a challenge. Macroemulsions, liquid crystals, and sponge phases are often encountered when formulating triglyceride microemulsions [78,79].

The search for environmentally safe surfactants from renewable sources has recently focused on new fatty acid derivatives such as ethoxylated fatty acid methyl esters and sulfomethyl esters [80].

The surfactant industry in the last decade has increasingly turned its attention to natural raw materials to replace petroleum-based products. This interest has been clearly demonstrated by the development of new sugar-based surfactants such as alkyl glucosides [21,81]. Efforts of finding new hydrophobic materials have also increased. This was partly because of increasing concerns with traditional products such as nonyl phenol. In addition, there is also awareness that natural hydrophobic compounds can yield properties not easily achieved through conventional synthesis from petrochemical products [80]. Another incentive is that of avoiding the health hazards and the high environmental cost associated with petrochemicals [82].

Ethoxylated fatty acid esters are well known as ether ester-type nonionic surfactants with multiple uses. For example, ethoxylated stearyl stearates are used as emulsifiers, dispersants, or oil phase adjusters in cosmetics or in industrial products. Also, ethoxylated methyl laureates have been studied as wetting agents. The hydrophilic–lipophilic balance (HLB) system proved to be a good tool to predict their emulsification properties. High-erucic rapeseed oil and castor oil-based polyoxyethylene glycol esters surfactants showed good water solubility, whereas soy-based surfactants were found to have lower water solubility [83].

Mixtures of ethoxylated fatty alcohols (EFA) and APG show better wettability than the corresponding pure surfactants. It was found that the wettability of the mixtures was not significantly influenced by water hardness. The chemical nature of these surfactants along with these properties makes these mixtures especially interesting for the formulation of biodegradable detergents or wetting agents [84].

Important nonionic surfactant families include the polyethoxylates based on fatty alcohols or alkyl phenols. Tertiary nonyl phenol ethoxylates have many industrial, commercial, institutional, and domestic uses because they are very efficient and cost-effective surfactants [85]. The use of cashew phenolic lipids for producing ethoxylated surfactants as well as their biodegradability has been documented [86].

Fatty alcohol ethoxylates can be synthesized in a range of homologs having the formula $C_x(EO)_y$. Different structures can be obtained by varying the carbon chain length of the fatty alcohol (x) and the number of EO units condensed on the fatty alcohol (y). If both these parameters are varied, a range of different molecular weights of alcohol ethoxylates having comparable hydrophilicity and HLB are obtained. On the other hand, if x or y is changed while holding the other constant, a range of ethoxylates with varying HLB are obtained [87].

Although a large variety of such surfactants are commercially available, their main application has been limited to emulsification in personal care, agricultural, or pharmaceutical products. Their commercial usage detergent has been limited primarily to lauryl alcohol ethoxylates with less than 9 mol of EO [88–91].

The high-EO ethoxylates can be used as effectively as low-EO ethoxylates in detergents, sometimes with better efficacy [92]. Detergents from fatty alcohol ethoxylates containing high levels of EO have limited applications due to the following reasons: (i) their cloud points (CPs) and, consequently, their phase inversion temperatures (PITs) are much higher than normal wash temperatures (typically above 100°C) and therefore unmeasurable; (ii) they are usually soft solids at ambient temperature (for greater than 11 EO units in the case of $C_{12}EO_x$ [93]) and are difficult to formulate; and (iii) they are difficult to manufacture since the viscosity in the ethoxylation reactors will reach very high values [94]. However, if the CPs are manipulated appropriately using salting-out electrolytes, they can offer certain distinct advantages in terms of their molecular and phase structures. For instance, in the case of solid detergent products, these materials can be of significant use since liquid ethoxylates cannot be incorporated at high levels due to processing constraints. Further, high-EO-containing alcohol ethoxylates are usually milder than low-EO counterparts [95] and are expected to be cheaper since the weight fraction of fatty alcohol in these surfactants is relatively lower. High-EO surfactants are also useful in formulating low-cost products with high salt concentrations, which are popular mass market products in some developing countries.

Goel [96] investigated phase structure, clouding, detergency, and related properties of tetradecyl EO mono dodecyl ether, $C_{12}EO_{14}$; this surfactant is a deformable solid at ambient conditions with a melting point of approximately 34°C. Since high CP is one of the major hurdles in formulating these materials, methods of lowering CP by incorporating appropriate levels of electrolytes are discussed and the effects of varying anions and cations in electrolytes have been elucidated [96].

Oily soil detergency was evaluated with hexadecane containing 10% w/w oleic acid as a model soil, which represents the hydrocarbon part of the oily soils present on fabrics. The oleic acid not only adds another important component to the model soil but also helps lower the phase inversion temperature, thus facilitating the observation of critical phenomena [97].

Commercial-grade ethoxylates have a broad distribution of EO depending on the catalyst used in the ethoxylation process. They also have a somewhat narrower distribution of alkyl chain length depending on the source of the alcohol [94].

In the past, matching of PIT with optimal detergency of commercial ethoxylates was accomplished using a simple technique. The method involves using an oil-soluble red dye and measuring the reflectance at 520 nm directly on the fabric before and after washing [91]. This technique developed some simple guidelines for selecting the optimal lauryl alcohol ethoxylate for maximizing detergency from an otherwise constrained formulation window [92].

Oskarsson et al. [98] synthesized the following nonionic surfactants: N,N-di-(3-aminopropyl) dodecyl amine, N-dodecyl-N,N-di-[(3-D-gluconylamido)propyl] amine (DGA), N-dodecyl-N,N-bis-[(3-lactobionylamido)propyl]amine (DLA), N-dodecyl-N-[(3-lactobionylamido)propyl]amine (LA), and ethoxylated N,N-di-(3-aminopropyl)dodecylamine (Y-amine with 4 and 8 EO units). They also measured adsorption at the air–water interface by the du Nouy ring method and at the solid surfaces consisting of self-assembled monolayers of alkane thiols on gold using surface plasmon resonance. The results from surface tension measurements showed that adsorption at the air–water interface was pH dependent. At low pH, the reduction in surface tension was less pronounced due to protonation of the amino groups of the surfactants. At the surface of the self-assembled monolayers, the highest amount of adsorbed surfactant was observed on a surface composed of a mixture of methyl and carboxyl groups. In general, the sugar-derived surfactants, DGA, DLA, and LA adsorbed less than the ethoxylates. The biodegradation test showed that the polyol surfactants were more biodegradable than the ethoxylate surfactants.

8.7.4 SUGAR-BASED SURFACTANTS AND THEIR APPLICATION

Several researches have been conducted about the cleaning action and properties of a series of alkyl glucosides (Figure 8.22), having high detergent power, emulsifying capacity, and biodegradability and low foaming capacity [99]. The effects of various factors on the cleaning action of the surfactants were investigated. Work is in progress at reducing the use of certain classes of surfactants polluting the environment and to replace them with new ones. Efforts are also being made to develop new classes of surfactants with reduced foaming capacity, with milder action on skin, effective at a lower temperature, and compatible with other components in detergent formulation. These new surfactants should be environmentally friendly and also cost effective.

The use of alkyl glucosides in the production of detergents started recently and, therefore, their properties are not well known. Some of these studies display good surface properties [100]. These surfactants are stable in both alkaline and acidic solutions. The use of alkyl glucoside surfactants for the production of detergent formulations promises good efficiency and environmental protection. Alkyl glucosides are characterized by high biodegradability since they are manufactured from natural raw materials. At the same time, they are mild, with low toxicity, and cause no skin irritation. In the past, the use of large-scale alkyl glucosides did not occur because of their high cost. However, in recent years, the cost has steadily decreased.

Researchers have mostly focused on long-chain alkyl glucosides with a broad distribution of carbon atoms in the chain (C_{10}–C_{18}). Such alkyl glucosides are of interest because they have low foaming and good emulsifying properties. A study of the cleaning effect and surfactant properties of a series of alkyl glucosides, C_6, C_8, C_{8-10}, C_{8-16}, and C_{12-14} revealed that with increasing chain length, the cmc decreases and the adsorption increases.

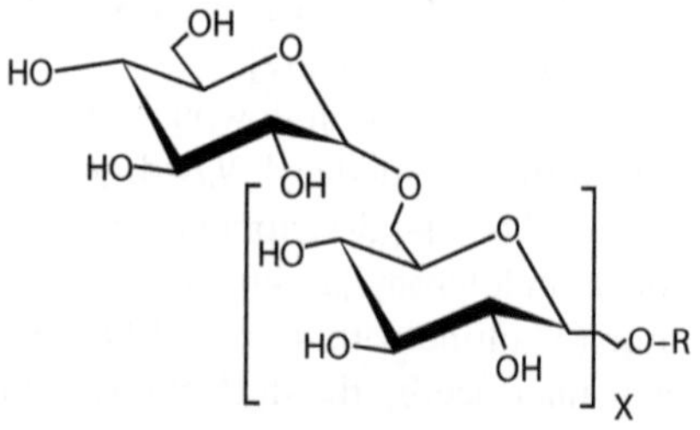

X = 0–8, R = Carbon alkyl chain range C_8–C_{16}

FIGURE 8.22 Structure of alkyl glycosides.

Zhang et al. [101] studied the adsorption behavior of sugar-based N-dodecyl-D-maltoside in mixtures with anionic sodium dodecyl sulfate at the solid/liquid interface. They found that synergism or antagonism between these two surfactants could be changed dramatically simply by changing the solution pH. To further explore the interfacial behavior of mixed surfactant systems, nonionic sugar-based N-dodecyl-D-maltoside was studied in mixtures with cationic dodecyl trimethyl ammonium bromide (DTAB) and nonionic ethoxylated surfactant, nonyl phenol ethoxylated decyl ether at the solid/liquid interfaces. Various interactions between the solid substrates and cationic–nonionic or nonionic–nonionic surfactant systems have been reported [102–104]. Although both N-dodecyl-D-maltoside and nonyl phenol ethoxylated decyl ether are nonionic surfactants, they have drastically different behaviors at solid/liquid interfaces. Sugar-based surfactants have been reported to adsorb strongly on alumina, hematite, and titania, but weakly on silica [105]. In contrast, ethoxylated surfactants are known to adsorb strongly on silica, but weakly on alumina [106].

Surfactants derived from sugar are widely used in food, cosmetic, and pharmaceutical formulations [107–109]. Physicochemical properties of these surfactants can be tailored to suit potential applications by varying the sugar head group size and the length and number of alkyl chains. Also, anionic sugar ester surfactants can be produced by incorporation of a sulfonate group. These anionic sugar esters are more water soluble than their nonionic counterparts and easily replace conventional anionic surfactants in product formulation. The effects of structural variations on the physicochemical properties of sugar ester surfactants have been reported [110,111]. Sucrose fatty acid esters are rapidly biodegradable [112–118]. However, the relationship between their biodegradability and chemical structure has not been comprehensively investigated.

In the study of Bakera et al. [119], the ultimate aerobic biodegradability of a series of sugar ester surfactants was determined using the international standards organisation test method [120]. The surfactants were nonionic sugar esters with different sugar head group size (formed from glucose, sucrose, or raffinose) and different lengths and numbers of alkyl chains (formed from lauric (C_{12}) or palmitic (C_{16}) acid). Analogous anionic sugar ester surfactants, formed by attaching sulfonyl group adjacent to the ester bond, and sugar esters with alkyl substituents were also studied. The researchers found that variations in sugar head group size or alkyl chain length and number did not significantly affect biodegradability. In contrast, the biodegradation rate of sugar esters with sulfonyl or alkyl groups was dramatically reduced compared with that of the unsubstituted sugar esters. The study showed a good correlation between the structure and biodegradability of sugar ester surfactants used in consumer products.

Amine surfactants with sugar-derived head groups are of interest as alternatives to traditional amine-based nonionic surfactants such as ethoxylated long-chain alkyl amines [121]. It was anticipated that an increased fraction of natural building blocks in the surfactant will be beneficial for reducing aquatic toxicity and improving biodegradation. Alkyl amine surfactants with polyoxyethylene head groups are toxic to aquatic organisms such as algae, and are not readily biodegradable [122]. Biodegradation of ethoxylated alkyl amines is initiated by degradation of the hydrophobic tail, followed by slower degradation of the amino polyoxyethylene chain [123]. Toxicity of the ethoxylated alkyl amine surfactants is believed to result from their

174   Surfactants in Tribology

tendency to adsorb at negatively charged surfaces. In nature, most surfaces carry a slight negative charge that attracts positively charged surfactants. Ethoxylated alkyl amines have a cationic character at neutral and low pH. Amine-based surfactants with sugar moieties polar head group are considered to be less toxic and have a higher rate of biodegradation [124].

Sugar esters, sugar amides, glycosides, and similar sugar-based surfactants can be synthesized in a range of structures. Some have a large array of useful surfactant properties. Furthermore, they often degrade relatively fast under natural conditions [125]. Sorbitan esters [126], sucrose esters [127], alkyl glucamides, and APGs [128–130] are commercially produced in significant quantities. Compared with analogous EO-based surfactants, surfactants with sugar polar head group have properties that are insensitive to temperature change [131].

Jönsson et al. [131] and Kameyama and Takagi [132] noted that sugar-based surfactants are often nontoxic and noncumulative. Waltermo et al. [133] and Matsumura et al. [134] showed that sugar surfactants are not sensitive to hard water and some have moderate foaming properties. They are also mild to the skin [135] and possess lower hemolytic activity than many other types of surfactants. However, these properties are also dependent on the nature of the hydrophobic moiety and, unfortunately, some sugar-based surfactants possess rather high hemolytic activity.

A new synthesis method was used to synthesize and characterize a set of glucose amine surfactants (Figure 8.23) [136]. These surfactants contain one or two amine groups that are protonated at low and moderate pH giving the surfactants a cationic character, whereas at high pH, they are considered as nonionic.

A range of nonionic d-gluconamides compounds were prepared from the monomer of chitosan (Figure 8.24). Three very common sugars are used as the basis for hydrophilic moiety of the surfactants: D-fructose, D-gluconic acid, and 2-deoxy-2-amino-D-glucose. These surfactants have been characterized for effectiveness as foaming agents, dispersion agents, emulsification agents, and wetting agents using a set of rapid tests suitable for screening surfactant properties. The properties of these surfactants are compared with that of commercial nonyl phenol ethoxylates type. The results may serve as a first guide to their use in various applications.

In most industrial applications, surfactants are used as processing aids and are disposed off following their usage. Issues of surfactant disposal have created interest in the use of more biologically acceptable alternatives and more focus on the use renewable resources as raw materials for surfactant synthesis nonrenewable resources. In this context, corn, potatoes, tapioca, and wheat are currently processed and constitute common sources of starch.

Polysaccharides are of interest because of the large number of reactive hydroxyl groups in their structures. Condensation of one of the reactive groups of a saccharide with a fatty acid produces surfactants. The saccharide provides the hydrophilic component, whereas the fatty acid provides the hydrophobic component. Various types of saccharides that have been used to provide the required hydrophilicity are outlined in Figure 8.25.

Carbohydrate-based surfactants have been used primarily in the cosmetic, detergent, food, and pharmaceutical industries [137–139] since they are physiologically, dermatologically, and biologically acceptable [140]. Since they are odorless,

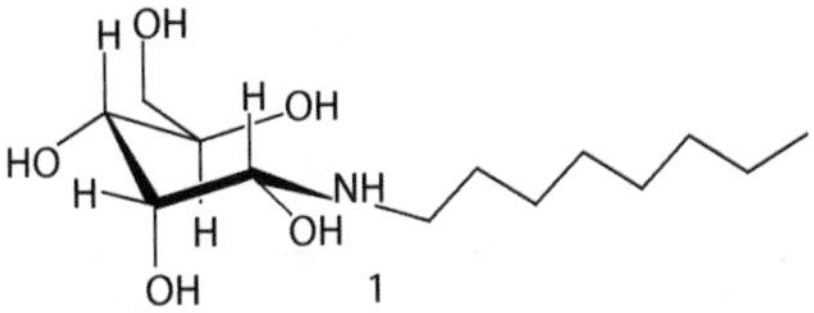

1
(2R,3R)-1-(octylamino)pentane-1,2,3,4,5-pentaol

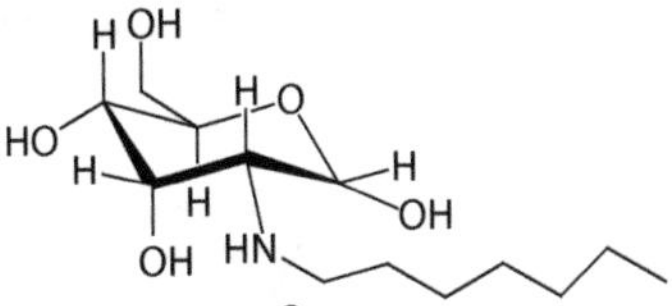

2
(2S,3R,4S,5S,6R)-3-(heptylamino)-6-(hydroxymethyl)
tetrahydro-2H-pyran-2,4,5-triol

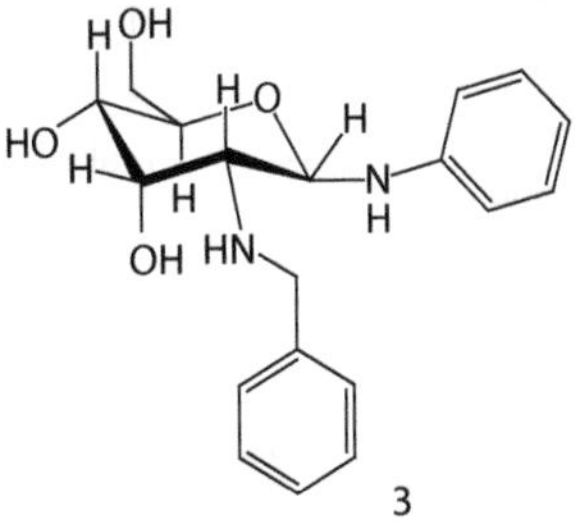

3
(2R,3S,4S,5R,6S)-5-(benzylamino)-2-(hydroxymethyl)-6-
(phenylamino)tetrahydro-2H-pyran-3,4-diol

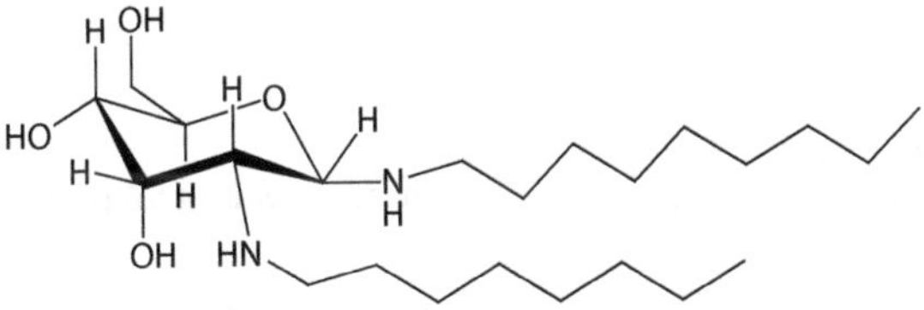

4
(2R,3S,4S,5R,6S)-2-(hydroxymethyl)-6-(nonylamino)-5-(octylamino)
tetrahydro-2H-pyran-3,4-diol

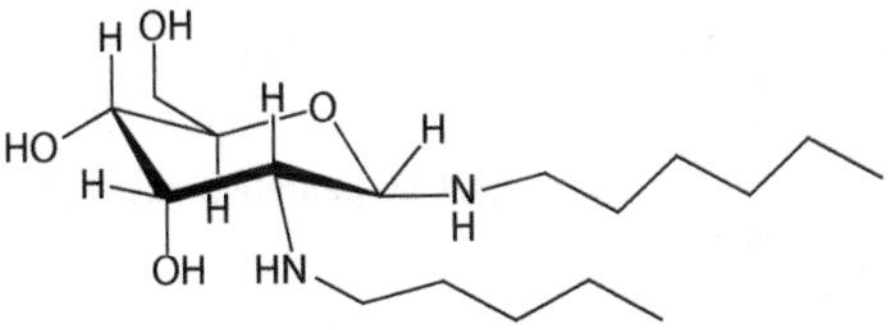

5
(2R,3S,4S,5R,6S)-6-(hexylamino)-2-(hydroxymethyl)-5-(pentylamino)
tetrahydro-2H-pyran-3,4-diol

FIGURE 8.23 Glucose amine surfactants.

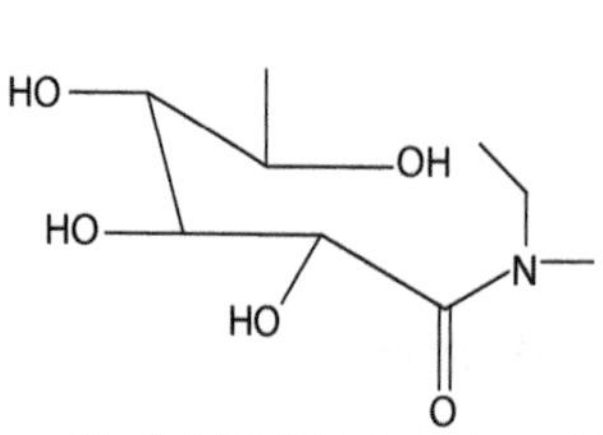

N-ethyl-2,3,4,5-tetrahydroxy-N-methylhexanamide

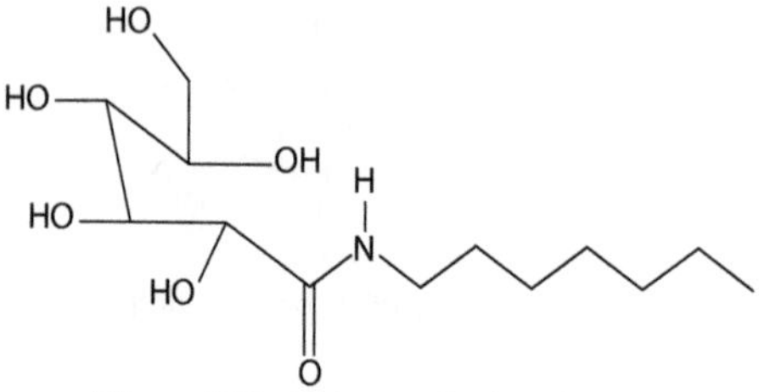

N-heptyl-2,3,4,5,6-pentahydroxyhexanamide

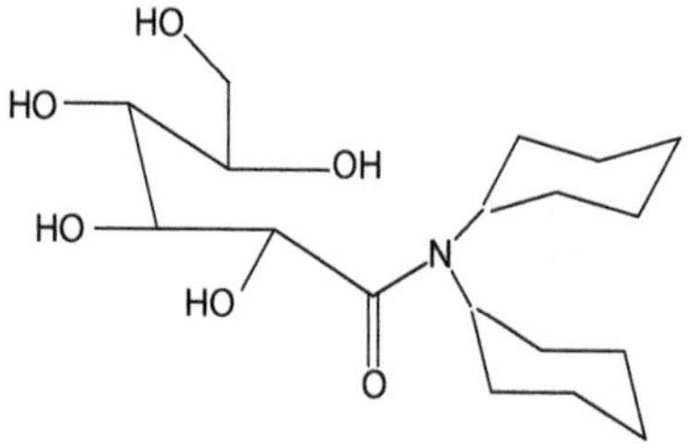

2,3,4,5,6-pentahydroxy-N,N-dimethylhexanamide compound with cyclohexane(1:2)

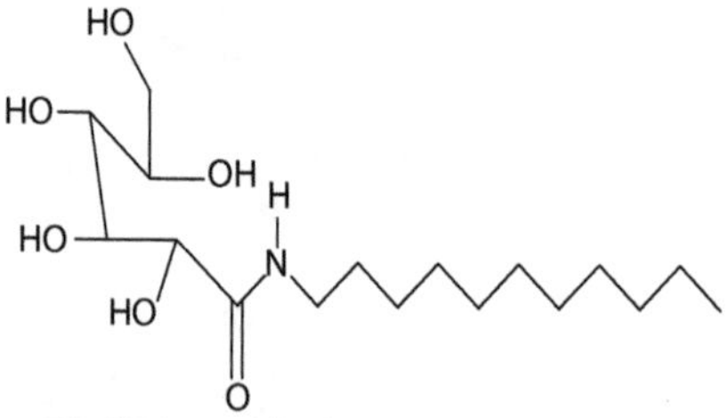

2,3,4,5,6-pentahydroxy-N-undecylhexanamide

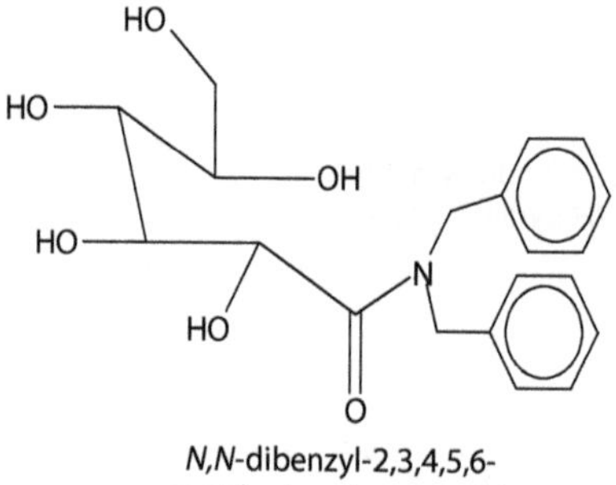

N,N-dibenzyl-2,3,4,5,6-pentahydroxyhexanamide

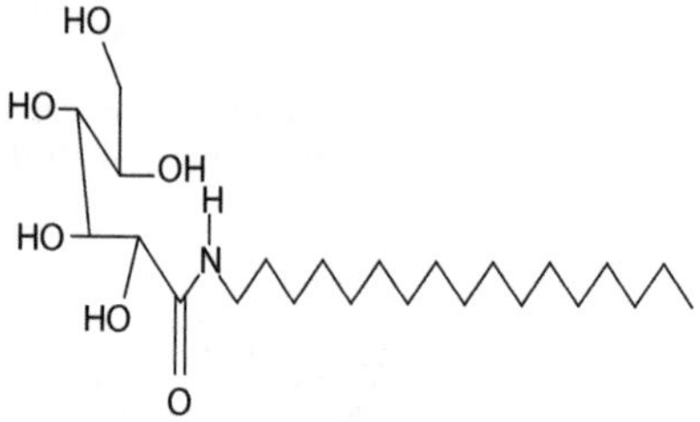

N-heptadecyl-2,3,4,5,6-pentahydroxyhexanamide

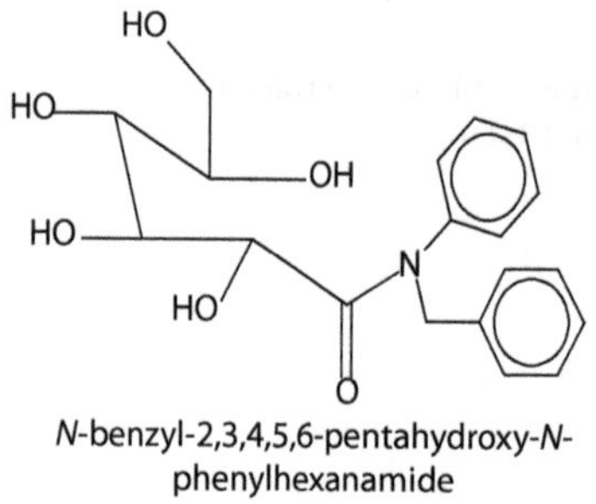

N-benzyl-2,3,4,5,6-pentahydroxy-N-phenylhexanamide

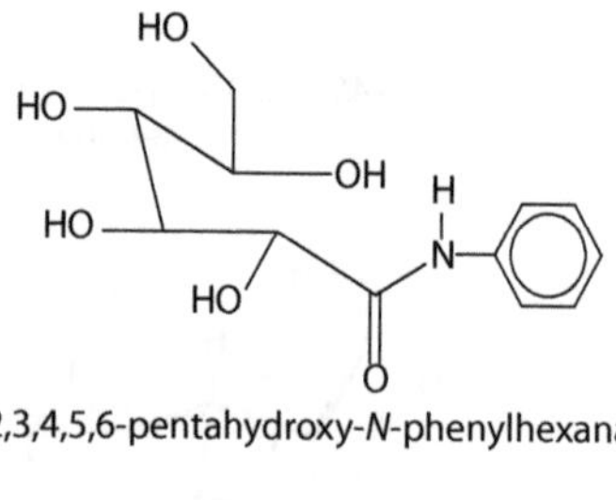

2,3,4,5,6-pentahydroxy-N-phenylhexanamide

FIGURE 8.24 Examples of nonionic D-gluconamides.

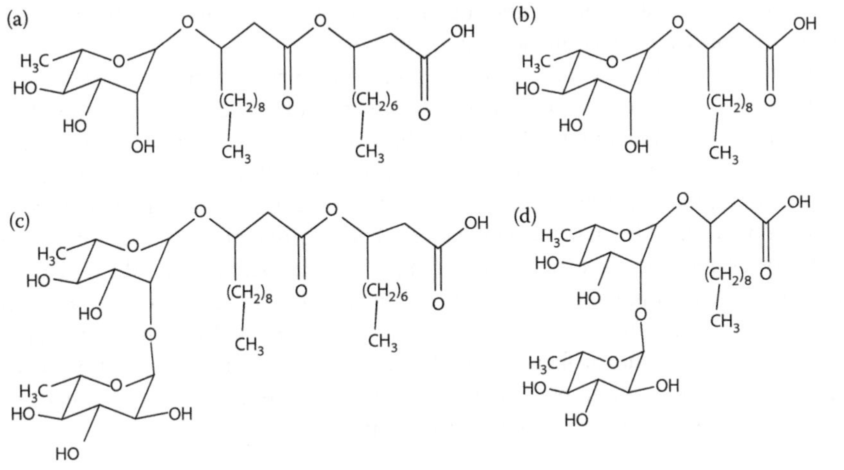

FIGURE 8.25 Various types of saccharides surfactants. (a) 3-(3-((3R,4R,5R,6S)-3,4,5-trihydroxy-6-methyltetrahydro-2H-pyran-2-yloxy)dodecanoyloxy)decanoic acid; (b) 3-((3R, 4R,5R,6S)-3,4,5-trihydroxy-6-methyltetrahydro-2H-pyran-2-yloxy)dodecanoic acid; (c) 3-(3-((3R,4R,5R,6S)-4,5-dihydroxy-6-methyl-3-((2S,3S,4R,5R,6S)-3,4,5-trihydroxy-6-methyltetrahydro-2H-pyran-2-yloxy)tetrahydro-2H-pyran-2-yloxy)dodecanoyloxy)decanoic acid; (d) 3-((3R,4R,5R,6S)-4,5-dihydroxy-6-methyl-3-((2S,3S,4R,5R,6S)-3,4,5-trihydroxy-6-methyltetrahydro-2H-pyran-2-yloxy)tetrahydro-2H-pyran-2-yloxy)dodecanoic acid.

tasteless, nonionic, and biodegradable, they compare well with other surface active compounds in overall performance (e.g., emulsification, detergency, foaming power, wetting power, and other related properties). Several major types of carbohydrate-based surfactants have been used worldwide.

In the study by Negm and Mohamad [141], two series of gemini cationic surfactants based on glucose and fructose molecules were prepared (Figure 8.26). The surface activities, adsorption, and free energies of micellization of these amphiphiles were measured in solutions. The results showed that these surfactants have a high tendency toward adsorption at interfaces. The biological activity of these amphiphiles showed high efficiency toward the growth of several microorganisms such as bacteria, fungi, and yeast at concentrations of 1, 2.5, and 5 mg/mL. These amphiphiles showed antimicrobial activity on Gram-positive and Gram-negative bacteria and antifungal activity comparable with commercially available controls. The study

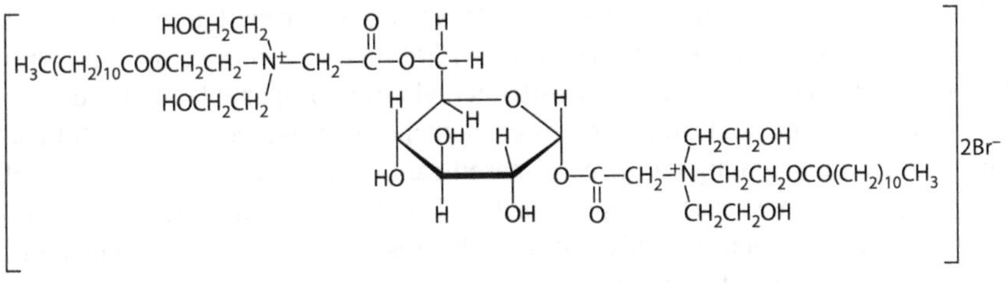

FIGURE 8.26 A gemini cationic surfactants based on glucose molecule.

allowed for correlation of the antimicrobial activity of the sugar-based surfactants with their surface activity and adsorption properties.

In the study by Kirk and coauthors [142], the surface active properties of six monoesters of 3-(trimethyl amino) propyl-D-glucopyranoside surfactants were found to be highly dependent on the fatty acyl chain length. The C_{10} and C_{12} esters exhibited the highest ability to lower surface tension as well as the lowest cmc values. Furthermore, the C_{10} ester had a good tendency toward foam formation and the stability of the formed foam in the applications that require foam as fire-fighting system and shower gel. The surfactants showed antimicrobial activity, and the dodecanoyl ester was able to inhibit the growth of both bacterial and fungal test strains. However, the antimicrobial effect was somewhat weaker than trimethyl benzyl ammonium chloride. The cationic surfactants exhibited a highly improved compatibility with anionic surfactants, as no precipitation occurred even in highly concentrated solutions, thereby providing a much more robust antimicrobial system.

8.7.5 Therapeutic Application of Green Surfactants

Pulmonary surfactants are complex mixtures of lipids and proteins that reduce the surface tension at the air–liquid interface. In addition to their biophysical function, some surfactant components play important roles in the innate and adaptive immunity of the lung. A negative modulation of the surfactant function was observed in allergic asthma, leading to the assumption that the therapeutic application of surfactant components might be beneficial in this disease. So far, there are a number of preclinical and clinical studies demonstrating various effects of the different pulmonary surfactant components administered for preventive or therapeutic purpose in allergic asthma [143].

In gene therapy, patients carrying identified defective genes are supplemented with copies of the corresponding normal genes. Many gene delivery reagents (transfection vectors), including retrovirus, adenovirus, positively charged polymers and peptides, and cationic amphiphilic compounds, are currently being used as carriers of genes in combating hereditary diseases by gene therapy. Reproducibility, low cellular and immunological toxicities, and the ease of preparation and administration associated with cationic transfection lipids are increasingly making them the transfection vector of choice in gene therapy [144,145]. Since the first report on cationic liposome-mediated gene delivery by Felgner and coworkers [146], an upsurge in global interest in the synthesis of efficient cationic transfection lipids has been witnessed (Figure 8.27) [147]. Many of the reported liposomal transfection vectors, for example, (1,2-dioleyl-3-N,N,N-trimethyl aminopropane chloride) [146], DMDHP (N,N-[bis(2-hydroxyethyl)]-N-[2,3-bis(tetradecanoyloxy) propyl] ammonium chloride) [148], DMRIE (1,2-dimyristyloxypropyl-3-dimethyl hydroxyethyl ammonium bromide) [149], and (1,2-dioleoyloxy-3-(trimethylamino)propane) [150], have a common element in their molecular structures, namely the presence of a glycerol backbone. Interestingly, among the glycerol-based cationic transfection lipids, the polar head group of the most efficient lipids, such as DMRIE and DMDHP, contain one or two hydroxy ethyl groups directly linked to the positively charged nitrogen atoms.

The developments of efficient nonglycerol-based liposomal transfection lipids have been reported, for example, DC–Chol synthesized by Gao and Huang [151].

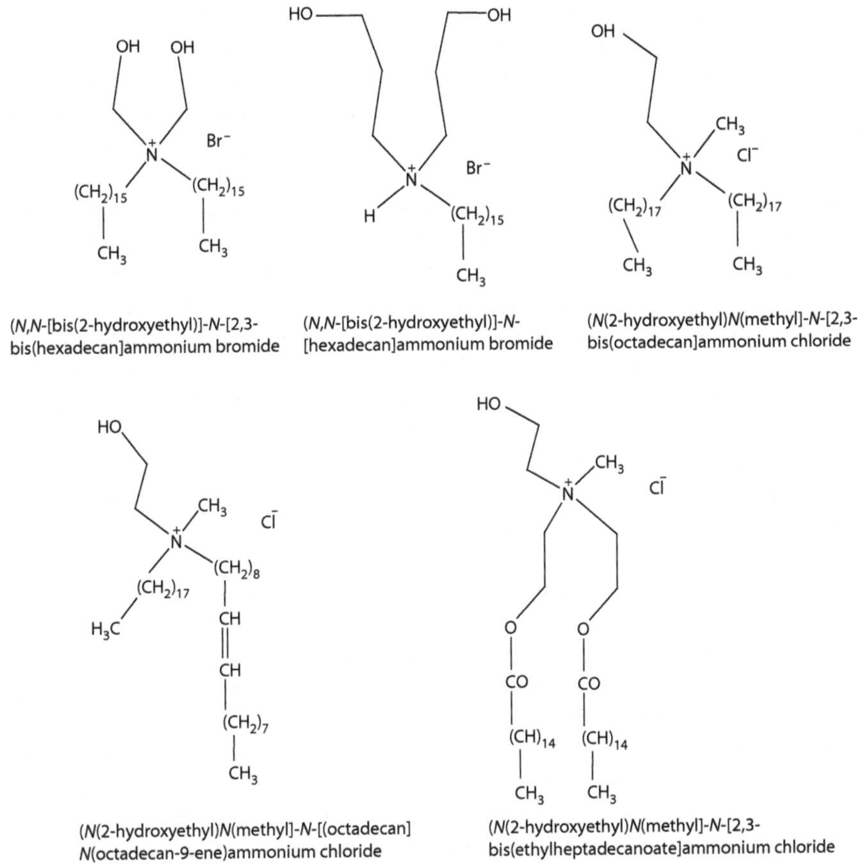

FIGURE 8.27 Structures of cationic lipids.

Long-chain alkyl acyl carnitine esters have also been designed by Szoka et al. [152]. These nonglycerol-based liposomal gene delivery reagents have no hydroxyethyl groups present in their polar head group regions. A detailed investigation of the transfection efficiencies of 1,4-diamino butane-based dicationic lipids containing hydroxyethyl groups has been reported [152].

8.7.6 Biodegradable Gemini-Type Cationic Surfactants and Their Application

Gemini-type cationic surfactants containing carbonate linkages as biodegradable and chemically recyclable segments were designed and synthesized. These surfactants consist of two long-chain alkyl groups, two quaternary ammonium groups, and a linker moiety. The surfactants are designed to be novel green and sustainable cationics with improved physicochemical and biological activities. These surfactants showed lower cmc compared with the corresponding nongemini cationics. They also showed strong antimicrobial activity. The biodegradability of gemini-type cationics was significantly improved when a carbonate linkage was introduced into the linker moiety. The maximum biochemical oxygen demand biodegradability of the gemini-type cationics containing a carbonate linkage in the linker moiety exceeded 70% after 28 days of incubation. Furthermore, gemini-type cationics containing both a carbonate linkage and a dodecyl group can be chemically recycled using a lipase enzyme [153].

Recently, the synthesis and properties of gemini-type surfactants consisting of two hydrophobic alkyl chains, two hydrophilic groups, and a linker have been extensively studied by many researchers [154]. It was found that the surface properties of gemini-type surfactants, such as a cmc and surface tension, were superior to those of the corresponding nongemini surfactants [155–162]. Thus, gemini-type surfactants can be regarded as green surfactants because their improved performance has led to a reduced surfactant consumption. Since they are water-soluble surfactants, they are generally difficult to recover or reuse. Therefore, they are discharged as drainage to the environment and are widely diffused, if they are not biodegradable. Thus, the development of biodegradable gemini-type surfactants is now needed because of environmental concerns. Biodegradable saccharide-derived gemini-type nonionic surfactants have been already reported [163,164]. Reduced sugar-based gemini-type nonionic surfactants have been synthesized and characterized [165–167].

Chemically cleavable gemini surfactants containing ester linkages showed good biodegradability [168]. These surfactants may become candidates for green surfactants. Furthermore, water-soluble surfactants should be chemically recycled as much as possible in order to establish their sustainability in the industrial field. Lipase-catalyzed chemical recycling may become one of the green methods since lipase is a catalyst with high catalytic activity. It is known that quaternary ammonium gemini-type cationic surfactants also show antimicrobial activity against a broad range of microorganisms [169,170]. However, there are only a few reports about biodegradable gemini-type cationic surfactants. Also, cationic surfactants are generally highly resistant to biodegradation due to the lack of a primary degradation site in the molecule [171]. Furthermore, sustainable chemical recycling may become an important issue for the next generation of surfactants, especially in the industrial field. Consequently, biodegradability and recyclability are now needed for the next generation of gemini-type cationic surfactants in terms of green and sustainable chemistry.

8.8 SYNTHETIC VERSUS NATURAL SURFACTANTS

Surfactants, or surface active agents, can be classified into two main groups: synthetic surfactants and natural surfactants. Natural surfactants are those obtained from plants or produced by biological processes, using excretions of microorganisms such as bacteria, fungi, and yeast. Natural surfactants have several advantages over synthetic surfactants, including high biodegradability, low toxicity, low irritancy, and compatibility with the human skin [172,173]. Owing to these superior properties, natural surfactants have potential application in petroleum, petrochemical, food, cosmetics, and pharmaceutical industries [174].

Increasing concern about the environment has led to consideration of natural surfactants as alternatives to synthetic surfactants. The development of cost-effective bioprocesses for production of natural surfactant is of great interest [175]. Synthetic surfactants are usually categorized according to the nature of their polar head group.

Major classes of natural surfactants include lipopeptides and lipoproteins, glycolipids, phospholipids, and polymeric surfactants [176]. Most of these compounds are either nonionic or anionic, but a few are cationic. Normally, the hydrophobic parts of natural surfactant molecules contain long-chain fatty acids, hydroxyl fatty acids, or alkyl hydroxy fatty acids. The hydrophilic parts can be carbohydrates, carboxylic acids, amino acids, cyclic peptides, phosphates, or alcohols [177].

The most common natural surfactants that have been isolated and studied are the glycolipids, which are composed of carbohydrates in combination with long-chain aliphatic acids or hydroxyl aliphatic acids. The best-known example of glycolipids are rhamnolipids, which is produced by certain species of *Pseudomonas*. In general, rhamnolipids are excreted as a heterogeneous mixture of several homologs. The chemical structure of each homolog in the mixture can be elucidated using liquid chromatography (LC) and mass spectrometry (MS). Rhamnolipids can be produced from various types of low-cost substrates and can be produced in high yields by controlling environmental factors and growth conditions. Rhamnolipids represent one of the most effective natural surfactants for commercial exploitation.

Rhamnolipids show good physicochemical and biological properties. Although these surface active compounds have many potential applications, they have been mainly investigated for environmental remediation.

Studies on emulsifier production by yeasts have been undetected by Carman et al. and Shepherd et al. [178,179]. The studies showed that biosurfactant can be produced by different yeast strains. For the last 10 years, several research groups have worked on biosurfactant molecules produced by yeasts, which are generally recognized as safe (GRAS). This is a concept used in the United States to identify substances that have been assessed as ingredients in many products and found in all cases to be safe. GRAS [180] is an American Food and Drug Administration (FDA) designation that a chemical or substance added to food is considered safe by experts [181], and so is exempted from the usual Federal Food, Drug, and Cosmetic Act (FFDCA) food additive tolerance requirements [182]. The GRAS lists acknowledged that many additives had existing scientific evidence of long and safe use in food. Among the additives on the list are sugar, salt, spices, and vitamins. Manufacturers can petition for GRAS status for new additives if the substances meet the criteria cited above. GRAS list additives are continually reevaluated based on current scientific evidence.

8.9   BIODEGRADATION PATHWAYS OF SURFACTANTS

Balson and Felix [183] described biodegradation as the destruction of a chemical by the metabolic activity of microorganisms. The literatures concerning the degradation of surfactants discuss two types of degradations: primary and ultimate biodegradation. Primary degradation occurs when the structure has changed sufficiently for a molecule to lose its surfactant properties. Ultimate degradation occurs when a surfactant molecule has completely converted into CO_2, CH_4, water, mineral salts, and biomass.

LAS is generally regarded as a biodegradable surfactant. Very high levels of biodegradation (97–99%) have been reported in some wastewater treatment plants

(WWTPs) using aerobic processes [184–186]. In contrast, APEs are less biodegradable and degradation rates of only 0–20% have been reported [187] based on oxygen uptake, and 0–9% based on spectroscopic analysis [187]. The mechanism of breakdown of LAS involves sequential degradation of first the straight alkyl chain, then the sulfonate group, and last, the benzene ring [188,189]. The breakdown of the alkyl chain starts with the oxidation of the terminal methyl group (ω-oxidation) to the alcohol, aldehyde, and finally to carboxylic acid as shown in Figure 8.28. The reactions are catalyzed by alkane monooxygenase and two dehydrogenase enzymes. The carboxylic acid then undergoes β-oxidation and the two carbon fragments enter the tricarboxylic acid cycle as acetyl CoA. At this stage, problems arise with branched alkyl chains. A methyl branch or a geminal dimethyl branch chain (geminal groups are two groups found on two neighbor carbon atoms in straight alkyl chain) as these cannot be β-oxidized by the microorganisms and must degrade by loss of one carbon atom at a time (α-oxidation). The second stage in LAS breakdown is the loss of the sulfonate group. There has been some discussion about the sequence of steps [188] and three mechanisms have been proposed for desulfonation reactions:

1. Hydroxyative desulfonation:

$$RSO_3H + H_2O \rightarrow ROH + 2H^+ + SO_3^{2-} \qquad (8.1)$$

2. Monooxygenase catalysis under acid conditions:

$$RSO_3H + O_2 + 2NADH \rightarrow ROH + H_2O + SO_3^{2-} + 2NAD^+ \qquad (8.2)$$

3. Reductive desulfonation:

$$RSO_3H + NADH + H^+ \rightarrow RH + NAD^+ + H_2SO_3 \qquad (8.3)$$

It is not clear which of these mechanisms are responsible for the breakdown of sulfonate to sulfate in the environment. The loss of the alkyl and the sulfonate groups from LAS leaves either phenyl acetic or benzoic acids. Microbial oxidation of phenyl acetic acid can result in fumaric and acetoacetic acids whereas benzene can be converted to catechol [188].

Studies on the biodegradation of LAS and other surfactants by biofilms of bacterial populations isolated from riverine [190] and estuarine [191] sites have been reported. A study of biodegradation of a range of anionic surfactants at a river site (river Ely, South Wales, UK) located near a sewage treatment plant outfall has been made [191]. Experiments were conducted in the laboratory using a population of bacteria isolated from river stone biofilms. Water samples collected at the beneath outflow (BO), beneath upstream (BU), and beneath downstream (BD) of the site were incubated with the isolated bacteria. The biodegradation tested using surfactant die-away time test. The die-away time test is the test that provides information about the rates of biodegradation under relevant environmental conditions using surface tension measurements. The biodegradation of surfactants follows the sequence: alkyl sulfates > alkyl ethoxy sulfates > secondary linear alkyl sulfates > linear alkyl sulfonates (LAS).

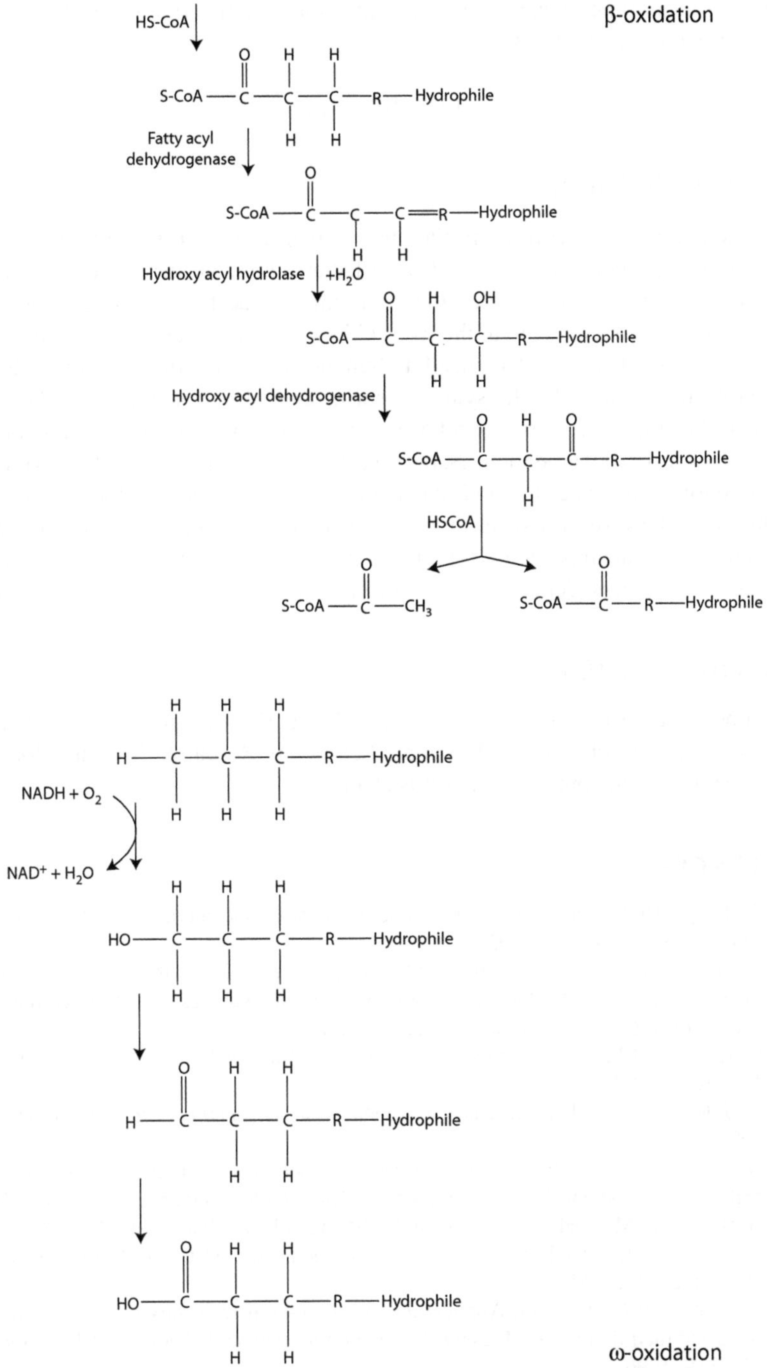

FIGURE 8.28 Reaction pathways of β- and ω-oxidation of the alkyl chain during surfactant degradation.

The die-away time of surfactants depended on the collection site of the bacteria evaluated in the sequence [187]:

$$BO > BD > BU$$

8.10 CONCLUSION

The accumulation of chemicals in the environment causes serious problems to the health of humans and other species. The surfactants are considered the most famous chemicals that have contaminated the environment due to their wide application. These compounds accumulate in the liver, kidney, and fat tissues, which causes dangerous health problems and even death to humans, fish, and other wild animals. The main problem is caused by the synthetic surfactants of petroleum origin. However, natural surfactants have no problem toward the environment. That is because of their tendency toward biodegradation. Natural surfactants of animal or plant origin are being developed to replace the traditional nondegrading surfactants for various applications. Nowadays, several surfactants have been tailored and developed to replace the synthetic surfactants in their applications, including domestic use, cosmetics, shampoos, emulsifiers, demulsifiers, corrosion inhibitors, and fire-fighting systems.

ACKNOWLEDGMENT

The authors would like to acknowledge Dr. A. M. Badawi and Dr. F. M. Ghuiba, Professors of Petrochemicals, Egyptian Petroleum Research Institute, for their encouragement in the preparation of this chapter.

REFERENCES

1. S. Lang, Biological amphiphiles (microbial biosurfactants), *Curr. Opin. Colloid Interface Sci.*, 7, 12–20, 2002.
2. L. Mu and S. S. Feng, Vitamin E TPGS used as emulsifier in the solvent evaporation extraction technique for fabrication of polymeric nano spheres for controlled release of paclitaxel, *J. Control. Release*, 80, 129–144, 2002.
3. K. Esumi and M. Ueno, *Structure-Performance Relationships in Surfactants*, Dekker, New York, 1997.
4. M. J. Rosen and D. J. Tracy, Gemini surfactants, *J. Surfactants Detergents*, 1, 547–554, 1998.
5. N. A. Negm, Solubilization, surface active and thermodynamic parameters of gemini amphiphiles bearing nonionic hydrophilic spacers, *J. Surfactants Detergents*, 10, 71–80, 2007.
6. H. Bagmann, M. Buehler, H. Fochem, F. Hirsinger, H. Zoebelein, and J. Falbe, Natural fats and oils: Renewable raw materials for the chemical industry, *Angew. Chem. Int. Ed. (Eng)*, 1, 27–41, 1988.
7. K. Urata and N. Takaishi, Applications of phase–transfer catalytic reactions to fatty acids and their derivatives: Present state and future potential, *J. Am. Oil Chem. Soc.*, 73, 831–839, 1996.
8. T. Maugard, M. Remaud Simeon, D. Petre, and P. Monsan, Lipase-catalyzed chemoselective *N*-acylation of amino–sugar derivatives in hydrophobic solvent: Acid–amine ion–pair effects, *Tetrahedron*, 53, 7587–7593, 1997.

9. T. Maugard, M. Remaud Simeon, D. Petre, and P. Monsan, Lipase-catalyzed synthesis of biosurfactants by transacylation of *N*-methyl-glucamine and fatty acid methyl esters, *Tetrahedron*, 53, 7629–7633, 1997.
10. M. Gelo-Pujic, E. Guibe-Jampel, and A. Loupy, Enzymatic glycosidations in dry media on mineral supports, *Tetrahedron*, 53, 17247–17254, 1997.
11. R. Rosu, M. Yasui, Y. Iwasaki, and T. Yamane, Enzymatic synthesis of symmetrical 1,3-diacyl glycerols by direct esterification of glycerol in solvent–free system, *J. Am. Oil Chem. Soc.*, 76, 839–844, 1999.
12. K. Hill, and O. Rhode, Sugar-based surfactants for consumer products and technical applications, *Lipid*, 101, 25–31, 1999.
13. G. Kruger, D. Boltersdorf, and H. Lewandowski, Preparation, applications, and bio-degradability, in *Novel Surfactants*, K. Holmberg, (Ed.), pp. 115–138, Marcel Dekker, New York, 1998.
14. N. A. Negm, F. M. Ghuiba, S. A. Mahmoud, and S. M. Tawfik, Biocidal, and anti-corrosive activities of benzoimidazol-3-ium cationic Schiff base surfactants, *Eng. Life Sci.*, 11, 496–510, 2011.
15. R. A. Thompson and S. Allenmark, Factors influencing the micellar catalyzed hydroly-sis of long-chain alkyl betainates, *J. Colloid Interface Sci.*, 148, 241–246, 1992.
16. E. N. Vulfson, Preparation, applications, and biodegradability, In *Novel Surfactants*, K. Holmberg, (Ed.), pp. 279–300, Marcel Dekker, New York, 1998.
17. I. Soderberg, C. J. Drummond, D. N. Furlong, S. Godkin, and B. Matthews, Non-ionic sugar-based surfactants: Self assembly and air/water interfacial activity, *Colloids Surf. A*, 102, 91–97, 1995.
18. A. Sokolowski, A. Bieniecki, K. A. Wilk, and B. Burczyk, Surface activity and micelle formation of chemodegradable cationic surfactants containing the 1,3-dioxolane moiety, *Colloids Surf. A*, 98, 73–82, 1995.
19. G. W. Wang, Y. C. Liu, X. Y. Yuan, X. G. Lei, and Q. X. Guo, Preparation, proper-ties, and applications of vesicle-forming cleavable surfactants with a 1,3-dioxane ring, *J. Colloid Interface Sci.*, 173, 49–54, 1995.
20. K. Holmberg, Surfactants with controlled half-lives, *Curr. Opin. Colloid Interface Sci.*, 1, 572–578, 1996.
21. W. Rybinski, Technology, properties and applications, In *Alkyl Polyglycosides*, K. Holmberg, (Ed.), pp. 31–85, Marcel Dekker, New York, 1998.
22. J. R. McElhanon, T. Zifer, G. M. Jamison, K. Rahimian, T. P. Long, S. R. Kline, D. A. Loy, D. R. Wheeler, and B. A. Simmons, Thermally cleavable surfactants based on furan maleimide diels—Alder adducts, *Langmuir*, 21, 3259–3266, 2005.
23. J. Small, D. A. Loy, D. R. Wheeler, J. R. McElhanon, and R. S. Saunders, Methods of making thermally removable polymeric encapsulants, U.S. patent 6271335 B1, 2001.
24. J. Small, D. A. Loy, D. R. Wheeler, J. R. McElhanon, and R. S. Saunders, Methods of making thermally removable polyurethanes, U.S. patent 6403753 B1, 2002.
25. J. Small, D. A. Loy, D. R. Wheeler, J. R. McElhanon, and R. S. Saunders, Methods of making thermally removable epoxies, U.S. patent 6337384 B1, 2002.
26. J. H. Aubert, Method of making thermally removable adhesives using diels–alder cyclo-addition reaction of bismaleimides and furan compounds, U.S. patent 2003116272 A1, 2003.
27. D. Ono, A. Masuyama, and M. Okahara, Preparation of new acetal type cleavable sur-factants from epichlorohydrin, *J. Org. Chem.*, 55, 4461–4464, 1990.
28. D. A. Jaeger, Y. M. Sayed, and A. K. Dutta, Second generation single-chain cleaveable surfactants, *Tetrahedron Lett.*, 31, 449–450, 1990.
29. D. A. Jaeger and S. G. G. Russell, Second generation double-chain cleavable surfac-tants, *Tetrahedron Lett.*, 34, 6985–6988, 1993.

30. K. Bergström and P. E. Hellberg, PCT International Patent Application PCT/SE97/00987, 1997.
31. D. A. Jaeger, V. B. Reddy, and D. S. Bohle, Cleavable double-chain surfactant Co(III) complexes, *Tetrahedron Lett.*, 40, 649–652, 1999.
32. S. A. Madison, M. Massaro, G. B. Rathinger, and C. J. Wenzel, PCT Patent Application WO 9514661 A1 950601, 1995.
33. D. A. Jaeger, J. Wettstein, and A. Zafar, Cleavable quaternary hydrazinium surfactants, *Langmuir*, 14, 1940–1941, 1998.
34. A. Masuyama, C. Endo, S. Takeda, M. Nojima, D. Ono, and T. Takeda, Ozone-cleavable gemini surfactants. Their surface-active properties, ozonolysis, and biodegradability, *Langmuir*, 16, 368–373, 2000.
35. C. L. Ferrenbach, and K. Hill, *Sugar-Based Surfactants for Consumer Products and Technical Applications, Sugar-Based Surfactants*, CRC Press, Boca Raton, FL, 2009.
36. K. Holmberg, *Cleavable Surfactants, Reactions and Synthesis in Surfactant Systems*, CRC Press, Boca Raton, FL, 2001.
37. D. Brault, A. Heyraud, V. Lognone, and M. Roussel, Methods for obtaining oligomannuronates and gluronates, WO 099870, 2003.
38. G. I. Keim, Treating fats and fatty oils, U.S. Patent No. 2,383,601, 1945.
39. T. Benvegnu, J. Franc, and O. Sassi, Oligomannuronates from seaweeds as renewable sources for the development of green surfactants, *Top Curr. Chem.*, 294, 143–164, 2010.
40. A. R. Tehrani Bagha and K. Holmberg, Cationic ester-containing gemini surfactants: Physical-chemical properties, *Langmuir*, 26, 9276–9282, 2010.
41. D. Lundberg, M. Stjerndahl, and K. Holmberg, Surfactants containing hydrolyzable bonds, *Adv. Polym. Sci.*, 218, 57–82, 2008.
42. K. Holmberg, Natural surfactants, *Curr. Opin. Colloid Interface Sci.*, 6, 148–159, 2001.
43. L. U. Ting and H. J. Bin, Synthesis and properties of novel gemini surfactant with short spacer, *Chin. Sci. Bull.*, 52, 2618–2620, 2007.
44. C. F. Jeffery, L. H. Lin, M. Y. Dong, W. S. Chang, and K. M. Chem, Preparation and properties of new ester-linked cleavable gemini surfactants, *J. Surf. Deterg.*, 14, 195–201, 2011.
45. K. Hill, Fats, and oils as oleochemical raw materials, *Pure Appl. Chem.*, 72, 1255–1264, 2000.
46. K. Shuwiwang, Y. Qiu, and R. Marchant, Poly(ethyleneoxide) surfactant polymers, *J. Biomater. Sci. Polymer Edn.*, 15, 95–110, 2004.
47. D. J. Wagner, R. G. Lappin, and J. R. Zietz, Alcohols, higher aliphatic, *Synth. Process.*, 2, 180–189, 2000.
48. H. Ferch and W. Leonhardt, Foam control in detergent products, *Surfactant Sci. Ser.*, Marcel Dekker, New York, 45, 22, 1992.
49. K. Lee, J. John, and R. George, Laundry detergents comprising modified alkylbenzene sulfonates, United States Patent 6514926, 2003.
50. Y. U. Yangxin, Z. Jin, and A. E. Bayly, Development of surfactants and builders in detergent formulations, *Chin. J. Chem. Eng.*, 16, 517–527, 2008.
51. T. F. Tadros (Ed.), *Applied Surfactants: Principles and Applications*, Wiley-VCH, Weinheim, Germany, 2005.
52. A. Cooper and M. W. Kennedy, Biofoams and natural protein surfactants, *Biophys. Chem.*, 151, 96–104, 2010.
53. F. T. Tadros, *Surfactants in Personal Care and Cosmetics, Applied Surfactants Principles and Applications*, Wiley-VCH Verlag, Weinheim, Germany, 7, 2005.
54. M. A. Kiani, M. F. Mousavi, S. Ghasemi, M. Shamsipur, and S. H. Kazemi, Inhibitory effect of some amino acids on corrosion of Pb–Ca–Sn alloy in sulfuric acid solution, *Corros. Sci.*, 50, 1035–1045, 2008.
55. G. Bereket and A. Yurt, The inhibition effect of amino acids and hydroxy carboxylic acids on pitting corrosion of aluminum alloy 7075, *Corros. Sci.*, 43, 1179–1195, 2001.

56. K. Olusegun, J. O. Otaigbe, O. J. Kio, and J. O. Abiola, L. *Gossipium hirsutum*: Extracts as green corrosion inhibitor for aluminum in NaOH solution, *Corros. Sci.*, 51, 1879–1881, 2009.

57. A. K. Satapathy, G. Gunasekaran, S. C. Sahoo, K. Amit, and P. V. Rodrigues, Corrosion inhibition by *Justicia gendarussa* plant extract in hydrochloric acid solution, *Corros. Sci.*, 51, 2848–2856, 2009.

58. S. Taj, *Development of Green Inhibitors for Oil and Gas Applications*, NACE International, San Diego, Ca, 2006.

59. M. A. Quraishi, Naturally occurring products as corrosion inhibitors, NACE meeting papers, 12–14, 2004.

60. N. A. Negm, N. G. Kandile, and M. A. Mohamad, Synthesis, characterization and surface activity of new eco-friendly Schiff bases vanillin derived cationic surfactants, *J. Surf. Deterg.*, 14, 325–331, 2011.

61. B. Ghosh, S. S Kundu, B. Senthilmurugan, and M. Haroun, Upstream scale inhibition in carbonate reservoir evaluation of a green chemistry, *Int. J. Petroleum Sci. Technol.*, 3, 51–64, 2009.

62. B. M. Folmer, K. Holmberg, E. G. Klingskog, and K. Bergströmc, Fatty amide ethoxylates: Synthesis and self-assembly, *J. Surf. Deterg.*, 4, 175–183, 2001.

63. P. V. Zimakova, and M. O. Dymenta, *Chapter III. Review of the Individual Reactions of Ethylene Oxide, Ethylene Oxide*, Khimiya Publication, Russia, Vol. 6, pp. 90–120, 1967.

64. M. Ionescu, Z. S. Petrovic, and X. Wan, Ethoxylated soybean polyols for polyurethanes, *J. Polym. Environ.*, 18, 1–7, 2010.

65. T. M. Schmitt (Ed.), Characterization of nonionic surfactants, *Analysis of Surfactants*, Marcel Dekker, New York, Vol. 96, pp. 58–107, 2001.

66. J. Smidrkal, R. Cervenkova, and V. Filip, Two-stage synthesis of sorbitan esters, and physical properties of the products, *Eur. J. Lipid Sci. Technol.*, 106, 851–855, 2004.

67. S. W. Medina, Ethoxylated acetylenic glycols having low dynamic surface tension, Patent 5650543, 1997.

68. S. P. Morell, Surfactants for waterborne coatings applications, *Coatings Technology Handbook*, CRC Press, Boca Raton, FL, 2005.

69. J. J. McKetta and W. A. Cunningham, *Encyclopedia of Chemical Processing and Design*, Marcel Dekker, New York, 259–260, 1984.

70. B. Stefan and A. Helga, Emulsifier mixture containing fatty alcohols, ethoxylated fatty alcohols and oil and wax components, U.S. Patent 7799333, 2010.

71. P. Sallay, L. Farkasa, Z. Szlováka, I. Rusznáka, P. Bakóa, M. Ahmed, A. Tungler, and G. Fogassyb, Novel general procedure for the preparation of homogeneous nonionic surfactants, *J. Surf. Deterg.*, 5, 353–357, 2002.

72. S. Kumar, D. Sharam, Z. A. Khan, and K. Din, Occurrence of cloud point in SDS–tetra-*n*-butylammonium bromide system, *Langmuir*, 17, 5813–5818, 2001.

73. K. Holmberg, B. Jonsson, B. Kronberg, and B. Lindman, *Surfactants and Polymers in Aqueous Solution*, John Wiley & Sons, Ltd., England, 2002.

74. S. K. Han, and B. H. Jhun, Effect of additives on the cloud point of polyethylene glycols, *Arch. Pharm. Res.*, 7, 1–9, 1984.

75. K. Alejski, E. Bialowas, W. Hreczuch, B. Trathnigg, and J. Szymanowski, Oxyethylation of fatty acid methyl esters. Molar ratio and temperature effects. Pressure drop modeling, *Ind. Eng. Chem. Res.*, 42, 2924–2933, 2003.

76. M. J. Rosen (Ed.), *Surfactants and Interfacial Phenomena*. Hoboken, John Wiley & Sons., New Jersey, p. 1, 2010.

77. Y. Xia, J. Mao, X. Lv, and Y. Chen, Synthesis of porphyrin endcapped water-soluble poly(phenylene ethynylene) and study of its optical properties, *Polym. Bull.*, 63, 37–46, 2009.

78. D. L. Do, A. Withayyapayanon, H. J. Harwell, and A. D. Sabatini, Environmentally friendly vegetable oil microemulsions using extended surfactants and linkers, *J. Surf. Deterg.*, 12, 91–99, 2009.

79. Y. A. El-Shattory, A. G. Abo El-Wafa, and M. S. Aly, Ethoxylation of fatty acids fractions of overused vegetable oils, *J. Surf. Deterg.*, 14, 151–160, 2011.

80. I. Johansson and M. Svensson, Surfactants based on fatty acids and other natural hydrophobes, *Curr. Opin. Colloid. Interface Sci.*, 6, 178–188, 2001.

81. O. Soderman and I. Johansson, Polyhydroxyl-based surfactants and their physicochemical properties and applications, *Curr. Opin. Colloid. Interface Sci.*, 6, 391–402, 1999.

82. M. F. Cox, Ethylene oxide-derived surfactants, In *Proceedings of the 3rd World Conference on Detergents*, A. Cahn (Ed.), pp. 141–146, 1994.

83. J. Anthony and J. O'Lenick, Evaluation of polyoxyethylene glycol esters of castor, high-erucic acid rapeseed, and soybean oils, *J. Surf. Deterg.*, 3, 201–206, 2000.

84. E. Jurado, M. Jose Vicaria, A. Fernandez-Arteaga, P. Chachalis, and J. Francisco García-Martın, Wetting power in aqueous mixtures of alkylpolyglucosides and ethoxylated fatty alcohols, *J. Surf. Deterg.*, 13, 497–501, 2010.

85. A. M. Soto, H. Justicia, J. W. Wray, and C. Sonnenschein, P-Nonylphenol, an estrogenic xenobiotic released from 'modified' polystyrene, *Environ. Health Persp.*, 92, 167–173, 1991.

86. J. H. P. Tyman and I. E. Bruce, *Biodegradable Surfactants Derived from Phenolic Lipids, in Surfactants in Lipid Chemistry: Recent Synthetic, Physical and Biodegradative Studies*, Royal Society of Chemistry, Cambridge, 159, 1992.

87. W. Delaware, The HLB system: A Time-saving guide to emulsifier selection, a catalogue of ICI surfactants, Wilmington, USA, p. 19897, 1992.

88. K. H. Raney, W. J. Benton, and C. A. Miller, Optimum detergency conditions with nonionic surfactants: I-Ternary water–surfactant–hydrocarbon systems, *J. Colloid Interface Sci.*, 117, 282–293, 1987.

89. S. Thompson, The role of oil detachment mechanisms in determining optimum detergency conditions, *J. Colloid Interface Sci.*, 163, 61–70, 1994.

90. S. Raney, Optimization of nonionic/anionic surfactant blends for enhanced oily soil removal, *J. Am. Oil Chem. Soc.*, 68, 525–534, 1991.

91. S. Goel, Measuring detergency of oily soils in the vicinity of phase inversion temperatures of commercial non-ionic surfactants using an oil-soluble dye, *J. Surf. Deterg.*, 1, 221–228, 1998.

92. S. Goel, Selecting the optimum linear alcohol ethoxylate for enhanced oily soil removal, *J. Surf. Deterg.*, 1, 213–220, 1998.

93. M. F. Cox, Effect of alkyl carbon chain length and ethylene oxide content on the performance of linear alcohol ethoxylates, *J. Am. Oil Chem. Soc.*, 66, 367–374, 1989.

94. C. L. Edwards, *Polyoxyethylene Alcohols in Nonionic Surfactants: Organic Chemistry*, N. M. van Os (Ed.), pp. 72–87, Surfactant Science Series, Marcel Dekker, New York, 1998.

95. J. S. Leal, M. T. Garcia, I. Ribosa, and F. Comelles, Environmental risk assessment of ethoxylated nonionic surfactants, In *Surfactants in Solution*, A. K. Chatto-padhyay, and K. L. Mittal (Eds.), No. 64, p. 379, Surfactant Science Series, Marcel Dekker, New York, 1996.

96. K. S. Goel, Phase behavior and detergency study of lauryl alcohol ethoxylates with high ethylene oxide content, *J. Surf. Deterg.*, 3, 221–227, 2000.

97. K. H. Raney and H. L. Benson, The effect of polar soil components on the phase inversion temperature and optimum detergency conditions, *J. Am. Oil Chem. Soc.*, 67, 722–729, 1990.

98. H. Oskarsson, M. Frankenberg, A. Annerling, and K. Holmberg, Adsorption of novel alkylaminoamide sugar surfactants at tailor-made surfaces, *J. Surf. Deterg.*, 10, 41–52, 2007.

99. J. Steber, W. Guhl, N. Stelter, and F. R. Schroder, Alkyl polyglycosides-ecological evaluation of a new generation of nonionic surfactants, *Tenside Surf. Deterg.*, 32, 51–52, 1995.

100. P. Veshchestva, S. Pletnev, and M. Yu (Eds.), *Surfactants: A Handbook*, Firma Klavel, Moscow, 2002.

101. L. Zhang, P. Somasundaran, and C. Maltesh, Adsorption of *n*-dodecyl-β-maltoside on solids, *J. Colloid Interface Sci.*, 191, 202–208, 1997.

102. L. Huang, C. Maltesh, and P. Somasundaran, Adsorption behavior of cationic and nonionic surfactant mixtures at the alumina–water interface, *J. Colloid Interface Sci.*, 177, 222–228, 1996.

103. F. Portet, P. L. Desbene, and C. Treiner, Non-ideality of mixtures of pure nonionic surfactants both in solution and at silica/water interfaces, *J. Colloid Interface Sci.*, 184, 216–226, 1996.

104. J. D. Hines, G. Fragneto, R. K. Thomas, P. R. Garrett, G. K. Rennie, and A. R. Rennie, Neutron reflection from mixtures of sodium dodecyl sulfate and dodecyl betaine adsorbed at the hydrophobic solid/aqueous interface, *J. Colloid Interface Sci.*, 189, 259–267, 1997.

105. P. Somasundaran, L. Zhang, and S. Lu, Adsorption of sugar-based surfactants at solid-liquid interfaces, *Sugar-Based Surfactants*, CRC Press, Boca Raton, FL, 2009.

106. P. Somasundaran, E. D. Snell, and Q. Xu, Adsorption behavior of alkyl aryl ethoxylated alcohols on silica, *J. Colloid Interface Sci.*, 144, 165–173, 1991.

107. E. N. Vulfson, Enzymatic synthesis of surfactants, In *Surfactants in Lipid Chemistry: Recent Synthetic, Physical and Biodegradative Studies*, J. H. P. Tyman (Ed.), p. 16, The Royal Society of Chemistry, London, 1992.

108. K. J. Parker, K. James, and J. Horford, Sucrose ester surfactants—A solventless process and the products thereof, In *Sucrochemistry*, ACS Symp. Series No 41, J. L. Hickson (Ed.), p. 97, American Chemical Society, Washington DC, 1977.

109. I. Söderberg, C. J. Drummond, D. N. Furlong, S. Godkin, and B. Matthews, Nonionic sugar-based surfactants: Self-assembly and air/water interfacial activity, *Colloids Surf. A.*, 102, 91–98, 1995.

110. S. Godkin, D. N. Furlong, C. J. Drummond, and B. Matthews, Sugar-based surfactants, In *Proceedings of the 3rd CESIO International Surfactants Congress*, London, 266, 1992.

111. T. M. Herrington, B. R. Midmore, and S. S. Sahi, Sucrose esters as emulsion stabilisers, In *Microemulsions and Emulsions in Foods*, ACS Symp. Series No. 448, M. El-Nokaly (Ed), p.82, American Chemical Society, Washington, DC, 1991.

112. P. C. Isaac and D. Jenkins, Biological oxidation of sugar-based detergents, *Chem. Ind.*, 976–977, 1958.

113. P. C. Isaac and D. Jenkins, A laboratory investigation of the breakdown of some of the newer synthetic detergents in sewage treatment, *J. Proc. Inst. Sewage Purif.*, 140, 314–318, 1960.

114. P. C. Isaac and D. Jenkins, The Biological breakdown of some newer synthetic detergents, In *Conference on Biological Waste Treatment: Advances in Biological Waste Treatment*, Macmillan, London, 5, 61, 1963.

115. C. H. Wayman and J. B. Roberts, Biodegradation of anionic and nonionic surfactants under aerobic and anaerobic conditions, *Biotechnol. Bioeng.*, 5, 367–384, 1963.

116. J. R. Cruz and M. C. D. García, The pollution of natural waters by synthetic detergents XIII: Biodegradation of nonionic surface active agents in river water and determination of their biodegradability by different test methods, *Grasas Y Aceitas J.*, 29, 1–9, 1978.

117. R. N. Sturm, Biodegradability of nonionic surfactants: Screening test for predicting rates and ultimate biodegradation, *J. Am. Oil Chem. Soc.*, 50, 159–165, 1973.

118. G. Brebion, R. Cabridenc, and A. Lerenard, Evaluation of the biodegradation of an eth-oxylated tallow sucro glyceride, *Rev. Fr. Corps. Gras*, 11, 191–198, 1964.
119. J. A. Bakera, B. Matthewsa, H. Suaresa, I. Krodkiewskaa, D. N. Furlong, F. Grieserb, and C. J. Drummond, Sugar fatty acid ester surfactants: Structure and ultimate aerobic biodegradability, *J. Surf.*, 3, 1–11, 2000.
120. International Standards Organisation, ISO 7827–1984(E), *Water Quality Evaluation in an Aqueous Medium of the Ultimate Aerobic Biodegradability of Organic Compounds Method by Analysis of Dissolved Organic Carbon* (DOC), Method no. ISO 7827–1984.
121. M. Hoey, and G. Berry, Polyoxyethylene alkyl amines, In *Nonionic Surfactants Organic Chemistry*, N.M. van Os (Ed.), Marcel Dekker, New York, 163, 1998.
122. C. G. Van Ginkel, Biodegradability of cationic surfactants, In *Biodegradability of Surfactants*, M. R. Porter and R. D. Karsa (Eds.), Blackie Academic & Professional, London, 183, 1995.
123. C. G. Van Ginkel, C. A. Stroo, and A. G. M. Kroon, Biodegradability of ethoxylated fatty amines and amides and the non-toxicity of their biodegradation products, *Tenside Surf. Deterg.*, 30, 213–219, 1993.
124. M. Stalmans, E. Matthijs, E. Weeg, and S. Morris, The environmental properties of glu-cose amide a new nonionic surfactant, *SOFW*, 13, 794–806, 1993.
125. D. R. Karsa and M. R. Porter (Eds.), *Biodegradability of Surfactants*, Blackie & Sons, London, 1995.
126. S. Osburn, The Manufacturing Confectioner, In *McCutcheon's Emulsifiers and Detergents*, International (Ed.), Glen Rock, New Jersey, 1986.
127. J. R. Hurford and C. K. Lee (Eds.), In *Developments in Food Carbohydrate*, Applied Science Publishers, London, 327, 1980.
128. H. Andree and B. Middelhauve, Möglichkeiten des einsatzesvon alkylpolyglucosiden in waschund spülmittein [Possibilities of the use of alkyl polyglucosides in washing up liquid], *Tenside Surf. Deterg.*, 28, 413–419, 1991.
129. D. Balzer, Alkylpolyglucosides, their physico-chemical properties and their uses, *Tenside Surf. Deterg.*, 28, 419–428, 1991.
130. M. A. Pes, K. Aramaki, N. Nakamura, and K. Hironobu, Temperature-insensitive micro-emulsions in a sucrose monoalkanoate system, *J. Colloid Interface Sci.*, 178, 666–671, 1996.
131. B. Jönsson, B. Lindman, K. Holmberg, and B. Kronberg, Introduction to surfactants, In *Surfactants and Polymers in Aqueous Solution*, John Wiley & Sons, Chichester, England, 2002.
132. K. Kameyama and T. Takagi, Micellar properties of octylglucoside in aqueous solutions, *J. Colloid Interface Sci.*, 137, 1–9, 1990.
133. Å. Waltermo, P. M. Claesson, E. Manev, S. Simonsson, I. Johansson, and V. Bergeron, Foam and thin-liquid-film studies of alkyl glucoside systems, *Langmuir*, 12, 5271–5278, 1996.
134. S. Matsumura, K. Imai, and S. Yoshikawa, Surface activities, biodegradability and antimicrobial properties of *n*-alkyl glucosides, mannosides, and galactosides, *J. Am. Oil Chem. Soc.*, 67, 996–1001, 1990.
135. P. S. Piispanen, An improved method of synthesis of 2-alkylamino-2-deoxy-D-gluco-pyranose and 1,2-Dialkylamino-1,2-dideoxy-D-(N)-glucoside, *J. Org. Chem.*, 68, 628–634, 2003.
136. P. A. Egan, Surfactants from biomass, *Chem. Tech.*, 12, 758–763, 1989.
137. H. Maag, Fatty acid derivatives: Important surfactants for household, cosmetic and industrial purposes, *J. Am. Oil Chem. Soc.*, 61, 259–265, 1984.
138. N. Ishler and F. D. Snell (Eds.), Uses of sucrose mono palmitate in food products, In *Sugar Esters*, Sugar Research Foundation, New York, 29, 1968.

139. H. Bertsch, F. Puschel, and E. Ulsperger, Manufacture and uses of fatty acid sugar esters, *Tenside Surf. Deterg.*, 2, 397–401, 1996.
140. R. Khan, The chemistry of sucrose, *Adv. Carbohydr. Chem. Biochem.*, 33, 268–274, 1976.
141. N. A. Negm and A. S. Mohamad, Synthesis, characterization and biological activity of sugar-based gemini cationic amphiphiles, *J. Surf. Deterg.*, 11, 215–221, 2008.
142. O. Kirk, F. Pedersen, and C. Fuglsang, Preparation and properties of a new type of carbohydrate-based cationic surfactant, *J. Surf. Deterg.*, 1, 37–40, 1998.
143. J. V. Erpenbeck, N. Krug, and M. J. Hohlfeld, Therapeutic use of surfactant components in allergic asthma, *Naunyn-Schmiedeberg's Arch. Pharmacol.*, 379, 217–224, 2009.
144. C. Stefaan De Smedt, J. Demeester, and E. W. Hennink, Cationic polymer based gene delivery systems, *Pharm. Res.*, 17, 2, 2000.
145. V. V. Kumar, R. S. Singhand, and A. Chaudhuri, Cationic transfection lipids in gene therapy: Successes, set-backs, challenges and promises, *Curr. Med. Chem.*, 10(14), 1297–306, 2003.
146. P. L. Felgner, T. R. Gadek, M. Holm, R. Roman, W. Chan, M. Wenz, J. P. Northorp, G. M. Ringold, and M. Danielsen, Lipofection: A highly efficient, lipid-mediated DNA-transfection procedure, *Proc. Natl. Acad. Sci. USA*, 84, 7413–7417, 1987.
147. R. Leventisand and J. R Silvius, Interactions of mammalian cells with lipid dispersions containing novel metabolizable cationic amphiphiles, *Biochim. Biophys. Acta.*, 1023, 124–132, 1990.
148. M. J. Bennett, A. M. Aberle, R. P. Balasubramaniam, J. G. Malone, R. W. Malone, and M. H. Nantz, *J. Med. Chem.*, 269, 4069–4078, 1997.
149. J. H. Felgner, R. Kumar, C. N. Sridhar, C. J. Wheeler, Y. J. Tsai, R. Border, P. Ramsey, M. Martin, and P. L. Felgner, Enhanced gene delivery and mechanism studies with a novel series of cationic lipid formulations, *J. Biol. Chem.*, 269, 2550–2561, 1994.
150. R. Leventis and J. R. Silvius, Interactions of mammalian cells with lipid dispersions containing novel metabolizable cationic amphiphiles, *Biochim. Biophys. Acta.*, 23, 124–132, 1990.
151. X. Gao and L. Huang, A novel cationic liposome reagent for efficient transfection of mammalian cells, *Biochem. Biophys. Res. Commun.*, 179, 280–285, 1991.
152. J. Wang, X. Guo, Y. Xu, L. Barron, and F. C. Szoka, Synthesis and characterization of long chain acyl carnitine esters. Potentially biodegradable cationic lipids for use in gene delivery, *J. Med. Chem.*, 41, 2207–2215, 1998.
153. T. Banno, K. Toshima, K. Kawada, and S. Matsumura, Synthesis and properties of gemini-type cationic surfactants containing carbonate linkages in the linker moiety directed toward green and sustainable chemistry, *J. Surf. Deterg.*, 12, 249–259, 2009.
154. F. M. Menger and C. A. Littuy, Gemini surfactants: A new class of self-assembling molecules, *J. Am. Chem. Soc.*, 115, 10083–10090, 1993.
155. M. J. Rosen, Geminis: A new generation of surfactants, *Chem. Tech.*, 23, 30–33, 1993.
156. F. M. Menger and C. A. Littauy, Gemini surfactants: A new class of self-assembling molecules, *J. Am. Chem. Soc.*, 115, 10083–10090, 1993.
157. R. Zana, Dimeric (gemini) surfactants: Effect of the spacergroup on the association behavior in aqueous solution, *J. Colloid. Interface Sci.*, 248, 203–220, 2002.
158. R. Zana, Dimeric, and oligomeric surfactants behavior at interfaces and in aqueous solution: A review, *Adv. Colloid Interface Sci.*, 97, 205–253, 2002.
159. T. Aisaka, T. Oida, and T. Kawase, A novel synthesis of succinic acid type gemini surfactant by the functional group interconversion of corynomicolic acid, *J. Oleo. Sci.*, 56, 633–644, 2007.
160. C. A. Bunton, L. Robinson, J. Schaak, and M. F Stam, Catalysis of nucleophilic substitutions by micelles of dicationic detergents, *J. Org. Chem.*, 36, 2346–2350, 1971.

161. F. Devinsky, L. Masarova, and I. Lacko, Surface activity and micelle formation of some new bisquaternary ammonium salts, *J. Colloid. Interface Sci.*, 105, 235–239, 1985.
162. F. Devinsky, I. Lacko, F. Bittererova, and L. Tomeckova, Relationship between structure-surface activity, and micelle formation of some new bisquaternary isosteres of 1,5-pentane diammonium dibromides, *J. Colloid. Interface Sci.*, 114, 314–322, 1986.
163. U. Laska, K. A. Wilk, I. Maliszewska, and L. Syper, Novel glucose-derived gemini surfactants with a 1,10-ethylene bisurea spacer: Preparation, thermotropic behavior, and biological properties, *J. Surf. Deterg.*, 9, 115–124, 2006.
164. K. A. Wilk, L. Syper, B. W. Domagalska, U. Komorek, and R. Maliszewska, Aldonamide-type gemini surfactants: Synthesis, structural analysis, and biological properties, *J. Surf. Deterg.*, 5, 235–244, 2002.
165. A. Wagenaar and J. B. Engberts, Synthesis of nonionic reduced-sugar based bola amphiphiles and gemini surfactants with an x-diamino-(oxa)alkyl spacer, *Tetrahedron*, 63, 10622–10629, 2007.
166. J. M. Pestman, K. R. Terpstra, M. C. A. Stuart, H. A. Doren, A. Brisson, and R. M. Kellogg, Nonionic bola amphiphiles and gemini surfactants based on carbohydrates, *Langmuir*, 13, 6857–6860, 1997.
167. M. L. Fielden, C. Perrin, A. Kremer, M. Bergsma, M. C. Stuart, and P. Camilleri, Sugar-based tertiary amino gemini surfactants with a vesicle-to-micelle transition in the endosomal pH range mediate efficient transfection in vitro, *Eur. J. Biochem.*, 268, 1269–1279, 2001.
168. D. Ono, S. Yamamura, M. Nakamura, and T. Takeda, Preparation and properties of bis (sodium sulfate) types of cleavable surfactants derived from diethyl tartrate, *J. Oleo. Sci.*, 54, 51–57, 2005.
169. P. Torres and M. Solans, Synthesis, aggregation, and biological properties of a new class of gemini cationic amphiphilic compounds from arginine, *Langmuir*, 12, 5296–5301, 1996.
170. E. Tsatsaroni, S. P. Koemtjopoulou, and G. Demertzis, Synthesis and properties of new cationic surfactants, *J. Am. Oil Chem. Soc.*, 64, 1444–1447, 1987.
171. P. Fernandez, M. Valls, and J. M. Bayona, Occurrence of cationic surfactants and related products in urban coastal environments, *Environ. Sci. Technol.*, 25, 547–550, 1991.
172. I. M. Banat, R. S. Makkar, and S. S. Cameotra, Potential commercial applications of microbial surfactants, *Appl. Environ. Microbiol.*, 53, 495–508, 2000.
173. S. S. Cameotra and R. S. Makkar, Recent applications of biosurfactants as biological and immunological molecules, *Curr. Opin. Microbiol.*, 7, 262–266, 2004.
174. J. D. Desai and I. M. Banat, Microbial production of surfactants and their commercial potential, *Microbiol. Mol. Biol. Rev.*, 61, 47–64, 1997.
175. M. E. Mercadé, M. A. Manresa, and M. Robert, Olive oil mill effluent (OOME) new substrates for biosurfactant production, *Bioresour. Technol.*, 43, 1–6, 1993.
176. M. G. Healy, C. M. Devine, and R. Murphy, Microbial production of biosurfactants, *Resour. Conserv. Recyc.*, 18, 41–57, 1996.
177. C. N. Mulligan, R. N. Yong, and B. F. Gibbs, Surfactant-enhanced remediation of contaminated soil: A review, *Eng. Geol.*, 60, 371–380, 2001.
178. G. M. Carman and M. C. Cirigliano, Purification and characterization of liposan, a bio-emulsifier from Candida lipolytica, *Appl. Environ. Microbiol.*, 50, 846–850, 1985.
179. R. Shepherd, J. Rockey, and I. W. Sutherland, Novel bioemulsifiers from microorganisms for use in foods, *J. Biotechnol.*, 40, 207–217, 1995.
180. T. Uniack, The citizen's wildlife refuge planning handbook: Charting the future of conservation on the National Wildlife Refuge near you: Defenders of Wildlife, Washington, D.C., accessed April 2010. Available at: http://www.defenders.org/resources/publications/programs_and_policy/habitat_conservation/federal_lands/citizen's_wildlife_refuge_planning_handbook.pdf

181. T. Balson and M. S. B. Felix, The biodegradability of non-ionic surfactants, In *Biodegradability of Surfactants*, D. R. Karsa and M. R. Porter (Eds.), Blackie Academic and Professional, Glasgow, UK, pp. 204–230, 1995.

182. P. H. Brunner, S. Capri, A. Marcomini, and W. Giger, Occurrence and behaviour of linear alkylbenzenesulfonates, nonylphenol, monophenol and nonylphenol diethoxylates in sewage and sewage-sludge treatment, *Water Res.*, 22, 1465–1472, 1988.

183. F. R. Bevia, D. Prats, and C. Rico, Elimination of LAS (linear alkylbenzene sulfonate) during sewage treatment, drying and composting of sludge and soil amending processes, In *Organic Contaminants in Waste Water*, D. Quaghebeur, I. Temmerman, and G. Angeletti (Eds.), Sludge, and Sediment, London, 1989.

184. H. Henau, E. Matthijs, and E. Namking, Trace analysis of linear alkylbenzene sulfonate (LAS) by HPLC. Detailed results from two sewage treatment plants, In *Organic Contaminants in Waste Water*, D. Quaghebeur, I. Temmerman, and G. Angeletti, (Eds.), Sludge, and Sediment, London, 1989.

185. R. D. Swisher, *Surfactant Biodegradation*, Marcel Dekker, New York, 1987.

186. M. A. Hashim, J. Kulandai, and R. S. Hassan, Biodegradability of branched alkylbenzene sulphonates, *J. Chem. Technol. Biotechnol.*, 54, 207–214, 1992.

187. J. A. Perales, M. A. Manzano, D. Sales, and J. M. Quiroga, Linear alkylbenzene sulphonates: Biodegradability and isomeric composition, *Bull. Environ. Contam. Toxicol.*, 63, 94–100, 1999.

188. C. Lee, N. J. Russell, and G. F. White, Modelling the kinetics of biodegradation of anionic surfactants by biofilm bacteria from polluted riverine sites: A comparison of five classes of surfactant at three sites, *Water Res.*, 29, 2491–2497, 1995.

189. S. Terzic, D. Hrsak, and M. Ahel, Primary biodegradation kinetics of linear alkylbenzene sulphonates in estuarine waters, *Water Res.*, 26, 585–591, 1992.

9 Lubricants and Functional Fluids from Lesquerella Oil*

Steven C. Cermak and Roque L. Evangelista

CONTENTS

* The mention of trade names or commercial products in this chapter is solely for the purpose of providing specific information and does not imply recommendation or endorsement by the U.S. Department of Agriculture. The USDA is an equal opportunity provider and employer.

Lesquerella fendleri is an oilseed crop belonging to the Brassicaceae (mustard) family that is native to the desert of southwestern United States. The interest in this crop is due to the high level of hydroxy fatty acids (HFAs) in the oil. The seed contains 33% oil, 23% protein, and 15% gums. The seed oil contains 54–60% lesquerolic (14-hydroxy-*cis*-11-eicosenoic) and 3–5% auricolic (14-hydroxy-11,17-eicosadienoic) acids. HFAs are used in a variety of industrial applications such as lubricants, corrosion inhibitors, engineering plastics, plasticizers, emulsifiers, and coatings. The current main source of HFAs is castor oil, which contains 90% ricinoleic (12-hydroxy-9-octadecanoic) acid. Because lesquerolic acid is very similar to ricinoleic acid, products derived from lesquerella oil would have comparable properties as those obtained from castor oil. Castor and lesquerella methyl esters have been shown to enhance lubricity in ultralow sulfur diesel at concentrations as low as 0.25%. Estolides synthesized from lesquerella and castor fatty acid esters with 2-ethylhexanoic acid have the best low-temperature properties of any estolides to date and have outperformed many commercial products even without additives.

9.1   INTRODUCTION

Lesquerella is a genus of annual, biennial, and perennial herbs belonging to the Brassicaceae (mustard) family. Of over 100 species identified, 83 are native to North America. Most of these species are found in the southwestern United States and northern Mexico [1–3]. Lesquerella seeds contain oil rich in one of three types of hydroxy fatty acids (HFAs): lesquerolic (14-hydroxy-*cis*-11-eicosenoic), auricolic (14-hydroxy-11,17-eicosadienoic), or densipolic (12-hydroxy-*cis*-9,15-octadienoic) acid (Figure 9.1) [1,4–7]. *L. lindheimeri* and *L. pallida*, both native to Texas, have the highest lesquerolic acid content (>80%). *L. densipila*, a native to Alabama and Tennessee, has oil containing 44% densipolic and 11% ricinoleic acids. *L. auriculata*, a native to Texas and Oklahoma, has oil with 40% auriculic and 11% lesquerolic acids, and 90% occur as estolides [5,6,8,9]. HFAs are used in a variety of industrial applications such as lubricants, corrosion inhibitors, engineering plastics, plasticizers, emulsifiers, and coatings. The current primary source of HFAs is castor oil, which contains 90% ricinoleic (12-hydroxy-9-octadecanoic) acid (Figure 9.1).

Among the wild plants evaluated for commercial cultivation, *L. fendleri* proved to be most promising for new crop development because of its abundant seed production, seed holding capacity, erect and compact growth, and ability to colonize disturbed soils, hardiness, and polymorphic characteristics [10,11]. It is widely distributed in southwestern Arizona, Texas, and Oklahoma. Wild stands have been found at elevations ranging from 610 to 1830 m in areas receiving 25–40 cm of annual precipitation [12]. The plant grows to a height of 45 cm with adequate water and warm temperature. *L. fendleri* can be distinguished from other *Lesquerella* species by its

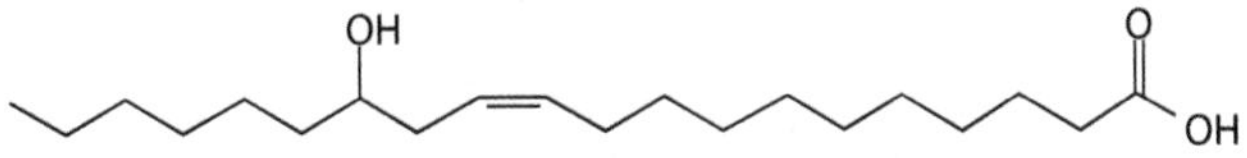

Lesquerolic acid (14-hydroxy-*cis*-11-eicosenoic)

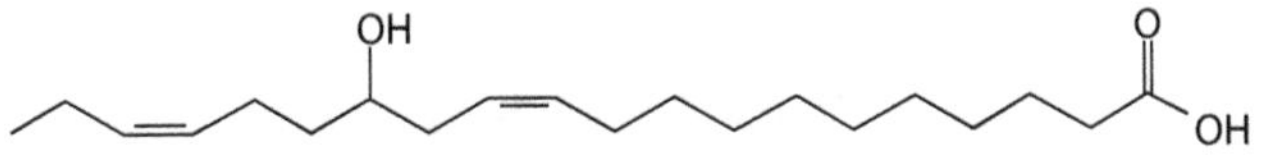

Auricolic acid (14-hydroxy-11,17-eicosadienoic)

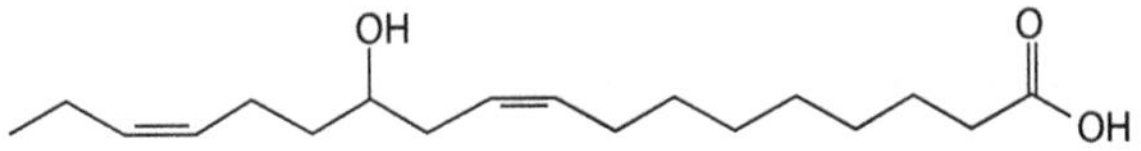

Densipolic acid (12-hydroxy-*cis*-9,15-octadienoic)

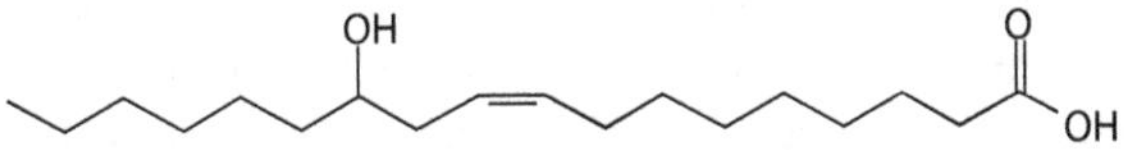

Ricinoleic acid (12-hydroxy-9-octadecanoic)

FIGURE 9.1 Structures and names of some common hydroxyl fatty acids found in castor and lesquerella oils.

glabrous siliques, fused trichomes, and yellow flowers. The siliques contain as many as 24 small seeds. The seeds vary in color from orange brown to yellow, measure up to 2 mm in diameter, and are flat to ovate in cross section. Seeds weigh about 0.63 g/1000 seeds. The seed contains 28% oil, 23% protein, and 15% gums [3,13]. The seed oil contains 54–60% lesquerolic and 3–5% auricolic acids [14]. *L. fendleri* variety (WCL-L03) with 33% oil content was released in 2005 [15]. Glucoiberin (3-(methylsulfinyl)propyl glucosinolate) is the principal glucosinolate in lesquerella. The seed contains 65 mg glucoiberin/g of defatted meal [16]. Glucosinolates are glucose and sulfur-containing organic anions (Figure 9.2) whose decomposition products are produced when plant cells are ruptured, and the glucosinolates present in vacuoles are hydrolyzed by the enzyme myrosinase (β-thioglucosidase glucohydrolase). Many glucosinolate degradation products are of interest because of their biological activities. Several of these hydrolysis products have biocidal activity against a wide variety of organisms, such as insects, plants, fungi, and bacteria, while others have human health benefits.

9.1.1 PRODUCTION OF LESQUERELLA

Domestication efforts on *Lesquerella* started in 1985. *L. fendleri* is the only species being cultivated at this time. The current production technique is best summarized

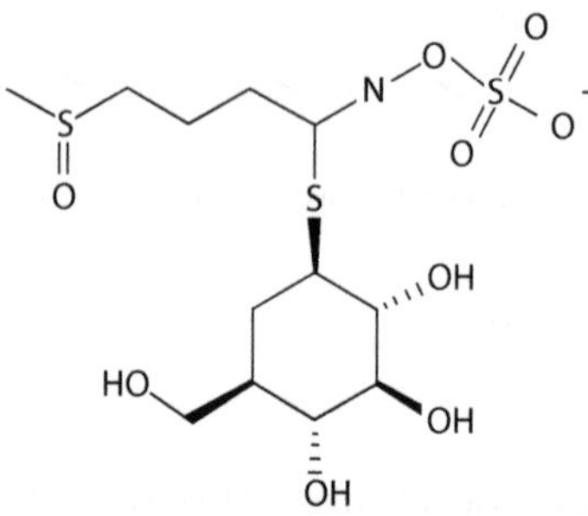

Glucoiberin (3-(methylsulfinyl)propyl glucosinolate)

FIGURE 9.2 Major lesquerella glucosinolate.

by Dierig [17]. The most important environmental factors affecting stand establishment are soil temperature, moisture, and depth of planting. The recommended planting date is late September in Texas and New Mexico, and October in Arizona. The temperature for germination ranges from 5°C to 35°C with 20°C as optimal. A level or slightly sloping field prepared for uniform flood irrigation is desirable. The surface is packed using a barbed roller, which makes small indentations on the soil that are desirable for germination of small seeded crops. The seeds are broadcast-planted at a rate of 8–12 kg/ha using a granule applicator and the planted field is rolled again so that the seeds have good contact with the soil. The field must be kept moist to avoid crusting until emergence is complete (about 10 days to 2 weeks after planting). *Lesquerella* requires 64–76 cm of water during the growing season; most of it is used between late February and May when 90% of growth occurs. Low water availability not only reduces yield but also decreases the oil content of the seeds and the percentage of lesquerolic acid in the oil [18]. Nitrogen fertilizer is applied at planting (73 kg/ha) and again at the onset of flowering. When the seeds reach maturity (May and June), irrigation is stopped to desiccate the plants for harvest. When the plants start to turn brown, a desiccant is applied to complete the process. The seeds are harvested about 3–4 weeks after the last irrigation using a conventional combine fitted with sieve screens designed for small seeds. Current seed yields are approximately 1800 kg/ha.

9.1.2 PROCESSING OF LESQUERELLA

9.1.2.1 Oil Extraction

Oil from lesquerella seeds has been successfully extracted using laboratory- and pilot-scale equipment. Like other mustard seeds, inactivation of β-thioglucosidase glucohydrolase (TGSase) is an important consideration in preparing the seed for oil extraction. When crushed, the glucosinolates in the seed can be hydrolyzed enzymatically (with sufficient moisture) by TGSase into isothiocyanates, nitriles, thiocyanates, and other undesirable sulfur-containing compounds. Thermal degradation of glucosinolates may also occur when seeds are subjected to high processing temperatures [19]. Some of these sulfur compounds end up in the oil during extraction. The sulfur compounds in the oil impart objectionable taste, odor, and can poison catalysts during hydrogenation.

Inactivation of TGSase is influenced largely by heating temperature, duration of heating, and seed moisture. In a study conducted by Carlson and coworkers [20], they reported that TGSase was rendered inactive when whole lesquerella seeds with 6% moisture content (MC) were cooked at 100°C for 60 min and for 15 min when seed MC was equal or greater than 10%. TGSase was also inactivated when cold-flaked seeds were extruded (with steam injection) at 77°C and 34 s residence time with expanded collets emerging at 10% MC. Dry extrusion of whole seeds with 6% MC at 111°C and 22 s residence time was also effective in inactivating TGSase [21].

Full solvent oil extraction typically is used for oilseeds with less than 25% oil content. The solvent, usually hexane, extracts the oil from flaked seeds or seeds that have been expanded into porous collets. Extrusion of cold-flaked lesquerella seeds produced collets acceptable for hexane extraction [20]. Residual oil in the defatted meal ranged from 0.8% in batch emersion to 2% in continuous extraction. For lesquerella seeds with oil content greater than 25%, prepressing is recommended to remove some of the oil before proceeding to hexane extraction.

Screw pressing is widely employed in older commercial operations when oil extraction facilities were still operating below 100 metric tons (MT)/day. Evangelista [21] conducted a study on screw pressing of lesquerella seeds and evaluated the quality of the oils obtained from uncooked seeds (cold pressing), seeds cooked and dried in the seed cooker, and seeds cooked by dry extrusion. Cold pressing unheated lesquerella seeds with 6% MC produced oil with very low phosphatides (<0.02%) and sulfur (9 ppm) (Table 9.1). Phosphatides (also called phospholipids) are lipid derivatives in which one fatty acid (FA) in the triglyceride has been replaced by phosphatidic acid or a phosphatidyl group such as lecithin, cephalin, phosphatidylserine, and phosphoinositides [22]. Sulfur is an indication of decomposition products of glucosinolates

TABLE 9.1

Analysis of Crude Lesquerella Oil Obtained by Screw Pressing

	Seed Treatment before Pressing[a]		
	Uncooked Seeds	**Cooked Seeds**	**Extruded Seeds**
Free fatty acid (%)	2.9	2.2	1.7
Phosphatides[b] (%)	0.01	<0.01	0.02
Sulfur (ppm)	9.0	71.6	44.6
Phosphorus (ppm)	4.0	2.1	6.8
Calcium (ppm)	1.9	1.3	3.0
Magnesium (ppm)	1.0	1.0	1.7

Source: Adapted from R.L. Evangelista, *Ind. Crops Prod.*, 29, 189–196, 2009.

[a] Starting seed moisture content = 6%.

[b] Calculated from phosphorus content: % phosphatides = $31.7 \times (P, ppm) \times 10^{-4}$.

present in the seed and coextracted with the oil during pressing. However, the press cake still contained 11.6% oil. The residual oil in the cake was reduced to 6.7% when the seeds were cooked in the cooker/conditioner (25–107°C, 50 min residence time) and dried to about 5% MC. The phosphatide in the extracted oil remained low but the sulfur level in the oil increased to 72 ppm. Cooking the seed by dry extrusion, a high temperature-short time cooking process, also resulted in oil with low phosphatides but sulfur content (45 ppm) was lower than that of seeds cooked in the seed cooker.

9.1.2.2 Oil Refining

Crude lesquerella oil is reddish brown (like molasses) and has a typical mustard odor. The amount of phosphatides and sulfur compounds in the oil will vary depending on the method used in oil extraction. In oil refining, phosphatides, free fatty acids (FFAs), sulfur compounds, metals, pigments, waxes, and low-boiling compounds are removed or reduced. The five major steps in oil refining include degumming, neutralization, bleaching, winterization, and deodorization. The extent of oil refining will depend on the intended application.

In laboratory refining, Carlson and Kleiman [23] used saturated NaCl solution to degum crude lesquerella oil because using water alone gave stubborn emulsion. Also, neutralization of the oil using alkali produced abundant and difficult emulsion. They also reported significant autodegumming during the initial storage of crude prepressed and solvent-extracted oils.

The crude oil used in the pilot-scale refining study was obtained by full-pressing cooked whole seeds. The crude oil had 11 ppm phosphorus (0.3% phosphatide) and 1.3% FFA. Because of the low phosphatide level, degumming was unnecessary. Bleaching with clay (6% w/w Tonsil 126FF, Sud-Chemie Inc., Louisville, KY) and activated carbon (2% w/w Darco KB, Norit Americas Inc., Marshall, TX) reduced the oil Gardner color from 13 to 11 [24]. Gardner color scale is a one-dimensional scale for grading color of transparent liquid. The sample is compared against 18 standards, which range from lightest yellow (1) to brownish red (18). To avoid dealing with difficult emulsion, the FFA was removed by physical refining (steam distillation). Deodorization was carried out using a packed column thin-film continuous deodorizer. Bleached oil and sparging steam were introduced into the deodorizer at 125 kg/h (about 40 min residence time) and 2.5 kg/h, respectively. The deodorization temperature was 270°C and vacuum pressure of less than 7 mm Hg. The exit temperature of the oil was set >80°C to maintain the flow of the oil out of the deodorizer. The Gardner color of the oil was reduced further from 11 to 4 due to heat bleaching in the deodorizer and the FFA content of the refined oil was 0.12%. However, the refined oil still had some detectable off odor.

9.2 ESTOLIDES

Estolides are a class of esters based on vegetable oils [25] made by the formation of a carbocation at the site of unsaturation that can undergo nucleophilic addition by another FA, with or without carbocation migration along the length of the chain, to form an ester linkage (Figure 9.3). These ester linkages are used to help characterize the structure of the estolide since the estolide number (EN) is defined as the average number of

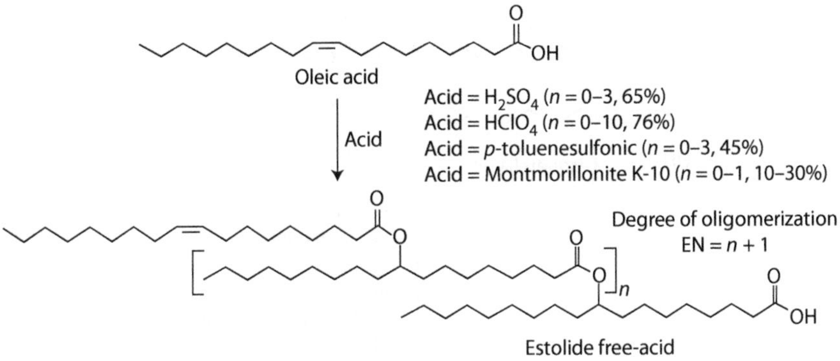

FIGURE 9.3 General scheme for oleic estolide free-acid synthesis. Degree of oligomerization (n) and yield (%) are dependent on the acid catalyst used for the synthesis.

FAs added to a base FA or the number of ester linkages (Figure 9.3, EN = n + 1). The secondary ester linkages of the estolide are more resistant to hydrolysis than those of triglycerides (TGs), and the unique structure of the estolide results in materials that have far superior physical properties (such as pour and cloud points, viscosity, color, and wear) for lubricant applications than vegetable and mineral oils [26,27].

Estolides have been developed from both FAs and directly from vegetable oil or TG (Figure 9.4). In order to obtain a wide variety of different estolides, oils from different crops (soy, coconut, meadowfoam, coriander, castor, lesquerella, and cuphea) have been explored to examine the full range of possibilities [28–31]. Unlike estolide FAs or estolide esters (Figure 9.3) that are formed when the carboxylic acid functionality of one FA links to the site of unsaturation of another FA to form oligomeric esters, the FA esters of lesquerella and castor have a hydroxy

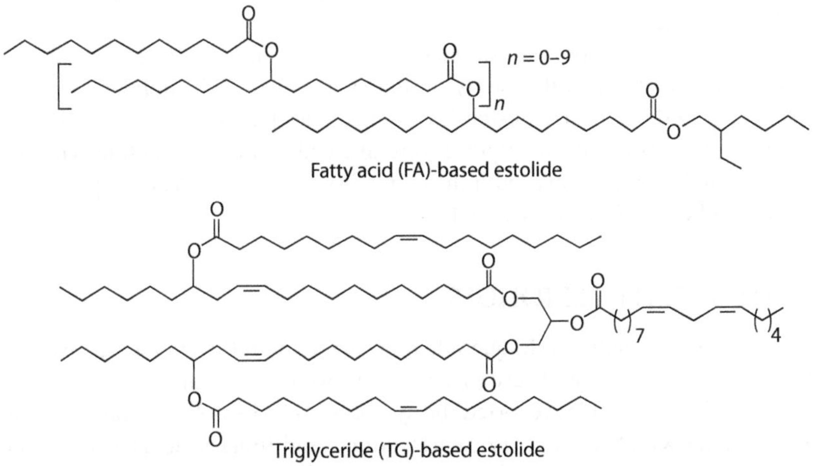

FIGURE 9.4 Fatty acid estolide 2-ethylhexyl esters versus TG-estolides.

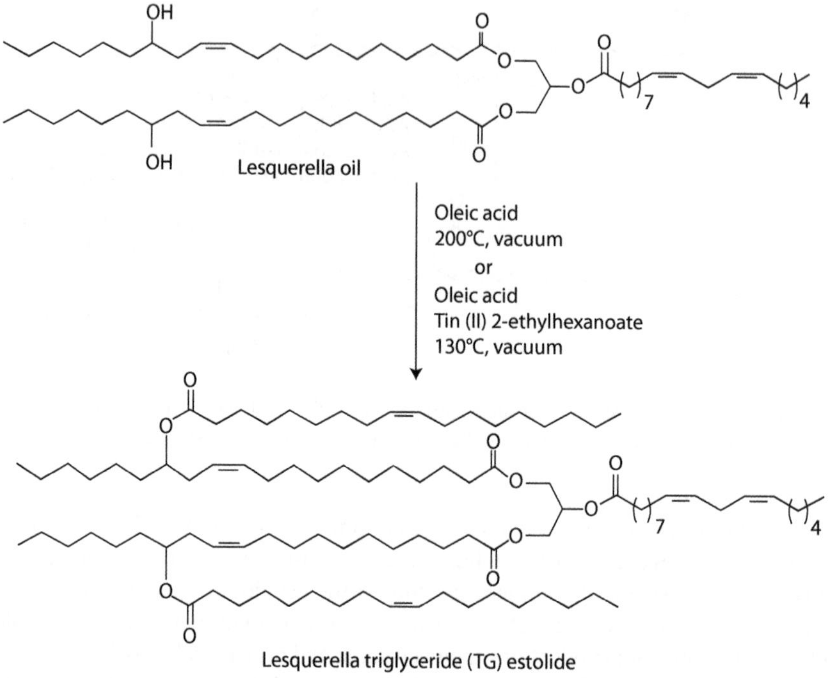

FIGURE 9.5   General scheme for lesquerella TG-estolide synthesis.

functionality that provides a site for a simple esterification to take place to produce TG-based estolides (Figure 9.5). HFAs such as lesquerolic can be readily converted into oil-based estolides either as TGs in the presence of FFA or from homopolymerization of the free lesquerolic FAs [32–35]. The synthesis of some estolides from castor oil and FAs has been reported [36,37], but not as a complete set with the physical properties [33].

Lesquerella and castor TG-estolides synthesized directly from the TGs were recently reported in a detailed study on the synthesis of TG-estolides [32]. Hayes and Kleiman [34] synthesized estolides from free lesquerolic and oleic acids using a lipase catalyst. However, no reports were available on either estolide types from lesquerella and castor capped with different FAs or their physical properties until reports by either Isbell and/or Cermak [32–33].

9.3 LESQUERELLA ESTOLIDES

The hydroxy functionality of lesquerella oil provides a useful site for reactive chemistry such as esterifications. These esterifications would result in estolide or estolide-type materials. Lawate [38] reported the synthesis of TG-estolides from both castor and lesquerella with heptanoic, isostearic, adipic, and fumaric acids as the capping FAs using p-toluenesulfonic acid as the catalyst (Figure 9.6). The reaction was carried out at 150°C while azeotropically removing water. The TG-estolide products

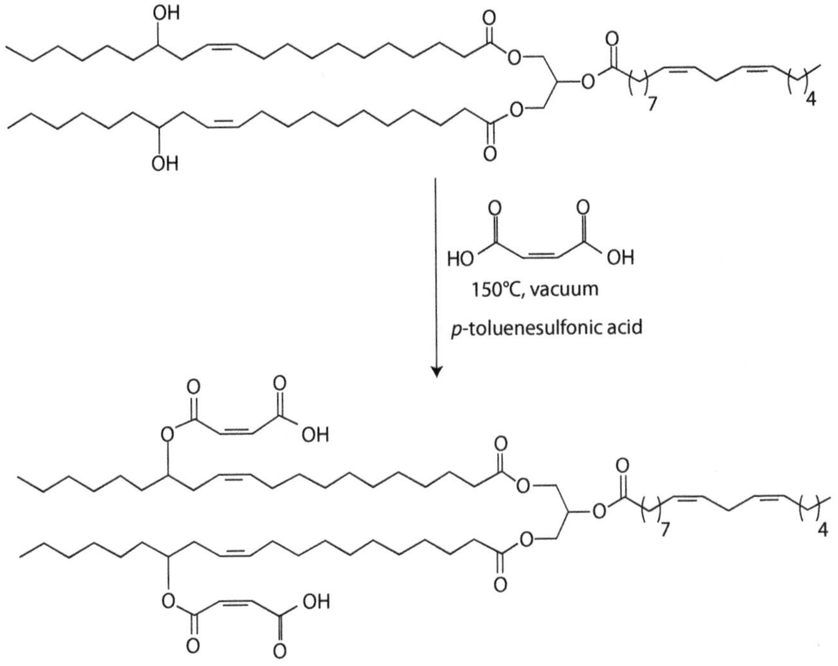

FIGURE 9.6 Synthesis scheme for lesquerella triglyceride estolides. (Adapted from S.C. Cermak and T.A. Isbell, *Ind. Crops Prod.*, 18, 223–230, 2003.)

were evaluated for their thickening performance by the change in viscosity in blends with high-oleic vegetable oil.

As of the beginning of 2012, three different types of estolides have been synthesized at the USDA laboratory in Peoria, Illinois: oleic FA-based estolides and two hydroxy-based estolides from TG and FA (Figure 9.4). The oleic-based estolides are synthesized [27,39] by the formation of a carbocation at the site of unsaturation on an FA, which can undergo nucleophilic addition by another FA, with or without carbocation migration along the length of the chain, to form an ester linkage (Figure 9.3). The simple estolide structure has been easily modified by the addition of a saturated FA using this same technology [40,41]. On the other hand, the TG-estolides require that the vegetable oil have a hydroxy functionality present that provides a site for simple esterification to take place with an FA to produce TG-based estolides (Figure 9.4). Finally, estolides have also been synthesized by taking advantage of these two different technologies where the starting material is an HFA but have used the TG-estolide technology to produce the final estolides (Figure 9.7).

Previously reported physical properties such as pour and cloud points, viscosity, color, and wear of estolide esters have compared favorably with commercially available industrial products such as petroleum-based hydraulic fluids, soy-based fluids, and petroleum oils [25–27,31,41–43]. Thus, lesquerella estolides should have very similar properties and industrial potential.

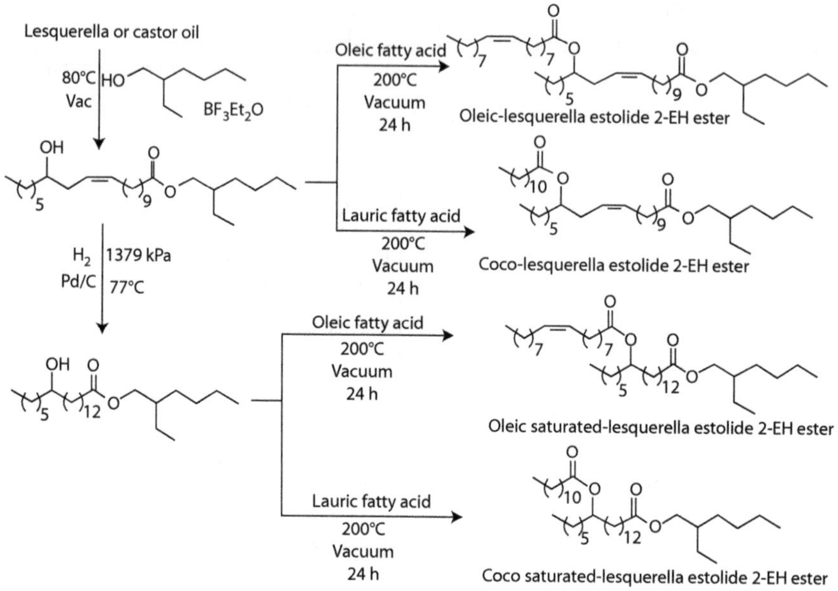

FIGURE 9.7 General scheme for lesquerella or castor estolide 2-EH ester synthesis.

9.3.1 SYNTHESIS OF LESQUERELLA AND CASTOR ESTOLIDES

Tables 9.2 and 9.3 outline a series of reactions that explore the formation of lesquerella and castor TG-estolides. In these reactions, series of different saturated FAs, acetic through stearic FAs, are used as the capping material to give the saturated-capped TG-estolides (Figure 9.5). Capping is defined as converting the hydroxy functionalities to esters in the lesquerella and castor cases. Thus, in a mono-capped estolide, just one of the hydroxy functionalities has been converted to the corresponding ester (one capping group per TG molecule) while in the fully capped estolide, all the hydroxy functionalities have been converted to the corresponding esters. Table 9.4 outlines a series of reactions that explores the formation of lesquerella and castor FA-based estolides where a series of different saturated and unsaturated FAs were used as the capping material to yield these lesquerella and castor estolides (Figure 9.7).

9.3.1.1 Lesquerella or Castor Methyl and 2-Ethylhexyl Esters

Lesquerella and castor oils underwent an acid-catalyzed transesterification reaction that was conducted in the presence of either methanol or 2-ethylhexanol via (Figure 9.7) Kugelrohr distillation under vacuum (6–13 Pa) at 90–110°C to remove the excess alcohol. The residue then underwent a second Kugelrohr distillation under vacuum (6–13 Pa) at 180–200°C to yield the purified esters.

9.3.1.2 Lesquerella or Castor Mono/Fully Capped Triglyceride Estolides

The hydroxyl-TGs of lesquerella and castor oil were converted into their corresponding mono- (one capping group per TG molecule) and fully capped (all hydroxy

TABLE 9.2
Physical Properties of Lesquerella TG-Estolides (See Figure 9.5 for Synthetic Scheme)

TG-Estolides	Fatty Acid[a]	EN[b]	Pour Point[c] (°C)	Cloud Point[d] (°C)	Vis[e] @40°C (cSt)	Vis[e] @100°C (cSt)	Viscosity Index[f]
Lesquerella	NA	NA	−21	−22	127.7	15.2	123
L2-M[g]	C2:0	0.90	−21	>r.t.	92.8	14.6	164
L2-F[g]	C2:0	1.70	−30	−18	79.7	14.2	186
L4-M	C4:0	0.88	−27	<−27	86.7	14.4	173
L4-F	C4:0	1.50	−33	−30	74.7	13.9	194
L6-M	C6:0	0.78	−33	>r.t.	103.0	15.9	165
L6-F	C6:0	1.23	−36	>r.t.	87.9	15.2	183
L8-M	C8:0	0.75	−27	−25	115.7	15.1	187
L8-F	C8:0	1.41	−33	−27	76.0	14.8	205
L10-M	C10:0	0.66	−27	−26	118.5	17.2	159
L10-F	C10:0	1.51	−30	−17	99.9	16.7	182
L12-M	C12:0	1.00	−27	−23	110.8	17.4	173
L12-F	C12:0	1.60	−18	−28	101.0	17.2	186
H₂-L12-M	H[h] L12-M	0.97	mp 28–38	N.A.	N.A.	N.A.	N.A.
H₂-L12-F	H[h] L12-M	1.61	mp 20–32	N.A.	N.A.	N.A.	N.A.
L14-M	C14:0	1.29	−18	1	129.9	19.2	168
L14-F	C14:0	1.46	3	21	118.6	18.4	174
L16-M	C16:0	0.83	0	15	135.2	20.5	176
L16-F	C16:0	1.75	6	27	114.2	18.7	184
L18-M	C18:0	1.46	9	28	137.4	20.5	173
L18-F	C18:0	1.75	24	45	Solid	34.3	NA
L18:1-M	C18:1	0.97	−27	−16	119.6	18.7	176
L18:1-F	C18:1	1.56	−27	−16	95.1	17.0	195

Note: r.t., room temperature; N.A., not available.

[a] Carbon length: # of unsaturation.
[b] Estolide number determined by NMR.
[c] ASTM D 97-96a [48].
[d] ASTM D 2500-99 [49].
[e] ASTM D 445-97 [50].
[f] ASTM D 2270-93 [51].
[g] M, mono-capped; F, fully capped.
[h] Hydrogenated material.

functionalities capped) TG-estolides according to the reaction conditions depicted in Figure 9.5 [32]. Triglyceride estolides (Tables 9.2 and 9.3) were synthesized from a series of even-numbered carbon saturated FAs from acetic (C2) to stearic (C18) and also unsaturated oleic acid. Owing to the differences in physical properties of the capping FAs, the TG-estolides required different procedures to utilize the wide range of capping materials (C2–C18).

TABLE 9.3

Physical Properties of Castor TG-Estolides (See Figure 9.5 for a General Synthetic Scheme)

TG-Estolides	Fatty Acid[a]	EN[b]	Pour Point[c] (°C)	Cloud Point[d] (°C)	Vis[e] @40°C (cSt)	Vis[e] @100°C (cSt)	Viscosity Index[f]
Castor	NA	NA	−15	−34	260.4	20.1	89
C2-M[g]	C2:0	0.82	−24	<−24	147.4	16.8	122
C2-F[g]	C2:0	2.59	−27	<−27	110.0	15.6	150
C4-M	C4:0	0.93	−27	<−27	117.4	15.4	138
C4-F	C4:0	2.70	−33	−30	82.2	14.0	177
C6-M	C6:0	0.80	−36	<−36	133.0	19.4	167
C6-F	C6:0	1.67	−45	<−45	79.0	17.0	234
C8-M	C8:0	0.98	−21	<−21	203.0	21.1	123
C8-F	C8:0	2.54	−36	<−36	105.9	16.8	172
C10-M	C10:0	1.04	−27	<−27	183.8	21.0	135
C10-F	C10:0	2.34	−36	<−36	91.8	16.2	191
C12-M	C12:0	0.92	−27	<−27	193.1	21.4	132
C12-F	C12:0	2.36	−33	<−33	120.0	19.1	181
H₂-C12-M	H[h] C12-M	1.19	mp 24–36	N.A.	N.A.	N.A.	N.A.
H₂-C12-F	H[h] C12-M	2.35	−3	21	161.2	22.1	164
C14-M	C14:0	1.11	−24	−17	223.4	25.6	146
C14-F	C14:0	2.69	−18	−7	155.9	23.2	179
C16-M	C16:0	0.81	−18	−3	220.6	24.1	137
C16-F	C16:0	2.69	3	21	177.8	26.1	182
C18-M	C18:0	2.10	9	27	226.8	26.6	151
C18-F	C18:0	2.42	18	24	174.5	25.7	182
C18:1-M	C18:1	1.55	−33	<−50	186.9	23.5	154
C18:1-F	C18:1	2.69	−27	−27	131.8	21.9	195

Note: N.A., not available.

[a] Carbon length: # of unsaturation.
[b] Estolide number determined by NMR.
[c] ASTM D 97-96a [48].
[d] ASTM D 2500-99 [49].
[e] ASTM D 445-97 [50].
[f] ASTM D 2270-93 [51].
[g] M, mono-capped; F, fully capped.
[h] Hydrogenated material.

9.3.1.3 Saturated and Unsaturated Lesquerella or Castor Estolide Esters

The hydroxyl-TGs (Table 9.4) of lesquerella and castor oil underwent acid-catalyzed transesterification reactions with 2-ethylhexanol to yield the 2-ethylhexyl (2-EH) lesquerella or castor esters [32]. The isolated hydroxy fatty esters underwent esterification with either saturated or unsaturated FAs in the presence of heat and reduced pressure, which generated a series of lesquerella and castor estolides (Figure 9.7).

TABLE 9.4

Physical Properties of Unsaturated and Saturated (Hydrogenated) Castor and Lesquerella-Based Estolide Esters (See Figure 9.7 for Synthetic Scheme)

Estolide	Capping FA	Hydroxy Fatty Acid Ester[a]	Product Name	Pour Point[b] (°C)	Cloud Point[c] (°C)	Vis[d] @40°C (cSt)	Vis[d] @100°C (cSt)	Viscosity Index[e]	Gardner Color
1	Oleic	Castor	Oleic-cas	−54	<−54	34.5	7.6	196	2+
2	Stearic	Castor	Stea-cas	3	23	41.7	8.6	191	8−
3	Coco[f]	Castor	Coco-cas	−36	−30	29.0	6.5	186	6−
4	2-EH acid[g]	Castor	2-EH-cas	−51	<−51	70.6	11.8	164	13−
5	Oleic	Lesquerella	Oleic-les	−48	−35	35.4	7.8	200	3+
6	Stearic	Lesquerella	Stea-les	3	12	38.6	8.2	195	4−
7	Coco[b]	Lesquerella	Coco-les	−24	<−24	40.4	8.4	192	17
8	2-EH acid[g]	Lesquerella	2-EH-les	−54	<−54	51.1	10.1	189	8−
9	Oleic	Saturated castor	Oleic-H-cas	−36	<−36	68.3	12.2	178	16+
10	Stearic	Saturated castor	Stea-H-cas	6	r.t.	43.6	8.7	186	7+
11	Oleic	Saturated lesquerella	Oleic-H-les	−12	−6	37.0	7.9	196	9+
12	Stearic	Saturated lesquerella	Stea-H-les	6	31	45.7	9.1	187	6+

Note: r.t., room temperature.

a 2-Ethylhexyl ester.
b ASTM D 97-96a [48].
c ASTM D 2500-99 [49].
d ASTM D 445-97 [50].
e ASTM D 2270-93 [51].
f FAs from coconut oil.
g 2-Ethylhexanoic.

TABLE 9.5

Lesquerella Estolides from Lipases

Lipase Type	Medium	% Estolide
C. rugosa	Biphasic	41.3
C. rugosa	Immobilized	9.6
C. rugosa	Reverse micelles	43.7
G. candidum	Biphasic	45.2
P. cyclopium	Biphasic	13.7
A. niger	Biphasic	12.3
Pseudomonas sp.	Biphasic	62.8

Source: Adapted from D.G. Hayes and R. Kleiman,
J. Am. Oil Chem. Soc., 72, 1309–1316, 1995.

9.3.1.4   Saturated and Unsaturated Lesquerella Estolide Esters from Lipases

Hayes and Kleiman reported in 1995 the synthesis of estolides from eight different lipases [34]. Table 9.5 shows the lipases that produced estolides as well as their yields. The lipase synthesis involved a two-step process of estolide formation followed by Lipozyme IM20 (*Mucor miehei* lipase immobilized on a weak anion exchange resin, Novo Nordisk A/S, Bagsveard, Denmark) catalyzed esterification.

9.3.2   Estolide Analysis and Identification

Chemical identification and analysis are very important techniques that make it possible to adequately describe the different series and size of estolide compounds. Under different synthetic conditions, estolides can vary greatly in molecular weight (Figure 9.3), and the ability to characterize these structures is very important to understand the physical properties of the estolides [25–26]. The estolides are easily analyzed by high performance liquid chromatography (HPLC) without the need for further chemical modifications [44]. Estolides can also be easily chemically modified to allow for chemical characterization that will help determine the size and connectivity of the estolides [44].

ENs were determined by gas chromatography (GC) from the SP-2380 column analysis (Supelco, Bellefonte, PA) [40,44] and/or by nuclear magnetic resonance (NMR) as previously described [44]. In the simplest case (Figure 9.3), general oleic estolides have an oleic acid backbone with a terminal FA. ENs are considered to be an average since molecules with a distribution of ENs are produced during the synthesis. Depending on reaction conditions, estolides of different sizes and shapes are possible that will all have different properties, thus, defining their exact size distribution is of utmost importance.

9.3.2.1   Estolide Physical Property Characterization

Estolides, all types, have certain physical characteristics that could help eliminate common problems, such as low resistance to thermal oxidative stability [45] and poor low-temperature properties [46], associated with vegetable oils as functional fluids. Simple oleic estolide esters, when formulated with a small amount of oxidative stability package, show better oxidative stability than both petroleum and vegetable oil-based fluids [47], but there is still room for improvement.

9.3.2.1.1 Gardner Color (AOCS Method Td 1a-64)

One of the most important physical properties to a consumer is the color of the oil or oil-based products. As a potential hydraulic fluid, the estolides need to meet the color requirements of currently used hydraulic fluids. Most consumers have become accustomed to the appearance of their oils, hydraulic fluids, and products. The measurement of the color of a material is designated as the Gardner color [24]. Gardner color scale is a one-dimensional scale for grading color of transparent liquid. The sample is compared against 18 standards that range from lightest yellow (1) to brownish red (18).

Gardner color was measured on a Lovibond three-field comparator from Tintometer Ltd. (Salisbury, England) using American Oil Chemists Society (AOCS) method Td 1a-64 [24]. In many cases, the Gardner color of materials can be susceptible to the interpretation of the recorder; thus, the + and − notation was employed (as in Table 9.4) to designate samples that did not match one particular Gardner color, with an upper limit of 18.

9.3.2.1.2 Pour Point (ASTM Method D 97-96a)

Pour points (PPs) were measured by the American Society for Testing Materials (ASTM) method D 97-96a [48] to an accuracy of ±3°C. The PPs were determined by placing a test jar with 50 mL of the sample into a cylinder submerged in a cooling medium. The sample temperature was reduced in 3°C increments at the top of the sample until the material stopped pouring. The sample no longer poured when the material in the test jar did not flow when held in a horizontal position for 5 s. The temperature of the cooling medium was chosen based on the expected PP of the material. Samples with PP that ranged from +9°C to −6°C, −6°C to −24°C, and −24°C to −42°C were placed in baths of temperatures −18°C, −33°C, and −51°C, respectively. The PP was defined as the lowest temperature at which the sample still poured. All PPs were determined in duplicate and average values were reported.

9.3.2.1.3 Cloud Point (ASTM Method D 2500-99)

Cloud points (CPs) were determined by ASTM method D 2500-99 [49] to an accuracy of ±1°C. The CPs were determined by placing a test jar with 50 mL of the sample into a cylinder submerged into a cooling medium. The sample temperature was reduced in 1°C increments until any cloudiness was observed at the bottom of the test jar. The temperature of the cooling medium was chosen based on the expected CP of the material. Samples with CPs that ranged from room temperature to 10°C, 9°C to −6°C, −6°C to −24°C, and −24°C to −42°C were placed in baths of temperatures 0°C, −18°C, −33°C, and −51°C, respectively. All CPs were determined in duplicate and average values were reported.

9.3.2.1.4 Viscosity and Viscosity Index (ASTM Methods D 445-97 and ASTM D 2270-93)

Viscosity measurements were made using calibrated Cannon-Fenske viscometer tubes purchased from Cannon Instrument Co. (State College, PA). Viscosity measurements were made in a Temp-Trol (Precision Scientific, Chicago, IL) viscometer bath set at 40.0°C and 100.0°C. Viscosity and viscosity index (VI) were calculated using ASTM methods D 445-97 [50] and D 2270-93 [51], respectively. All viscosity measurements were run in duplicate and the average value was reported.

9.3.2.1.5 Oxidative Stability (ASTM Method D 2272-98)

Rotating bomb oxidation tests were conducted on a rotating pressurized vessel oxidation test (RPVOT) apparatus manufactured by Koehler (Bohemia, NY) using ASTM method D 2272-98 [52]. Estolides and commercial products were tested at 150°C. Samples were measured to $50.0 + 0.5$ g with 5.0 mL of reagent water added to the sample. The copper catalyst used was 3 m long and polished with 220 grit silicon carbide sandpaper produced by Abrasive Leaders and Innovators (Fairborn, OH) and was used immediately. The wire was wound to have an outside diameter of 44–48 mm and a weight of $55.6 + 0.3$ g and a height of 40–42 mm. The bomb was assembled and slowly purged with oxygen twice. The bomb was charged with $90.0 + 0.5$ psi (620 kPa) of oxygen and then tested for leaks by immersing in water. The test was completed after the pressure dropped more than 25.4 psi (175 kPa) from the maximum pressure. All the samples were tested in duplicate runs and the average time is reported here.

9.3.2.1.6 High-Frequency Reciprocating Rig Test (ASTM D 6079-11)

High-frequency reciprocating rig (HFRR) tests were done using ASTM D 6079-11 [53]. The HFRR test uses a weighted steel ball and a stationary steel disk that is completely submerged in a test sample. The ball and disk are heated to 60°C and brought into contact with each other and the entire apparatus is vibrated at 50 Hz for 75 min. The diameter of the wear scar left on the ball is measured under a microscope; this value is reported as the HFRR test result.

9.4 POTENTIAL APPLICATIONS OF LESQUERELLA-BASED OILS

Estolides have certain physical characteristics that could help eliminate common problems associated with the application of vegetable oils as functional fluids, such as low thermal oxidative stability [45] and poor low-temperature properties [46,54]. Simple oleic estolide esters, when formulated with a small amount of oxidative stability package, show better oxidative stability than both petroleum or vegetable oil-based fluids [47], but there is still room for improvement. There are a number of ways to improve the oxidative stability of oils. Akoh [55] reported that refined soybean oil has an oxidative stability index (OSI) of 9.4 h at 110°C, but once the oil is partially hydrogenated, the OSI increases to 15.3 h at 110°C, an improvement of more than 60%. The same approach was taken with the oleic estolides, where hydrogenation with 2% w/w of 10% palladium on activated carbon as catalyst gave completely saturated estolides [26]. The saturated oleic estolides were expected to be more oxidatively stable than the unsaturated estolides (assuming the same trend displayed by soybean oil held true). However, as was shown by Isbell and coworkers [26], saturated estolide synthesized via hydrogenation had a PP (ASTM D 97-96a [48]) of −9°C, which was unsatisfactory for many functional fluids. Cermak and Isbell [40] envisioned a new class of saturated estolides with superior oxidative stability and low-temperature properties.

To date, some of the estolides synthesized that best address low-temperature properties have been made from a mixture of saturated and unsaturated FAs [25,31,33]. When saturated FAs are added to the estolide synthesis, a saturated-capped estolide is formed. These estolides have an oleic acid backbone with a terminal saturated tallow FA acting as a capping group. Cermak and Isbell [25] theorized that by varying the capping material on the estolide, the crystal lattice structure of the material was disrupted as it approached its PP, which led to estolide esters with excellent low-temperature properties: PP of −36°C and CP (ASTM D 2500-99 [49]) of −41°C. These saturated-capped estolide 2-EH esters thus have eliminated the common problems associated with the use of vegetable oils as functional fluids. To date, all the different types of estolide FAs and estolide esters have PP that compared favorably with commercially available industrial products such as petroleum-based hydraulic fluids, soy-based fluids, and petroleum oils [25,31,33].

9.4.1 BIODIESEL APPLICATION

Vegetable oils, animal fats, or waste oils are converted into biodiesel through methanolysis or transesterification. Figure 9.8 shows the flowchart for the production of biodiesel from vegetable oil by transesterification [56]. Biodiesel is an attractive and environmentally friendly alternative to petrodiesel for combustion in compression–ignition

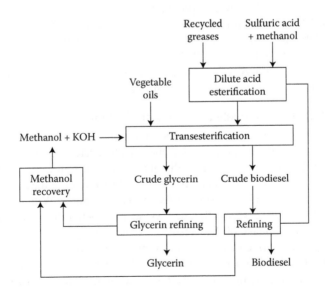

FIGURE 9.8 Flowchart for biodiesel production through methanolysis. (Adapted from U.S. Department of Energy, viewed January 6, 2012. http://www.afdc.energy.gov/afdc/fuels/biodiesel_production.html.)

(diesel) engines because it is biodegradable and nontoxic, renewable, and reduces over-all exhaust emissions. In addition, biodiesel has a higher flash point but similar viscos-ity, cetane number, and gross heat of combustion as petrodiesel [57–59].

Increasingly strict regulations to reduce the sulfur content of commercial petro-leum diesel fuels have resulted in reduced lubricity of these fuels. This can be dam-aging to the engine and fuel injection systems that use low-sulfur fuels. Vegetable oil-based diesel fuel additives, derivatives, and substitutes may be a potential solution to this emerging problem. Previous studies [60–62] have shown that acid esters of TGs derived from vegetable oils have increased diesel fuel lubricity at concentrations of less than 1%. These mixtures of fatty acid methyl esters (FAMEs), commonly known as biodiesel, provide a clean effective fuel for diesel engines. In addition, when methyl esters of vegetable oil are added in concentrations of less than 5%, significant increases in the lubricity of diesel have been observed [62,63]. The lubric-ity property has become more and more of an issue as the United States Federal Government has mandated regulations on the composition of on-highway diesel fuel. The regulation requires a decrease in the sulfur content of diesel fuels in ultralow sulfur diesel fuel (ULSD) to ≤ 15 ppm. This reduction of sulfur in the fuel leads to reduced lubricity and low-temperature performance of the fuel as compared with the previous fuels with higher sulfur content [61]. Reduced lubricity leads to increased wear and damage on the engine, thus shortening its life and increasing repair costs to the consumer. Finding fuels or additives that will help solve this lubricity problem without increasing sulfur exhaust emissions has become an important research area [59,60,63,64]. These problems with ULSD are currently addressed through the use of additives. Unfortunately, additives usually increase costs but provide only slight improvement in lubricity.

In the past, castor oil has been reported to have improved lubricity over other oils with similar FA carbon chain length [65]. Castor and lesquerella methyl esters were explored as lubricity enhancers for diesel fuel and were compared with other methyl esters by Goodrum and Geller [63]. The HFRR test (ASTM D 6079-11 [53]) was used in the investigation. A series of tests were performed using different concentra-tions of castor and lesquerella oil esters in mixtures with diesel fuel.

9.4.1.1 Methyl Lesquerolate as a Biodiesel Additive

Goodrum and Geller [63] previously reported the lubricity of mixtures of diesel fuel with castor, lesquerella, rapeseed, and soybean methyl esters. Each ester was added to a reference diesel fuel, 0.05% sulfur diesel type 2, at the following concentrations: 0.25%, 0.50%, and 3.00% on a mass basis. The 100% reference diesel was tested in each analysis and the value is given in the respective figures. With the HFRR test (ASTM D 6079-11 [53]), the smaller the wear scar, the better the performance of the lubricant/fuel. A summary of their results is shown in Figure 9.9. They found that castor oil methyl esters enhanced lubricity most effectively at all concentrations. It was also noted that lesquerella oil methyl esters exhibited very similar lubricity behavior to castor oil methyl esters. Both showed improvement consistent with the recommended ISO 450 µm wear scar limit at concentrations of 0.25%. For soybean oil or rapeseed oil methyl esters, the recommended wear scar limit was met when the amounts added were >0.5%.

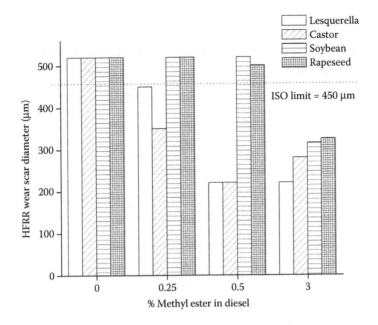

FIGURE 9.9 Effect of concentration of castor, lesquerella, soybean, and rapeseed methyl esters on wear scar diameter from HFRR ASTM D 6079 [53,63].

The lubricity properties between castor and lesquerella oils versus soy and rapeseed oils are very dramatic but explainable. The FA compositions of these oils are compared in Table 9.6. Rapeseed and soy oils have relatively similar chemical compositions. This is especially obvious when one looks at the concentration of unsaturated compounds such as oleic acid. In contrast to this, castor and lesquerella oils are significantly different due to the presence of HFAs such as ricinoleic acid and dihydroxystearic acid in castor oil and lesquerolic acid in lesquerella oil. Not only is ricinoleic acid unique to castor oil, but it also comprises nearly all of the oil. Likewise, lesquerolic acid comprises over half of lesquerella oil composition. In addition, both of these major components are unsaturated as well as hydroxylated. This difference in the chemical composition probably had a significant effect on the lubricating properties of these esters [65]. HFRR values for castor and lesquerella oil ester-enhanced diesel fuel were consistently lower than those for fuel with soy and rapeseed ester additives, indicating that these oils perform better as lubricity enhancers.

9.4.1.2 Lesquerella Estolides and Methyl Esters as a ULSD Additive

The utilization of ULSD fuel in the United States has necessitated further development of lubricity-enhancing additives to satisfy wear requirements under ASTM D 975 [66] method specifications (520 µm maximum wear scar diameter by HFRR at 60°C, Table 9.7). Since vegetable oils and derivatives thereof, such as CP #1–CP #4 shown in Figure 9.10, exhibit excellent lubrication characteristics (Table 9.8), CP #1–CP #4 were evaluated in ULSD as potential lubricity enhancers by Moser and coworkers [59]. As shown in Table 9.9, all materials imparted significantly enhanced lubricity

TABLE 9.6
Fatty Acid Profile (% w/w) of Methyl Esters from Different Vegetable Oils

Fatty Acid		Vegetable Oils			
Carbon:unsaturation	Name	Castor	Lesquerella	Rapeseed	Soybean
C14:0	Myristic	n.d.	n.d.	n.d.	0.56
C14:1	Myristoleic	n.d.	n.d.	n.d.	0.18
C16:0	Palmitic	0.86	1.00	2.70	14.17
C16:1	Palmitoleic	n.d.	0.60	n.d.	1.27
C16:2	Hexadecanoic	n.d.	n.d.	n.d.	0.24
C18:0	Stearic	1.01	1.70	0.90	5.19
C18:0, 2OH	Densipolic	0.70	n.d.	n.d.	n.d.
C18:1	Oleic	2.63	16.70	12.60	48.20
C18:1, OH	Ricinoleic	89.54	0.50	n.d.	n.d.
C18:2	Linoleic	4.10	6.80	12.10	22.19
C18:3	Linolenic	0.36	11.40	8.00	1.45
C18:4	Stearidonic	0.29	n.d.	n.d.	n.d.
C20:0	Arachidic	0.16	0.80	n.d.	0.28
C20:1	Eicosenoic	0.35	n.d.	7.40	n.d.
C20:1, OH	Lesquerolic	n.d.	56.30	n.d.	n.d.
C20:2, OH	Auricolic	n.d.	3.50	n.d.	n.d.
C22:0	Behenic	n.d.	n.d.	0.70	n.d.
C22:1	Erucic	n.d.	n.d.	49.80	n.d.

Source:　Adapted from J.W. Goodrum and D.P. Geller, *Bioresources Technol.*, 96, 851–855, 2005.
Note:　n.d., not detected.

TABLE 9.7
Selected Biodiesel and Ultralow Sulfur Diesel (ULSD) Fuel Specifications

	Biodiesel		ULSD
	ASTM D 6751	EN 14214	ASTM D 975
Kinematic viscosity[a] (mm²/s, 40°C)	1.9–6.0	3.5–5.0	1.9–4.1
Cloud point[b] (°C)	Report	—	Depends[e]
Pour point[c] (°C)	—	—	Depends[e]
Wear scar diameter[d] (μm)	—	—	520 max

Source:　Adapted from B.R. Moser, S.C. Cermak, and T.A. Isbell, *Energy Fuels*, 22, 1349–1352, 2008.
[a]　ASTM D 445-97 [50].
[b]　ASTM D 2500-99 [49].
[c]　ASTM D 97-96a [48].
[d]　ASTM D 6079-11 [53].
[e]　On location.

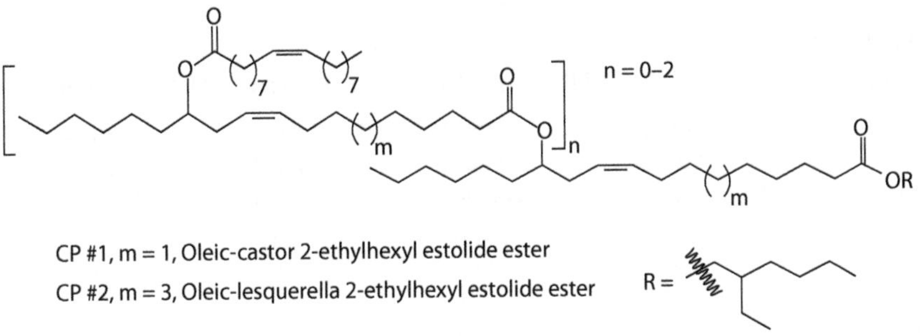

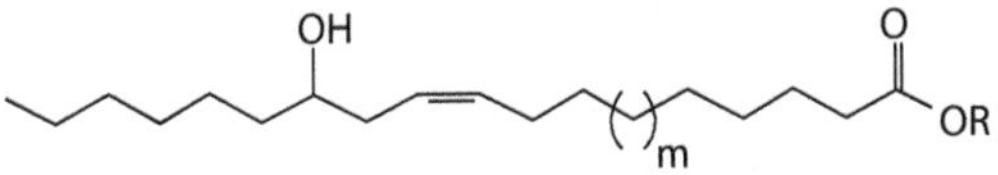

CP #1, m = 1, Oleic-castor 2-ethylhexyl estolide ester

CP #2, m = 3, Oleic-lesquerella 2-ethylhexyl estolide ester

CP #3, m = 1,2-Ethylhexyl castorate ester

CP #4, m = 3,2-Ethylhexyl lesquerellate ester

FIGURE 9.10 Structures of lesquerella and castor estolides and methyl esters used in fuel property testing.

TABLE 9.8
Fuel Properties[a] of Neat Materials (See Figure 9.10)

Material Number	Material Name	CP[b] (°C)	PP[c] (°C)	Visc[d] (mm²/s)	WSD[e] (μm)
Castor estolide, CP #1	Oleic-castor 2-ethylhexyl estolide ester	−45	−54	36.24	180
Lesquerella estolide, CP #2	Oleic-lesquerella 2-ethylhexyl estolide ester	−35	−48	26.56	133
Castor ester, CP #3	2-Ethylhexyl castorate ester	−27	−33	22.03	201
Lesquerella ester, CP #4	2-Ethylhexyl lesquerellate ester	−25	−27	15.41	161
ULSD	Ultralow sulfur diesel	−12	−21	2.32	551

Source: Adapted from B.R. Moser, S.C. Cermak, and T.A. Isbell, *Energy Fuels*, 22, 1349–1352, 2008.

[a] CP, cloud point; PP, pour point; Visc, kinematic viscosity at 40°C; WSD, wear scar diameter measurement from HFRR test.

[b] ASTM D 2500-99 [49].

[c] ASTM D 97-96a [48].

[d] ASTM D 445-97 [50].

[e] ASTM D 6079-11 [53].

TABLE 9.9

Effect of Compound (CP) in ULSD on Wear Scar (μm) for HFRR[a] Measurement

wt% (Material #)[b]	CP#1 Estolide Castor	CP#2 Estolide Lesquerella	CP#3 FAME Castor	CP#4 FAME Lesquerella
0	551	551	551	551
0.1	399	349	393	420
0.5	311	306	267	312
1.0	272	270	189	198
2.0	243	208	180	188

Source: Adapted from ASTM (D 445-97), *Standard Test Method for Kinematic Viscosity of Transparent and Opaque Liquids (the Calculation of Dynamic Viscosity)*, 1997.

[a] ASTM D 6079-11 [53].
[b] See Figure 9.10 for structures.

to the ULSD. In particular, esters CP #3 and CP #4 yielded wear scar data (HFRR, 60°C) at ≥1 wt% in ULSD (<200 μm). Solids CP #1 and CP #2 at similar concentrations provided wear scar diameters between 200 and 280 μm, which was a substantial improvement over neat ULSD (551 μm, Table 9.9). The neat ULSD failed to meet the lubricity requirement specified in ASTM D 975 (Table 9.7). Although pure estolides (CP #1 and CP #2) exhibit superior lubricity compared with pure esters (CP #3 and CP #4) (Table 9.8), this trend was reversed when blended (wt%) with ULSD (Table 9.9), which is attributable to the considerably higher molecular weights of estolides CP #1 and CP #2 in comparison with esters CP #3 and CP #4. In effect, although similar weight percentages of materials were used, the concentrations in moles/L of estolides CP #1 and CP #2 were far less than esters CP #3 and CP #4 in ULSD due to differences in the molecular weight of the estolides as compared with the esters. Thus, fewer molecules of CP #1 and CP #2 were available to influence the lubricity of ULSD when compared with CP #3 and CP #4 at similar weight percentages.

The low-temperature performance and kinematic viscosity (40°C) of CP #1 through CP #4 in ULSD were also measured and reported [59]. Although the neat CP #1 through CP #4 displayed considerably lower CP and PP values than ULSD, at the concentration tested in ULSD (>2 wt%), these materials failed to produce a positive impact on the low-temperature performance of ULSD, that is, unchanged CP and PP values as reported. Also, addition of up to 2% CP #1 through CP #4 in ULSD resulted in minor increase in kinematic viscosity. However, all samples remained within ASTM D 975 kinematic viscosity specifications for ULSD given in Table 9.7.

From the results reported above, the following conclusions were made by Moser and coworkers [59]. The estolide and 2-EH esters of both castor and lesquerella significantly improved the lubricity of ULSD at low (≤2%) blend concentrations. The estolides and methyl esters were superior to both soybean oil methyl esters and rapeseed methyl esters as lubricity enhancers at the same weight percent concentrations [59]. The estolide and 2-EH esters were ineffective as either CP or PP depressants in ULSD. However, CP #1 through CP #4 did not adversely affect the CP or PP of the ULSD; they may prove useful as substitutes for biodiesel as lubricity enhancers in ULSD. Finally, the neat CP #1 through CP #4 exhibited high kinematic viscosities. However, at low blend levels of CP #1 through CP #4 in ULSD, the kinematic viscosity remained within ULSD specification. As the concentration of CP #1 through CP #4 was increased in ULSD, only small increases in kinematic viscosity were observed.

9.4.2 FUNCTIONAL FLUID APPLICATIONS

Consumers today are demanding more from their automobile lubricants than ever before. The use of renewable lubricants can meet these demands and at the same time lessen the demand for foreign oil. All types of estolides as functional fluids have shown great promise as cosmetics, coatings, and biodegradable lubricants. Estolides and their coproducts compare favorably with commercially available industrial products such as petroleum-based hydraulic fluids, soybean-based fluids, and petroleum oils, and usually outperform the competition.

9.4.2.1 Cold Weather Lubricants

9.4.2.1.1 Pour Point Evaluation of Castor and Lesquerella Estolide 2-Ethylhexyl Esters

Castor and lesquerella estolide 2-EH esters, where the castor and lesquerella base units were unsaturated (Figure 9.7), produced estolides that had the lowest PP when capped with oleic acid (PP = −54°C and PP = −48°C, respectively) or with a branched material, 2-ethylhexanoic acid (PP = −51°C and PP = −54°C, respectively) (Table 9.4). By capping the FA esters of castor and lesquerella, the compounds no longer had the opportunity to undergo either intra- or intermolecular hydrogen bonding interactions, thus yielding a lower PP. As the capping material was changed to a saturated FA, dramatic changes occurred in the physical properties, stearic (PP = 3°C for castor and PP = 3°C for lesquerella) and coco (PP = −36°C for castor and PP = −24°C for lesquerella), respectively (Table 9.4). The stearic group had higher PPs than the base FA esters because the long saturated alkyl group allows for sufficient alkyl stacking. With the shorter branched-chain 2-EH, the opposite trend was observed with both castor and lesquerella estolides (Table 9.4, estolide entry 4 and 8). The shorter branched chains disrupt the stacking interactions and produce PPs that are considerably lower than the underivatized FA esters [33]. It has been previously demonstrated that coco-capped estolides have beneficial effects on the low-temperature properties of estolides [25,31].

As the base of the estolide was hydrogenated, castor and lesquerella FA esters had PPs of 9°C and 15°C, respectively [33]. These saturated base unit esters were combined with either oleic or stearic acid. The oleic-capped estolide esters had reasonable PPs (PP = −36°C for castor and PP = −12°C for lesquerella) whereas the stearic-capped materials allowed for sufficient alkyl stacking, thus producing higher PPs (Table 9.4).

Some of the estolide esters synthesized from castor and lesquerella FAs have very low PPs (<−50°C) (Figure 9.11). With respect to the castor- and lesquerella-based estolides, the general trend for the lowest PP increases in the

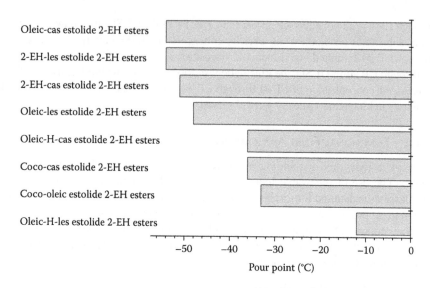

FIGURE 9.11 Pour point (ASTM D 97-96a [48]) comparison of hydrogenated (H) and non-hydrogenated lesquerella (Les) and castor (Cas) estolides 2-EH esters.

order oleic-cas ~ 2-EH-cas ~ 2-EH-les ~ les-cas < coco-cas ~ oleic-H-cas < coco-les < oleic-H-les as shown in Table 9.4.

9.4.2.1.2 Pour Points of Castor and Lesquerella TG-Estolides

The hydroxy TGs of lesquerella and castor oil were converted into their corresponding mono- and fully capped estolides according to the reaction conditions depicted in Figure 9.5. Lesquerella and castor vegetable oils have PPs of −21°C and −15°C, respectively [37]. The fully capped lesquerella TG-estolides had slightly lower PPs than the mono-capped estolides in all cases. The lowest PPs were observed for both the mono- (PP = −33°C) and fully capped lesquerella TG-estolides (PP = −36°C) when the capping group was hexanoic (Table 9.2, L6-M and L6-F). When the saturated capping chain was stearic, the materials were semisolid. The fully capped castor TG-estolides also had lower PPs than the mono-capped materials for all saturated FA groups shorter than lauric. The lowest PPs observed for both mono- (PP = −36°C) and fully capped castor TG-estolides (PP = −45°C) were also seen when the capping group was hexanoic (Table 9.3, C6-M and C6-F). The stearic fully capped castor TG-estolides were also solids at room temperature. The oleic-capped TG-estolides for both lesquerella and castor had PPs of about −30°C.

The longer-chain capping groups provided sufficient alkyl stacking to yield PPs that were higher than the TGs (lesquerella and castor oils). The shorter-chain groups appear to disrupt the TG stacking interactions to yield PPs that are considerably lower than the underivatized oils (les, cas; PP = −21°C and −15°C, respectively). Interestingly, the acetic-capped TG-estolides with their two carbon capping units do not appear to significantly disrupt the stacking interactions in the oil and have the same PP (−21°C) as lesquerella oil. Fully capping the lesquerella oil removed both the intra- and intermolecular hydrogen bonding interactions, yielding a lower

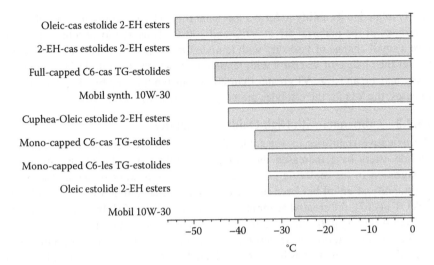

FIGURE 9.12   Pour point (ASTM D 97-96a [48]) comparison of various estolides and commercial engine oils.

PP. When oleic was used as the capping FA, the unsaturation of the oleate improved the PP, resulting in considerably lower PP temperatures compared with the stearic. Thus, the PP decreased from 24°C for stearic fully capped to −27°C for the oleic fully capped lesquerella TG-estolides (Table 9.2, L18-F and L18:1-F).

The lesquerella and castor TG-estolides compared favorably with previously synthesized estolides [37]. In terms of PPs, the TG-estolides were equal to the saturated-capped estolide 2-EH ester series. Figure 9.12 shows the PPs of the TG-estolides that are not some of the best low-temperature performers to date, this allows the lesquerella and castor estolide 2-EH esters to be the estolides with the lowest PPs to date.

9.4.2.2   Lubricants with Superior Viscosity Properties

9.4.2.2.1   Viscosities of Castor and Lesquerella TG-Estolides

The two main factors in the TG-estolide series of homologous compounds that affect viscosity are hydrogen bonding by the hydroxyl groups and the steric bulk of the molecules. Intermolecular hydrogen bonding between the hydroxyl TGs increases the steric bulk of the molecules by creating weakly associated dimers, trimers, and so on of the original TGs, thereby increasing the viscosity of the oil. In all cases, the fully capped TG-estolides have lower viscosities for all chain lengths than the mono-capped estolides (Tables 9.2 and 9.3). Capping of the hydroxyl moiety reduces or eliminates hydrogen bonding and, consequently, reduces the viscosity of the oil. The viscosity trend shows that viscosity increases linearly with increasing chain length of the capping FA.

9.4.2.2.2   Viscosity Index

VI is a term used for a lubricating oil quality indicator. It is a measure of the change of kinematic viscosity with change in temperature. The viscosity of a lubricant is closely related to its ability to form a lubricant film. The VI represents how close the viscosities of the material are at 40°C and 100°C. Thus, the ideal lubricant would have a small change of viscosity with temperatures. The lower the VI numbers, the greater the change of viscosity with temperature, which is not desirable where viscosity may be an issue. VI greater than 200 is outstanding and highly preferable. In all cases, the VIs for the estolides are very high (>150 to 200+).

9.4.2.3 Lubricants with High Oxidative Stability

The oxidative stabilities of vegetable oils in general are very poor, whereas past estolide esters have increased the oxidative stability by reducing the amount of unsaturation in the molecule [33]. There are numerous ways by which the oxidative stability of an oil has been measured. Some of the common ways are OSI [55], RPVOT [47,67], differential scanning calorimetry (DSC) [68], Indiana stirring oxidation test (ISOT) [69], and the thin-film microoxidation test [70].

The original estolides were developed as an industrial base oil or as an industrial oil additive, so the materials had to be evaluated under conditions commonly associated with industrial-based materials. These industrial oils will replace petroleum oils and by-products for which the recommended tests are generally microoxidation or RPVOT. The ASTM has developed detailed test procedures for these materials. For the RPVOT, the time to failure is reported in minutes; failure is identified as a pressure drop of 175 kPa from the maximum recorded pressure. The longer the RPVOT time, the better the oxidative stability of the material. Typical commercial engine oils have RPVOT times in the range of 210–250 min (Table 9.10). The RPVOT test method calls for the oil to be tested with materials that would be present in most applications such as water, copper, and oxygen. The test has been accepted for bio-based material producers as a suitable method to test the oxidative stability of these fluids. With vegetable-based materials, the RPVOT method tests both thermal oxidative stability and hydrolytic stability [67].

9.4.2.3.1 Oxidative Stability: Lesquerella and Castor Estolide 2-EH Esters

Estolide esters synthesized from HFAs (Figure 9.7) do not decrease the unsaturation in the molecule; therefore, improvements in the oxidative stability from the formation of estolides are unlikely. At the time of this study, lesquerella oil was available only in a limited quantity, so most large-scale destructive tests were conducted on castor versions. Castor estolides have somewhat similar FA profiles as lesquerella (Table 9.6) in regard to the effects on oxidative stabilities. This hypothesis proved to be correct when pure castor-oleic estolide 2-EH ester was tested for oxidative stability and displayed RPVOT time of 15 min (Figure 9.13). The RPVOT times for the estolides tested increased as the concentration of the antioxidant package was increased (Figure 9.13). The unsaturated castor-based oleic estolide ester with an antioxidant stability package concentration of 3.5% compared favorably with commercially available petroleum-based oils (Table 9.10). Increased RPVOT times were observed when the "capping" unit was replaced with saturated FAs (e.g., coco) (Figure 9.7). Addition of 3.5% oxidative stability package in coco–castor estolide 2-EH ester resulted in RPVOT time that exceeded most commercially available oils (Figure 9.13).

TABLE 9.10
RPVOT[a] Values of Select Common Functional Fluids

Fluid	Time (min)
Aeroshell® 15W-50 aviation oil	552
Biosoy®	28
Castrol® synthetic 10W-30	246
Crambe oil[b]	13
Environlogic®-132 Terrsolve	67
Environlogic-146 Terrsolve	51
Environlogic-168 Terrsolve	71
Meadowfoam oil[b]—crude	20
Soybean oil[b]	13
Soylink®	83
Traveller® All Season H.F.[c] IVG-46	274
Traveller Premium Universal H.F.[c] IVG-46	464
Valvoline® 5W-30	228
Valvoline 10W-30	223
Valvoline 10W-40	224
Valvoline 20W-50	214
Valvoline SAE-30	224

[a] ASTM D 2272-98 [52].
[b] Unformulated/research sample.
[c] Hydraulic fluid.

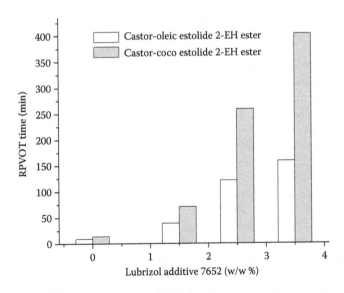

FIGURE 9.13 Effect of antioxidant package concentration on oxidative stability (RPVOT-ASTM D 2272-98 [52]) of castor estolide esters.

TABLE 9.11

Effect of Antioxidant Concentration on RPVOT[a] Values of Lesquerella and Castor TG-Estolides

TG	Lubrizol 7652 Antioxidant Added (wt%)				
Estolide[b]	0	0.5	1	2	4
L12-M	13	17	29	66	118
L12-F	14	22	47	106	158
H_2L12-M	56	n.d.	439	559	541
H_2L12-F	37	n.d.	476	688	668
L18:1-M	16	18	29	34	121
L18:1-F	14	20	31	60	127
C12-M	18	30	44	102	241
C12-F	16	32	46	124	318
H_2C12-M	47	n.d.	142	289	396
H_2C12-L	28	n.d.	242	410	515
C18:1-M	15	23	52	90	197
C18:1-F	14	26	48	103	180

Note: n.d., not determined.

[a] ASTM D 2272-98 [52] and all RPVOT times are reported in minutes.

[b] Estolides from Tables 9.2 and 9.3.

9.4.2.3.2 Oxidative Stability: Lesquerella and Castor TG-Estolides

The oxidative stabilities of vegetable oils are very poor due to the high amount of unsaturation inherently present in the molecule [45]. The formation of estolides from oleic acid reduces the amount of unsaturation in the molecule, which results in improved oxidative stability [35]. The synthesis of TG-estolides from lesquerella and castor oils, however, does not remove any unsaturation (Figure 9.6). Consequently, the RPVOT times for the TG-estolides are still short without antioxidants, approximately 15 min (Table 9.11). The addition of an antioxidant package can improve the oxidative stability of the TG-estolides measured with the RPVOT. However, even at 4% concentration of antioxidant package, the TG-estolides displayed RPVOT times only in the range of 200 min (typical of most mineral oil-based automotive crankcase fluids, Table 9.10).

Hydrogenation of the lauric-capped TG-estolides from both lesquerella and castor greatly improved their RPVOT oxidative stability performance with no antioxidant added, where the lesquerella mono-capped TG-estolide had an RPVOT of 56 min. When 1% antioxidant was added to this material, the RPVOT time increased to 439 min. The castor TG-estolide performance was also improved at the same antioxidant concentration (1%), but the improvement in RPVOT time was smaller. Unfortunately, the PPs of these hydrogenated oils were negatively impacted, since they all formed solids at room temperature.

9.5 CONCLUSIONS

Lesquerella offers a unique oil product along with several coproducts that makes commercialization attractive. Agronomics will play an important role in improving seed yield while reducing production costs. A lesquerella hydroxy oil would reduce dependence on imported castor oil, provide a new crop for diversification in the United States, and provide oil for biobased products.

Estolides have been synthesized from different sources of FAs without catalysts and solvent, to produce materials that have good low-temperature properties. Estolides synthesized from castor and lesquerella FA esters with oleic acid (PP = −54°C and PP = −48°C) and 2-ethylhexanoic acid (PP = −51°C and PP = −54°C), respectively, yielded the best cold flow performing estolides to date. These new estolides from castor and lesquerella FA esters have outperformed commercial products in low-temperature properties even without additives. Generally, as the saturation in the molecule increased, the PPs increased, whereas the RPVOT oxidative stability increased. The key approach to designing high-performing estolides is to maintain some degree of unsaturation to maintain acceptable low-temperature properties without sacrificing oxidative stability.

ACKNOWLEDGMENTS

The authors are extremely grateful to Girma Biresaw for his work and insight in the oxidative stability package for estolides. The authors are also extremely grateful to Kendra B. Brandon, Billy Deadmond, Amy B. Deppe, Amber L. Durham, Jeffery Forrester, Natalie A. LaFranzo, Benjamin A. Lowery, Jonathan L.A. Phillips, Alex L. Skender, Melissa L. Winchell, and other new crop researchers for their assistance with sample preparation, data collection, and synthesis. The authors are grateful to Terry A. Isbell, Bryan R. Moser, and David A. Dierig for their work on the development of lesquerella as an industrial crop.

REFERENCES

1. K.L. Mikolajczak, F.R. Earle, and I.A. Wolff, Search for new industrial oils. VI. Seed oils of genus *Lesquerella*, *J. Am. Oil Chem. Soc.*, 39, 78–80, 1962.
2. A.S. Barclay, H.S. Gentry, and Q. Jones, The search for new industrial crops II: Lesquerella (Cruciferae) as a source of new oilseeds, *Econ. Bot.*, 16(2), 95–100, 1962.
3. T.P. Abbott, D.A. Dierig, M. Foster, J.M. Nelson, W. Coates, H.B. Frykman, K.D. Carlson, and J.D. Arquette, Status of lesquerella as an industrial crop, *Information*, 8, 1169–1175, 1997.
4. C.R. Smith, T.L. Wilson, T.K. Miwa, H. Zobel, R.L. Labmar, and I.A. Wolff, Lesquerolic acid: A new hydroxy acid from *Lesquerella* seed oil, *J. Org. Chem.*, 26, 2903–2905, 1961.
5. C.R. Smith, T.L. Wilson, R.B. Bates, and C.R. Scholfielf, Densipolic acid: A unique hydroxydienoic acid from *Lesquerella densipila* seed oil, *J. Org. Chem.*, 27, 3112–3117, 1962.
6. R. Kleiman, G.F. Spencer, F.R. Earle, H.J. Nieschlag, and A.S. Barckay, Tetra-acid triglycerides containing a new hydroxy eicosanoyl moiety in *Lesquerella auriculata* seed oil, *Lipids*, 7, 660–665, 1972.

7. D.A. Dierig, Lesquerella, *New Crop Factsheet*, Center for New Crops and Plant Products, Purdue University, W. Lafayette, Indiana, 1995. http://www.hort.purdue.edu/newcrop/cropfactsheets/Lesquerella.html.

8. T.A. Isbell, US effort in the development of new crops (lesquerella, pennycress, coriander and cuphea), *Proc. Journées Chevreul.*, 16(4), 205–210, 2009.

9. M.M. Janderek, D.A. Dierig, and T.A. Isbell, Fatty-acid profile of lesquerella germplasm in the National Plant Germplasm System collection, *Ind. Crops Prod.*, 29, 154–164, 2009.

10. H.S. Gentry and A.S. Barclay, The search for new industrial crops III: Prospectus of *Lesquerella fendleri*, *Econ. Bot.*, 16(3), 206–211, 1962.

11. D.A. Dierig, A.E. Thompson, and F.S. Nakayama, Lesquerella commercialization efforts in the United States, *Ind. Crops Prod.*, 1, 280–293, 1993.

12. A.E. Thompson, D.A. Dierig, and E.R. Johnson, Yield potential of *Lesquerella fendleri* (Gray) Wats., a new desert plant resource for hydroxyl fatty acids, *J. Arid Environ.*, 16, 331–336, 1988.

13. R.W. Miller, C.H. Van Etten, and I.A. Wolff, Amino acid composition of lesquerella seed meals, *J. Am. Oil Chem. Soc.*, 39, 115–117, 1962.

14. D.G. Hayes, R. Kleiman, and B.S. Phillips, The triglyceride composition, structure, and presence of estolides in the oils of lesquerella and related species, *J. Am. Oil Chem. Soc.*, 72, 559–569, 1995.

15. D.A. Dierig, G.H. Dahlquist, and P.M. Tomasi, Registration of WCL-L03 high oil *Lesquerella fendleri* germplasm, *Crop Sci.*, 46, 1832–1833, 2006.

16. S.F. Vaughn and M.A. Berhow, Glucosinolate hydrolysis products from various plant sources: pH effects, isolation, and purification, *Ind. Crops Prod.*, 21, 193–202, 2005.

17. D.A. Dierig, G. Wang, W.B. McCloskey, K.R. Thorp, T.A. Isbell, D.T. Ray, and M.A. Foster, Lesquerella: New crop development and commercialization in the U.S., *Ind. Crops Prod.*, 34, 1381–1385, 2011.

18. Lesquerella Task Force, *Lesquerella as a Source of hydroxy Fatty Acids for Industrial Products*, Growing Industrial Materials Series, U.S. Department of Agriculture, Washington, D.C., 1991.

19. A.J. MacLeod, S.S. Panesar, and V. Gil, Thermal degradation of glucosinolates, *Phytochemistry*, 20, 977–980, 1981.

20. K.D. Carlson, R. Kleiman, and L.R. Watkins, Pilot-scale extrusion processing/solvent extraction of lesquerella seed, In *Proceedings of the First International Conference on New Industrial Crops and Products*, H.H. Naqvi, A. Estilai, and I.P. Ting (Eds.), Association for the Advancement of Industrial Crops, Riverside, CA, Oct 8–12, 1990.

21. R.L. Evangelista, Oil extraction from lesquerella seeds by dry extrusion and expelling, *Ind. Crops Prod.*, 29, 189–196, 2009.

22. M.C. Erickson, Chemistry and function of phospholipids, In *Foof Lipids: Chemistry, Nutrition and Biotechnology*, 3rd ed., C.C. Akoh and D.B. Min (Eds.) pp. 39–62, CRC Press, Boca Raton, FL, 2008.

23. K.D. Carlson and R. Kleiman, Degumming and bleaching of *Lesquerella fendleri* seed oil, *J. Am. Oil Chem. Soc.*, 70, 579–582, 1993.

24. Firestone, D. (Ed.), *Official and Tentative Methods of the American Oil Chemists' Society*, 4th ed., AOCS, Champaign, IL, 1994.

25. S.C. Cermak and T.A. Isbell, Physical properties of saturated estolides and their 2-ethylhexyl esters, *Ind. Crops Prod.*, 16, 119–127, 2002.

26. T.A. Isbell, M.R. Edgcomb, and B.A. Lowery, Physical properties of estolides and their ester derivatives, *Ind. Crops Prod.*, 13, 11–20, 2001.

27. S.C. Cermak and T.A. Isbell, Biodegradable oleic estolide ester having saturated fatty acid end group useful as lubricant base stock, U.S. Patent 6,316,649 B1, 2000.

28. R.H. Purdy and C.D. Craig, Meadowfoam: New source of long-chain fatty acids, *J. Am. Oil Chem. Soc.*, 64, 1493–1497, 1987.

29. T.A. Isbell, K. Kleiman, and S.M. Erhan, Characterization of monomers produced from thermal high-pressure conversion of meadowfoam and oleic acids into estolides, *J. Am. Oil Chem. Soc.*, 69, 1177–1183, 1992.
30. R.W. Miller, F.R. Earle, I.A. Wolff, and Q. Jones, Search for new industrial oils. IX. Cuphea, a versatile source of fatty acids, *J. Am. Oil Chem. Soc.*, 41, 279–280, 1964.
31. S.C. Cermak and T.A. Isbell, Synthesis and physical properties of cuphea-oleic estolides and esters, *J. Am. Oil Chem. Soc.*, 81, 297–303, 2004.
32. T.A. Isbell and S.C. Cermak, Synthesis of triglyceride estolides from lesquerella and castor oils, *J. Am. Oil Chem. Soc.*, 79, 1227–1233, 2002.
33. S.C. Cermak, K.B. Brandon, and T.A. Isbell, Synthesis and physical properties of estolides from lesquerella and castor fatty esters, *Ind. Crops Prod.*, 23, 54–64, 2006.
34. D.G. Hayes and R. Kleiman, Lipase-catalyzed synthesis and properties of estolides and their esters, *J. Am. Oil Chem. Soc.*, 72, 1309–1316, 1995.
35. B.H. Zoleski and F.J. Gaetani, Low foaming railroad diesel engine lubricating oil composition, U.S. Patent 4,428,850, 1984.
36. L.A. Nelson, C.M. Pollock, and G.J. Achatz, Secondary alcohol esters of hydroxyacids and uses thereof, U.S. Patent 6,407,272 B1, 2002.
37. T.A. Isbell, B.A. Lowery, S.S. DeKeyser, M.L. Winchell, and S.C. Cermak, Physical properties of triglyceride estolides from lesquerella and castor oils, *Ind. Crops Prod.*, 23, 256–263, 2006.
38. S.S. Lawate, Triglyceride oil thickened with estolides of hydroxyl-containing triglycerides, U.S. Patent 5,427,704, 1995.
39. T.A. Isbell, T.P. Abbott, S. Asadauskas, and J.E. Lohr, Biodegradable oleic estolide ester base stocks and lubricants, U.S. Patent 6,018,063, 2000.
40. S.C. Cermak and T.A. Isbell, Synthesis of estolides from oleic and saturated fatty acids, *J. Am. Oil Chem. Soc.*, 78, 557–565, 2001.
41. T.A. Isbell, Chemistry and physical properties of estolides, *Grasas Y Aceites*, 62, 8–10, 2011.
42. S.C. Cermak, Estolides: Biobased lubricants, In *Surfactants in Tribology*, Volume 2, G. Biresaw and K.L. Mittal (Eds.) pp. 269–320, CRC Press, Boca Raton, FL, 2011.
43. S.C. Cermak, A.L. Skender, A.B. Deppe, and T.A. Isbell, Synthesis and physical properties of tallow-oleic estolide 2-ethylhexyl esters, *J. Am. Oil Chem. Soc.*, 84, 449–456, 2007.
44. T.A. Isbell and R. Kleiman, Characterization of estolides produced from the acid-catalyzed condensation of oleic acid, *J. Am. Oil Chem. Soc.*, 71, 379–383, 1994.
45. R. Becker and A. Knorr, An evaluation of antioxidants for vegetable oils at elevated temperatures, *Lubr. Sci.*, 8, 95–117, 1996.
46. S. Asadauskas and S.Z. Erhan, Depression of pour points of vegetable oils by blending with diluents used for biodegradable lubricants, *J. Am. Oil Chem. Soc.*, 76, 313–316, 1999.
47. S.C. Cermak and T.A. Isbell, Improved oxidative stability of estolide esters, *Ind. Crops Prod.*, 18, 223–230, 2003.
48. ASTM (D 97-96a), *Standard Test Method for Pour Point of Petroleum Products*, 1996.
49. ASTM (D 2500-99), *Standard Test Method for Cloud Point of Petroleum Products*, 1999.
50. ASTM (D 445-97), *Standard Test Method for Kinematic Viscosity of Transparent and Opaque Liquids (the Calculation of Dynamic Viscosity)*, 1997.
51. ASTM (D 2270-93), *Standard Practice for Calculating Viscosity Index from Kinematic Viscosity at 40 and 100°C*, 1993.
52. ASTM (D 2272-98), *Standard Test Method for Oxidation Stability of Steam Turbine Oils by Rotating Pressure Vessel*, 1998.
53. ASTM (D 6079-11), *Standard Test Method for Evaluating Lubricity of Diesel Fuels by the High-Frequency Reciprocating Rig (HFRR)*, 2011.

54. G.R. Zehler, Performance tiering of biodegradable hydraulic fluids, *Lubr. World*, 9, 22–26, 2001.

55. C.C. Akoh, Oxidative stability of fat substitutes and vegetable oils by the oxidative stability index method, *J. Am. Oil Chem. Soc.*, 71, 211–216, 1994.

56. U.S. Department of Energy, viewed January 6, 2012. http://www.afdc.energy.gov/afdc/fuels/biodiesel_production.html.

57. G. Knothe, Analysis of oxidized biodiesel by ^{1}H-NMR and effect of contact area with air, *Eur. J. Lipid Sci. Technol.*, 108, 493–500, 2006.

58. B.R. Moser, M.J. Haas, J.K. Winkler, M.A. Jackson, S.Z. Erhan, and G.R. List, Evaluation of partially hydrogenated methyl esters of soybean oil as biodiesel, *Eur. J. Lipid Sci. Technol.*, 109, 17–24, 2007.

59. B.R. Moser, S.C. Cermak, and T.A. Isbell, Evaluation of castor and lesquerella oil derivates as additives in biodiesel and ultralow sulfur diesel fuels, *Energy Fuels*, 22, 1349–1352, 2008.

60. D.P. Geller and J.W. Goodrum, Effects of specific fatty acid methyl esters on diesel fuel lubricity, *Fuel*, 83, 2351–2356, 2004.

61. G. Anastopoulos, E. Lois, F. Zannikos, S. Kalligeros, and C. Teas, Influence of aceto acetic esters and di-carboxylic acid esters on diesel fuel lubricity, *Tribol. Int.*, 34, 749–755, 2001.

62. J.H. Van Gerpen, S. Soylu, and E.T. Mustafa, Evaluation of the lubricity of soybean oil-based additives in diesel fuel, ASAE Paper No. 996134, 1999.

63. J.W. Goodrum and D.P. Geller, Influence of fatty acid methyl esters from hydroxylated vegetable oils on diesel fuel lubricity, *Bioresources Technol.*, 96, 851–855, 2005.

64. B.R. Moser, G. Knothe, and S.C. Cermak, Biodiesel from meadowfoam (*Limnanthes alba* L.) seed oil: Oxidative stability and unusual fatty acid composition, *Energy Environ. Sci.*, 3, 318–327, 2010.

65. D.C. Drown, K. Harper, and E. Frame, Screening vegetable oil alcohol esters as fuel lubricity enhancers, *J. Am. Oil Chem. Soc.*, 78, 579–584, 2001.

66. ASTM (D 975-11), *Standard Specification for Diesel Fuel Oils*, 2011.

67. S.C. Cermak, G. Biresaw, and T.A. Isbell, Comparison of a new estolide oxidative stability package, *J. Am. Oil Chem. Soc.*, 85, 879–885, 2008.

68. W.F. Bowman and G.W. Stachowiak, Application of sealed capsule differential scanning calorimetry. Part I: Predicting the remaining useful life of industry-used turbine oils, *Lubr. Eng.*, 54, 19–24, 1998.

69. D.C. Du, S.S. Kim, J.S. Chun, C.M. Suh, and W.S. Kwon, Antioxidation synergism between ZnDTC and ZnDDP in mineral oil, *Tribol. Lett.*, 13, 21–27, 2002.

70. S. Asadaushas, J.M. Perez, and J.L. Duda, Lubrication properties of castor oil—Potential basestock for biodegradable lubricants, *Lubr. Eng.*, 53, 35–41, 1997.

10 Green Tannic Acid-Based Surfactants and Their Metal Complexes for Inhibiting Microbial Growth

Nabel A. Negm, Mohamed M. El Sukkary,
and Dina A. Ismail

CONTENTS

This chapter is divided into two sections: The first part reviews the occurrence, extraction, and modification of tannic acid as a green compound for producing surfactants that have the ability to reduce or prevent microbial growth. The second part describes an experimental work dealing with the preparation, characterization, and application of some tannic acid surfactants and their metal complexes as antimicrobial agents. The study also rationalizes the surface activity of the synthesized compounds with their antimicrobial efficacy.

10.1   INTRODUCTION

Tannic acid is a special commercial form of tannin, a type of polyphenol. It is a weak acid ($pK_a \approx 6$) due to the numerous phenol groups in the structure. The chemical formula for commercial tannic acid is often given as $C_{76}H_{52}O_{46}$ (Scheme 10.1), which corresponds with decagalloyl glucose. But, in fact, it is a mixture of polygalloyl glucoses or polygalloyl quinic acid esters with the number of galloyl moieties per molecule ranging from 2 up to 12, depending on the plant source used to extract the tannic acid. The structure of gallic acid is given in Scheme 10.2. Commercial tannic acid is usually extracted from the following plant parts: tara pods (*Caesalpinia spinosa*, a small leguminous tree), gall nuts from *Rhus semialata* and *Quercus infectoria*, or Sicilian Sumac leaves (*Rhus coriaria*, a deciduous shrub to small tree). Tannic acid hydrolyzes into glucose and gallic or ellagic acid units. Tannic acid is odorless but has a very astringent taste. Pure tannic acid is a light yellowish and amorphous powder.

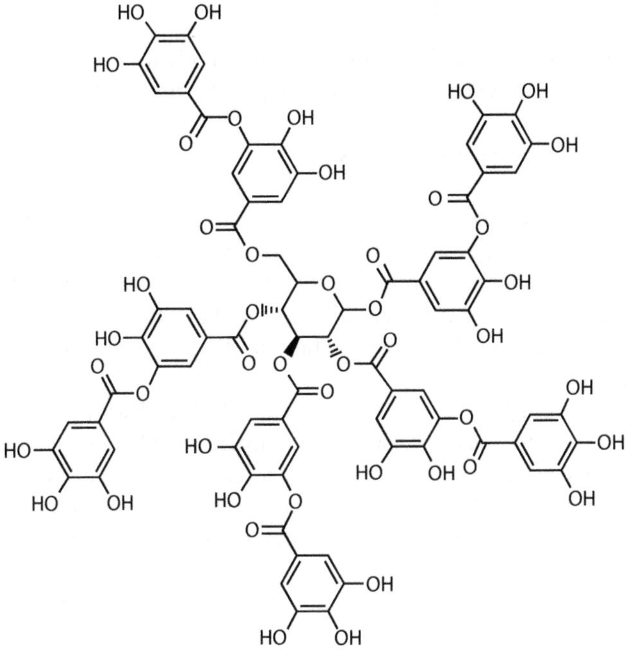

SCHEME 10.1 Structure of tannic acid.

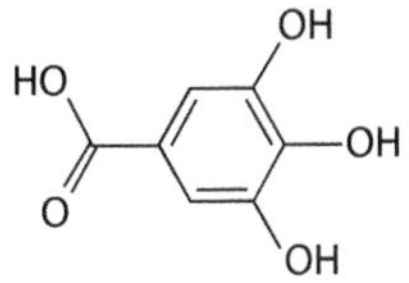

SCHEME 10.2 Structure of gallic acid.

10.1.1 DEFINITION

The name tannin comes from the leather industry where production of leather from hide is called tanning. The plant materials used in tanning contain polyphenols that react with proteins; hence the name tannin. In reality, most of the commercial tannins are developed for leather tanning application.

Plant tannins are water-soluble phenolic compounds with a molar mass between 300 and 3000, showing the usual phenol reactions (e.g., blue color with iron (III) chloride and precipitating alkaloids, gelatin, and other proteins) [1]. The molecular weight of tannins can reach 5000, and they contain sufficient phenolic hydroxyl groups to permit the formation of stable cross-links with proteins and, as a result, cross-linking enzymes may be inhibited. Tannins have also been described as oligomeric compounds with multiple structure units with free phenolic groups, with molecular weight ranging from 500 to >20,000 and are soluble in water except those with very high-molecular-weight structures.

10.1.2   Classification of Tannins

Tannins are a group of phenol compounds found in plants, which produce a group of chemicals called "polyphenols." These polyphenols are, for the most part, soluble in water.

There are two main types of tannins:

1. Hydrolyzable tannins: Tannic acid is a particular type of hydrolyzable (basically means that it can be split up and be broken down by interaction with water) tannin commonly found in the bark and wood of oaks and other plants. It is used commercially in tanning leather and in certain dying processes. The term "hydrolyzable tannin" refers to both ellagitannins and gallotannins.
   a.   Gallotannins: these consist of a sugar substituted with galloyl groups.
   b.   Ellagitannins: these are esters of hexahydroxy diphenoyl (HHDP) groups with a sugar core (usually glucose) and often contain galloyl groups.
   Ellagitannins differ from gallotannins in that their galloyl groups are linked through C–C bonds, whereas the galloyl groups in gallotannins are linked by deepside bonds.
2. Nonhydrolyzable (condensed) tannins: nonhydrolyzable tannins are called condensed tannins (also called "proanthocyanidins"). Condensed tannins are often found in plant sources such as tea, pomegranates, grape seeds, and grape skins.
   Condensed tannins are further subclassified into three subcategories:
   a.   Red skin tannins
   b.   White skin tannins
   c.   Seed tannins

10.1.3   Structure of Tannic Acids

Tannic acids are composed of a central glucose molecule, which is derivatized at its hydroxyl groups with one or more galloyl residues. It was reported that the polyphenolic nature of tannic acid (hydrophobic core and hydrophilic shell) was the feature responsible for its antioxidant properties. Tannins, also referred to as tannic acid, have a structure consisting of a central glucose and 10 galloyl groups. They are a type of water-soluble polyphenol present in the bark and fruits of many plants, and particularly found in bananas, grapes, raisins, sorghum, spinach, red wine, persimmons, coffee, chocolate, and tea.

10.1.4   Applications of Tannins

10.1.4.1   Tanning

Tannins are applied widely for uses ranging from tanning, known over millennia since ca. 1500 BC, to medical uses and to uses in the food industry. The biological significance and current and promising new applications of tannins rely on their complexation with other biopolymers [2]. The structure, stereochemistry, and reactivity of some natural condensed tannins have been described in detail by Roux et al. [3].

The leather tanning industry is one of the oldest industries known. Although the technology of leather has evolved over the years, the basic principles have remained the same. Hide proteins, mainly collagen, are rendered insoluble and dimensionally more stable by treatment with chemical products such as natural vegetable tannins. This produces leather, which is more resistant to mechanical wear and less suscep- tible to biological and other types of attack. In view of the continuing interest in producing high-quality leathers, a new formulation of sulfonated melamine–urea– formaldehyde resin containing different vegetable tannins have been developed. This resin is able to produce leather with the same good characteristics as leather prepared with chrome salts [4]. In addition, Pizzi et al. [5] have developed a cor- relation between the antioxidant property of tannins and the problem of color variation of leather. More recently, a novel transposed process for making leather that provides a tanning process without pickling and basification steps has been developed using basic chromium sulfate, vegetable tannins, aluminum syntan, and chromium–silica.

The pickling process is the lowering of the pH value to the acidic region in the presence of salts. Pickling is normally done to help with the penetration of certain tanning agents, for example, chromium (and other metals) and aldehydic and some polymeric tanning agents, whereas in the basification process, the pH of the float is raised slowly. This basification process fixes the tanning material to the leather.

Sivakumar et al. [6] utilized ultrasound to improve the extraction of myrobalan tannins, leading to higher shrinkage temperature, better diffusion, and subsequent tanning. On the other hand, a high shrinkage temperature could also be achieved by first adding mimosa tannins with oxazolidine. Shrinkage temperature is the temper- ature at which the energy input (heat) exceeds the energy bound in existing hydro- gen bonding of the collagen structure resulting in the decomposition of the helical structure.

10.1.4.2 Adhesives

Applications of tannins as adhesives have been the subject of investigations since the 1950s and the 1960s. It is perhaps the second most important bulk application of tannins. Their application as wood adhesives is based mainly on the reaction, gelling, and hardening of these tannins with formaldehyde. New formulations have been developed and investigated over the years [7]. Recent studies on tannin-based adhesives have included:

1. Rheological studies of various tannin extracts [7]
2. The development of time–temperature–transformation curing diagrams for understanding the behavior of thermosetting resins under isothermal cure conditions and continuous heating transformation [8]
3. Condensed tannin of Douglas fur and polyethylenimine as formaldehyde- free wood adhesives [9]
4. Bonding quality of *Eucalyptus globulus* plywood using tannin–phenol– formaldehyde adhesives [10]

5. Resorcinol–tannin–formaldehyde resin prepared from Taiwan Acacia and China fir bark that had cold setting capability comparable with that of resorcinol formaldehyde resin
6. Solid-phase microextraction (SPME) to analyze birch bark tar
7. Use of mimosa tannins to reduce gelation time

10.1.4.3 Food

Tannic acid has always been used as a food additive. Its safe dosage ranges from 10 to 400 mg per kilogram of food depending on the type of food to which it is added [11]. Also, several authors have demonstrated that tannic acid and other polyphenols have antimutagenic and anticarcinogenic activities.

Recently, the consumption of polyphenol-rich fruits, vegetables, and beverages (tea and red wine) has been credited with inhibitory and preventive effects to various human cancers and cardiovascular diseases. This may be related at least in part to the antioxidant activity of polyphenols [12]. In other studies, tannic acid inhibited skin, lung, and fore stomach tumors induced by polycyclic aromatic hydrocarbon carcinogens and N-methyl-N-nitrosourea. Gulcin et al. [13] investigated the antioxidant and radical scavenging properties of tannic acid using different analytical methods such as:

1. Ferric thiocyanate method
2. Total reducing capacity using the potassium ferricyanide reduction method
3. Scavenging activity of free radicals (e.g., 1,1-diphenyl-2-picryl hydrazyl free radical, 2,20-azinobis(3-ethylbenzthiazoline-6-sulfonic acid))
4. Superoxide anion and chelation capacity of ferrous ions (Fe^{2+})
5. Hydrogen peroxide scavenging

Furthermore, an important goal of this investigation was to show the *in vitro* antioxidative effects of tannic acid relative to commercial and standard antioxidants commonly used by the food and pharmaceutical industry (e.g., butylated hydroxyanisole, butylated hydroxytoluene, α-tocopherol, and trolox).

10.1.4.4 Medical

Recent reports suggest that a wide range of herbal medicines and foodstuffs may be credited for prevention of chronic diseases due to their radical scavenging activities or antioxidant properties. The radical scavenging properties or antioxidant capability of tannins have been confirmed by several authors by using colorimetric measurements [14], electron spin resonance (ESR) [15], chemiluminescence (CL), and rancimat and oxidograph methods [16].

10.1.4.5 Environmental Remediation and Other Applications of Tannins

Lately, there has been a growing interest of various low-cost adsorbents derived from agricultural or biomass wastes for removing heavy metal ions and dyes from industrial effluents [17]. The use of tannin-based adsorbents for such application has been investigated [18]. Reports have shown that polyphenolic compounds or tannins are capable of complexing with metal ions and adsorbing significant quantities of dyes.

Tannins have also been used or are being developed for the following industrial applications: (a) corrosion inhibitors in the formulation of pigments in paint coatings [19]; (b) flocculants; (c) depressants; (d) viscosity modifier agent; (e) chemical cleaning agents (for removing iron-based deposits), and (f) oxygen scavengers for boiler water treatment system [20]. These reflect their importance as industrial raw materials relative to the synthetic phenols.

10.1.4.6 Corrosion Inhibitors

Tannic acid is used in the conservation of ferrous metal objects to passivate and inhibit corrosion. Tannic acid reacts with the corrosion products to form a more stable compound, thus preventing further corrosion from taking place. After treatment, the tannic acid residue remains on the surface so that if the moisture reaches the surface, the tannic acid will be hydrated and this prevents or retards further corrosion. Tannic acid treatment for conservation of iron is very effective and widely used. However, it does have a significant visual effect on the treated object, turning the corrosion products black and any exposed metal dark blue. Tannic acid should be used with care on copper alloy components since it has a slightly etching effect.

The behavior of copper in 1 M solutions of HCl, HNO_3, and HCl–HNO_3 mixture containing various amounts of tannic acid was investigated. Since tannic acid is a mixture of galloyl glucose, these same mineral acids containing gallic acid or glucose have also been studied. Glucose inhibits the corrosion of copper in HNO_3 solution to a very slight extent and has no effect in the other acidic solutions. Gallic and tannic acids had no effect in HCl at a concentration of 50%, but at the highest concentrations tested, they showed an inhibitive efficiency of 80% and 70%, respectively, in HNO_3 and in 65–70% HCl–HNO_3 mixtures [21]. Gallic acid may be considered a cathodic inhibitor that does not interfere directly with the cathodic process but modifies the corrosive environment. Tannic acid behaves both as a cathodic and an anodic inhibitor due to the formation of an oxidizing compound that is adsorbed on the copper surface. Tannic acid is also part of commercial iron/steel corrosion treatments such as Hammerite Kurust, which is a mixture of inorganic fillers, a styrene–acrylate copolymer used to remove rust.

Most of the well-known corrosion inhibitors used in the protection of carbon steel or copper metal in acidic media are organic compounds that contain nitrogen, sulfur, oxygen, and multiple bonds in the molecules. These compounds adsorb on the metal surface to inhibit acidic corrosion of the metals and have continued to provoke research interest [22,23].

Tannins as corrosion inhibitors are applied both in nonaqueous solvent and waterborne pretreatment formulations. These formulations could be applied on partially rusted substrates, reducing the need for cleaning the surface by sandblasting or other methods, which are expensive and may not be possible in some situations. Tannins have been called rust converters since their presence converts active rust into compounds that are stable and corrosion resistant. A rapid reaction was found to occur between rusty iron and natural tannins. The transformation of rusty iron into a blue-black coating layer has been attributed to the complexation of the polyphenolic moiety of the tannin with iron oxides and oxyhydroxides. Although other complexation products were undoubtedly formed, the ferric–tannate complex has been cited as the major product. However, the protective efficiency of ferric–tannate against further corrosion generated contradictory results.

According to Jamroz et al. [24] and Matamala et al. [25], tannins are more effective when used in conjunction with phosphoric acid. On the other hand, Des Lauriers [26] found that the efficiency of this type of pretreatment was inadequate. Owing to the diversity of the materials used in the different studies, different explanations of the corrosion inhibitory mechanisms of tannins have been suggested.

Tannins form chelates with iron and other metallic cations due to the presence of hydroxyl groups on the aromatic rings. Rust protection properties result from the reaction of polyphenolic parts of the tannin molecule with ferric ions, thereby forming a highly cross-linked network of ferric–tannate [27]. Gust [28] has reported via Mossbauer spectroscopy that a mixture of mono- and bicomplexes was formed as a result of the reaction between the iron rust phase components and oak tannins in an aqueous solution.

Similar mixtures were also observed during the reaction of extracts from mangrove barks of Panama [29] with ferrous and ferric salts.

A rust-modifying or rust-stabilizing action, resulting in the formation of more dense and pore-free corrosion or rusted layers, may affect the protective property of tannins [30]. In addition, the composition of the rust layer and its kinetics of formation at the corroded steel surface are other factors that influence the converter protection efficiency. Infrared spectroscopy has shown that the reaction of iron oxides with phosphoric acid decreased in the order: lepidocrocite > magnetite > goethite [31]. Owing to the diversity of the materials used in different studies, different explanations on the inhibitory mechanisms of tannins have been suggested. Phytic acid and tannins are reported as nontoxic corrosion inhibitors for metals in aggressive media. Hence, the anticorrosion activity of CO extract (CO, *Chromolaena odorata*, is a species of flowering shrub in the sunflower family) can be attributed to phytic acid and tannins. It is well known that H_2 gas is frequently evolved when metals in service degrade in the presence of acids. The rate at which H_2 gas is evolved is related to the rate of metal dissolution in the environment as shown by weight loss. Thus, if there is a way to measure the volume of H_2 gas evolved as a function of time with the metal mass loss, the rate of corrosion of the metal within the environment can be predicted without causing plant downtime. This will allow engineers to have knowledge about the degree of conversion required to carry out repair [32].

Carlin and Keith [33] studied the MOP-30 coating system for the protection for ferrous artifacts not exposed to the elements. MOP-30 is a new vapor-barrier coating derived from a combination of octyl phenol ethoxylate containing 30 ethylene oxide units and tannic acid. They contained three components: (a) tannic acid solution with increased penetration properties; (b) clear coating derived from tannic acid that provides good compatibility when applied over a tannic acid inhibitor coating identified as MOP-30 (this coating has good vapor-barrier properties); and (c) graphite pigment added into the MOP-30 coating to produce a gray matt appearance. This mixture imparts an aesthetically pleasing finish. Like MOP-30, the coating can be removed in an alcohol bath.

Natural products such as Henna extract (*Lawsonia inermis*) and its main constituents (lawsone, gallic acid, δ-glucose, and tannic acid) were investigated as corrosion inhibitors for corrosion of mild steel in 1 M HCl solution using electrochemical and surface analysis methods [34]. Polarization measurements indicated that all the examined compounds acted as mixed inhibitor (mixed inhibitor is the inhibitor that retards both the cathodic and anodic reaction of the metal in the corrosive medium) and inhibition efficiency increased by increasing the inhibitor concentration.

10.1.4.7 Medicinal Application of Tannic Acid

Several studies have shown that tannic acid possesses antiviral, antimicrobial, and antibacterial properties. In many cases, it acts directly on the organism to inactivate it. Tannins have also been implicated in hyaluronidase system, that is, they destroy hyaluronidase in much the same manner as echinacea does, thereby defending the cells against viral invasion. Tannins are also toxic to fungi, bacteria, and viruses and inhibit their growth [35]. A 1:50,000 solution of acerin and similarly a 1:50,000 solution of mimosa, quebracho, canaigre, and babul were found to destroy viruses within 5 min, while a milder effect was observed for a chestnut wood, valonia, and sumac solution [36]. With reference to Colak et al., Carson and Frmsch [37,38] stated that tannic acid and digallic acid inactivate influenza viruses, while gallic acid in carob has antibacterial, antifungal, and antioxidant properties.

Antibacterial properties of vegetable tannins have been exploited in various fields, particularly in medicine for their antioxidant, antimutagenic, and anticarcinogenic properties [39–42]. Hydrolyzable tannins such as tannic acid and epigallocatechin gallate have been reported to have natural antioxidant, antimicrobial, and antiviral activities [43].

As an antioxidant compound, tannic acid was also shown to prevent lipid oxidation and radical-mediated DNA cleavage by scavenging oxygen and oxygen-derived radicals [44]. Tannic acid inhibited the growth of *Aeromonas hydrophila, A. sobria, Escherichia coli, Klebsiella pneumoniae, Listeria monocytogenes, Staphylococcus aureus, Helicobacter pylori, Cytophaga columnaris*, human immuno-deficiency virus (HIV), and the influenza virus. As their names infer, hydrolyzable tannins can be hydrolyzed under certain hydrolytic conditions. Previous studies [45,46] have shown that thermally processed muscadine grapes (their seeds and juice) enhanced antibacterial activity in their water-soluble fractions. These contained tannic acid as the major antimicrobial substance. These results suggest that the thermal hydrolysis of tannic acid might increase its antioxidant capacity and antimicrobial activity.

Rancimat and disc diffusion tests showed that thermally processed tannic acid has stronger antioxidant capacity and antimicrobial activity [47] than fresh tannic acid. Soybean oil treated with thermally processed tannic acid had a 33–84% longer induction period of oxidation (8.0–14.7 h) than fresh tannic acid-treated oil (6–8 h), while untreated oil showed only 5.9 h. Thermally processed tannic acid samples had strong antimicrobial activity (inhibition zone diameter) against 10 human pathogens while fresh tannic acid showed antimicrobial activity only on two strains, *Salmonella typhimurium* (ATCC 14028) and *Enterobacter sakazakii* over 1–2 day incubation periods. ^{13}C-NMR spectra showed that thermally processed tannic acid had a higher content of hydrolyzed aromatic carboxylic acid groups than fresh tannic acid. The results showed that tannic acid thermally processed for 15 min had about 67% higher antioxidant capacity and about 50% higher antimicrobial activity than its fresh counterpart. Thermal processing could be useful for enhancing antioxidant capacity and antimicrobial activity of hydrolyzable polyphenols in natural plants.

Tannic acids with different concentrations in water were tested and showed a significant effect on the decline in the growth of *Pectobacterium chrysanthemi*; 91% inhibition was observed at a concentration of 200 mg/mL [48], while the other concentrations were not studied. This phenolic acids exhibit antimutagenic properties.

Tannins and tannic drugs are used in many medical applications to stop minor bleeding, protect the skin and mucosal infections and inflammatory reactions in many medical applications such antidiarrheals, and as antidotes for poisoning by heavy metals and alkaloids. Special attention was paid to the antimutagenic and antitumorigenic properties of tannins. Tannins from immature fruit mangosteen pericarp offer protection from insects, fungi, plant viruses, and bacteria. All effects of tannins on animal and human cells and tissues are related to their ability to chemically react with proteins and form insoluble complexes, which is manifested by the precipitation of proteins in the surface layers (astringent effect). Scalbert [49] found that tannins exhibit an inhibitory effect on filamentous algae, yeasts, and bacteria. Condensed tannins are active toward *Streptococcus mutans*. It is proven that the condensed tannins bind to the cell wall of bacteria in ruminants, preventing growth and protease activity [50].

Tannins may be formed by polymerization of quinone. This group of compounds has received a great deal of attention in recent years, since it was suggested that the consumption of tannin-containing beverages, especially green teas and red wines, can cure or prevent a variety of illnesses.

Many human physiological activities, such as stimulation of phagocytic cells, host-mediated tumor activity, and a wide range of anti-infective actions, have been assigned to tannins [51]. One of their molecular actions is to complex with proteins through so-called nonspecific forces such as hydrogen bonding and hydrophobic effects, as well as by covalent bond formation [51,52]. Thus, their mode of antimicrobial action may be related to their ability to inactivate microbial adhesions, enzymes, cell envelope transport proteins, and so on. The antimicrobial significance of this particular activity has not been explored. There is also evidence for direct inactivation of microorganisms: low tannin concentrations modify the morphology of germ tubes of *Crinipellis perniciosa*. Tannins in plants inhibit insect growth and disrupt digestive events in luminal animals [53].

Scalbert [49] reviewed the antimicrobial properties of tannins in 1991 and listed 33 studies up to that point which had documented the inhibitory activity of tannins. According to these studies, tannins can be toxic to filamentous fungi, yeasts, and bacteria. Condensed tannins have been determined to bind to cell walls of luminal bacteria, preventing growth and protease activity [54]. Although this is still speculative, tannins are considered, at least partially, responsible for the antibiotic activity of methanolic extracts of the bark of *Terminalia alata* found in Nepal. This activity was enhanced by UV light activation. At least two studies have shown tannins to be inhibitory to viral reverse transcriptases [55,56].

Segura et al. [57] evaluated the biological activity of alkaloids and tannins extracted from roots of pomegranate on axenic cultures from *Entamoeba histolytica* and *E. invadens*. Two milliliters of the aqueous extract had higher activity on cultures from *E. histolyticathan* and *E. invadens* strains, producing growth inhibitions of about 100% and 40%, respectively. Alkaloid concentrations of 1 mg/mL had no amoebicide activity, but tannins at concentrations of 10 μg/mL for *E. histolytica* and 100 μg/mL for *E. invadens* were sufficient to produce growth inhibition of about 100%. Tannic acid was also tested on the cultures of *E. histolytica* observing a high inhibitory activity on the growth. Its effect at 0.01 mg/mL was similar to that observed with the tannin mixture. Dried pomegranate peels were also used for curing acute enteritis and dysentery [58]. The major components in the extract of the fruits of pomegranate were tannins or polyphenols [59]. Vasconcelos et al. [60] reported that the gel prepared from the extract of the pericarp of fresh pomegranate fruits is an antifungal agent against *C. albicans*. The mechanism of action of tannins against *Candida* is unclear, although it is suggested that tannins may act on the cell membrane because they can precipitate proteins. Tannins also inhibit many enzymes such as glycosyl transferases in *Streptococcus mutans* affecting its ability to attach to dental surfaces. The adhesion of *Candida* to acrylic surfaces is probably related to the presence of *S. mutans*. Hence, polyphenols certainly interfere in salivary proteins and some oral bacterial enzymes. In addition, they may affect bacterial membranes and disturb bacterial aggregation [61].

Tannic acid was considered as one of the more important therapeutics used in the topical treatment of burn patients in the twentieth century. Its use, however, changed dramatically over time, and many positive reactions regarding its effectiveness emerged shortly after its introduction. However, the tannic acid method became obsolete after the appearance of several reports on hepatotoxicity in the early 1940s. Only in present times, with advances in the fields of burn research, does a better understanding of the pathophysiology of the burn syndrome, the involvement of the liver therein and tannic acid research, occur. As a result, with the availability of highly purified tannic acids, the usefulness of tannic acid as an adjuvant therapy for burns has again attracted interest. This rise and fall in popularity of tannic acid is also reflected in review articles on this subject that have been published over the years. Lee and Rhoads in 1944 [62] stated that tannic acid was applied successfully and, in fact, had produced a decrease in mortality rate despite the occurrence of liver necrosis. In subsequent reviews, the latter aspect was emphasized [63–65]. Conversely, in 1995, Hupkens et al. [66] once more made the necessary differentiation in the reports on hepatotoxicity of tannic acid and evaluated its use for better cosmetic results.

In this respect, a parallel may be drawn between tannic acid and silver, another therapeutic regimen for the local treatment of burns. After a period in which the use of silver was reviled, it has again occupied an important place in the contemporary treatment of burn wounds [67,68]. Similarly, highly purified tannic acids gained interest as a means to improve wound healing and to reduce scar tissue formation. However, prior to the reintroduction of tannic acid in burn treatment, a thorough benefit–risk analysis should be carried out. This will, in the near future, require controlled prospective studies comparing tannic acid with current standard therapeutic regimens, such as the local application of silver sulfadiazine, with rigorous toxicological assessment in which any detrimental effects of tannic acid on the liver are to be excluded.

In anticipation of such tests, it was attempted in this review to tentatively answer the question: whether or not tannic acid actually is toxic to the liver. From the collected data, it can be concluded that the evidence is yet inconclusive. Thus, many clinical and experimental studies in the past were indicative of a potential hepatotoxic effect of tannic acid. However, tannic acid preparations that were ill defined and of poor quality were used, often in extremely high concentrations. This could have negatively affected the outcome of these studies. Moreover, liver damage and impairment of liver function also occurred in patients who did not receive tannic acid treatment at all, and this phenomenon is now considered part of the burn syndrome. On the basis of these considerations, the suggestion by some that tannic acid has been a contributing factor to death among burn patients seems to be untenable. This is supported by the low mortality rate observed in most clinical studies. It is our opinion that the local application of tannic acid to burn wounds causes no serious hepatic damage, at least no more than the thermal injury does in itself. A prerequisite for using tannic acid for burn treatment is that it be used under strictly controlled conditions, that a highly purified tannic acid product be applied in moderate concentrations (at maximum 2.5–5% tannic acid), and that an adequate pharmaceutical formulation be used, which minimizes decomposition and creates a favorable environment for wound healing [69].

10.2   EXPERIMENTAL

10.2.1   Materials

The following chemicals were obtained from Sigma-Aldrich chemical company (Germany): tannic acid, glycine, p-toluene sulfonic acid, benzaldehyde, ferric chloride, manganese chloride, and cobalt chloride with high-grade purity (99.99%). The following chemicals were obtained from El-Gomhoria chemicals company (Egypt): toluene (ADWIC, Cairo, Egypt), ethanol, iso-propanol, and petroleum ether (60–80°C).

10.2.1.1   Synthesis of Tannic Acid–Glycine Derivatives (TG_1, TG_3, TG_5)

Tannic acid (T) was reacted with glycine (G) in different molar ratios (T:G = 1:1, 1:3, and 1:5, respectively) in toluene as a solvent and in the presence of 0.1% by weight of p-toluene sulfonic acid as a dehydrating agent. The reaction was conducted in a dean-stark trap under reflux condition until the theoretical amount of water for each ratio of reactants was obtained. The reaction product was filtered and the solid was recrystallized twice from acetonitrile. The resulting tannic acid–glycine derivatives (TG_1, TG_3, and TG_5) were dried under vacuum (Schemes 10.3 through 10.5).

10.2.1.2   Synthesis of Tannic Acid–Glycine–Benzaldehyde Derivatives (TGB_1, TGB_3, TGB_5)

Each of the three tannic acid–glycine derivatives (TG_1, TG_3, and TG_5) were reacted with 1, 3, and 5 molar ratio of benzaldehyde (B) in ethanol under reflux condition for 6 h. The reaction product was filtered and washed with iso-propanol and then dried under vacuum. The resulting tannic acid–glycine Schiff bases were designated as TGB_1, TGB_3, and TGB_5 (Schemes 10.3 through 10.5).

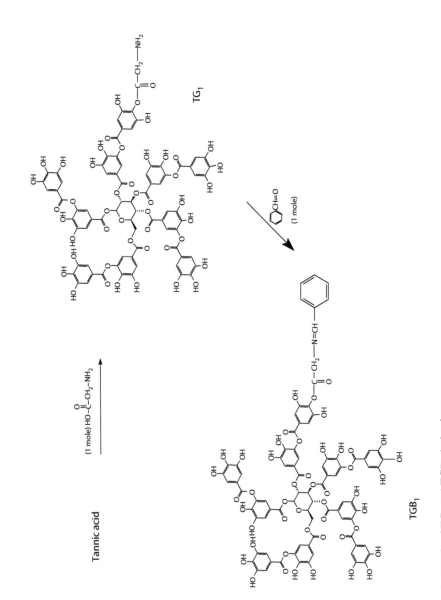

SCHEME 10.3 Synthesis of TG₁ and TGB₁ derivatives.

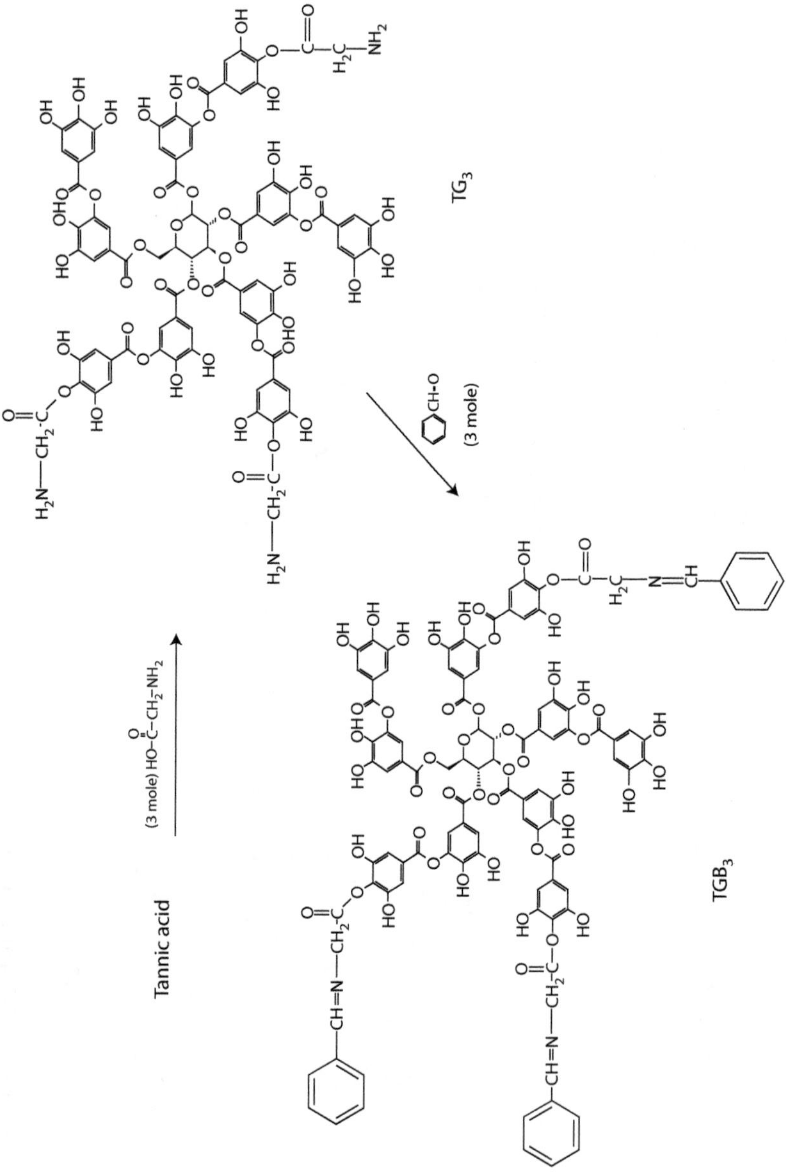

SCHEME 10.4   Synthesis of TG₃ and TGB₃ derivatives.

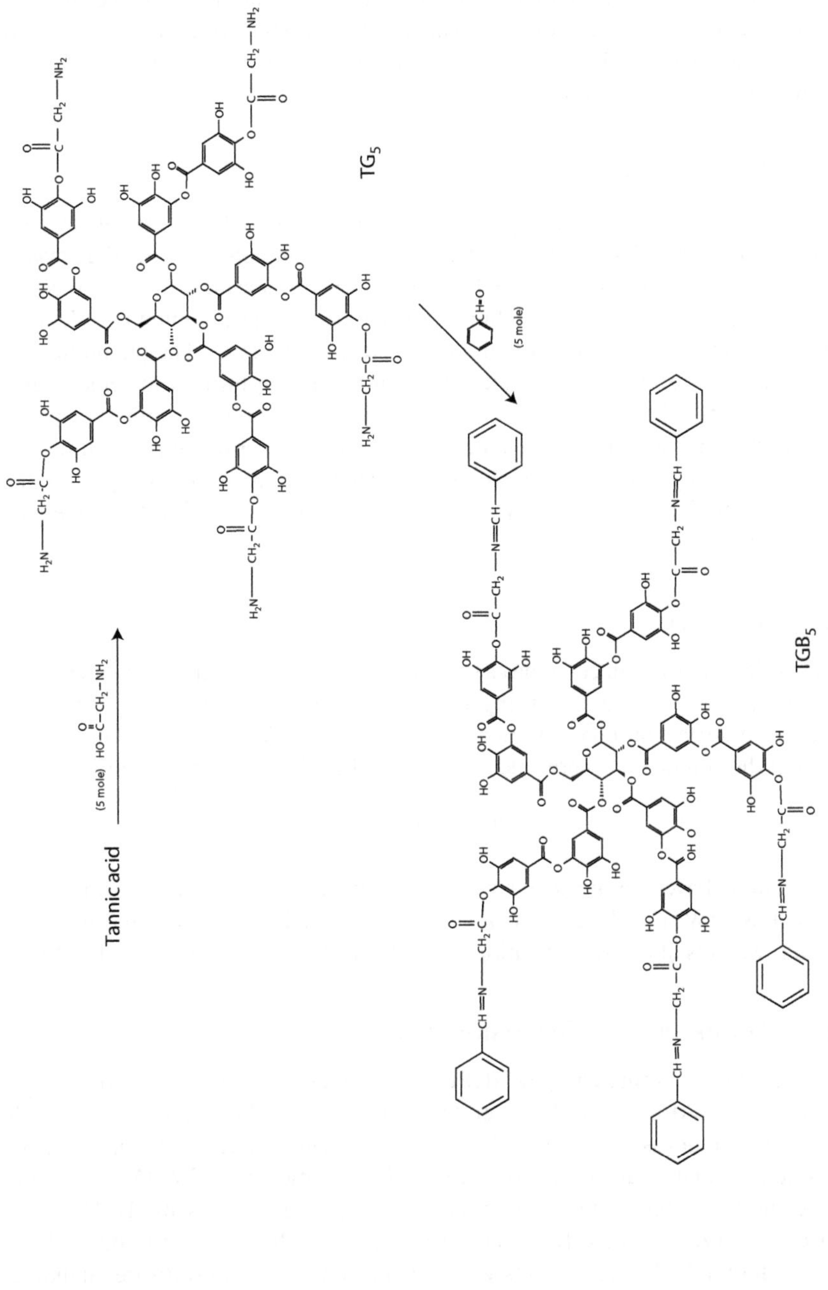

SCHEME 10.5 Synthesis of TG$_5$ and TGB$_5$ derivatives.

10.2.1.3 Synthesis of Metal Complexes

The tannic acid–glycine Schiff base derivatives (TGB$_1$, TGB$_3$, and TGB$_5$) were complexed with different metal ions, namely, FeCl$_3$, MnCl$_2$, and CoCl$_2$, by a ratio double to the azomethine groups in each compound in ethyl alcohol as solvent. The reaction mixture was refluxed for 16 h. On cooling, the different metal complexes were precipitated and filtered under vacuum. The resulting crystals were washed with cold ethanol/petroleum ether (50:50 vol) and finally with diethyl ether and kept in a desiccator over fused CaCl$_2$ [70,71].

10.2.2 SURFACE TENSION

The surface tension (γ) of the synthesized tannic acid Schiff base surfactants (TGB$_1$, TGB$_3$, and TGB$_5$) and their metal complex solutions in bidistilled water (in concentration range of 0.1 to 0.0001 mol/L) were measured using Du–Noüy tensiometer (Krüss model K6, Hamburg, Germany) using the platinum ring detachment method and was calibrated with deionized water at 25°C. The platinum ring was cleaned before each measurement by flame (Bunsen burner) to remove any residual deposits. The surface tension measurements were taken 1 min after pouring the surfactant solution into the measuring container to ensure equilibrium. The measurements for each solution were taken in triplicate to check for repeatability and the measured surface tension values were within an error of less than or equal to ±1 mN/m [72–74].

10.2.3 VISCOSITY

The intrinsic viscosities in centipoises (η) of the prepared compounds were measured in double-distilled water at 25°C using a capillary viscometer (Übbelhode suspended level type) at surfactant concentrations in the range 0.005–5.0 g/L. The molecular weights of the compounds were calculated using the following equation [75]:

$$\eta = 3.38 \times 10^{-3}\, M^{0.43} \tag{10.1}$$

The obtained average molecular weights obtained from viscosity measurements (indicated as MW^V) of different synthesized compounds were compared with the molecular weights of standard solutions of gallic acid and are listed in Table 10.1.

10.2.4 GEL PERMEATION CHROMATOGRAPHY

Gel permeation chromatography (GPC) experiments were carried out on a Supremamax 3000 column (Polymer Standard Service, Mainz, Germany) with water (HPLC grade) as eluent (at 1 mL/min). The system comprised a pump (Hitachi, Darmstadt, Germany), an autosampler (Merck Hitachi model AS-2000A), and an in-line vacuum degreaser. The volume of injected sample per run was 40 μL. The samples were analyzed with a differential refractive index (RI) detector (model RI-71, made by Merck). Molecular weights were calculated using Astra software. Standard compounds (MW with 1000, 2000, 3000, and 6000) were used for calibration. The

TABLE 10.1

Theoretical and Measured Molecular Weights of Tannic Acid–Glycine–Benzaldehyde (TGB) Surfactants and Their Metal Complexes

Surfactant	Expected Molecular Weight[a] (g/mol)	Molecular Weight[b] (g/mol)	Molecular Weight[c] (g/mol)
TGB$_1$	1846.4	1790	1809
TGB$_3$	1991.5	1950	1952
TGB$_5$	2136.7	2090	2094
TGB$_1$-Mn	3818.6	3735	3742
TGB$_3$-Mn	4108.9	4019	4027
TGB$_5$-Mn	4399.3	4303	4311
TGB$_1$-Co	3822.6	3739	3746
TGB$_3$-Co	4112.9	4022	4031
TGB$_5$-Co	4403.3	4306	4315
TGB$_1$-Fe	3854.9	d	d
TGB$_3$-Fe	4145.3	d	d
TGB$_5$-Fe	4435.6	d	d

[a] Calculated from the chemical formula of expected structure.
[b] Calculated from gas chromatography.
[c] Calculated from intrinsic viscosity measurements.
[d] No values were recorded as the compounds were insoluble in water.

average molecular weights of the synthesized compounds obtained from GPC measurements (indicated as MW^{GPC}) were listed in Table 10.1.

10.2.5 ANTIMICROBIAL ACTIVITY

10.2.5.1 Microorganisms

The biocidal activity of the synthesized tannic acid derivatives and their metal complexes were tested against different bacterial strains (American Type Culture Collection (ATCC)) as follows: *Staphylococcus aureus* ATCC 29213, *Escherichia coli* ATCC 25922, *Pseudomonas aeruginosa* ATCC 27853, *Bacillus subtilis* ATCC 55422, *Desulfomonas pigra* ATCC 29098 (SRB), and *Staphylococcus typhimurium* ATCC 27948. SRB is a sulfate-reducing bacteria and grows in oil fields and reduces petroleum components that contain sulfur to hydrogen sulfide (H_2S). H_2S is corrosive in nature and toxic; hence, these bacteria are a problem in the oil field systems and should be protected. On the other hand, the selected species cause dangerous diseases for humans and animals.

10.2.5.2 Growing of Microorganisms

The bacterial strains were cultured according to the standards of the National Committee for Clinical Laboratory (NCCL) [76]. The bacterial species were grown on nutrient agar. The nutrient agar medium consisted of beef extract (3.0 g/L), peptone

(5.0 g/L), sodium chloride (5.0 g/L), and agar (20.0 g/L), with the volume made up to 1 L with distilled water. The mixture was heated until boiling and sterilized in an autoclave. The bacterial strains were kept on the nutrient agar medium and showed no inhibition zones.

10.2.5.3 Resistance and Susceptibility

For preparation of discs and inoculation, 1.0 mL of inocula was added to 50 mL of the agar medium (40°C) and mixed. The agar was poured into 120 mm petri dishes and allowed to cool to room temperature. Wells (6 mm in diameter) were cut in the agar plates using proper sterile tubes and filled up to the surface of agar with 0.1 mL of the synthesized tannic acid Schiff base surfactants (TGB$_1$, TGB$_3$, and TGB$_5$) and their metal complexes dissolved in dimethyl formamide (DMF) (2.5 mg/mL). The plates were placed on a leveled surface, incubated for 24 h at 30°C, and then the diameters of the inhibition zones were measured. The inhibition zone formed by the tested compounds against a particular test bacterial strain determined qualitatively the antibacterial activities of these compounds. The mean value obtained for three replicates was used to calculate the zone of growth inhibition of each sample. The antimicrobial activity was calculated as a mean of the three replicates. The tested compounds were completely compatible with the medium of agar and no turbidity was observed during the mixing process [76,77].

10.2.5.4 Minimum Inhibitory Concentration

The biocidal activity of the synthesized tannic acid Schiff base surfactants (TGB$_1$, TGB$_3$, and TGB$_5$) and their metal complexes against the tested strains were expressed as the minimum inhibitory concentration (MIC) values, defined as the lowest concentration of compounds inhibiting the development of visible growth after 24 h of incubation. The MIC values were determined by the dilution method [77]. The compounds tested were dissolved in a mixture of distilled water/alcohol (3/1; v/v) at various concentrations and 1 mL aliquot of the cationic surfactant solution was added to 14 mL agar medium. The final concentrations of the tested compounds in the medium were 1.25, 0.65, 0.325, 0.156, 0.1, and 0.098 mg/mL.

10.2.6 BIODEGRADABILITY

Biodegradability test in river water of the synthesized tannic acid Schiff base surfactants (TGB$_1$, TGB$_3$, and TGB$_5$) was determined by the surface tension method using Du–Noüy tensiometer (Krüss type K6) using platinum ring detachment method [78–80]. In this method, each surfactant was dissolved in river water at a concentration of 100 ppm and incubated at 38°C. A sample was withdrawn daily (for 30 days), filtered, and the surface tension measured. The biodegradation percent (D%) was calculated as follows:

$$D\% = (\gamma_t - \gamma_o)/(\gamma_{bt} - \gamma_o) \times 100 \qquad (10.2)$$

where γ_t is the surface tension at time t, γ_o is the surface tension at time 0 (initial surface tension), and γ_{bt} is the surface tension of river water without the addition of surfactant at time t.

10.2.7 CRITICAL MICELLE CONCENTRATION

The critical micelle concentration (cmc) is defined as the concentration of surfactant in the solution at which the micelles start to form. It can be determined from the extrapolation of the pre- and postmicellar regions in the surface tension versus $-\log C$ profile [81].

10.2.8 EFFECTIVENESS

The effectiveness (π_{cmc}) is the difference between the surface tension of the twice-distilled water and the surfactant solution at the cmc according to the following equation:

$$\pi_{cmc} = \gamma_o - \gamma_{cmc} \tag{10.3}$$

where γ_o is the surface tension of the twice-distilled water (71.8 mN/m) and γ_{cmc} is the surface tension of the surfactant solution at the cmc [82].

10.2.9 MAXIMUM SURFACE EXCESS

The maximum surface excess (Γ_{max}) is defined as the maximum concentration of surfactant molecules that can be attained at the air–solution interface and can be calculated using the following equation [83]:

$$\Gamma_{max} = (\partial\gamma/\partial\ln C)/RT \tag{10.4}$$

where R is the gas constant (8.314 J/K mol), T is the absolute temperature (K), and $(\partial\gamma/\partial\ln C)$ is the slope of γ versus $\ln C$ plots.

10.2.10 MINIMUM SURFACE AREA

Minimum surface area (A_{min}) is the area occupied by one surfactant molecule at the air/solution interface at the maximum saturation condition and can be calculated using the following equation [83]:

$$A_{min} = 10^{16}/(N_{av}\Gamma_{max}) \tag{10.5}$$

where Γ_{max} is the maximum surface excess, N_{av} is the Avogadro's number (6.023×10^{23} molecule per mol), and A_{min} is the area per molecule given in nm^2 per molecule.

10.2.11 EMULSIFICATION POWER

The emulsification power of the prepared surfactants (TGB_n) was determined as follows: 10 mL of the surfactant solution (1% by weight) and 10 mL of light paraffin oil were placed in a graduated glass tube (50 mL capacity), which was then placed in a controlled heating apparatus at 25°C. Then, the glass tube was shaken vigorously for 2 min [84] and allowed to settle. The emulsification power is the time in seconds after shaking for separation of 9 mL of pure surfactant solution.

10.3 RESULTS AND DISCUSSION

10.3.1 STRUCTURE OF TANNIC ACID SCHIFF BASE SURFACTANTS AND THEIR METAL COMPLEXES

The chemical formula for commercial tannic acid is often given as $C_{76}H_{52}O_{46}$, which corresponds with decagalloyl glucose. In fact, it is a mixture of polygalloyl glucose or polygalloyl quinic acid esters with the number of galloyl moieties per molecule ranging from 2 up to 12, depending on the plant source used to extract the tannic acid form. In our study, the tannic acid used was a fine chemical purchased from Aldrich Chemical Company (Berlin, Germany) and extracted from tara pods (*Caesalpinia spinosa*) plant. Hence, analysis of the produced surfactants focused on the functional groups expected to be present after the applied chemical reactions and the molecular weight of the products.

The three tannic acid Schiff base surfactants were prepared by the reaction of tannic acid with glycine at different molar ratios (1:1, 1:3, and 1:5) in toluene. The esters were then condensed with benzaldehyde in absolute ethanol in good yield and purity. The different metal complexes were prepared by the reaction of the Schiff bases and metal ions in molar ratio 1:2 (metal:base) in absolute ethanol. The prepared compounds were characterized by elemental analysis, FTIR, and H^1-NMR spectroscopy. The complexes were stable at room temperature in air in the solid state.

10.3.1.1 IR Spectra

The FTIR spectra of the prepared compounds were made by Mattson Genies Fourier transformer infrared spectroscopy, France, using a KBr disc. The discs were prepared by pressing 0.01 g of different compounds in a standard amount of dry KBr at 10 bars. The infrared spectroscopic analysis of tannic acid Schiff base surfactants exhibited the absorption bands at 3515, 3010, 2850, 1737, 1634, 1563, and 1210 cm^{-1}, which were, respectively, attributable to the stretching vibration of O–H, C–H aromatic, CH$_2$, C=O, C=N, C=C, and C–O. IR spectra of the prepared compounds confirmed the reaction between tannic acid and glycine as no absorption band appeared at 3400 cm^{-1} assigned to the N–H linkage of glycine. A strong band at 1634 cm^{-1} in the free ligand was assigned to the azomethine frequency (N=C–CH), which shifts by ±5–10 cm^{-1} in the spectra of metal complexes. The shift is an indication of coordination through nitrogen atom, which reduces the double bond character of the carbon–nitrogen bond of the azomethine group [85–89]. The formation of the metal–nitrogen bond is further supported by the appearance of a band in the far region of 540–480 cm^{-1}.

10.3.1.2 ¹H-NMR Spectroscopy

¹H-NMR spectra were performed using Bruker DRX-300 (Bruker Biospin, Germany) spectrometer with trimethyl silane as an internal standard. The ¹H-NMR spectra

revealed that synthesized surfactants (TGB$_1$ surfactant was taken as a representative sample for the synthesized surfactants) showed the following signals: $\delta = 1.28$ ppm (m, 2H, CH$_2$), 2.64 ppm (t, 2H, OCOCH$_2$), 5.34 ppm (s, 1H, OH).

10.3.2 MOLECULAR WEIGHT MEASUREMENT

The molecular weight of the synthesized surfactants and their cobalt and manganese complexes were measured using two different methods: viscosity and GPC. The measured molecular weights were comparable with the calculated molecular weights within 98% (Table 10.1), which confirms high purity of the synthesized compounds as represented in Scheme 10.1.

10.3.3 SURFACE ACTIVITY OF THE TANNIC ACID SCHIFF BASE SURFACTANTS AND THEIR METAL COMPLEXES

Tannic acid is a water-soluble natural compound with maximum solubility of 2850 g/L in water and weak acidity of p$K_a = 10$. The solution of tannic acid in water has very low surface activity due to the large number of hydroxyl groups in its structure. The surface tension of the tannic acid solution in double-distilled water is 65 mN/m and this value did not change considerably when the concentration of tannic acid was increased to values up to 0.1 mol/L. The introduction of side chains in its chemical structure is expected to increase its surface activity and decrease its surface tension in solution. The lowering of surface tension comes from the increasing adsorption tendency of these molecules at the interfaces because the chains are hydrophobic. Consequently, increasing the number of attached chains to the tannic acid molecules gradually decreases the surface tension of its aqueous solutions.

Figure 10.1 represents the dependence of surface tension of the three synthesized tannic acid Schiff base derivatives on their concentration in double-distilled water at 25°C. For each substitution ratio, that is, TGB$_1$, TGB$_3$, or TGB$_5$, there is a gradual decrease in the surface tension values with increasing concentration of tannic acid in solution. This decrease in surface tension is attributed to the gradual increase in adsorption of the surfactant molecules with increasing concentration. Also, increasing the degree of substitution at constant concentration from one to five side chains decreases surface tension. The lowest surface tensions were obtained for TGB$_5$ surfactants at 25°C. This has to do with the high adsorption tendency of TGB$_5$ at the interface relative to TGB$_1$ and TGB$_3$ surfactants.

Figure 10.1 also shows a typical profile of surface tension as a function of concentration of the surfactants in solution. The profile shows two characteristic regions, one at low concentration (premicellar region) characterized by a sharp decrease in surface tension due to a small increase in surfactant concentration. The slope of this region represents the tendency of these molecules to adsorb at the interfaces. The second region is at high concentration (postmicellar region), which shows a very low change in the surface tension values with increasing surfactant concentration. Extrapolation of these two regions determines the cmc of the synthesized surfactants, which are summarized in Table 10.2.

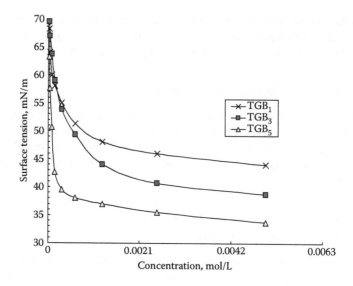

FIGURE 10.1 Surface tension versus concentration of tannic acid Schiff base surfactants at 25°C.

TABLE 10.2

Critical Micelle Concentrations, Effectiveness, Efficiency, Maximum Surface Excess, and Minimum Surface Area Values of the Synthesized Surfactants and Their Metal Complexes at 25°C

Surfactant	$cmc \times 10^{-5}$ (mol/L)	Effectiveness (π_{cmc}) (mN/m)	Efficiency (Pc_{20}) $\times 10^{-4}$ (mol/L)	Maximum Surface Excess (Γ_{max}) $\times 10^{-10}$ (mol/cm²)	Minimum Surface Area (A_{min}) (nm²/ molecule)
TGB$_1$	64.5	23.3	5.8	1.95	84.96
TGB$_3$	40.5	28.3	5.5	2.83	58.73
TGB$_5$	18.1	33.8	0.7	3.67	45.26
TGB$_1$-Mn	64.8	11.3	1.6	0.58	129.42
TGB$_3$-Mn	51.2	17.8	1.5	1.97	84.43
TGB$_5$-Mn	23.5	25.8	0.6	2.20	75.61
TGB$_1$-Co	39.5	20.4	1.5	1.61	102.86
TGB$_3$-Co	31.0	24.6	0.8	2.28	72.93
TGB$_5$-Co	22.5	33.8	0.5	2.31	71.82
TGB$_1$-Fe	a	a	a	a	a
TGB$_3$-Fe	a	a	a	a	a
TGB$_5$-Fe	a	a	a	a	a

a No values were recorded as the compounds were insoluble in water.

The measured cmc values for TGB_1, TGB_3, and TGB_5 were 0.00065, 0.00041, and 0.00018 mol/L, respectively, which are very low compared with conventional nonionic surfactants [90,91].

On complexation of the tannic acid Schiff base surfactants with cobalt and/or manganese ions, the order of surface tension variation of the different metal complexes was identical to that for their uncomplexed parent surfactants (see Figures 10.2 and 10.3). This can be attributed to the fact that the metal complexes retain their ligand entities in the solutions. The surface tension is mainly affected by the ligands of the complex molecules.

The influence of complexation on surface tension compared with the parent surfactants is compared in the surface tension profiles shown in Figures 10.4 through 10.6. In general, the surface tension of the metal complexes is higher than that of the parent surfactants.

Determination of cmc values of the different metal complexes (Table 10.2) showed two points: first, the cmc of manganese complexes is higher than that of the parent surfactants and second, the parent surfactants have higher cmc values than that of cobalt complexes. This difference was attributed to the difference in ionic radii and electronegativity of these two transition metal ions and their ability to attract ligands to their outer shells to form stable complexes. This difference affects the volume of the complexes and consequently its effective area at the air–water interface.

The effectiveness is the difference between the surface tension of the surfactant solution at cmc and that of the double distilled water. The effectiveness values (π_{cmc}) of the synthesized tannic acid Schiff base surfactants were calculated at 25°C, and the results are listed in Table 10.2. The effectiveness values are useful for comparison of surfactants in the same series, that is, having the same functional groups and

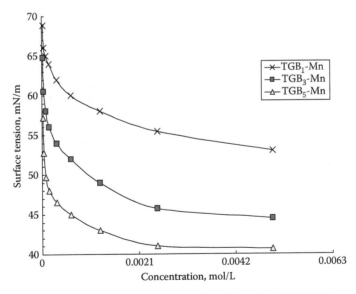

FIGURE 10.2 Surface tension versus concentration of tannic acid Schiff base surfactant manganese complexes at 25°C.

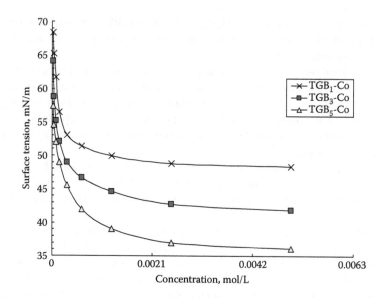

FIGURE 10.3 Surface tension versus concentration of tannic acid Schiff base surfactant cobalt complexes at 25°C.

different alkyl chains. It is clear that the effectiveness values of the tannic acid Schiff base derivatives (TGB$_1$, TGB$_3$, and TGB$_5$) increased with increasing number of substituents on the tannic acid moiety. The most surface active homolog is TGB$_5$ with maximum effectiveness, that is, lowest surface tension at the cmc (Table 10.2). A similar order is observed for the cobalt or manganese complexes of these surfactants.

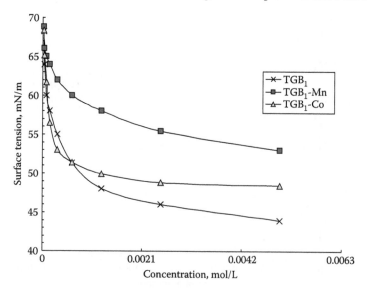

FIGURE 10.4 Comparison of surface tension versus concentration of TGB$_1$ surfactant and its metal complexes at 25°C.

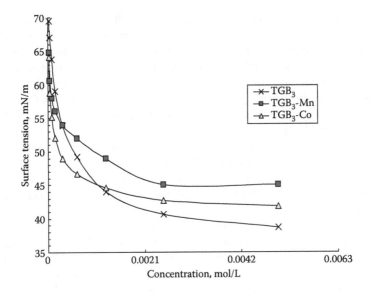

FIGURE 10.5   Comparison of surface tension versus concentration of TGB$_3$ surfactant and its metal complexes at 25°C.

The effectiveness of manganese or cobalt–tannic acid Schiff base surfactant complexes decreased in the same order as the surfactants. Cobalt complexes generally exhibit higher surface tension reduction than manganese complexes (Table 10.2).

Another effective factor for describing the surface activity of surfactants is efficiency (Pc$_{20}$). Efficiency is the concentration of the surfactant that reduces the

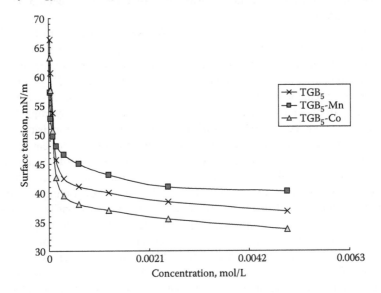

FIGURE 10.6   Comparison of surface tension versus concentration of TGB$_5$ surfactant and its metal complexes at 25°C.

surface tension of the solution by 20 surface tension units (mN/m) at the desired temperature. Efficiency is an important factor that can be used to compare surfactants with different functional groups and different alkyl chains. The more surface active compound shows a lower efficiency value, and vice versa. Comparison of the synthesized surfactants showed that increasing the degree of substitution on the tannic acid nucleus decreases its efficiency. In addition, complexation decreases the efficiency values considerably. The most efficient surfactants at the interface were the cobalt complexes (Table 10.2).

The overall observation of effectiveness and efficiency values of the synthesized tannic acid Schiff bases and their metal complexes revealed that tannic acid Schiff base derivatives are more surface active than their metal complexes in the bulk of the solutions. On the other hand, the metal complexes were more surface active at the solution–air interface.

The maximum surface excess (Γ_{max}) is the concentration of the surfactant molecules at the interface per unit area. It describes the tendency of surfactants to adsorb at the air–liquid interface at complete saturation of the interface. The Γ_{max} value depends on the slope of surface tension versus concentration curve at premicellar regions. The slope determines the variation of surface tension at finite change in concentration, and shows the degree of surfactant molecules diffusing to the interface when the bulk concentration increased. Γ_{max} values of the synthesized surfactants and their metal complexes are given in Table 10.2. It is clear from the data that the main factor that influences Γ_{max} values is the degree of substitution on the tannic acid molecules. The highest Γ_{max} value was obtained for TGB$_5$ surfactants, which indicates that the molecules are more strongly adsorbed at the air–aqueous interface than TGB$_1$ and TGB$_3$ surfactant molecules. This result is in good agreement with the variation of surface tension as a function of concentration values shown in Figure 10.1.

The order of Γ_{max} variation of the metal complexes is also identical to TGB$_1$, TGB$_3$, and TGB$_5$ surfactant molecules (Table 10.2). The complexation of the tannic acid Schiff base surfactants by manganese and cobalt ions showed moderate influence on Γ_{max} values compared with their parent molecules. The lowering in Γ_{max} values of the metal complexes can be attributed to increasing hydrophobicity of these complexes. Increasing hydrophobicity breaks down the water shell around the hydrophobic parts of the surfactant molecules [92,93], which increases the mutual repulsion between the complex molecules and the water phase. Water molecules adjacent to organic (nonpolar) liquid and hydrophobic parts of molecules form a shell-like structure via hydrogen bonding. Frank and Evans [92] have called this an "iceberg structure" because of its similarity to the structure of ice. This structure building is associated with loss of entropy, and therefore, this effect is partially responsible for the low solubility of nonpolar liquids in water [93]. This effect increases the tendency of the molecules to form micelles in the bulk solutions at comparatively low concentrations, which is obvious from the cmcs of the different metal complexes.

The minimum surface area (A_{min}) is defined as the average area occupied by one surfactant molecule at the interface under saturated conditions. This value determines the geometrical position of the surfactant molecules in the adsorbed layer. A_{min} values of the synthesized tannic acid Schiff bases decreased with increasing degree of substitution on the tannic acid molecule in the order: $TGB_1 > TGB_3 > TGB_5$. The introduction of one side chain ($O-CO-CH_2-N=CH-Ph$) gives an area of molecule (TGB_1) of 84.96 nm^2 per molecule. This large value indicates that the molecules are planar at the interface. The gradual increase of substitution on the tannic acid molecule decreases the average area per molecule to reach the minimum value for the case of five substituents, TGB_5. This is attributed to the overlap between side chains, which twists the surfactant molecules and decreases the actual area at the interface (Table 10.2). Complexation of the synthesized surfactants is expected to increase the average area per molecule at the interface. Complexation reactions involve participation of two surfactant molecules with one transition metal ion. This suggests approximate doubling of the area of the complex at the interface compared with the parent surfactant (with minor deviation). The metal ion attracts ligands to its outer shell, which decreases their area, and that explains the minor deviation in A_{min}.

The tannic acid Schiff base surfactants were evaluated as emulsifying agents for light paraffin oil (1% by weight surfactant concentration). The results obtained showed that the average time for breaking the emulsion formed was relatively lower than that for conventional emulsifiers (in the range of several hours or even days). The stability of the emulsions (in seconds) increased with increasing degree of substitution in the order (Table 10.3): $TGB_1 < TGB_3 < TGB_5$. Complexation decreased the emulsion stability considerably, which was attributed to an increase in the molecular polarity of the complexes, that is, increase in the hydrophilicity of complex molecules [91].

TABLE 10.3
Emulsification Properties of the Synthesized Surfactants and Their Metal Complexes toward Light Paraffin Oil

Surfactant	Emulsification Power[a] (s)
TGB_1	240
TGB_3	540
TGB_5	1200
TGB_1-Mn	120
TGB_3-Mn	150
TGB_5-Mn	1080
TGB_1-Co	102
TGB_3-Co	120
TGB_5-Co	960
TGB_1-Fe	b
TGB_3-Fe	b
TGB_5-Fe	b

[a] Time for 90% of surfactants solution to separate from emulsion. Bigger the number, more the emulsification power.
[b] No values were recorded as the compounds were insoluble in water.

10.3.4 Antimicrobial Assay

Antibacterial activities (zone of growth inhibition and MICs) of the tannic acid Schiff base surfactants and their cobalt, manganese, and iron complexes and the used standard compound are shown in Tables 10.4 and 10.5. Amikacin™ (Amikacin is an aminoglycoside antibiotic used to treat different types of bacterial infections, Scheme 10.4) was used as standard compound. The organisms used in the present investigation included *Bacillus subtilis* (NCIB-3610) and *Staphylococcus aureus* (NCTC-7447) as Gram-positive bacteria and *Pseudomonas aeruginosa* (NCIB-9016), *Escherichia coli* (NCTC-10416) as Gram-negative bacteria, and *Candida albicans* (ATCC 14053) and *Aspergillus niger* (Ferm-BAM C-21) as fungi. The results indicate high activity of the three tannic acid Schiff base surfactants against Gram-negative bacteria, *P. aeruginosa* and *E. coli*, and moderate activity toward Gram-positive bacteria, *B. subtilis* and *S. aureus* (compared with the standard used). On the other hand, the three surfactants showed high activity against the fungi, *C. albicans* and *A. niger*. The activity toward different microorganisms decreased with increasing number of substituents on the tannic acid molecules in the following order: $TGB_1 > TGB_3 > TGB_5$.

TABLE 10.4

Antimicrobial Assay Results in Term of Zone of Growth Inhibition of the Free Tannic Acid Schiff Base Surfactants and Their Metal Complexes

	Microorganisms					
	Gram-Positive Bacteria		Gram-Negative Bacteria		Fungi	
	Bacillus subtilis, NCIB-3610	*Staphylococcus aureus,* NCTC-7447	*Pseudomonas aeruginosa,* NCIB-9016	*Escherichia coli,* NCTC-10416	*Candida albicans,* ATCC 14053	*Aspergillus niger,* Ferm-BAM C-21
Surfactant	Diameter of Zone of Growth Inhibition					
TGB_1	17	18	19	19	20	24
TGB_3	15	17	17	19	19	22
TGB_5	15	16	16	16	15	20
TGB_1-Mn	29	28	28	27	28	30
TGB_3-Mn	27	27	27	27	28	29
TGB_5-Mn	26	26	27	26	27	25
TGB_1-Co	28	28	27	28	27	28
TGB_3-Co	25	26	27	26	26	25
TGB_5-Co	24	27	26	25	26	24
TGB_1-Fe	26	29	26	27	26	34
TGB_3-Fe	24	22	22	24	22	26
TGB_5-Fe	24	0	0	21	21	29
Amikacin	14	13	13	14	15	0

Note: Indicated values are in mm. Bigger the number, more effective the antimicrobe. The standard used is Amikacin (see Scheme 10.6).

TABLE 10.5

Antimicrobial Assay Results in Terms of Minimal Inhibitory Concentrations (MIC) of the Free Tannic Acid Schiff Base Surfactants and Their Metal Complexes

	Microorganisms					
	Gram-Positive Bacteria		Gram-Negative Bacteria		Fungi	
Surfactant	*Bacillus subtilis,* NCIB-3610	*Staphylococcus aureus,* NCTC-7447	*Pseudomonas aeruginosa,* NCIB-9016	*Escherichia coli,* NCTC-10416	*Candida albicans,* ATCC 14053	*Aspergillus niger,* Ferm-BAM C-21
TGB$_1$	1.25	0.312	0.312	0.312	0.312	0.312
TGB$_3$	0.625	0.625	0.625	0.625	0.625	0.625
TGB$_5$	1.25	1.25	1.25	1.25	1.25	0.625
TGB$_1$-Mn	0.097	0.097	0.097	0.097	0.097	0.097
TGB$_3$-Mn	0.097	0.097	0.156	0.097	0.156	0.156
TGB$_5$-Mn	0.156	0.156	0.312	0.156	0.156	0.312
TGB$_1$-Co	0.097	0.097	0.097	0.097	0.097	0.097
TGB$_3$-Co	0.156	0.156	0.097	0.156	0.156	0.156
TGB$_5$-Co	0.156	0.156	0.156	0.312	0.156	0.312
TGB$_1$-Fe	0.156	0.097	0.156	0.097	0.156	0.097
TGB$_3$-Fe	0.312	0.312	0.312	0.312	0.312	0.156
TGB$_5$-Fe	0.312	0.625	0.625	0.625	0.625	0.156
Standard	0.625	0.625	0.625	0.625	0.625	>0.625

Note: Indicated values are in mg/mL. Bigger the number, less the antimicrobe. The standard used is Amikacin (see Scheme 10.6).

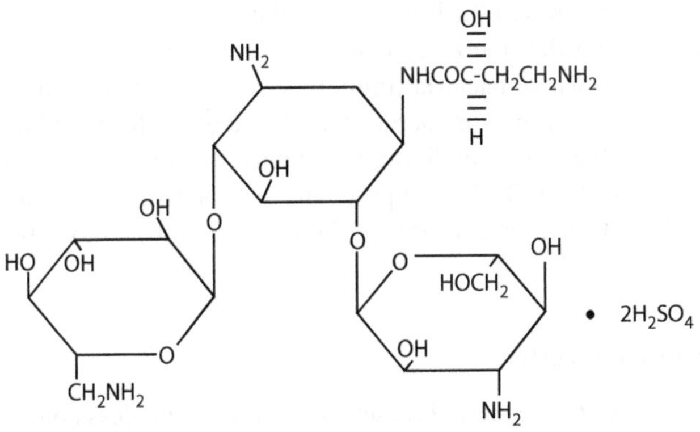

SCHEME 10.6 Structure of Amikacin.

The generally accepted mode of action of the surface active biocides in biocidal action is that the biocide molecules adsorb onto cell membranes, which will then lead to decrease in the osmotic stability of the cell and leakage of intracellular constituents [94–96]. The adsorption onto the cellular membranes mainly occurs due to the presence of polar functional groups in the chemical structure of the biocide molecules. The antimicrobial efficacy of the synthesized tannic acid Schiff base surfactants might be due to the presence of azomethine groups ($-CH=N-$) and several hydroxyl groups. These groups increase the adsorption tendency of different molecules on the cellular membrane, which may play an effective role in their antibacterial activity [97–100]. However, the exact mechanism of antimicrobial action is still unknown. Several other mechanisms may contribute to antimicrobial action, including the formation of an impermeable coating on the bacterial surface [101], uptake of low-molecular-weight biocides that will interact with electronegative substances in the cell [101], and inhibition of bacterial growth through chelation of trace metals [95]. The mechanisms of interaction for Gram-positive bacteria could be different from that for Gram-negative bacteria. Gram-negative bacteria are generally more resistant than Gram-positive bacteria to antimicrobial agents. This can be explained by the different cell membrane structures of the two types of bacteria [94]. The external layer of the outer membrane of Gram-negative bacteria is almost entirely composed of lipopolysaccharides and proteins that restrict the entry of biocides and amphiphilic compounds [94,102]. The perturbation of this outer membrane requires fine tuning of the hydrophobic–hydrophilic balance. In the current results, the tested biocides were more efficient against Gram-negative bacteria than Gram-positive bacteria. These compounds may be useful where conventional biocides have failed in defeating the Gram-negative bacterial strains growth.

The antimicrobial efficacies of the three tannic acid Schiff base surfactants considerably increased with complexation. The metal complexes of TGB_1 were more efficient than those of TGB_3 and TGB_5. The difference in activity of metal complexes may be explained on the basis of chelation theory. Chelation reduces the polarity of the metal atom mainly because of partial sharing of its positive charge with the donor groups and possible π-electron delocalization within the whole chelation. Also, chelation increases the lipophilic character of the central atom, which subsequently favors its permeation through the lipid layer of the cell membrane [103–105]. Tables 10.4 and 10.5 indicate that the three complexes are very active against all Gram-negative and Gram-positive bacteria (inhibitory zones <25 mm), whereas TGB_3–Fe and TGB_5–Fe complexes show moderate activities toward different bacterial strains (inhibitory zones <21 mm). In addition, no activity was observed against *S. aureus* and *P. aeruginosa* with TGB_5–Fe complex. The quantitative assays gave MIC values in the range 0.097–1.25 mg/mL (Table 10.5), which confirmed the above-discussed results.

10.3.5 Biodegradability

Biodegradation is the conversion of organic compounds into less complex structures under the influence of microorganisms. In aerobic conditions, this process results in the formation of water, carbon dioxide, and biomass. A surfactant molecule is ultimately converted into carbon dioxide, water, and low-toxicity smaller molecules. If the surfactant does not undergo natural biodegradation, then it is stable and persists in the environment. It was reported that the biodegradation rate of a surfactant molecule depends on its chemical structure, concentration, pH, and temperature. In traditional surfactants, the rate of biodegradation varied from 1–2 h for fatty acid-derived surfactants, 1–2 days for linear alkyl benzene sulfonates, and several months for branched alkyl benzene sulfonates. The temperature effect is particularly important, since the rate of biodegradation can vary by a factor of five between summer and winter [106]. Two criteria are important when testing for biodegradation: first, primary degradation that results in loss of surface activity, and second ultimate biodegradation, that is, conversion into carbon dioxide, which can be measured using closed bottle tests.

The biodegradability of the surfactants synthesized in this work was evaluated using the Die-away test. In this test, a standard solution of the surfactant is prepared in river water and the surface tension is measured at different interval times of 0, 1–7 days. Then, the surface tension of the used river water is measured at the same interval times. The biodegradation extents of the different surfactants were calculated using the methodology described in Ref. [78]. Since all the surfactants prepared in this investigation have the same hydrophilic part, only the number of substituents will affect biodegradability. The results of biodegradation using Die-away test in river water reflected the fact that lowering of the surface tension is a reverse function of biodegradation.

It is clear from the data in Table 10.6 that the biodegradation rate of the prepared compounds was in the range 80–100% after 28 days. In addition, it is clear that there is a direct relationship between the number of substituents (Ph–CH=N–COO–) attached to the tannic acid moiety and the percent biodegradation. Compounds TGB_1, TGB_3, and TGB_5 showed gradual increase in the degradation rate after 28 days. In case of Mn complexes, the biodegradation extent reached 80% after 40 days. While in case of Co complexes, the biodegradation extent reached 70% after 40 days. These values qualify the synthesized tannic acid Schiff base surfactants as biodegradable compounds surpass the international standard values (70% after 28 days) [107]. The metal complexes required longer time to reach the 70% biodegradation value. The metal complexes reached 70% biodegradation after 40 days, which is much better compared with that of branched alkyl of benzene sulfonates (several months). This allows us to classify the synthesized surfactants as readily biodegradable compounds, while their metal complexes can be classified as moderate biocides, which degraded in the environment after longer time. These good results are in contrast to the low biodegradation of classic nonionic surfactants [108]. In fact, the structures of the synthesized surfactants were designed from naturally occurring compounds (tannic acid and glycine), which can degrade by the action of environmental microorganisms. The biodegradation of tannic acid was confirmed by evidence of its degradation products (gallic acid and glucose) as reported by Mingshu et al. [109].

The reported pathway of biodegradation of the synthesized compounds is as follows: the microorganisms first attach to the tannic acid moiety and then the substituents are degraded through the β-oxidation pathway. This includes chain-shortening.

TABLE 10.6
Biodegradation of the Synthesized Surfactants in River Water[a]

Surfactant	Biodegradation (%)
TGB$_1$	100[b]
TGB$_3$	94[b]
TGB$_5$	80[b]
TGB$_1$-Mn	86[c]
TGB$_3$-Mn	83[c]
TGB$_5$-Mn	80[c]
TGB$_1$-Co	79[c]
TGB$_3$-Co	75[c]
TGB$_5$-Co	70[c]
TGB$_1$-Fe	_[d]
TGB$_3$-Fe	_[d]
TGB$_5$-Fe	_[d]

[a] Calculated using Equation 10.2: Biodegradation $(\%) = (\gamma_t - \gamma_o / \gamma_{bt} - \gamma_o) \times 100$. Bigger the number, more biodegradable.
[b] The indicated biodegradation percent was obtained after 28 days.
[c] The indicated biodegradation percent was obtained after 40 days.
[d] No values were recorded as the compounds were insoluble in water.

Finally, the microorganism completely degrades the hydrocarbon chain to carbon dioxide and water [110].

10.4 SUMMARY

Tannic acid and its derivatives can be used in several applications, including corrosion inhibition, emulsification, and medical and foodstuff additives. The practical part of the chapter showed that three tannic acid Schiff base derivatives and their metal complexes had good cytotoxicity against several bacterial and fungal strains. Complexation of the ligands increased cytotoxicity against the tested strains. The complexes containing one Schiff base moiety showed higher antimicrobial activity than those containing three or five moieties. The inhibition mechanism toward Gram-positive or Gram-negative bacteria was explained according to the complexation and adsorption theory. The compounds showed noticeable biodegradability in the environment as a comprise feature of their parent compound, tannic acid.

ACKNOWLEDGMENT

Nabel A. Negm is greatly thankful to Professor Dr. Satish V. Kailas, Department of Mechanical Engineering and Center for Product Design and Manufacturing, Indian Institute of Science, Bangalore, India for helping him in novel applications of the tannic acid derivatives.

REFERENCES

1. E. C. Smith and T. Swain, Flavonoid compounds, H. S. Mason and A. M. Florkin (Eds.), pp. 755–809. In: *Comparative Biochemistry*, Academic Press, New York, 1962.
2. P. J. Hernes, R. Benner, G. L. Cowie, M. A. Goni, B. A. Bergamashi, and J. I. Hedges, Tannin diagnosis in mangrove leaves from a tropical estuary: A novel molecular approach, *Geochim. Cosmochim. Acta*, 65, 3109–3122, 2001.
3. D. G. Roux, D. Ferreira, H. K. L. Hundt, and E. Malan, Structure, stereochemistry, and reactivity of natural condensed tannins as basis for their extended industrial application, *Appl. Polym. Symp.*, 28, 335–353, 1975.
4. C. Simon and A. Pizzi, Tannins/melamine-urea-formaldehyde (MUF) resins substitution of chrome in leather and its characterization by thermomechanical analysis, *J. Appl. Polym. Sci.*, 88, 1889–1903, 2003.
5. A. Pizzi, C. Simon, B. George, D. Perrin, and M. C. Triboulot, Tannin antioxidant characteristics in leather versus light stability, *Model J. Appl. Polym. Sci.*, 91, 1030–1040, 2004.
6. V. Sivakumar, V. Ravi Verma, P. G. Rao, and G. Swaminathan, Studies on the use of power ultrasound in solid-liquid myrobolan extraction process, *J. Clean Prod.*, 15, 1813–1818, 2007.
7. S. Garnier, A. Pizzi, O. C. Vorster, and L. Halasz, Rheology of polyflavanoid tannin-formaldehyde reactions before and after gelling II. Hardener influence and comparison of different tannins, *J. Appl. Polym. Sci.*, 86, 864–871, 2002.
8. S. Garnier and A. Pizzi, CHT, and TTT curing diagrams of polyflavonoid tannin resins, *J. Appl. Polym. Sci.*, 81, 3220–3230, 2001.
9. K. Li, X. Geng, J. Simonsen, and J. Karchesy, Novel wood adhesives from condensed tannins and polyethylenimine, *Int. J. Adhesion Adhesives*, 24, 327–333, 2004.
10. G. Vazquez, J. Gonzalez-Alvarez, F. Lopez-Suevos, and G. Antorrena, Effect of veneer side wettability on bonding quality of *Eucalyptus globulus* plywoods prepared using a tannin-phenol-formaldehyde adhesive, *Bioresour. Technol.*, 87, 349–353, 2003.
11. S. C. Chen and K. T. Chung, Mutagenicity and antimutagenicity of tannic acid and its related compounds, *Food Chem. Toxicol.*, 38, 1, 2000.
12. R. G., andrade, L. T. Dalvi, J. M. C. Silva, G. K. B. Lopes, A. Alonso, and M. H. Lima, The antioxidant effect of tannic acid on the *in vitro* copper-mediated formation of free radicals, *Arch. Biochem. Biophys.*, 437, 1–9, 2005.
13. I. Gulcin, Z. Huyut, M. Elmastas, and H. Y. Aboul-Enein, Radical scavenging and anti-oxidant activity of tannic acid, *Arab J. Chem.*, 3, 43–53, 2010.
14. R. C. Minussi, M. Rossi, L. Bologna, L. Cordi, D. Rotilio, G. M. Pastore, and N. Durán, Phenolic compounds and total antioxidant potential of commercial wines, *Food Chem.*, 82, 409–416, 2003.
15. M. Noferi, E. Masson, A. Merlin, A. Pizzi, and X. Deglise, Antioxidant characteristics of hydrolysable and polyflavonoid tannins, *J. Appl. Polym. Sci.*, 63, 475–482, 1997.
16. E. L. Szczapa, J. Korczak, M. N. Kalucka, and R. Z. Wojtasiak, Antioxidant properties of lupin seed products, *Food Chem.*, 83, 279–285, 2003.
17. D. Sud, G. Mahajan, and M. P. Kaur, Agricultural waste material as potential adsorbent for heavy metal ions from aqueous solutions, *Bioresour. Technol.*, 99, 6017–6027, 2008.
18. G. Palma, J. Freer, and J. Baeza, Removal of metal ions by modified *Pinus radiata* bark and tannins from water solutions, *Water Res.*, 37, 4974–4980, 2003.
19. O. R. Pardini, J. I. Amalvy, A. R. D. Sarli, R. Romagnoli, and V. F. Vetere, Formulation and testing of a waterborne primer containing chestnut tannin, *J. Coat Technol.*, 73, 99–111, 2001.
20. A. A. Rahim and J. Kassim, Recent development of vegetal tannins in corrosion protection of iron and steel, *Rec. Pat. Mater. Sci.*, 1, 223–231, 2008.

21. E. B. Yang, L. Wei, K. Zhang, Y. Z. Chen, and W. N. Chen, Tannic acid, a potent inhibitor of epidermal growth factor receptor tyrosine kinase, *J. Biochem.*, 139, 495–502, 2006.

22. R. Ravichandran, S. Nanjundan, and N. Rajendran, Effect of benzotriazole derivatives on the corrosion of brass in NaCl solutions, *Appl. Surface Sci.*, 236, 241–250, 2004.

23. E. S. Ferreira, C. Giacomelli, F. C. Giacomelli, and A. Spinelli, Evaluation of the inhibitor effect of L-ascorbic acid on the corrosion of mild steel, *Mater. Chem. Phys.*, 83, 129–134, 2004.

24. D. Jamroz, A. Wiliczkiewicz, J. Skorupińska, J. Orda, J. Kuryszko, and H. Tschirch, Effect of sweet chestnut tannin (SCT) on the performance, microbial status of intestine and histological characteristics of intestine wall in chickens, *Br. Poult. Sci.*, 50, 687–99, 2009.

25. G. Matamala, W. Smeltzer, and G. Droguett, Comparison of steel anticorrosive protection formulated with natural tannins extracted from *Acacia* and from *pine bark*, *Corros. Sci.*, 42, 1351–1362, 2000.

26. P. J. Des Lauriers, Rust conversion coatings, *Mater. Perform.*, 26, 35–39, 1987.

27. J. Gust, Application of infrared spectroscopy for investigation of rust phase component conversion by agents containing *oak* tannin and phosphoric acid, *NACE*, 47, 453–457, 1991.

28. J. A. Jaen, G. Saldana, and C. Hernandez, Characterization of reaction products of iron and aqueous plant extracts, *Hyperfine Interact.*, 122, 139–145, 1999.

29. A. A. Rahim, E. Rocca, J. Steinmetz, and M. J. Kassim, Inhibitive action of mangrove tannins and phosphoric acid on pre-rusted steel via electrochemical methods, *Corros. Sci.*, 50, 1546–1550, 2008.

30. J. Mabrour, M. Akssira, M. Azzi, M. Zertoubi, N. Saib, A. Messoudi, A. Albizane, and S. Tahiri, Effect of vegetal tannin on anodic copper dissolution in chloride solutions, *Corros. Sci.*, 46, 1833–1847, 2004.

31. C. A. Barrero, L. M. Ocampo, and C. E. Arroyave, Possible improvements in the action of some rust converters, *Corros. Sci.*, 43, 1003–1018, 2001.

32. O. O. Ajayi, O. A. Omotosho, K. O. Ajanaku, and B. O. Olawore, Degradation study of aluminum alloy in 2 M hydrochloric acid in the presence of *Chromolaena odorata*, *J. Eng. Appl. Sci.*, 6, 10–17, 2011.

33. W. Carlin and D. H. Keith, An improved tannin-based corrosion inhibitor-coating system for ferrous artifacts, *Int. J. Naut. Arch.*, 25, 38–45, 1996.

34. A. Ostovari, S. M. Hoseinieh, M. Peikari, S. R. Shadizadeh, and S. J. Hashemi, Corrosion inhibition of mild steel in 1 M HCl solution by henna extract: A comparative study of the inhibition by *henna* and its constituents, *Corros. Sci.*, 51, 1935–1944, 2009.

35. M. M. Cowan, Plant products as antimicrobial agents, *Clin. Microbiol. Rev.*, 12, 564–582, 1999.

36. K. H. Gustavson, *Chemistry of Tanning Processes*, Academic Press, New York, 403, 1956.

37. S. M. Colak, B. M. Yapici, and A. Yapici, Determination of antimicrobial activity of tannic acid in pickling process, *Romanian Biotech. Lett.*, 15, 5325–5330, 2010.

38. R. S. Carson and A. W. Frmsch, The inactivation of influenza viruses by tannic acid and related compounds, *J. Bact.*, 66, 572–575, 1953.

39. J. G. Melo, T. A. S. Araújo, V. T. N. Castro, D. L. Cabral, M. Rodrigues, S. Nascimento, E. L. Amorim, and U. P. Albuquerque, Antiproliferative activity, antioxidant capacity and tannin content in plants of semi-arid northeastern Brazil, *Molecules*, 15, 8534–8542, 2010.

40. L. R. Ferguson, Role of plant polyphenols in genomic stability, *Mutat. Res.*, 75, 89–111, 2001.

41. S. C. Chen and K. T. Chung, Mutagenicity and antimutagenicity studies of tannic acid and its related compounds, *Food Chem. Toxicol.*, 38, 1–5, 2000.

42. H. Rodriguez, B. Rivas, C. G. Cordoves, and R. Munoz, Degradation of tannic acid by cell-free extracts of *Lactobacillus plantarum*, *Food Chem.*, 107, 664–670, 2008.

43. K. T. Chung, T. Y. Wong, C. I. Wei, Y. W. Huang, and Y. Lin, Tannins and human health, *Crit. Rev. Food Sci. Nutr.*, 38, 421–464, 1998.

44. N. S. Khan, A. Ahmad, and S. M. Hadi, Anti-oxidant, pro-oxidant properties of tannic acid and its binding to DNA, *Chem. Biol. Interact.*, 125, 177–189, 2000.

45. T. J. Kim, J. L. Silva, W. L. Weng, W. W. Chen, M. Corbitt, Y. S. Jung, and Y. S. Chen, Inactivation of *Enterobacter sakazakii* by water-soluble *muscadine* seed extracts, *Int. J. Food Microbiol.*, 129, 295–299, 2009.

46. T. J. Kim, W. L. Weng, J. Stojanovic, Y. Lu, Y. S. Jung, and J. L. Silva, Antimicrobial effect of water-soluble *muscadine* seed extracts on *Escherichia coli* O157:H7, *J. Food Protect.*, 71, 1465–1468, 2008.

47. T. J. Kim, J. L. Silva, M. K. Kim, and Y. S. Jung, Enhanced antioxidant capacity and antimicrobial activity of tannic acid by thermal processing, *Food Chem.*, 118, 740–746, 2010.

48. R. Z. Yahiaoui, F. Z. Yahiaoui, R. F. Zaidi, A. Zaidi, and A. A. Bessai, Influence of gallic and tannic acids on enzymatic activity and growth of *Pectobacterium chrysanthemi*, *African J. Biotechnol.*, 7, 482–486, 2008.

49. A. Scalbert, Antimicrobial properties of tannins, *Phytochemistry*, 30, 3875–3883, 1991.

50. R. K. Upadhyay, Plant natural products: Their pharmaceutical potential against disease and drug resistant microbial pathogens, *J. Pharm. Res.*, 4, 1179–1185, 2011.

51. E. Haslam, Natural polyphenols (vegetable tannins) as drugs: Possible modes of action, *J. Nat. Prod.*, 59, 205–215, 1996.

52. J. L. Stern, A. E. Hagerman, P. D. Steinberg, and P. K. Mason, Phlorotannin protein interactions, *J. Chem. Ecol.*, 22, 1887–1899, 1996.

53. H. E. Brownlee, A. R. McEuen, J. Hedger, and I. M. Scott, Antifungal effects of *cocoa* tannin on the witches broom pathogen *Crinipellis perniciosa*, *Physiol. Mol. Plant Pathol.*, 36, 39–48, 1990.

54. G. A. Jones, T. A. McAllister, A. D. Muir, and K. J. Cheng, Effects of sainfoin (*Onobrychis viciifolia scop.*) Condensed tannins on growth and proteolysis by four strains of *ruminal* bacteria, *Appl. Environ. Microbiol.*, 60, 1374–1378, 1994.

55. R. S. L Taylor, F. Edel, N. P. Manandhar, and G. H. N. Towers, Antimicrobial activities of southern Nepalese medicinal plants, *J. Ethnopharmacol.*, 50, 97–102, 1996.

56. G. I. Nonaka, I. Nishioka, M. Nishizawa, T. Yamagishi, Y. Kashiwada, G. E. Dutschman, A. J. Bodner, R. E. Kilkuskie, Y. C. Cheng, and K. H. Lee, Anti-AIDS agents. 2. Inhibitory effects of tannins on HIV reverse transcriptase and HIV replication in H9 lymphocyte cells, *J. Nat. Prod.*, 53, 587–595, 1990.

57. J. J. Segura, L. H. Morales-Ramos, J. Verde-Star, and D. Guerra, Growth inhibition of *Entamoeba histolytica* and *E. invadens* produced by *pomegranate* root (*Punica granatum L.*), *Arch. Invest. Medic.*, 21, 235–239, 1990.

58. Z. Shoutang, Preparation and processing method for enterocleaning drug, Chinese Patent CN 929-21096-15, 1994.

59. J. V. Pereira, Atividade antimicrobianado estrato hidroalcoolico da P. granatum Linn, sobre microorganismos *Formadores daplaca* bacterium, Joao Pessoa, Federal University of Paraiba, Post-graduation in Dentistry. Thesis, 91, 1998.

60. L. C. S. Vasconcelos, M. C. C. Sampaio, F. C. Sampaio, and J. S. Higino, Use of *Punica granatum* as an antifungal agent against *candidosis* associated with denture stomatitis, *Mycoses*, 46, 192–196, 2003.

61. S. Petti and C. Scully, Polyphenols, oral health and disease: A review, *J. Dentis.*, 37, 413–423, 2009.

62. W. E. Lee and J. E. Rhoads, The present status of the tannic acid method in the treatment of burns, *J. Am. Med. Assoc.*, 125, 610–612, 1944.

63. H. Rosenquist, The primary treatment of extensive burns, *Acta Chir. Scand.*, 95, 1–128, 1947.

64. H. N. Harkins, The treatment of burns and freezing, In: D. Lewis, W. Walters, A. Blalock, et al. (eds). *Lewis' Practice of Surgery*, Volume 1. WF Prior Company Inc., pp. 1–177, 1955.

65. O. Cope, The end of the tannic acid era, *Harvard Med. Alumn. Bull.*, 65, 17–19, 1992.

66. P. Hupkens, H. Boxma, and J. Dokter, Tannic acid as a topical agent in burns: Historical considerations and implications for new developments, *Burns*, 21, 57–61, 1995.

67. H. J. Klasen, A historical review of the use of silver in the treatment of burns. I. Early uses, *Burns*, 26, 117–130, 2000.

68. H. J. Klasen, A historical review of the use of silver in the treatment of burns. II. Renewed interest for silver, *Burns*, 26, 131–138, 2000.

69. S. B. Halkes, Use of tannic acid in the local treatment of burn wounds: Intriguing old and new perspectives, wounds, *Health Management Publications, Inc.*, 13, 144–158, 2001.

70. N. A. Negm and M. F. Zaki, Surfactants in Solution, Paper presented at the 17th International Conference, Berlin, Germany, 17–22, 2008.

71. N. A. Negm, S. M. I. Morsy, and A. M. Badawi, Biological activities of some novel cationic metallomicelles, *Egypt. J. Chem.*, 48, 645–652, 2005.

72. N. A. Negm, Solubilization, surface active and thermodynamic parameters of gemini amphiphiles bearing nonionic hydrophilic spacer, *J. Surf. Deterg.*, 8, 71–80, 2007.

73. N. A. Negm, M. A. I. Salem, A. M. Badawi, and M. F. Zaki, Synthesis, surface and thermodynamic properties of some novel methyl diethanol ammonium bromide as cationic surfactants, Paper presented at the 7th International Conference of Chemical Engineering, Cairo, Egypt, 29, 2004.

74. N. A. Negm and A. S. Mohamed, Synthesis, characterization and biological activity of sugar-based gemini cationic amphiphiles, *J. Surf. Deterg.*, 11, 215–221, 2008.

75. R. X. Yan, *Water-Soluble Polymers*, Chemical Industry Press, Beijing, pp. 192–193, 1998.

76. N. A. Negm, I. A. Aiad, and S. M. Tawfik, Screening for potential antimicrobial activities of some cationic uracil biocides against wide spreading bacterial strains, *J. Surf. Deterg.*, 13, 503–511, 2010.

77. National Committee for Clinical Laboratory Standards (1997) Methods for dilution antimicrobial susceptibility tests for bacteria that grow aerobically. Approved standard M7-A4. Methods for dilution antimicrobial susceptibility tests for bacteria that grow aerobically. 4th Villanova, Pennsylvannia, National Committee for Clinical Laboratory Standards, 1997.

78. C. G. Naylor, J. B. Williams, P. T. Varineau, R. P. Yunick, K. Serak, C. Cady, and D. J. Severn, Biodegradation of the C_{14} ring label nonylphenol ethoxylate in activated sludge and in river water, Presented at the 19th Annual Society Environmental Toxicology and Chemistry, 1998.

79. S. A. Wakasman and H. A. Lechevalver, *The Actinomycetes Antibiotic of Actinomycetes*, Baltimore, Williams & Wilkins Co., USA, 430, 1962.

80. M. M. A. EL-Sukkary, N. A. Sayed, I. Aiad, S. M. Helmy, and W. I. M. EL-Azab, Aqueous solution properties, biodegradability, and antimicrobial activity of some alkyl-glycosides surfactants *Tenside Surf. Det.*, 46, 312–317, 2009.

81. M. Z. Mohamed, D. A. Ismail, and A. S. Mohamed, Synthesis and evaluation of new amphiphilic polyethylene glycol-based triblock copolymer surfactants, *J. Surf. Det.*, 8, 97–105, 2005.

82. N. A. Negm, Solubilization, surface active and thermodynamic parameters of Gemini amphiphiles bearing nonionic hydrophilic spacers, *J. Surf. Deterg.*, 10, 71–80, 2007.

83. N. A. Negm and I. A. Aiad, Synthesis and characterization of multifunctional surfactants in oil–field protection applications, *J. Surf. Deterg.*, 10, 87–92, 2007.

84. A. M. Alsabagh, D. R. K. Harding, N. G. Kandile, A. M. Badawi, and A. E. El-Tabey, Synthesis of some novel nonionic ethoxylated surfactants based on α-amino acids and investigation of their surface active properties, *J. Disp. Sci. Technol.*, 30, 427–438, 2009.

85. S. Gaur and B. Sharma, Synthesis of polystyrene anchored Schiff base its complexes with some transition metals *J. Ind. Chem. Soc.*, 80, 841–842, 2003.

86. K. B. Gudasi, P. B. Maravalli, and T. R. Goudar, Thermokinetic and spectral studies of niobium (V) complexes with 3-substituted-4-amino-5-mercapto-1,2,4-triazole Schiff bases, *J. Serb. Chem. Soc.*, 70, 643–650, 2005.

87. K. Singh, M. S. Barwa, and P. Tyagi. Synthesis and characterization of cobalt(II), nickel(II), copper(II) and zinc(II) complexes with Schiff base derived from 4-amino-3-mercapto- 6-methyl-5-oxo-1,2,4-triazine, *Eur. J. Med. Chem.*, 42, 394–402, 2007.

88. L. J. Bellamy, *The Infrared Spectra of Complex Molecules*, 3rd ed. Methuen, London, 1996.

89. R. M. Silverstein, G. Bassler, and G. B. Morril, *Spectrometric Identification of Organic Compounds*, 5th ed. John Wiley and Sons, New York, 1991.

90. N. A. Negm and A. S. Mohamed, Surface and thermodynamic properties of diquaternary bola-form amphiphiles containing aromatic spacer, *J. Surf. Deterg.*, 7, 23–30, 2004.

91. M. J. Rosen, *Surfactants and Interfacial Phenomena*, 2nd ed. John Wiley & Sons, New York, 1989.

92. H. S. Frank and M. W. Evans, Free volume and entropy in condensed systems. III. Entropy in binary liquid mixtures; partial molal entropy in dilute solutions; structure and thermodynamics in aqueous electrolytes, *J. Chem. Phys.*, 13, 507, 1945.

93. C. Tanford, *The Hydrophobic Effect: Formation of Micelles and Biological Membranes*, 2nd ed. Wiley, New York, 1980.

94. L. Pérez, A. Pinazo, M. T. García, M. Lozano, A. Manresa, M. Angelet, M. P. Vinardell, M. Mitjans, R. Pons, and M. R. Infante, Cationic surfactants from lysine: Synthesis, micellization and biological évaluation, *Eur. J. Med. Chem.*, 44, 1884–1892, 2009.

95. E. I. Rabea, M. E. T. Badawy, C. V. Stevens, G. Smagghe, and W. Steurbaut, Chitosan as antimicrobial agent: Applications and mode of action, *Biomacromolecules*, 4, 1457–1465, 2003.

96. F. Devlieghere, A. Vermeulen, and J. Debevere, Chitosan: Antimicrobial activity, interactions with food components and applicability as a coating on fruit and vegetables, *Food Microbiol.*, 21, 703–714, 2004.

97. (a) L. Shi, H. M. Ge, S. H. Tan, H. Q. Li, Y. C. Song, H. L. Zhu, and R. X. Tan, Synthesis and antimicrobial activities of Schiff bases derived from 5-chloro-salicylaldehyde, *Eur. J. Med. Chem.* 42, 558–564, 2007; (b) J. Lv, T. Liu, S. Cai, X. Wang, L. Liu, and Y. Wang, Synthesis, structure and biological activity of cobalt (II) and copper (II) complexes of valine-derived Schiff bases, *J. Inorg. Biochem.*, 100, 1888–1896, 2006.

98. G. G. Mohamed and Z. H. Abd El-Wahab, Mixed ligand complexes of bis-(phenylimine) Schiff base ligands incorporating pyridinium moiety: Synthesis, characterization and antibacterial activity, *Spectrochim. Acta A*, 61, 1059–1068, 2005.

99. N. Sari, S. Arsalan, E. Logoglu, and I. Sakiyan, Antibacterial activities of some Amino acid Schiff bases, *G.U.J. Sci.*, 16, 283–288, 2003.

100. F. C. de Silva, M. C. B. V. Souza, I. I. P. Frugulhetti, H. C. Castro, S. L. Souza, T. M. L. Souza, D. Q. Rodrigues et al., Synthesis, HIV–RT inhibitory activity and SAR of 1-benzyl-1H-1, 2,3-triazole derivatives of carbohydrates, *Eur. J. Med.*, 44, 373–383, 2009.

101. L. Y. Zheng and J. A. F. Zhu, Study on antimicrobial activity of chitosan with different molecular weights, *Carbohyd. Polym.*, 54, 527–530, 2003.

102. G. Oros, T. Cserhati, and E. Forgacs, Separation of the strength and selectivity of the microbiological effect of synthetic dyes by spectral mapping technique, *Chemosphere*, 52, 185–193, 2003.

103. A. S. Gaballa, M. S. Asker, A. S. Barakat, and S. M. Teleb, Synthesis, characterization and biological activity of some platinum (II) complexes with Schiff bases derived from salicylaldehyde, 2-furaldehyde and phenylenediamine, *Specrochim. Acta Part A: Mol. Biomol.*, 67, 114–121, 2007.
104. K. B. Chew, M. T. H. Tarafder, K. A. Crouse, A. M. Ali, B. M. Yamin, and H. K. Fun, Synthesis, characterization and bio-activity of metal complexes of bidentate N-S isomeric Schiff bases derived from S-methyldithiocarbazate (SMDTC) and the X-ray structure of the bis-[S-methyl-β-N-(2-furyl-methylketone) dithiocarbazato] cadmium (II) complex, *Polyhedron*, 23, 1385–1392, 2004.
105. K. Singh and P. P. Dharampal, Synthesis and spectroscopic studies of some new organometallic chelates derived from bidentate ligands, *Turk. J. Chem.*, 34, 499–507, 2010.
106. T. F. Tadros, *Applied Surfactants: Principles and Applications*, Chapter 10. Microemulsions, Wiley-VCH Verlag, Weinheim, 2005.
107. N. A. Negm, N. G. Kandile, and M. A. Mohamad, Synthesis, characterization and surface activity of new eco-friendly schiff bases vanillin derived cationic surfactants, *J. Surf. Deterg.*, 14, 325–331, 2011.
108. J. S. Leal, J. J. Gonzalez, K. L. Kaiser, V. S. Palabrica, F. Comelles, and M. T. Garcia, On the toxicity and biodegradation of cationic surfactants, *Acta Hydrochim. Hydrobiol.*, 22, 13–21, 1994.
109. L. Mingshu, Y. Kai, H. Qiang, and J. Dongying, Biodegradation of gallotannins and ellagitannins., *J. Bas. Microbiol.*, 46, 68–84, 2005.
110. C. Ratledge, Biodegradation of oils, fats and fatty acids, In: Ratledge C. (Ed.), *Biochemistry of Microbial Degradation.* Kluwer, Amsterdam, p. 89, 1994.

11 Elastohydrodynamics of Farm-Based Blends Comprising Amphiphilic Oils*

Grigor B. Bantchev and Girma Biresaw

CONTENTS

---

* The mention of trade names or commercial products in this publication is solely for the purpose of providing specific information and does not imply recommendation or endorsement by the U.S. Department of Agriculture. The USDA is an equal opportunity provider and employer.

Vegetable oils contain nonpolar hydrocarbon chains and polar ester groups (and possibly also other functional groups such as hydroxyl groups in castor oil). The presence of polar and nonpolar groups within the same molecule gives vegetable oil an amphiphilic character. The density, refractive index, viscosity, pressure–viscosity coefficient, and elastohydrodynamic film thickness of neat oils, and binary blends of vegetable oils or estolides with synthetic esters, polyalpha olefins, and polyglycols are discussed. Several literature models for predicting blend properties from neat oil properties are described and compared with experimental data. The effect of vegetable oil amphiphilicity and aging on elastohydrodynamic film thickness of lubricating blends are discussed. The effect is most pronounced in the ultralow film thickness (below 20 nm) regime. This lubrication regime is very critical since wear rate starts increasing rapidly with decreasing film thickness.

11.1   INTRODUCTION

11.1.1   Amphiphilicity of Vegetable Oils

Small molecules can be classified into nonpolar (e.g., CH_4) and polar (e.g., H_2O), based on the difference in the electronegativities between the bonded atoms. The nonpolar molecules contain atoms with similar electronegativities (like C and H), whereas in polar molecules, atoms with a wide difference of electronegativities (like C and O, H and O, or Al and Cl) are present. Polarity influences the miscibility between compounds. Polar molecules are soluble in polar solvents while nonpolar molecules are miscible in nonpolar solvents (like hexane). On the other hand, polar molecules are insoluble in nonpolar solvents and vice versa.

Lubricants typically contain base oils, consisting of molecules with a large number of atoms (50–1000). Such molecules can be amphiphilic, which is a term used to describe that the molecule has both polar and nonpolar segments. Figure 11.1 shows a triglyceride, a type of molecule that is the main component of vegetable oils (VO). The polar bonds between highly electronegative oxygen atoms and less electronegative carbon atoms are shown in bold. It can be seen that the polar bonds are clustered close to each other, away from most of the nonpolar bonds. Thus, the triglyceride molecules show an amphiphilic character—different parts of the molecules have distinct affinities toward surfaces, based on their polarities.

Most metals are covered with their oxides. Since metals have low electronegativity and oxygen is highly electronegative, the oxides, and correspondingly, the metal

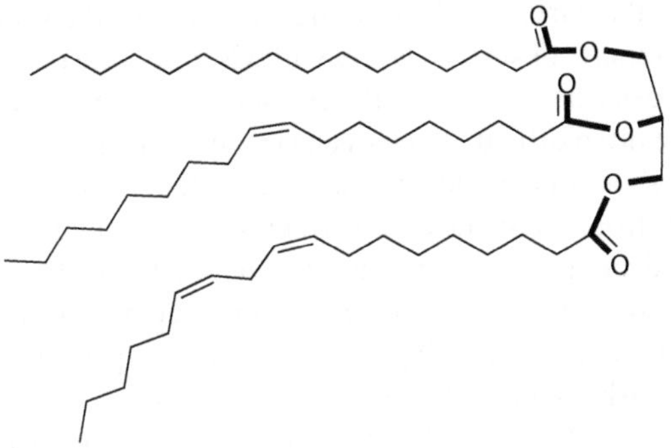

FIGURE 11.1   An example of a triglyceride molecule. The polar groups are displayed in bold.

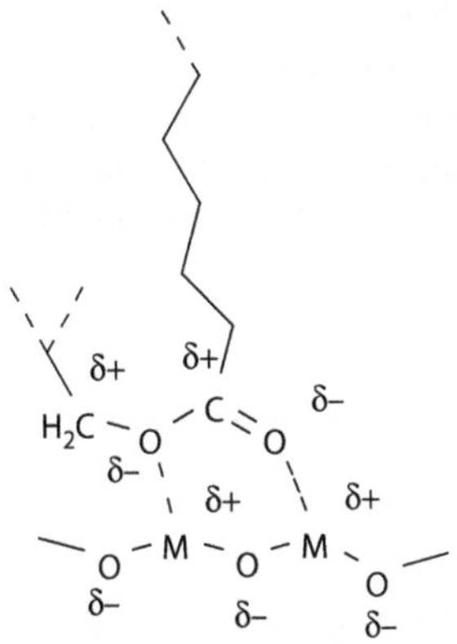

FIGURE 11.2   Interaction between dipoles in one of the ester groups of a triglyceride molecule and the metal atoms at the surface. The oxygen atoms have partial negative charge (denoted by $\delta-$) and metal atoms have partial positive charge ($\delta+$). This leads to attraction between the two, represented by the dashed lines.

surfaces, have a polar character. This will cause the polar parts of a triglyceride molecule to have a higher attraction toward the metal surfaces (Figure 11.2). These attractive forces are dipole–dipole, and have a short range of action, on the order of nanometers or less. These forces help the molecules to stay on the surface and not be squeezed out when another surface is approaching. This means that a direct metal-to-metal contact will be prevented. Thus, the presence of a polar segment in the molecule leads to better boundary lubrication.

11.1.2 ADSORPTION

If a lubricant is composed of a mixture of molecules with different polarities, then more polar molecules will displace less polar molecules from the surface. The adsorption of polar molecules on surfaces leads to their immobilization, which means that the layers close to the surface will have higher viscosity than the molecules away from it. There are reports of surface layers with viscosity orders of magnitude higher than the bulk viscosity or even behaving like solids [1,2]. Such layers are generally very thin and quite often difficult to characterize. As already mentioned, the dipole–dipole forces are short range and the increase in the viscosity is detectable only within nanometers from the surface. Nevertheless, these nanometer-thick films play a crucial role in reducing the wear and coefficient of friction (COF) in the boundary and mixed lubrication regimes, which are characterized by various degrees of direct metal-to-metal contact of the moving parts. As an example, Figure 11.3 shows the COF in a ball-on-disk tribometer. The experiments were in the boundary regime (12.7 mm diameter ball, 181 kg load, 6.22 mm/s). The lubricant was hexadecane with varying amounts of soybean oil (SBO). It can be seen that even a small concentration of SBO substantially decreased the COF. This is because the triglyceride molecules adsorb on the steel surface and form a protective layer. Increasing the concentration of triglyceride leads to higher coverage of the surface and, correspondingly, lower friction. At approximately 20 mM SBO in hexadecane, the surface has boundary layer coverage close to the maximum amount. This leads to the COF reaching values close to the minimum. Further increase in the SBO concentration leads to only small changes in the adsorption amount and COF. More details can be found in the literature [3].

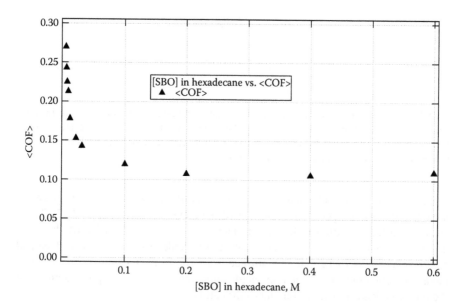

FIGURE 11.3 Coefficient of friction (COF) between a moving ball and a stationary disc, lubricated by hexadecane with dissolved soybean oil (SBO). Small amounts of SBO in hexadecane lead to a big decrease of COF between the moving parts, due to the formation of a boundary layer. (With kind permission from Springer Science+Business Media: *J. Am. Oil Chem. Soc.*, Friction and adsorption properties of normal and high-oleic soybean oils, 79, 2002, 53–58, G. Biresaw et al.)

11.1.3   ELASTOHYDRODYNAMIC FILM THICKNESS

The optimum lubrication regime for moving parts is the regime in which there is no direct contact between the moving parts, but are separated by a thin lubricating film. A challenging problem in lubrication is the elastohydrodynamic (EHD) regime. In this regime, the film thickness is strongly influenced by two phenomena, caused by the contact pressure between the parts. The first is the increase of the lubricant density and viscosity, and the second is the elastic deformation of the contacting parts.

Figure 11.4 shows an EHD film thickness measurement apparatus. The essential components are a rotating glass disc and a freely rolling ball pressed against it. There is also an optical system that illuminates the contact zone and collects the reflected light. Figure 11.5 shows a schematic of the contact zone between the moving glass disc and the steel ball. In the area of apparent contact, the pressure is high and it

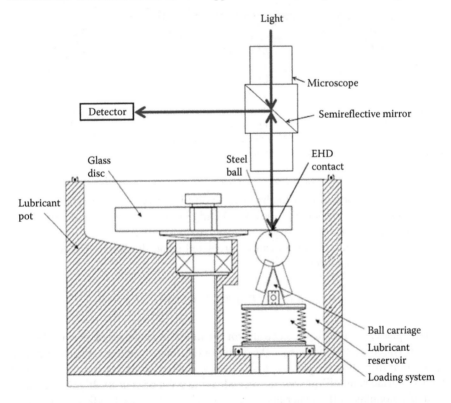

FIGURE 11.4   Illustration of the elastohydrodynamic instrument used in this study. A steel ball is pressed against a glass disc by the loading system. The ball carriage allows free rotation of the ball. The glass disc is rotated by a motor and gear system (not shown), allowing precise control of speed. When the system is engaged, the motion of the disc causes the ball to rotate. The contact zone is illuminated through a microscope. A semireflective mirror allows a microscope to be used for both illumination and observation of the contact zone. The detector contains a diffraction grating, CCD detector, and computer with software for analyzing the signal.

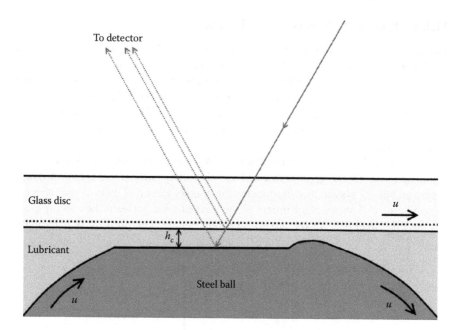

FIGURE 11.5 Schematic of the contact zone in the elastohydrodynamic system (not to scale). A steel ball is pressed against a glass disc, which rotates with speed u. The ball moves with the same speed, due to traction. Interference pattern analysis of the reflected light allows determining the lubricant film thickness h_c.

causes elastic deformation of the ball (shown in Figure 11.5) and the glass disc (the deformation of the glass is not shown). The high pressure in this region also compresses the lubricant and increases its density and viscosity.

The lubricant film thickness can be predicted using the Hamrock and Dowson (H–D) numerical solution [4]:

$$h_c = c_{H-D} R'(u\eta/(E'R'))^{0.67}(\alpha E')^{0.53}(W/(E'R'^2))^{-0.067} \qquad (11.1)$$

In Equation 11.1, h_c is the central film thickness in nanometers, that is, the distance between the two surfaces in the center of the contact zone. u is the entrainment speed in millimeters per second, $u = (u_1 + u_2)/2$, where u_1 and u_2 are the speeds of the two surfaces in millimeters per second, at the point of contact. E' is the reduced Young's modulus of the materials in pascals. R' is the reduced radius of curvature in meters. η is viscosity of the lubricant at ambient pressure in millipascals–seconds (mPa s). W is the load in newtons [4]. α is the pressure–viscosity coefficient (PVC) in GPa^{-1}, and it describes how the viscosity changes with pressure. c_{H-D} is a constant dependent on the geometry of the instrument. According to Equation 11.1, the only lubricant properties influencing film thickness are the viscosity and PVC. In the H–D model and in the current study, the system under investigation contains sufficient amount of lubricant (fully flooded contact).

In deriving Equation 11.1, Hamrock and Dowson used Roelands' formula [5] for the viscosity dependence on pressure:

$$\log\eta_p + 1.2 = (\log\eta + 1.2)(1 + p/1.9609 \times 10^8)^Z \tag{11.2}$$

$$\alpha = 1.151 \times 10^{-3} Z (\log\eta + 1.2) \tag{11.3}$$

In Roelands' formula, η_p is the viscosity (in mPa s) at pressure p (in Pa) and Z is a dimensionless parameter called the viscosity–pressure index. The strict definition of α is

$$\alpha = (\partial\ln\eta/\partial p)_T \tag{11.4}$$

Equations 11.2 and 11.3 have been derived for mineral oil and Roelands claims that it fits well with the data of many lubricating oils [5]. The validity of Equations 11.2 and 11.3 for VO has not been extensively studied.

In this chapter, we discuss the effect of the amphiphilic character of VO on the film thickness of lubricating blends in EHD contact. The variation of viscosity with blend composition is discussed. Film thickness measurement of moderately thin films (20–500 nm) is reported. The data from the measurements are used to estimate the PVC. Finally, the film thickness in ultrathin regime (below 20 nm), where the adsorption effects can be detected, is reported.

11.2 EXPERIMENTAL

11.2.1 MATERIALS

The chemical structures of the oils, used in this study, are illustrated in Figure 11.6. SBO and canola oil were obtained from the local supermarket and used as received. High-oleic sunflower oil (HOSO) was a sample from KIC Chemicals (New Paltz, NY). The oils contain triglycerides of fatty acids (FA). Most FA have 18 carbon atoms and a varying number of double bonds (stearic—no double bonds, oleic—one double bond, linoleic—two double bonds, and linolenic—three double bonds), as shown in Figure 11.6 [6]. Castor oil was obtained from Fisher Scientific (Pittsburgh, PA, USA). The difference in the FA structure between castor oil and the rest of VOs is that the major FA component (89%) of castor oil is ricinoleic acid, which has 18 carbons with a cis-double bond between carbons 9 and 10 and a hydroxy group located at carbon 12 (Figure 11.6).

Polyalphaolefins (PAO), with kinematic viscosity at 100°C of approximately 2 (PAO2), 4 (PAO4), 6 (PAO6), or 40 mm²/s (PAO40), were free samples from INEOS Oligomers (League City, TX, USA).

The polyol esters were from Hatco Corp. (Fords, NJ, USA). They were penta-erythritol esters (PE), dipentaerythritol esters (DPE), and complexed ester of tri-methylolpropane (TMP) with linear or branched short-chain FA.

Oleic estolides (ESTs) are oligomers with repeating units of hydroxy C18 FA (see Figure 11.6d). The hydroxy group is in the middle of the fatty acid chain. The ESTs

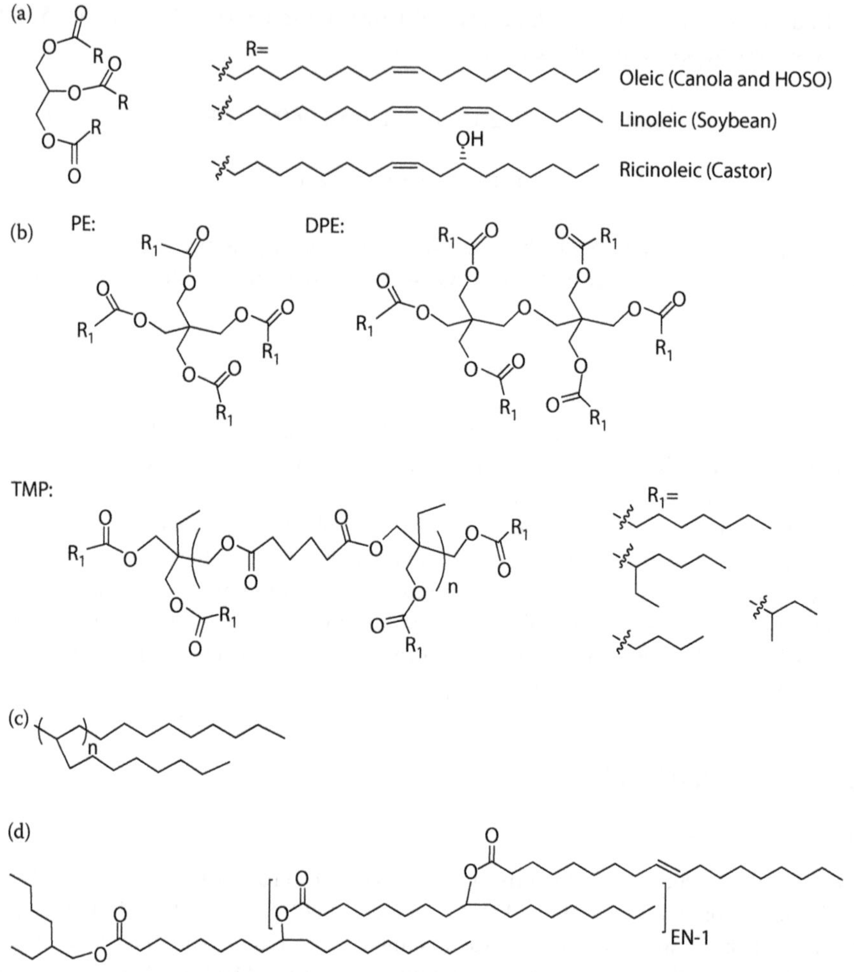

FIGURE 11.6 Chemical structures of investigated oils. (a) Triglycerides (with some examples of fatty acids), (b) polyol esters (with some examples of short branched and linear fatty acids), (c) polyalphaolefin (PAO), and (d) 2-ethylhexyl ester of an oleic estolide (EST) (the average estolide number (EN) is 2).

are synthesized by oligomerization of oleic acid in the presence of acid catalyst [7]. During the reaction, the acid group of an oleic acid molecule reacts with the double bond from another molecule, forming an ester group (Figure 11.7). For lubricant applications, the acid group of the estolide is esterified with an alcohol, to obtain an estolide ester. In the current study, an estolide ester, obtained by esterification of an oleic estolide with 2-ethylhexyl alcohol (2-EH) was used. It will be denoted as EST. The estolide number (EN) is a measure of estolide oligomerization. For EST, the average EN was 2, which means that in an EST molecule, there are on an average three ester groups (two hydroxy FA and one 2-EH ester group).

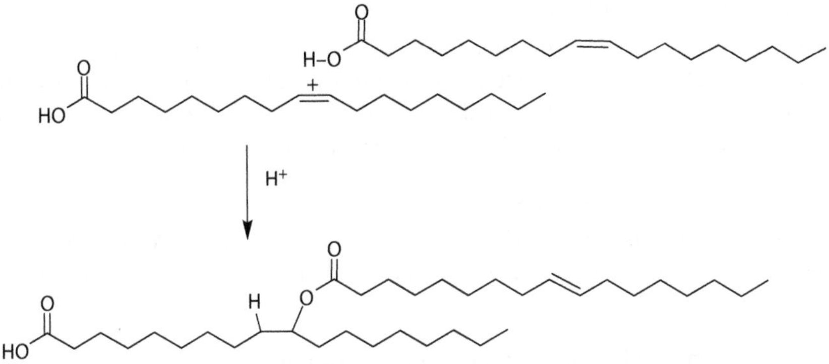

FIGURE 11.7 The reaction of formation of an estolide. During the reaction, some of the double bonds migrate and/or isomerize from *cis* to *trans* configuration. The migration results in some of the ester groups being formed at other atoms (6, 7, 11, and 12) rather than expected from the original formula (atoms 9 or 10).

Hexane, ethanol, isopropyl alcohol, and potassium hydroxide, used for cleaning the glassware and the EHD instrument (described below), were from Fisher Scientific (Pittsburgh, PA, USA). Florisil, used for the purification of VO, was from Fisher Scientific (Pittsburgh, PA, USA). Silica and alumina, used for the purification of PAO, were from Sorbent Technology (Atlanta, GA, USA). 2,6-Di-*tert*-butyl-4-methylphenol (BHT) was purchased from Fisher Scientific, and was a product of Acros Organics (New Jersey, USA).

11.2.2 METHODS

11.2.2.1 Purification of Oils

Oils were purified by filtration through an adsorbent to remove traces of surface active compounds that may be present due to contamination during manufacturing or oxidation during storage. Approximately 1 L of VO was filtered through 40–50 g of florisil. The rest of the oils were purified using 40–50 g silica plus 40–50 g alumina. The purity of the oil was confirmed by measurements of the interfacial tension between water droplet and oil, using axy-symmetric drop-shape analysis method. The oils before purification had a dynamic interfacial tension that decreased with time, whereas the interfacial tensions with the purified oils were higher and also did not change significantly with time. The purified oils were protected against oxidation with addition of 0.1 wt.% BHT.

11.2.2.2 Cleaning of Glassware and Instruments

Glassware was cleaned in an alkaline bath, containing ethanol, deionized water, and potassium hydroxide. It was immersed in the solution overnight, rinsed afterwards with a copious amount of deionized water, and left to dry. The instruments were cleaned thoroughly with alcohol, allowed to dry, cleaned with hexane, and allowed to dry again. It is important to use solvents that do not have nonvolatile contaminants.

11.2.2.3 Refractive Index

The refractive index (RI) of the blends was measured to four-digit precision using a model Mark II Plus Abbe Refractometer from Reichert Inc. (Depew, NY, USA). The measurement temperature was varied from ambient to 70°C. Linear regression of RI versus temperature data was used to obtain the RI value at 100°C.

11.2.2.4 Viscosity and Density

Dynamic viscosity and density of the oils were measured on an SVM-3000 Stabinger viscometer from Anton Paar (Graz, Austria). The data were used to calculate kinematic viscosities.

11.2.2.5 Film Thickness: Measurement Instrument

Lubricant film thickness was measured by optical interferometry method between a rotating glass disc and a stainless-steel ball using an EHD ultra thin-film measurement system (PCS Instruments, London, England). The ball is pressed with a controlled force against the coated side of the disc (Figure 11.4). The disc is rotated by a motor and the ball rotates due to traction with the disc. The ball is partially immersed in the lubricant and the rotation of the ball and the disc causes a thin lubricant film to form between the two. From the top, the contact between the disc and the ball is illuminated with a white light and observed with a microscope and a charge-coupled device (CCD) camera. The glass disc, on the side of the contact with the ball, is coated with two layers—a semitransparent chromium layer and an approximately 500-nm-thick silica spacer layer. Part of the light is reflected from the chromium layer part from the lubricant–glass interface, and the rest from the ball. The reflected light beams interfere and a pattern is formed, which depends on the optical distance between the reflecting surfaces. The interference pattern is analyzed by splitting it with a diffraction grating, capturing it with a CCD camera, and determining the wavelength of maximum intensity using a computer and appropriate software. The thickness of the film is calculated from the value of the wavelength of maximum intensity. The system allows for measurement of film thickness with precision down to 1 nm [8]. A detailed explanation of the instrument and the procedure has been given before [8,9].

11.2.2.6 Film Thickness Measurements: Procedure

All film thickness measurements were carried out in a laboratory with controlled temperature (25°C) and humidity (50%). Prior to use in film thickness measurements, the glass disc and the balls were thoroughly cleaned with hexane and allowed to dry. For each blend, a new unused track on the glass disc and a new stainless-steel ball were used. The ball was placed into the carriage, about 100 mL oil was poured into the instrument pot, and the glass disc was mounted. The pot temperature was set to the required value (30°C, 40°C, 70°C, or 100°C). The time to heat the instrument to the desired temperature varied between 15 and 60 min, and afterwards the system was allowed to equilibrate thermally for 15 more min. The ball was then pressed against the glass disc with 20 N load (0.53 GPa, contact radius 135 µm) and the thickness of the silica spacer layer was determined. The spot on the disc, where the thickness of the silica spacer is measured is called "the trigger point," and is the location where the film thickness is measured. The load was then released; the rotation of the disc started at the lowest speed, and the ball engaged against the disc with the selected load (10, 20, 30, or 40 N). Most measurements were conducted at 20 N load. The software automatically takes three consecutive measurements of film thickness at the trigger point, averages the measured values, and rejects them if the standard deviation is more than 5 nm. The film thickness measurement was carried out, and the speed was changed to the next value, usually 20% higher than the previous. The measurement procedure was repeated until the maximum speed (~2.5 m/s) was reached. After the film thicknesses at various speeds were scanned, the ball was disengaged, the disc stopped, and the spacer layer thickness at the trigger point remeasured. The next scan was carried out at the required new experimental conditions, which may be at different load, higher temperature, or after the system was left to age.

After each scan, the thickness of the spacer layer at the trigger point was checked again by pressing the ball against the stationary disc. If the spacer layer thickness decreased by the amount larger than the precision of the instrument (1 nm), it was assumed that the disc had been damaged and the measured values were corrected accordingly.

In some experiments, the viscosity and the density of the tested oil were remeasured to determine whether deterioration had occurred due to oxidation, shear degradation, or contamination.

The instrument measures optical film thickness, which is the product of film thickness and RI, n. In order to determine the actual film thickness, n of the fluid is required. The n at elevated pressure can be calculated, using the Lorentz–Lorenz equation [10]:

$$(n^2 - 1)/(n^2 + 2) = \rho(n_r^2 - 1)/(\rho_r(n_r^2 + 2)) \tag{11.5}$$

where ρ is the density. The n_r and ρ_r are the RI and the density at the reference pressure and temperature. Since the data for n or ρ under pressure are often unavailable, the change in n with pressure is often neglected. This leads to some error in the film thickness calculations. However, since a reference fluid is also used in the film thickness measurement, the errors due to changes in the refractive indices with pressure for the reference and test fluids cancel each other to a large extent.

11.2.2.7   Calculation of Effective Pressure–Viscosity Coefficient

For experiments conducted on the same instrument, where the instrument parameters (load, ball and disc dimensions, and materials) are constant, the H–D equation simplifies to

$$h_c = c(u\eta)^{0.67}\alpha^{0.53} \tag{11.6}$$

where c is a constant. c can be determined by measuring the film thickness of a reference fluid with known viscosity and PVC. In the current studies, PAO6 [11] or PAO4 [12] were used as reference fluids. After the instrument constant c is determined, α of the test fluids can be calculated using the film thickness versus entrainment speed and the viscosity data for the fluids [9,13–15]. In the current manuscript, the PVC calculated by this method will be referred to as the "effective PVC" or α. In our experiments, the film thickness data in the range of 20 (or even 50) to 200 nm were selected for calculations of α. Lower and higher film thicknesses were not used because of potential contribution by factors unaccounted for in the H–D model.

The method for determining the α was used with slight modification, since some other researchers have derived equations, similar to Equation 11.1, but with slightly different powers (see e.g., [16]). If the actual expression is $h_c \sim (u\eta)^a$, where a is different from 0.67, the use of Equation 11.6 will lead to some error in α determination. To avoid this type of error, α is calculated from the ratios of the film thickness at the condition

$$u\eta = u_r\eta_r = \text{const} \qquad (11.7)$$

When the condition (11.7) is satisfied, it can be combined with Equation 11.6 for the reference and the test liquids [10] to give

$$(\alpha/\alpha_r)^{0.53} = h_c/(h_c)_r \qquad (11.8)$$

In Equation 11.8, the subscript "r" indicates reference fluid.

11.3 RESULTS AND DISCUSSION

11.3.1 DENSITY

11.3.1.1 Dependence of Density on Temperature

The change in the density of a liquid is described by the coefficient of thermal expansion, β_V:

$$\beta_V = (1/V)(\partial V/\partial T) = -(1/\rho)(\partial \rho/\partial T) \qquad (11.9)$$

For almost all fluids (water between 0°C and 4°C being one exception), β_V is positive, that is, density decreases with increase of the temperature. In general, β_V is a function of temperature. For nonassociating fluids, Rackett [17] proposed an equation for density as a function of temperature:

$$\ln(\rho) = \ln(\rho_c) - (1 - T/T_c)^{2/7}\ln(Z_c) \qquad (11.10)$$

$$\rho_c = M_W R_G T_c/P_c \qquad (11.11)$$

where the subscript "c" refers to the properties of the liquid at a critical point: pressure (P_c), temperature (T_c), and density (ρ_c), R_G is the universal gas constant, and M_W is the molecular weight. Z_c is the critical compressibility, a material constant with typical values between 0.2 and 0.3.

The Rackett model requires knowledge of the critical temperature of a liquid. For lubricating oils, it is usually above the temperature of oil degradation, which is above the temperature range of their practical use. The use of the Rackett model or its modifications [18,19] requires theoretical calculations of the critical properties. Such methods yield for typical VO values of T_c above 500°C and P_c above 1.2 MPa [18,19]. These values indicate that, for lubricating oils, Equation 11.12 approximates Equation 11.10 with good precision:

$$\ln(\rho) = \ln(\rho_{ref}) - (T - T_{ref})\beta_V \tag{11.12}$$

ρ_{ref} is the density of the VO at the reference temperature, T_{ref}, T is the temperature at which the density ρ is needed, and β_V is assumed to be independent of temperature.

For liquid with $T_c = 500°C$, Equation 11.12 approximates Equation 11.10 in the temperature range $-30°C$ to $120°C$ with a relative error of less than 0.3%. The relative error decreases further if a narrower temperature range is used, or if $T_c > 500°C$. This reasoning, and the simplicity of Equation 11.12, justified its use for describing our experimental data.

Our experimental density data for all the neat oils and blends gave excellent fit to Equation 11.12. The data for ρ_{ref}, β_V, and the experimental temperature interval, used for their determination, are summarized in Table 11.1. In the lubrication literature, 15°C is often chosen for reporting densities of lubricants. We, thus, used a reference temperature (T_{ref}) of 15°C in the linear regression. The R^2 of the fits was at least 0.999, in most cases above 0.9998. Table 11.1 as well as Table 11.2 contain the maximum absolute deviation (MAD), in percent. MAD is calculated from the maximum absolute difference between experimental and predicted data obtained using the parameters in the Tables. The assumptions that the density or the volume (instead of their logarithm, used in Equation 11.12) should be a linear function of the temperature fit the data less accurately, but still with R^2 above 0.998 and, in most cases, above 0.999. These alternative fits are not shown.

TABLE 11.1
Density (ρ_T, $T = 15°C$) and the Coefficient of Thermal Expansion (β_V) of Base Oils

Base Oil Name and (Abbreviation)	$\rho_{15°C}$, g/cm^3	β_V, °C^{-1} ×10^4	T_{min},°C	T_{max},°C	MAD,%
Soybean (SBO)	0.9228	7.43	30	100	0.029
Canola	0.9204	7.47	30	100	0.010
High-oleic sunflower (HOSO)	0.9160	7.46	−5	100	0.047
Castor	0.9632	7.28	20	100	0.018
2-EH ester of oleic estolide (EST)	0.9081	7.38	30	100	0.015
Polyalphaolefin (PAO2)	0.7968	8.73	30	100	0.011
Polyalphaolefin (PAO4)	0.8190	7.98	10	100	0.010
Polyalphaolefin (PAO40)	0.8493	7.20	30	100	0.018
Pentaerythritol ester (PE)	0.9651	7.65	20	100	0.020
Dipentaerythritol ester (DPE)	0.9730	7.21	20	100	0.021
Trimethylolpropane complexed ester (TMP)	1.0198	7.14	30	100	0.024
Oil-soluble polyalkyl glycol (PAG32)	0.9445	7.94	10	100	0.034
Oil-soluble polyalkyl glycol (PAG220)	0.9795	7.17	10	100	0.038

Note: The temperature range $(T_{min}-T_{max})$ used for their determination and the deviation between the experimental data and Equation 11.12 are also given.

There was an overall trend of higher density at 15°C for oils with more polar groups per molecule. VO showed slight increase of density with decrease in the oleic acyl group content (HOSO—81% oleic, canola—61%, and SBO—28%). This may be due to the fact that in the oils under study, the oleic acyl groups are replaced by linoleic and linolenic acyl groups, which have more double bonds. The ricinoleic acyl group, which is the main component of castor oil, differs from the oleic acyl group due to the presence of OH (hydroxyl) group in the middle of the fatty acid chain. The hydrogen-bond-forming ability of the hydroxyl group contributes to the increase of density since castor oil has a much higher density than HOSO.

The data in Table 11.1 show a general trend in which the higher the density, the lower the thermal expansion coefficient. However, there are exceptions, even for the same type of oils.

11.3.1.2 Dependence of Density on Composition

An ideal mixture has a volume equal to the sum of the volumes of its components. This definition, combined with the law of conservation of mass, yields for a two-component system

$$1/\rho = x_1/\rho_1 + x_2/\rho_2 \tag{11.13}$$

In Equation 11.13, ρ is the density of the mixture, ρ_i and x_i are the densities and mass fractions of the components. The density of all the blends in the current investigation showed good agreement with Equation 11.13 (see Figure 11.8 and Table 11.2).

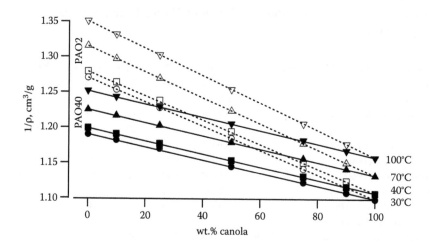

FIGURE 11.8 Inverse densities of canola–PAO blends as a function of blend composition at different temperatures. The markers show experimental data points; lines show prediction of Equation 11.13. Open symbols and dashed lines are canola–PAO2 blends; solid symbols and lines are canola–PAO40 blends. Circles are for 30°C, squares are for 40°C, up triangles are for 70°C, and down triangles are for 100°C.

TABLE 11.2

List of the Blends and the Maximum Absolute Difference (MAD) between Measured Blend Density and That Predicted using Equation 11.13

	Blend	MAD Difference (%)
Castor	Soybean (SBO)	0.08
Castor	Pentaerythritol ester (PE)	0.05
Castor	Trimethylolpropane complexed ester (TMP)	0.06
Castor	Dipentaerythritol ester (DPE)	0.05
Canola	Polyalphaolefin (PAO2)	0.16
Canola	Polyalphaolefin (PAO40)	0.23
Soybean (SBO)	Polyalphaolefin (PAO2)	0.16
Soybean (SBO)	Polyalphaolefin (PAO40)	0.14
High-oleic sunflower (HOSO)	Oil-soluble polyalkyl glycol (PAG32)	0.05
High-oleic sunflower (HOSO)	Oil-soluble polyalkyl glycol (PAG220)	0.05
2-EH ester of oleic estolide (EST)	Polyalphaolefin (PAO40)	0.02
2-EH ester of oleic estolide (EST)	Polyalphaolefin (PAO2)	0.04

11.3.2 REFRACTIVE INDEX

The RI versus temperature data were analyzed using the following empirical equation:

$$n = n_{ref} - [c_n(T - T_{ref})] \tag{11.14}$$

In Equation 11.14, n is the RI (dimensionless) at temperature T (°C), c_n is the RI-temperature coefficient (°C^{-1}), and n_{ref} is the RI at the reference temperature (T_{ref}).

The minus sign in Equation 11.14 indicates that the RI decreased with increase of the temperature.

Lorentz–Lorenz equation (Equation 11.5) implies correlation of change in density and change in RI as a function of temperature. An approximation for c_n can be derived from Equations 11.5, 11.12, and 11.14 as follows:

$$c_n \approx n_{ref}(1 + n_{ref}^2 - 2/n_{ref}^2)\beta_V/6 \tag{11.15}$$

The experimental refractive indices of blends were analyzed using Equation 11.14 and the n_{ref} and c_n values were obtained (Table 11.3). The values of c_n, calculated using Equation 11.15, are also included in Table 11.3 for comparison. While there is an agreement between c_n values determined by the two methods, there is a systematic trend for Equation 11.15 to overpredict c_n by ~7.5%.

If we examine the data within one type of oil (e.g., only vegetable oils), we observe a correlation between density and RI as follows: higher the density, higher the RI. This correlation does not hold when oils from different types are compared (e.g., SBO vs. PAO2).

TABLE 11.3

Refractive Indices (n_{20}) at 20°C (T_{ref}), and the Refractive Index-Temperature Coefficient (c_n) Obtained Using Equations 11.14 and 11.15

	n_{20}, from Equation 11.14	c_n, °C^{-1} ×10^4 from Equation 11.14	c_n, °C^{-1} ×10^4 from Equation 11.15	% c_n Difference Equation 11.14 versus Equation 11.15
SBO	1.4753	3.70	4.01	8.2
HOSO	1.4698	3.74	3.97	5.9
Castor	1.4798	3.65	3.97	9.0
Estolide ester	1.4655	3.67	3.88	5.8
PAO2	1.4435	3.96	4.32	9.1
PAO4	1.4555	3.91	4.08	4.3
PAO40	1.4683	3.58	3.82	6.8
PE	1.4555	3.63	3.92	8.1
DPE	1.4582	3.45	3.73	8.1
TMP	1.4653	3.47	3.76	8.2
PAG32	1.4490	3.68	4.00	8.7
PAG220	1.4529	3.61	3.87	7.3

Note: The temperature range (T_{min}–T_{max}) of the data used for their determination was 24–70°C.

The compositional dependence of the RI of blends is well described by the equation

$$n = x_1 n_1 + x_2 n_2 \tag{11.16}$$

where x_i is the mass fraction, n_i is the RI of component i, and n is the RI of the blend. Equation 11.14 and the experimental values for c_n were used to extrapolate n to the experimentally inaccessible 100°C. The linear fit of n on composition and temperature yielded numbers that agreed with the experimental data to four digits after the decimal point, whereas for the film thickness measurements three- or even two-digit precision is sufficient. No further comparison of the RI data with theoretical models was conducted.

11.3.3 VISCOSITY

11.3.3.1 Compositional Dependence of Viscosity

The viscosity of blends is said to obey a simple mixing rule if the blend viscosity can be predicted from the viscosity of the pure components and their fractions. For a two-component blend, the mixing equation is

$$\ln(\eta) = x_1 \ln(\eta_1) + x_2 \ln(\eta_2) = x_1 \ln(\eta_1) + (1 - x_1)\ln(\eta_2) \tag{11.17}$$

where x_i indicates the mole fraction of component i with viscosity η_i. However, most blends show nonideal behavior. Several models are available to account for the nonideality of blends. The simplest model is due to Grunberg and Nissan (G–N) given as follows [20]:

$$\ln(\eta) = x_1\ln(\eta_1) + (1 - x_1)\ln(\eta_2) + x_1(1 - x_1)d \tag{11.18}$$

where d is an interaction parameter related to blend vapor pressure. The authors claim that Equation 11.18 is obtained from theoretical derivation.

Another widely used equation, based on a three-body interaction model, is due to McAllister [21]:

$$\ln v = x_1^3 \ln(v_1/M_1) + 3x_1^2x_2 \ln(v_{12}/M_{12}) + 3\ x_1x_2^2 \ln(v_{21}/M_{21})$$

$$+ x_2^3 \ln(v_2/M_2) - \ln(x_1\ M_1 + x_2\ M_2) \tag{11.19}$$

where x_i, v_i, and M_i are, respectively, the mole fraction, kinematic viscosity (cSt), and the molecular weight (Da) of the component i ($i = 1$ or 2), $M_{12} = (2\ M_1 + M_2)/3$, $M_{21} = (M_1 + 2M_2)/3$, and v_{12} and v_{21} are adjustable parameters with units of cSt. At fixed temperature, the McAllister equation has two adjustable parameters. If the temperature dependence is taken into consideration, then it has four adjustable parameters. The interesting thing about the McAllister model is that it does not assume "ideal mixing" blends. v_{12} and v_{21} in the McAllister model cannot be set to zero, neither is there a combination of their values, to simplify the McAllister formula to Equation 11.17. The McAllister model has not been applied to lubricant oils since they are mixtures of many components, and quite often, with unknown properties of the individual components. In addition, the derivation of the McAllister formula restricts it to molecules with spherical shape, similar sizes, and known molecular weights. Such molecules are rarely used as lubricating oils.

There have been attempts at finding a mixing rule for the viscosity of lubricating blends. In this work, we will discuss the method of Lederer (as cited by Roelands [5]). In the Lederer method, the ideal mixing model (Equation 11.17) is modified to

$$\ln(\eta) = \phi_1\ln(\eta_1) + \phi_2\ln(\eta_2) = \phi_1\ln(\eta_1) + (1 - \phi_1)\ln(\eta_2) \tag{11.20}$$

$$\phi_1 = mx_1/(mx_1 + 1 - x_1) \tag{11.21}$$

where x_1 is the mass fraction of component 1, and m is a fitting parameter and corresponds to the ratio of the molecular weights of the segments constituting a unit of flow. In this model, an assumption is made that the flow behavior is determined not by the movement of whole molecules, but rather by the movement of their segments. Thus, Equation 11.21 provides a method to calculate the molar fraction of the molecular segments (ϕ_1) in component 1. This model is proposed for large and flexible lubricant molecules. In this model, m is a fitting parameter. Akei and Mizuhara [22] demonstrated that Equation 11.20 describes well the behavior of binary mixtures of several refrigerant fluids with lubricating oils such as alkylbenzene, polyol ester, polyalkyl glycol, organic carbonate, and naphthenic mineral oil.

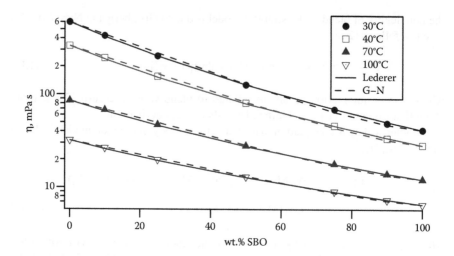

FIGURE 11.9   Viscosity of SBO–PAO40 blends. The solid lines are the best fit to Equation 11.20 (Lederer model, $m = 1.39$ for all temperatures). The dashed line is the best fit to Equation 11.18 (Grunberg–Nissan model, $d(30°C) = -2.16$, $d(40°C) = -1.89$, $d(70°C) = -1.48$, and $d(100°C) = -1.18$).

We have found that Equation 11.20 describes well the viscosity of blends of SBO–PAO2, SBO–PAO40, canola–PAO2, and canola–PAO40 [9]. An example is shown in Figure 11.9. We found that, for each combination of oils, only a single value of m was sufficient to fit the data of each blend at all test temperatures (30°C, 40°C, 70°C, and 100°C). Thus, it can be interpreted that the number of the segments per molecule did not vary with temperature. The values of the fitting parameters m for the above blends are given in Table 11.4.

Figure 11.9 also shows predicted blend viscosity using the G–N model (Equation 11.18). A different fitting parameter d is required at each temperature. As shown in Figure 11.9, the value of fitting parameter d varied with temperature. The temperature dependence of d should be expected, since the authors relate it to vapor pressure.

TABLE 11.4
Fitting Parameter m Used in Prediction of Viscosity of Binary Blends (Equations 11.20 and 11.21)

Component 1	Component 2	m
SBO	PAO2	0.85
SBO	PAO40	1.39
Canola	PAO2	0.89
Canola	PAO40	1.33

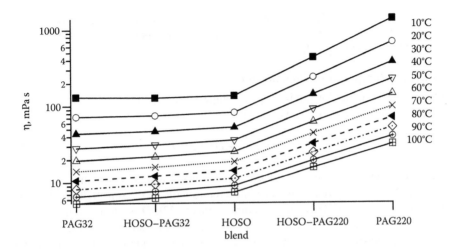

FIGURE 11.10   Viscosity of PAG32, PAG220, HOSO, and 50–50 wt.% PAG–HOSO blends at different temperatures. Points are experimental, lines are predicted using the simple mixing rule (Equation 11.17).

HOSO–polyalkyl glycol (PAG) blends showed viscosity results that were very close to the ideal blend predictions of Equation 11.17 (see Figure 11.10). The Lederer formula yielded slight improvements [14].

Castor–PE, castor–DPE, and castor–TMP blends showed viscosities that could not be predicted satisfactorily with either Lederer or G–N models (Figure 11.11).

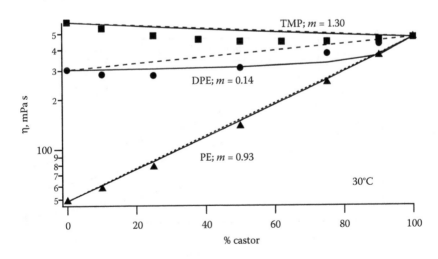

FIGURE 11.11   Experimental versus predicted viscosity of castor–polyol blends at 30°C. The dashed lines are fits with the simple mixing rule (Equation 11.17); the solid lines are fits with the Lederer model (Equations 11.20 and 11.21), where only one value of m was used to fit the data at all temperatures (other temperatures are not shown).

Castor–PE, castor–DPE, and castor–TMP blends showed mean square error between the experimental data and the best fits with Lederer model of 3.9%, 7.6%, and 6.9%, respectively [13].

11.3.3.2 Temperature Dependence of Viscosity

One common model for predicting the temperature dependence of viscosity is the Vogel–Tammann–Fulcher (VTF) equation

$$\ln\eta = \ln\eta_0 + DT_0/(T - T_0) \tag{11.22}$$

where η_0, fragility D (dimensionless), and T_0 (in kelvins) are constants, specific for a blend. The fragility of a liquid is considered to indicate the sensitivity of its short- and medium-range order to temperature changes [23]. D is always > 0, since the liquid viscosity always decreases with increase of temperature. T_0 is a temperature, related to, and slightly lower than the glass transition temperature. η_0 is the limit of viscosity, when the temperature T tends to infinity. There are still disputes about the fundamental nature of Equation 11.22. Some researchers have derived Equation 11.22 from theoretical models for interactions between the molecules [24]. Others claim that it is just a fitting formula, whose parameters have no fundamental meaning [25]. We used the VTF formula since it is well known.

From the VTF equation, the following can be derived:

$$d(\ln\eta)/dT = - DT_0/(T - T_0)^2 \tag{11.23}$$

Figure 11.12 shows how the viscosity of some oils and blends depends on temperature and how Equation 11.22 fits the data. Table 11.5 summarizes the coefficients of

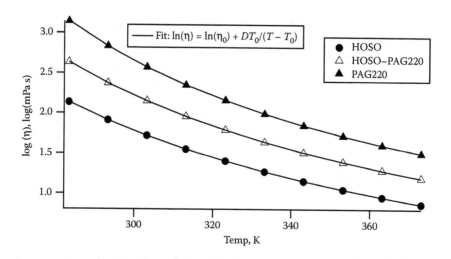

FIGURE 11.12 Viscosity versus temperature of HOSO, PAG220, and their 50–50 wt.% blend. Points are experimental, lines are fit with the VTF model (Equation 11.22). Fitting parameters for various oils are summarized in Table 11.5.

TABLE 11.5
Fitting Parameters Used with the VTF (Equation 11.22) Model to Describe the Temperature Dependence of Viscosity of Neat Base Oils

Base Oil	$\ln(\eta_0)$, mPa s	D	T_0, K
SBO	−1.76	4.63	164
Canola	−1.99	5.42	157
HOSO	−2.12	5.72	156
Castor	−2.76	5.82	183
EST	−2.42	7.00	155
PAO2	−3.27	6.39	132
PAO4	−2.97	5.83	154
PAO40	−2.50	7.94	160
PE	−2.51	5.43	164
DPE	−2.58	6.00	176
TMP	−2.03	6.87	167
PAG32	−2.15	4.31	176
PAG220	−0.91	4.72	179

Equation 11.22 for the base oils used in this study. The raw data are taken from our previous publications [9,13,14].

11.3.3.3 Kinematic Viscosity

Kinematic viscosity is equal to dynamic viscosity divided by density (Equation 11.24):

$$\mu = \eta/\rho \qquad (11.24)$$

Its units are mm²/s, or centistokes (cSt). Combining Equations 11.22 and 11.12 gives for the temperature dependence of kinematic viscosity

$$\ln(\mu) = \ln(\eta_0) + DT_0/(T - T_0) - \ln(\rho_{ref}) + (T - T_{ref})\beta_V \qquad (11.25)$$

Kinematic viscosity is very popular in the lubrication literature for the description of lubricants. The kinematic viscosity data, as well as viscosity index (VI), calculated from the kinematic viscosity using ASTM D2270 [26], for the neat oils studied here, is given in Table 11.6.

11.3.4 EHD FILM THICKNESS

Typical results for EHD film thickness are shown in Figure 11.13 [9]. In general, the film thickness increases with increasing entrainment speed. If the data are presented in a log–log plot, the slope is approximately constant and close to the value predicted by the H–D equation (0.67). The decrease of viscosity of an oil or blend,

TABLE 11.6

Kinematic Viscosity (KV, in mm²/s) at 40°C and 100°C

Base Oil	KV40°C, mm²/s	KV100°C, mm²/s	VI
SBO	31.65	7.63	224
Canola	35.88	8.20	214
HOSO	39.73	8.61	203
Castor	253.4	19.51	87
EST	95.41	15.08	167
PAO2	5.23	1.71	—
PAO4	17.27	3.92	124
PAO40	396.6	40.02	151
PE	33.69	6.36	143
DPE	174.0	17.46	109
TMP	332.0	35.43	152
PAG32	33.27	6.91	175
PAG220	243.0	37.52	206

Note: The viscosity index (VI, dimensionless) is calculated from the KV using the ASTM method 2270 [26]. The VI is not defined for oils with KV at 100°C below 2 cSt.

due to an increase in temperature or the amount of the lower-viscosity component, leads to decrease in film thickness. In Figure 11.13, it is illustrated that the film thickness of the PAO40–SBO blends decreased with increase of the SBO concentration, since the SBO had lower viscosity than PAO40. For the SBO–PAO2 blends,

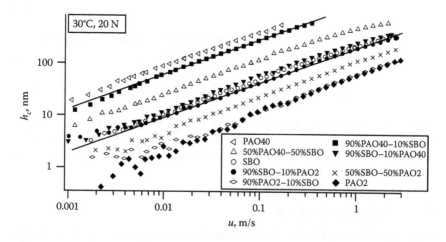

FIGURE 11.13 EHD film thickness, h_c, of PAO–SBO blends at 30°C, 20 N load as a function of entrainment speed, u. The solid lines have a slope of 0.67 as predicted by the H–D equation (Equation 11.6).

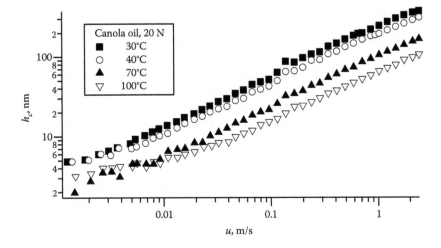

FIGURE 11.14   Effect of temperature on EHD film thickness, h_c, of canola oil as a function of the entrainment speed, u.

the film thickness decreased with increase of the PAO2 concentration, since PAO2 had lower viscosity than SBO. This was the same trend as the trend of the viscosities of the blends. Figure 11.14 illustrates how film thickness decreases when temperature increases.

The change in the load had a weak effect on film thickness, as expected from the H–D formula (see Figure 11.15) [13]. In most cases, film thickness was lower under higher load.

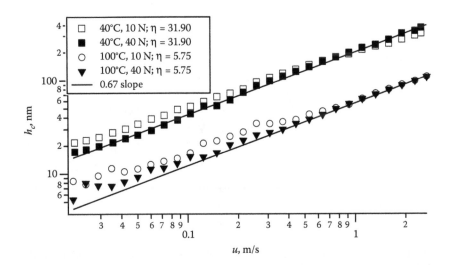

FIGURE 11.15   The film thickness, h_c, of PE at two temperatures and two loads as a function of the entrainment speed, u. In the legend, the viscosity, η, is in mPa s.

11.3.5 PRESSURE–VISCOSITY COEFFICIENT

11.3.5.1 Dependence of PVC on Composition

As outlined in Section 11.2.2, the film thickness of a blend can be used to determine the effective PVC, α, if the film thickness of a reference fluid on the same instrument is known. The results for α of SBO–PAO blends are shown in Figure 11.16, and the results for canola–PAO blends are shown in Figure 11.17. The Lederer model can be used to predict the PVC of blends from the PVC of the neat oils as follows:

$$\alpha = \alpha_1\phi_1 + \alpha_2(1 - \phi_1) \tag{11.26}$$

where ϕ_1 is the segment fraction of component 1, calculated using Equation 11.21. The parameter m, used in Equation 11.21, is determined by fitting the data for the ambient-pressure viscosity of blends to Equation 11.20. The segment fraction is calculated from the mole fraction and determined by fitting the ambient-pressure viscosity versus composition data using Equation 11.20. Figures 11.16 and 11.17 indicate that Equation 11.26 overpredicted the α values for PAO–vegetable oil blends [9]. This demonstrates that the good fit of the ambient-pressure data to Lederer model does not guarantee its applicability to high-pressure data. This is contrary to the observations of Akei and Mizuhara for other oils [22].

The α of blends of castor oil with polyol esters are shown in Figures 11.18 through 11.20. The error bars show the standard error as determined from independent experiments, averaged over all values. The error bars are included only for 40°C and omitted for 70°C and 100°C in order to avoid crowding of the plot. It can be seen that for such systems, the α was a nonmonotonic function of composition at 40°C. At higher

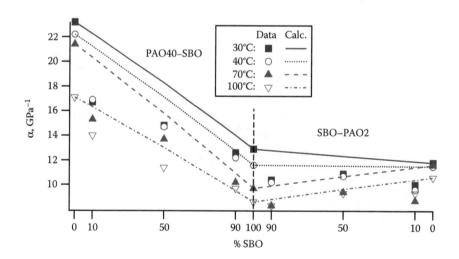

FIGURE 11.16 Effective pressure–viscosity coefficients, α, versus composition for SBO–PAO blends. Points are experimental and lines are predictions of the Lederer mixing rule (Equation 11.26) using segment fraction (ϕ) fitting parameters derived from viscosity versus composition data (Table 11.4 and Equation 11.20).

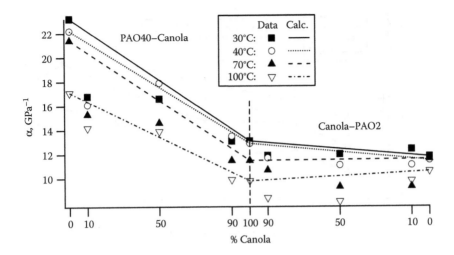

FIGURE 11.17 Effective pressure–viscosity coefficients, α, versus composition for canola–PAO blends. Points are experimental and lines are predictions of the Lederer mixing rule (Equation 11.26) using segment fraction (ϕ) fitting parameters derived from viscosity versus composition data (Table 11.4 and Equation 11.20).

temperatures, the α values were closer to a linear function of the composition. A common feature of all blends was that the α of castor oil decreased with the addition of 10% polyol ester, even if the neat ester (DPE) had higher α value than castor oil. All base oil blends showed a maximum in α values, the most prominent one being for the castor oil–TMP system. The order of the α values (DPE > TMP $\approx$ castor > PE) did not follow the order of viscosity of the oils (TMP > castor > DPE > PE). The

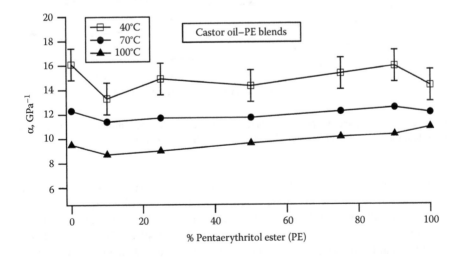

FIGURE 11.18 Effective pressure–viscosity coefficients, α, determined from film thickness data for castor oil–pentaerythritol ester blends as a function of composition.

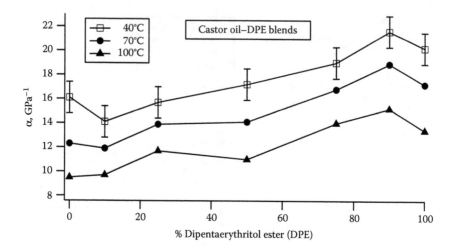

FIGURE 11.19 Effective pressure–viscosity coefficients, α, determined from film thickness data for castor oil–dipentaerythritol ester blends as a function of composition.

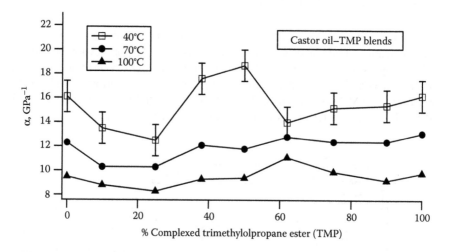

FIGURE 11.20 Effective pressure–viscosity coefficients, α, determined from film thickness data for castor oil–TMP ester blends as a function of composition.

Lederer model cannot be applied for prediction of α of castor–polyol ester blends, since it failed to describe the ambient-pressure data for viscosity [13].

The α results for HOSO, PAG, and their blends are shown in Figure 11.21. The PVC of neat HOSO was lower than that of neat PAG at 40°C. The same holds true at 100°C, but the differences were smaller. For PAG–HOSO blends, α values of the blends were equal to those predicted by the Lederer mixing rule (Equation 11.26) within experimental error [14].

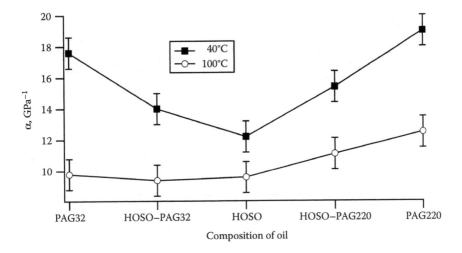

FIGURE 11.21   Effective pressure–viscosity coefficients, α, of oil-soluble PAG, HOSO, and their 50–50 wt.% blends, calculated from film thickness data.

11.3.5.2   Dependence of PVC on Temperature

The α values of the studied neat and blended oils decrease with temperature. Spikes [27], based on thermodynamic considerations, derived the following dependence of PVC on temperature:

$$\alpha = -bT(\partial(\ln\eta)/\partial T)_P \qquad (11.27)$$

where b is a positive constant that is almost independent of temperature. Combination of Equation 11.23 with Equation 11.27 yields

$$\alpha = bTDT_0/(T - T_0)^2 \qquad (11.28)$$

Equation 11.28 predicts the temperature dependence of PVC. It can be used to fit to the experimental data, using only one fitting parameter, b. Figure 11.22 illustrates such a fit for SBO and canola oil. It can be seen that Equation 11.28 predicted a sharper decrease of α with temperature than observed experimentally [9]. Similar results were observed for most VO–PAO and castor oil–polyol ester blends. A possible explanation for the discrepancy is that b may not be constant but increase with temperature.

11.3.6   FILM THICKNESS IN THE ULTRATHIN (BELOW 20 NM) REGIME

The H–D equation is derived using the assumption that the properties of the fluid are not influenced by the proximity between the surfaces of the lubricated parts. As already mentioned in Section 11.2, this is not totally correct, especially when molecules with different polarities are present in the blend. The more polar molecules form boundary layers, and will have different properties close to the surface than in the bulk, due to changes in composition and the interactions with the surface.

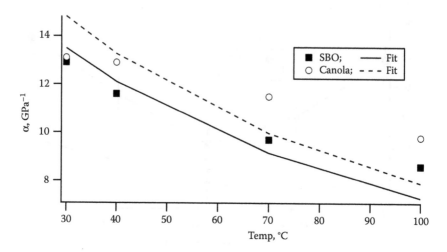

FIGURE 11.22 Effective pressure–viscosity coefficients, α, of soybean and canola oils. Points are experimental; the lines are predicted by Equation 11.28.

Figure 11.23a compares the film thickness of SBO, PAO2, and 10% SBO in PAO2 [15]. PAO2 is less viscous and less polar than SBO. The measurements were taken as soon as the lubricant reached the target temperature. Lines with slope equal to 0.67, predicted by the H–D equation (Equation 11.6), are drawn through the data points. It can be seen that SBO obeyed quite closely the H–D equation down to 3–4 nm. PAO2 and the blend of 10% SBO in PAO2 also showed agreement with the H–D equation down to 1–2 nm, despite some slightly larger deviations of some data points.

The film thickness results changed after the system was aged for ~100 min (Figure 11.23b). PAO2 still showed film thickness close to or slightly below the H–D line, while SBO and 10% SBO in PAO2 oils showed about 4 nm thicker films. This can be explained by the presence of about 2-nm-thick boundary layers on both contacting surfaces. These boundary layers were strongly attached to the surfaces and could not be easily squeezed out of the contact zone. Similar results were observed with these blends when the experiments were carried out at 100°C. At this temperature, the boundary layers seemed to be thinner, probably because higher temperature promotes desorption and melting.

Figure 11.24 shows the EHD film thicknesses for SBO, PAO40, and blend of 20% SBO in PAO40. PAO40 is less polar but more viscous than SBO. Initially (Figure 11.24a), the film thickness of the oils followed the H–D prediction. After aging (Figure 11.24b), we can see that the film thickness of SBO increased, of PAO40 remain unchanged, and of SBO–PAO40 blend decreased.

The fact that the film thickness of PAO40 remained unchanged with time suggests that no boundary layer was formed by the nonpolar oil. On the other hand, the increase of the film thickness of SBO with time demonstrates that the polar molecules adsorb within 100–200 min to form boundary layers of higher viscosity than the bulk viscosity.

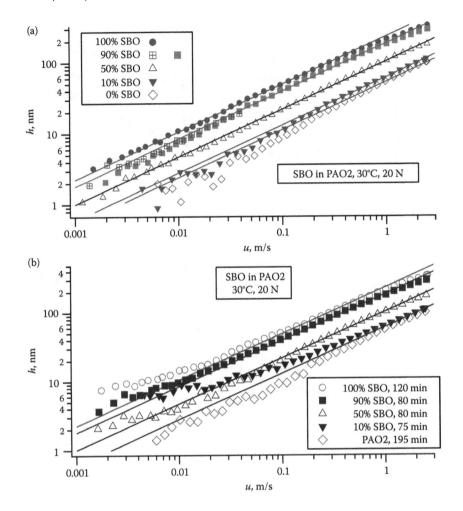

FIGURE 11.23 EHD film thickness, h_c, of SBO, PAO2, and 10% SBO in PAO2 as a function of entrainment speed, u. The lines have slopes of 0.67, as predicted by the H–D equation (Equation 11.6). (a) Measurements were taken without aging the sample; (b) measurements were taken after ~100 min aging. (With kind permission from Springer Science+Business Media: *J. Am. Oil Chem. Soc.*, Elastohydrodynamic study of bio-based esters–polyalphaolefin blends in the low-film thickness regime, 89, 2012, 1091–1099, G.B. Bantchev, G. Biresaw, and S.C. Cermak.)

At low speeds, the film thickness of aged SBO–PAO40 blend was close to the film thickness of neat SBO. The decrease of the SBO–PAO40 film thickness with time can be explained with the formation of a fractionation layer, which is a different mechanism than the boundary layer. The fractionation layer can be considered as a type of boundary layer where the polar molecules are close to the surface, but are not attached strongly to it and retained their mobility. The more polar component in this experiment was SBO. If it formed fractionation layers, the layers should have viscosity, close to SBO, which is lower than the viscosity of SBO–PAO40 blend. This lower

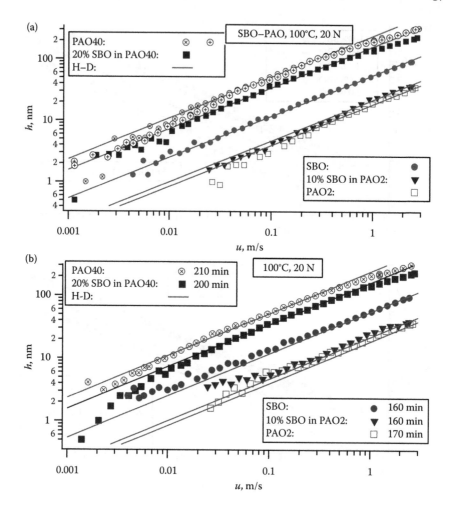

FIGURE 11.24 EHD film thickness, h_c, of SBO, PAO40, and 20% SBO in PAO40 as a function of entrainment speed, u. The lines have slopes of 0.67, as predicted by the H–D equation (Equation 11.6). (a) Measurements were taken without aging the sample; (b) measurements were taken after ~200 min aging. (With kind permission from Springer Science+Business Media: *J. Am. Oil Chem. Soc.*, Elastohydrodynamic study of bio-based esters–polyalphaolefin blends in the low-film thickness regime, 89, 2012, 1091–1099, G.B. Bantchev, G. Biresaw, and S.C. Cermak.)

viscosity of the fractionation layer led to lower film thickness for the SBO–PAO40 blend, compared with the nonaged sample, when there was not enough time for fractionation layer formation.

When examined in more detail, the behavior of the SBO–PAO40 blend fractionation layer was even more complicated. The evolution of the film thickness with aging time for the SBO–PAO40 blend is shown in Figure 11.25. It can be seen that the film thickness below 10 nm initially decreased and afterwards started to recover. This can be explained with a slow conversion of the fractionation layer into boundary

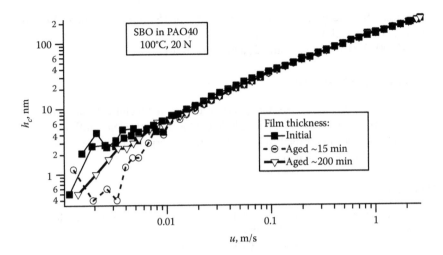

FIGURE 11.25 EHD film thickness, h_c, of 20% SBO in PAO40 as a function of entrainment speed, u, at different times elapsed since the beginning of the experiment.

layer; the polar molecules that had been concentrated close to the surface, but without adsorbing to it, have started attaching and losing their mobility. This could lead to an increase in the viscosity of the SBO molecules close to the surface, and the measured film thickness started growing from the initially depressed values [15].

11.4 CONCLUSIONS

Blends of base oils with different polarities show diverse and difficult-to-predict EHD behavior. Theories for predicting viscosity of blends work satisfactorily only in some cases. The temperature–viscosity dependence of neat base oils and their blends can be predicted using the VTF equation.

The film thickness of the blends in the 20–200 nm range can be described satisfactorily using the H–D equation. The film thickness of blends increases with increasing viscosity and entrainment speed.

Predicting the PVC of blends from the value for the neat base oils is a challenging task. The Lederer model showed good agreement for PVC to the experimental data only for high-oleic sunflower oil–polyalkyl glycol blends at 40°C and 100°C and for castor–polyol ester blends at elevated temperatures (70°C and 100°C). For PAO–SBO, PAO–canola blends, and castor–polyol ester blends at low temperature (40°C), there were deviations in the predictions of the Lederer model from the experimental PVC values. On the other hand, the Lederer model described well the ambient-pressure viscosity of PAO–vegetable oil blends.

At ultralow film thickness (below 20 nm), when a polar component was present in the oils, the H–D equation proved to be inadequate for describing their behavior. Without aging, the base oils and blends showed film thicknesses close to the H–D prediction. However, with aging time, discrepancies appeared between measured and predicted film thickness for VO and their blends. The experimental data can be explained by assuming an initial (in the order of 15 min) formation of SBO fractionation layer, which was fluid, but converted into boundary layer with time (on the order of 100–200 min) and attained reduced mobility.

ACKNOWLEDGMENTS

The authors thank Linda Cao and Andrew Ruddy for technical help, Dr. Steven Cermak for the estolide sample, KIC Chemical for the high-oleic sunflower oil sample, INEOS Oligomers for the PAO samples, Hatco Corp. for the polyol samples, and Dow Chemical for PAG samples.

REFERENCES

1. D.Y.C. Chan and R.G. Horn, The drainage of thin liquid films between solid surfaces, *J. Chem. Phys.*, 85, 5311–5324, 1985.
2. J.M. Georges, S. Millot, J.L. Loubet, and A. Tock, Drainage of thin liquid films between relatively smooth surfaces, *Chem. Phys.*, 98, 7345–7360, 1993.
3. G. Biresaw, A. Adhvaryu, S.Z. Erhan, and C.J. Carriere, Friction, and adsorption properties of normal and high-oleic soybean oils, *J. Am. Oil Chem. Soc.*, 79, 53–58, 2002.
4. B.J. Hamrock and D. Dowson, Isothermal elastohydrodynamic lubrication of point contacts. Part III—Fully flooded results, *Trans. ASME: J. Lubr. Technol.*, 99, 264–276, 1977.
5. C.J.A. Roelands, Correlational aspects of the viscosity-temperature-pressure relationship of lubricating oils, PhD thesis, The Technological University of Delft, Holland, 1966.
6. S.S. Lawate, K. Lal, and C. Huang, Vegetable oil—structure and performance, In: *Tribology Data Handbook*, E.R. Booser (Ed.), pp. 103–116, CRC Press, Boca Raton, FL, 1997.
7. S.C. Cermak, G. Biresaw, and T.A. Isbell, Comparison of a new estolide oxidative stability package, *J. Am. Oil Chem. Soc.*, 85, 879–885, 2008.
8. G.J. Johnson, R. Wayte, and H.A. Spikes, The measurement and study of very thin lubricant films in concentrated contacts, *Tribol. Trans.*, 34, 187–194, 1991.
9. G.B. Bantchev and G. Biresaw, Elastohydrodynamic study of vegetable oil-polyalphaolefin blends, *Lubr. Sci.*, 20, 283–297, 2008.
10. C.A. Foord, W.C. Hammann, and A. Cameron, Evaluation of lubricants using optical elastohydrodynamics, *ASLE Trans.*, 11, 31–43, 1968.
11. S. Gunsel, S. Korcek, M. Smeeth, and H.A. Spikes, The elastohydrodynamic friction and film forming properties of lubricant base oils, *Tribol. Trans.*, 42, 559–569, 1999.
12. S. Bair and F. Qureshi, Accurate measurements of pressure-viscosity behavior in lubricants, *Tribol. Trans.*, 45, 390–396, 2002.
13. G.B. Bantchev and G. Biresaw, Film–forming properties of castor oil–polyol ester blends in elastohydrodynamic conditions, *Lubr. Sci.*, 23, 203–219, 2011.
14. G.B. Bantchev and G. Biresaw, Film-forming properties of blends of high-oleic sunflower oil with polyalkyl glycol, *J. Am. Oil Chem. Soc.*, DOI: 10.1007/s11746-012-2120-0, 2012.
15. G.B. Bantchev, G. Biresaw, and S.C. Cermak, Elastohydrodynamic study of bio-based esters—Polyalphaolefin blends in the low film thickness regime, *J. Am. Oil Chem. Soc.*, 89, 1091–1099, 2012.
16. R.S. Fein, High-pressure viscosity and EHL pressure-viscosity coefficients, In: *Tribology Data Handbook*, E.R. Booser (Ed.), pp. 638–644, CRC Press, Boca Raton, FL, 1997.
17. H.G. Rackett, Equation of state for saturated liquids, *J. Chem. Eng. Data*, 15, 514–517, 1970.

18. K. Anand, A. Ranjan, and P.S. Mehta, Predicting the density of straight and processed VO from fatty acid composition, *Energy Fuels*, 24, 3262–3266, 2010.
19. J.D. Halvorsen, W.C. Mammel, Jr., and L.D. Clements, Density estimation for fatty acids and vegetable oils based on their fatty acid composition, *J. Am. Oil Chem. Soc.*, 70, 875–880, 1993.
20. L. Grunberg and A.H. Nissan, Mixture law for viscosity, *Nature*, 164, 799–800, 1949.
21. R.A. McAllister, The viscosity of liquid mixtures, *AIChE J.*, 6, 427–431, 1960.
22. M. Akei and K. Mizuhara, The elastohydrodynamic properties of lubricants in refrigerant environments, *Tribol. Trans.*, 40, 1–10, 1997.
23. C.A. Angell, Formation of glasses from liquids and biopolymers, *Science*, 267, 1924–1935, 1995.
24. A.V. Granato, A derivation of the Vogel–Fulcher–Tammann relation for supercooled liquids, *J. Non-Cryst. Solids*, 357, 334–338, 2011.
25. F. Mallamace, C. Branca, C. Corsaro, N. Leone, J. Spooren, S.-H. Chen, and H. E. Stanley, Transport properties of glass-forming liquids suggest that dynamic crossover temperature is as important as the glass transition temperature, *Proc. Natl. Acad. Sci.*, 107, 22457–22462, 2010.
26. Anon., Standard practice for calculating viscosity index from kinematic viscosity at 40 and 100°C, *ASTM Stand.*, D2270, 769–774, 2001.
27. H.A. Spikes, A thermodynamic approach to viscosity, *Tribol. Trans.*, 33, 140–148, 1990.

Part III

Tribological Properties of Aqueous and Nonaqueous Systems

12 Ionic Liquids as Novel Lubricant Bases

Marian W. Sulek and Marta Ogorzalek

CONTENTS

The excellent tribological characteristics of ionic liquids are the main reason for their application as lubricant bases. Their most important characteristics include a wide temperature interval in which they remain in the liquid state, excellent thermo-oxidative stability, nonflammability, and low vapor pressure that is virtually unmeasurable in many ionic liquids. These properties are of interest for potential applications in devices operating under conditions of high vacuum, high temperatures, fire hazard, and under cosmic space conditions.

The subject of this study is the tribological properties of three phosphonium-based liquids. They have a common tetradecyl(trihexyl)phosphonium-cation [P_{66614}] but different anions: benzotriazole [Bt], 1,2,4-triazole [Tr], and 3-aminotriazole [3AT]. Paraffin oil, which is commonly applied in the lubrication of machines in the food, cosmetics, and pharmaceutical industries, was used as the standard substance.

The study of the tribological properties of ionic liquids and paraffin oil was carried out using a ball-on-disk tribometer (T-11) and a four-ball machine

(T-02) using steel–steel and steel–aluminum pairs under constant load. The friction coefficient (μ) and wear scar diameter (d) were determined.

The tribological test results have shown that resistance to motion and wear are lower for the ionic liquids than for paraffin oil. It can thus be claimed that the phosphonium-based ionic liquids can function as both effective and efficient lubricant bases.

12.1 INTRODUCTION

Mineral, synthetic, semisynthetic, and vegetable oils are most commonly used as lubricant bases. For specific applications, the properties of the bases are modified by incorporating specific additives, such as viscosity modifiers, oxidation and corrosion inhibitors, friction modifiers, detergents, dispersants, pour-point depressants, and so on. Currently, formulated lubricants contain from a fraction of 1% to as much as 60% of additives, the rest being a base. For example, the concentration of the base oil in transformer and turbine oils exceeds 99 (w/w%) and in compressor and hydraulic oils it ranges from 98 (w/w%) to 99 (w/w%), whereas the concentration of base oil in motor oils may range from 60 to 90 (w/w%) [1–3]. It can thus be claimed that the quality of the base oil determines, to a large extent, performance characteristics of lubricants. Moreover, a properly selected base oil can reduce the number of performance-enhancing additives. Therefore, a search for new base oils and modifications of existing ones have been continuing. The driving forces include increasing environmental and market demands, technological advances in natural raw materials processing and in the synthesis of new compounds, as well as economic factors.

In recent years, there have been reports that discuss the application of ionic liquids as lubricants [4–10]. The interest in this subject is connected with their beneficial properties that are highly desirable in tribology. The current tribological research on ionic liquids can be treated as fundamental research. The analysis focuses on the effect of the structure of the liquid on selected tribological properties. The main drawback of a number of publications is the research equipment used. The authors' own procedures are often used and, therefore, the various results obtained are difficult to compare.

12.2 CHARACTERISTICS OF IONIC LIQUIDS

Ionic liquids (ILs) are liquids composed exclusively of dissociated salt ions. The salts forming ionic liquids dissociate into ions under the influence of thermal energy. The number of long-range interactions decreases above the melting point and a liquid is formed. Most salts melt at high temperatures, for example, sodium chloride at 800°C. The subject of our interest are ionic liquids whose melting point is below 100°C or comparable with room temperature (low-temperature ionic liquids—LTILs). They will also be called ionic liquids, for short. Their liquid state at these temperatures (below RT) is due to a relatively low crystal lattice energy, lower ion packing density, and an asymmetry of liquid-forming ions. As an example, replacing sodium in NaCl with a large imidazolium cation may lead to a decrease in the melting point from 800°C to 80°C [11–13].

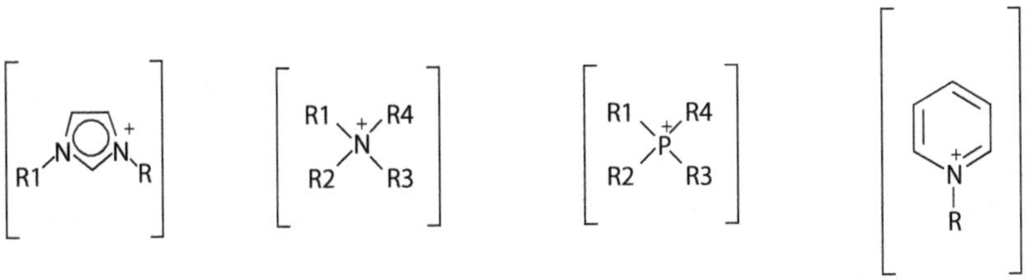

1-Alkyl-3-   Tetraalkylammonium   Tetraalkylphosphonium   N-alkylpyridinium
methylimidazolium

FIGURE 12.1   Examples of cations in ionic liquids.

The salts forming ionic liquids consist of various kinds of cations and anions. The cations are characterized by low symmetry. The most common cations are imidazolium, ammonium, phosphonium, and pyridinium. Their chemical structures are shown in Figure 12.1.

In Figure 12.1, R, R_1, R_2, R_3, and R_4 denote alkyl chains with different numbers of carbon atoms. The anions can be organic or inorganic [12]. The number of possible combinations of cations and anions is estimated to be of the order of 10^{18}. Owing to such a large number of possible combinations of salts, they can have a variety of physicochemical properties. Therefore, by appropriate selection of ions, it is possible to obtain ionic liquids with preset properties. This is an interesting example of materials engineering at the molecular level.

Ionic liquids exhibit low vapor pressure at room temperature, and therefore, are non-flammable and harmless for the environment and living organisms. They are liquid in a wide temperature range from below room temperatures up to about 400°C [11,12]. The most important parameter characterizing ionic liquids is their melting point (t_M), which varies from low temperatures to a few hundred degrees and depends on the chemical structure of cations and anions [14]. Ionic liquids are stable in a broad temperature range of their occurrence, of the order of a few hundred degrees Celsius, that is, from melting point to boiling point. Some of them undergo decomposition above 400°C [14,15]. The viscosity of ionic liquids is similar to that of typical oils and ranges from 10 to 500 mPa.s [11,16]. Ionic liquids exhibit high surface activity. The values of surface tension for ionic liquids are higher than those for conventional solvents (e.g., acetone 24 mN/m, hexane 19 mN/m, toluene 28 mN/m) but lower than for water (72 mN/m) [17]. The density of ionic liquids ranges from 1.05 g/cm³ to 1.64 g/cm³ at 293 K [11,12]. They can possess both hydrophobic properties (oil solubility) and hydrophilic properties (water solubility) [18]. Thus, it is relatively easy to select a performance-enhancing additive package.

It follows from the properties presented above that ionic liquids are an attractive alternative for traditional and synthetic lubricant bases.

12.3   EXPERIMENTAL

12.3.1   MATERIALS

Three ionic liquids (phosphonium salts) were selected for this study: tetradecyl(trihexyl) phosphoniumbenzotriazole [P$_{66614}$][Bt], tetradecyl(trihexyl)phosphonium 1,2,4-triazole

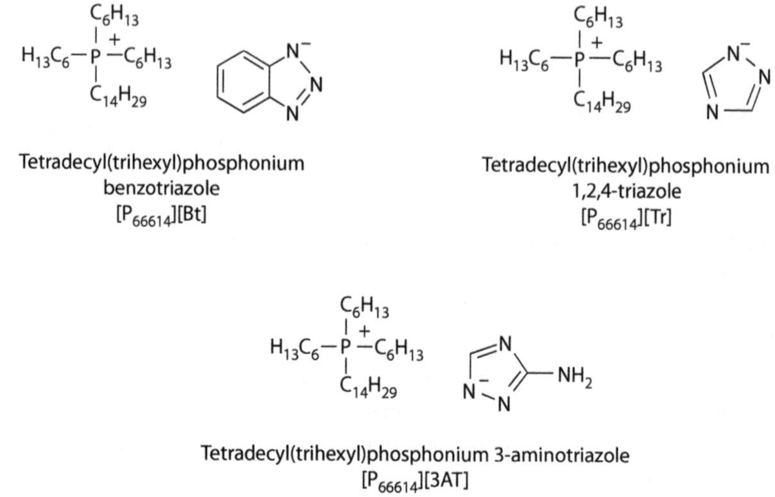

Tetradecyl(trihexyl)phosphonium
benzotriazole
$[P_{66614}][Bt]$

Tetradecyl(trihexyl)phosphonium
1,2,4-triazole
$[P_{66614}][Tr]$

Tetradecyl(trihexyl)phosphonium 3-aminotriazole
$[P_{66614}][3AT]$

FIGURE 12.2 Structures of ionic liquids investigated in this work.

$[P_{66614}][Tr]$, and tetradecyl(trihexyl)phosphonium 3-aminotriazole$[P_{66614}][3AT]$. They possess a common tetradecyl(trihexyl)phosphonium cation and three kinds of anions: benzotriazole, 1,2,4-triazole, and 3-aminotriazole. The ionic liquids were synthesized at Poznan University of Technology in Poland by Professor J. Pernak's team [19]. The structures of the ionic liquids are shown in Figure 12.2.

Paraffin oil used in this study was obtained from Augmed PPUH Warsaw (Poland). The reason for its selection was the fact that it has been applied in machines used in the food, cosmetics, and pharmaceutical industries. The oil exhibits good lubricating properties and has a limited negative effect on living organisms and the environment.

12.3.2 Methods

12.3.2.1 Ball-On-Disk Tribometer (T-11) and Procedure

A T-11 device produced at the Institute for Sustainable Technologies in Radom was used to evaluate motion resistance and wear of friction pair elements under concentrated contact of the ball-on-disk type. The upper element of the friction pair was a 6.35 mm diameter steel ball. The lower elements of the pair were either steel or aluminum with the following specifications:

- Disks 25.4 mm in diameter and 8 mm in thickness made of 100Cr6 steel (surface roughness $R_a = 0.043$ μm)
- Disks 25.4 mm in diameter and 8 mm in thickness made of aluminum (surface roughness $R_a = 0.009$ μm)

The experimental conditions were as follows: load 50 N; linear velocity 0.1 m/s; test duration 900 s; path length 90 m; temperature 20°C, 50°C, and 80°C. Based on friction force measurements, the coefficient of friction was calculated using the formula:

$$\mu = \frac{F_T}{P},$$ (12.1)

where F_T is the friction force [N] and P is the load [N].

Wear scar diameters (d) of the balls were measured parallel and perpendicular to the direction of sliding using a Polar reflection microscope with a measuring accuracy of 0.01 mm produced by PZO-Warszawa (Poland).

12.3.2.2 Four-Ball Machine: Tribological Tester (T-02)

A four-ball machine produced at the Institute for Sustainable Technologies in Radom was used in the tribological tests [20]. Bearing balls of 0.5″ in diameter made of 100Cr6 bearing steel (surface roughness $R_a = 0.32$ μm, hardness 60 ÷ 65HRC) were used. The temperature was measured by means of a thermocouple placed in the lubricant. The tests were carried out at a constant load (3 kN) and at a constant rotational speed (200 rpm). The duration of the tests was 3600 s. The coefficient of friction (μ) was calculated on the basis of the load (P) and frictional torque (M_T) measured. The following formula was used:

$$\mu = 222.47 \frac{M_T}{P},$$ (12.2)

where M_T is the frictional torque [N · m] and P is the load [N], the coefficient 222.47 resulting from the distribution of forces among the four balls in contact.

Wear scar diameters (d) of the balls were measured parallel and perpendicular to the direction of sliding using a Polar reflection microscope with a measuring accuracy of 0.01 mm produced by PZO-Warszawa (Poland).

12.3.2.3 Profilometer

A TOPO 01P profilometer produced at the Institute of Advanced Manufacturing Technology, Kraków, Poland was used for the investigation of profiles and roughnesses of wear tracks on aluminum disks after the tests were carried out with the T-11 apparatus. Changes in surface profiles and roughness of wear tracks whose measure is a mean arithmetic profile deviation from the middle line (R_a) were analyzed.

12.4 RESULTS AND DISCUSSION

The tribological properties of the ionic liquids and paraffin oil were evaluated using a T-11 tribotester with a ball-on-disk configuration (see Figure 12.3) and a T-02 four-ball machine (see Figure 12.4). The results obtained are presented separately for each of the two tribotesters.

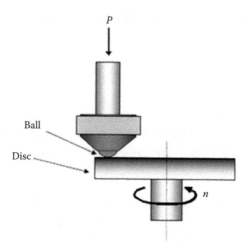

FIGURE 12.3   Schematic of the T-11 ball-on-disk tribometer.

FIGURE 12.4   Schematic of the T-02 four-ball tribometer.

12.4.1   BALL-ON-DISK TRIBOLOGICAL TESTS (T-11 TRIBOMETER)

12.4.1.1   (Steel 100Cr6–Steel 100Cr6)

Typical variations in the coefficient of friction (μ) as a function of time (t) for paraffin oil and tetradecyl(trihexyl)phosphonium 3-aminotriazole [P_{66614}][3AT] are presented in Figure 12.5. The course of $\mu(t)$ changes for the other phosphonium ionic liquids was of a similar nature.

The friction coefficient (μ) value for paraffin oil was about 0.1, whereas the value for the ionic liquid [P_{66614}][3AT] was about 0.08. There was no significant influence of friction time and temperature on the coefficient of friction.

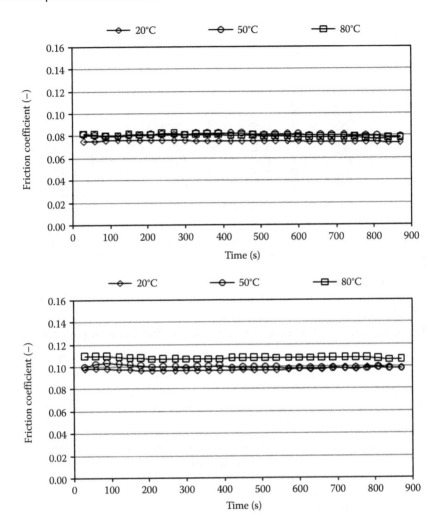

FIGURE 12.5 Dependence of friction coefficient on time at various temperatures for (a) tetradecyl(trihexyl)phosphonium 3-aminotriazole [P_{66614}][3AT] and (b) paraffin oil. T-11 tester. Ball-on-disk pair (100Cr6 steel ball–100Cr6 steel disk). Test duration 900 s, load 50 N, sliding velocity 0.1.

Figure 12.6 and Table 12.1 show the averaged values of the coefficient of friction from 900-s tests for all the ionic liquids.

The values of the coefficient of friction for paraffin oil at various temperatures ranged from 0.10 ± 0.01 to 0.11 ± 0.02 while the values for ionic liquids ranged from 0.07 ± 0.02 to 0.09 ± 0.02. No significant effect of temperature and anion structure on the coefficient of friction was detected.

The wear scar diameter (d) value is a measure of wear, and the results obtained are given in Figure 12.7 and Table 12.2.

The wear scar diameter value for paraffin oil after the tests at various temperatures was 0.40 ± 0.1 mm. A decrease in wear was observed in the case of phosphonium

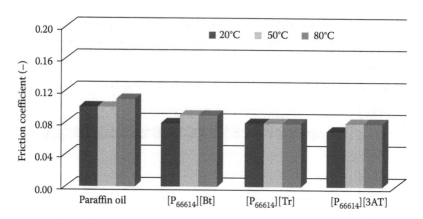

FIGURE 12.6 Friction coefficient for paraffin oil and phosphonium ionic liquids. T-11 tester. Ball-on-disk pair (100Cr6 steel ball–100Cr6 steel disk). Test duration 900 s, load 50 N, sliding velocity 0.1 m/s.

TABLE 12.1
Friction Coefficient Values for Paraffin Oil and Phosphonium Ionic Liquids

	20°C	50°C	80°C
Paraffin oil	0.10 ± 0.01	0.10 ± 0.01	0.11 ± 0.02
$[P_{66614}][Bt]$	0.08 ± 0.01	0.09 ± 0.02	0.09 ± 0.02
$[P_{66614}][Tr]$	0.08 ± 0.01	0.08 ± 0.01	0.08 ± 0.01
$[P_{66614}][3AT]$	0.07 ± 0.02	0.08 ± 0.01	0.08 ± 0.01

Note: T-11 tester. Ball-on-disk pair (100Cr6 steel ball–100Cr6 steel disk). Test duration 900 s, load 50 N, sliding velocity 0.1 m/s. $[P_{66614}][Bt]$, tetradecyl(trihexyl)phosphoniumbenzotriazole; $[P_{66614}][Tr]$, tetradecyl(trihexyl)phosphonium 1,2,4-triazole; $[P_{66614}][3AT]$, tetradecyl(trihexyl) phosphonium 3-aminotriazole.

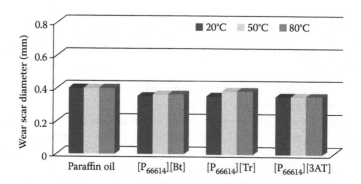

FIGURE 12.7 Wear scar diameters for paraffin oil and phosphonium ionic liquids. T-11 tester. Ball-on-disk pair (100Cr6 steel ball–100Cr6 steel disk). Test duration 900 s, load 50 N, sliding velocity 0.1 m/s.

TABLE 12.2
Wear Scar Diameter Values for Paraffin Oil and Phosphonium Ionic Liquids

	20°C	50°C	80°C
Paraffin oil	0.40 ± 0.02	0.40 ± 0.02	0.40 ± 0.03
$[P_{66614}][Bt]$	0.35 ± 0.02	0.36 ± 0.03	0.36 ± 0.03
$[P_{66614}][Tr]$	0.35 ± 0.02	0.38 ± 0.04	0.38 ± 0.04
$[P_{66614}][3AT]$	0.35 ± 0.02	0.35 ± 0.02	0.35 ± 0.02

Note: T-11 tester. Ball-on-disk pair (100Cr6 steel ball–100Cr6 steel disk). Test duration 900 s, load 50 N, sliding velocity 0.1 m/s. $[P_{66614}][Bt]$, tetradecyl(trihexyl)phosphoniumbenzo-triazole; $[P_{66614}][Tr]$, tetradecyl(trihexyl)phosphonium 1,2,4-triazole; $[P_{66614}][3AT]$, tetradecyl(trihexyl)phosphonium 3-aminotriazole.

ionic liquids compared with the oil. Wear scar diameter ranged from 0.35 ± 0.1 to 0.38 ± 0.2 mm in the case of a significant majority of the ionic liquids.

12.4.1.2 (Steel 100Cr6–Aluminum)

Figure 12.8 shows changes in the coefficient of friction (μ) as a function of time (t) for paraffin oil and phosphonium ionic liquids in a steel ball-on-aluminum disk test.

The friction coefficient values for phosphonium ionic liquids were lower than those for paraffin oil. The data for the ionic liquids were independent of the kind of anion. The coefficient of friction remained at a constant value of about 0.05 during the test. It can thus be claimed that phosphonium ionic liquids exhibit lower values of μ.

The averaged μ values from three independent measurements are given in Figure 12.9 and Table 12.3.

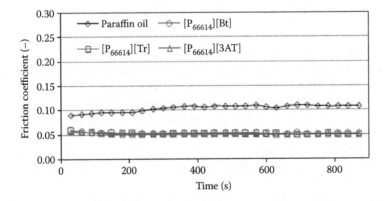

FIGURE 12.8 Dependence of friction coefficient on time for paraffin oil and phosphonium ionic liquids. T-11 tester. Ball-on-disk pair (100Cr6 steel ball–aluminum disk). Test duration 900 s, temperature 20°C, load 50 N, sliding velocity 0.1 m/s.

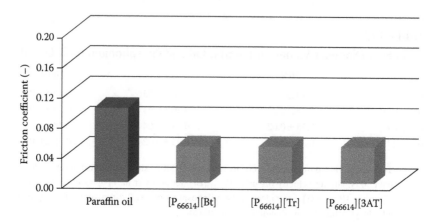

FIGURE 12.9 Friction coefficient for paraffin oil and phosphonium ionic liquids. T-11 tester. Ball-on-disk pair (100Cr6 steel ball–aluminum disk). Test duration 900 s, temperature 20°C, load 50 N, sliding velocity 0.1 m/s.

TABLE 12.3
Friction Coefficient Values for Paraffin Oil and Phosphonium Ionic Liquids

Paraffin oil	0.10 ± 0.01
$[P_{66614}][Bt]$	0.05 ± 0.01
$[P_{66614}][Tr]$	0.05 ± 0.01
$[P_{66614}][3AT]$	0.05 ± 0.01

Note: T-11 tester. Ball-on-disk pair (100Cr6 steel ball–aluminum disk). Test duration 900 s, temperature 20°C, sliding velocity 0.1 m/s. $[P_{66614}][Bt]$, tetradecyl(trihexyl)phosphoniumbenzotriazole; $[P_{66614}]$ [Tr], tetradecyl(trihexyl)phosphonium 1,2,4-triazole; $[P_{66614}][3AT]$, tetradecyl(trihexyl)phosphonium 3-aminotriazole.

The friction coefficient value for paraffin oil was 0.10 ± 0.02. Resistance to motion for phosphonium ionic liquids remains at a constant level of $\mu = 0.05 \pm 0.01$ and this value is two times lower compared with paraffin oil.

In the steel ball-on-aluminum disk test, the harder material (steel) did not display wear and the wear scar diameter on the ball was unmeasurable. Therefore, a profilometer was used to measure the wear track on the aluminum disk (see Figure 12.10). A change in the profile of the disk surface is given in μm on the *Y*-axis and wear scar diameter is given in mm on the *X*-axis. The results obtained are presented in Figure 12.10.

The profile changes observed are small for paraffin oil and practically unidentifiable for ionic liquids (Figure 12.10). Therefore, R_a values for the aluminum disk after friction test are a good measure for the roughness changes observed. The value for paraffin oil is 0.19 μm and for ionic liquids it is about 10 times lower. These results

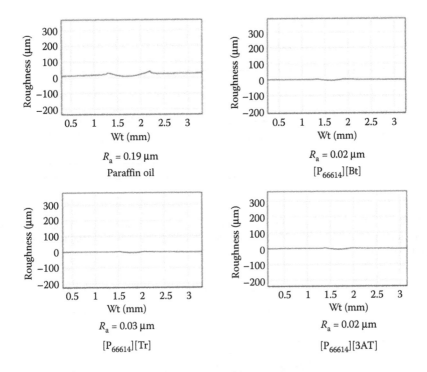

FIGURE 12.10 Profiles and roughness (R_a) of wear tracks on aluminum disks after 900 s tests on a steel ball-on-aluminum disk device (100Cr6 steel ball, aluminum disk) at rotational speed of 95 rpm, load 50 N, for paraffin oil and phosphonium ionic liquids (TOPO 01P profilometer).

clearly indicate that ionic liquids have a strong effect on the reduction in material wear in a steel-on-aluminum friction pair.

12.4.2 TRIBOLOGICAL TESTS AT A CONSTANT LOAD (T-02 FOUR-BALL MACHINE)

A T-02 four-ball machine (see Figure 12.4) was used in order to evaluate the effect of ionic liquid structure on resistance to motion and wear at a considerably higher load (3000 N) than that used on the T-11 tribometer (50 N) (see Figure 12.3) with a different friction geometry. The quantities measured were mean friction coefficient (μ) and wear scar diameter of the ball (d).

The plots of changes in the coefficient of friction as a function of time for the ionic liquids and paraffin oil are shown in Figure 12.11.

The friction coefficient value for ionic liquids decreased to about 0.1 after 360 s whereas in the case of paraffin oil it decreased to about 0.2 after 720 s. After these times, the μ values remained constant or showed a slight decrease.

The arithmetic mean of friction coefficients was calculated for individual runs in order to analyze the results more easily. The dependence of the mean friction coefficients on the chemical structure of the phosphonium ionic liquid used is compared in Figure 12.12 and Table 12.4.

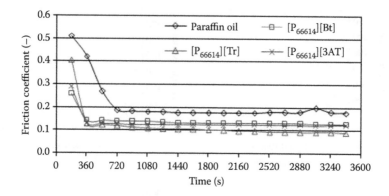

FIGURE 12.11 Dependence of friction coefficient on time for paraffin oil and phosphonium ionic liquids. Four-ball tester (T-02), rotational speed of the spindle 200 rpm, load 3 kN, test duration 3600 s.

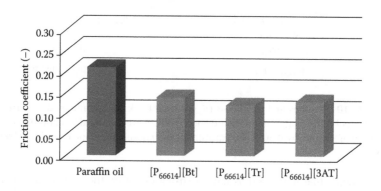

FIGURE 12.12 Friction coefficient for paraffin oil and phosphonium ionic liquids. Four-ball tester (T-02), rotational speed of the spindle 200 rpm, load 3 kN, test duration 3600 s.

TABLE 12.4
Friction Coefficient Values for Paraffin Oil and Phosphonium Ionic Liquids

Paraffin oil	0.21 ± 0.02
$[P_{66614}][Bt]$	0.14 ± 0.01
$[P_{66614}][Tr]$	0.12 ± 0.02
$[P_{66614}][3AT]$	0.13 ± 0.01

Note: Four-ball tester (T-02), rotational speed of the spindle 200 rpm, load 3 kN, test duration 3600 s. $[P_{66614}][Bt]$, tetradecyl(trihexyl)phosphoniumbenzotriazole; $[P_{66614}][Tr]$, tetradecyl(trihexyl)phosphonium 1,2,4-triazole; $[P_{66614}][3AT]$, tetradecyl(trihexyl)phosphonium 3-aminotriazole.

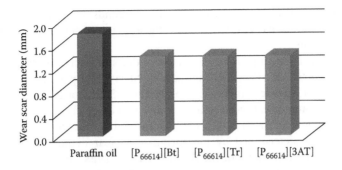

FIGURE 12.13 Wear scar diameters for paraffin oil and phosphonium ionic liquids. Four-ball tester (T-02), rotational speed of the spindle 200 rpm, load 3 kN, test duration 3600 s.

TABLE 12.5
Wear Scar Diameter Values for Paraffin Oil and
Phosphonium Ionic Liquids

Paraffin oil	1.8 ± 0.1
[P$_{66614}$][Bt]	1.4 ± 0.1
[P$_{66614}$][Tr]	1.4 ± 0.1
[P$_{66614}$][3AT]	1.4 ± 0.1

Note: Four-ball tester (T-02), rotational speed of the spindle 200 rpm, load 3 kN, test duration 3600 s. [P$_{66614}$][Bt], tetradecyl(trihexyl) phosphoniumbenzotriazole; [P$_{66614}$][Tr], tetradecyl(trihexyl)phosphonium 1,2,4-triazole; [P$_{66614}$][3AT], tetradecyl (trihexyl)phosphonium 3-aminotriazole.

The averaged friction coefficient value for paraffin oil from the whole 3600-s test is 0.21 ± 0.03. The μ values for the three phosphonium liquids [P$_{66614}$][Bt], [P$_{66614}$][Tr], and [P$_{66614}$][3AT] are 0.14 ± 0.01, 0.12 ± 0.02, and 0.13 ± 0.01, respectively. This means that the coefficients of friction obtained after tests in the presence of phosphonium liquids decrease by over 40% relative to paraffin oil.

Wear scar diameter values (d) from four-ball experiments for individual lubricants are given in Figure 12.13 and Table 12.5.

The wear scar diameter value for paraffin oil was 1.8 ± 0.2 mm. It has been found that the values of d are lower (about 1.4 ± 0.1 mm) for phosphonium ionic liquids. No significant effect of the anion chemical structure on wear was observed.

12.5 SUMMARY

The application of ionic liquids as novel lubricant base oils has been investigated and an analysis of their physicochemical properties has been conducted. Since they are described as "green chemistry raw materials," they should also satisfy ecological criteria. Another advantage for using ionic liquids as base oils is the fact that they are good solvents for a variety of compounds and, thus, their tribological properties can be easily modified with the addition of performance-enhancing additive packages.

Three types of ionic liquids were selected for this study: tetradecyl(trihexyl)phosphoniumbenzotriazole [P$_{66614}$][Bt], tetradecyl(trihexyl)phosphonium 1,2,4-triazole [P$_{66614}$][Tr], and tetradecyl(trihexyl)phosphonium 3-aminotriazole [P$_{66614}$][3AT], which have a common cation structure but different anion chemical structures.

The coefficient of friction (μ) was determined by measuring friction force while wear was determined by measuring wear scar diameter of the ball (d) or changes in the roughness of the disk (R_a). The results were compared relative to paraffin oil as a model base oil. In the case of the steel ball–steel disk friction pair (T-11), the values of μ and d for ionic liquids decreased relative to paraffin oil by 30% and 13%, respectively. Even more beneficial properties of ionic liquids were observed for the steel ball–aluminum disk friction pair (T-11) for which there was a twofold reduction in the μ value and about a 10-fold reduction in surface roughness (R_a) compared with paraffin oil. Tests on the four-ball tribometer (T-02) showed more beneficial properties for phosphonium liquids. The values of the coefficient of friction (μ) and wear scar diameter (d) for the ionic liquids decreased by about 40% and 20%, respectively, relative to the paraffin base oil, at a considerably higher load (3000 N) than the one used in the T-11 tribometer (50 N).

It can thus be concluded that the potential benefits for applications of ionic liquids as novel ecological lubricant bases have been demonstrated. More work is required to verify the benefits of using other ionic liquids with differing chemical structure of both cation and anion.

ACKNOWLEDGMENT

This research was financed by the Ministry of Science and Higher Education, Warsaw, Poland in 2008–2010—Grant No. N N507 373535.

REFERENCES

1. S. M. Hsu, Molecular basis of lubrication, *Tribol. Int.*, 37, 553–559, 2004.
2. M. Ratoi, V. Anghel, C. Bovington, and H. A. Spikes, Mechanisms of oiliness additives, *Tribol. Int.*, 33, 241–247, 2000.
3. H. A. Spikes, Mixed lubrication—an overview, *Lubr. Sci.*, 9, 221–253, 1997.
4. I. Minami, Ionic liquids in tribology, *Molecules*, 14, 2286–2305, 2009.
5. M. D. Bermúdez, A. E. Jiménez, J. Sanes, and F. J. Carrón, Ionic liquids as advanced lubricant fluids, *Molecules*, 14, 2888–2908, 2009.
6. H. Arora and P. M. Cann, Lubricant film formation properties of alkyl imidazoliumtetrafluoroborate and hexafluorophosphate ionic liquids, *Tribol. Int.*, 43, 1908–1916, 2010.
7. D. Jiang, L. Hu, and D. Feng, Crown-type ionic liquids as lubricants for steel-on-steel system, *Tribol. Lett.*, 41, 417–424, 2011.
8. M. W. Sulek, T. Wasilewski, M. Ogorzalek, J. Pernak, and F. Walkiewicz, The influence of anion type in imidazolium ionic liquids on tribological properties, *Tribologia*, 6, 107–115, 2010, (in Polish).

9. M. W. Sulek, T. Wasilewski, M. Ogorzalek, A. Bak, P. Skrzek, J. Pernak, and F. Walkiewicz, Tribological properties of selected ionic liquids in the material pairs: steel–PA6 and steel–PMMA, *Tribologia*, 4, 215–222, 2009, (in Polish).

10. M. W. Sulek, T. Wasilewski, M. Ogorzalek, A. Bak, J. Pernak, and F. Walkiewicz, Tribological characteristics of ionic liquids with ammonium cation, *Tribologia*, 4, 207–214, 2009, (in Polish).

11. D. Rooney, J. Jacquemin, and R. Gardas, Thermophysical properties of ionic liquids, *Top. Curr. Chem.*, 290, 185–212, 2009.

12. B. Clare, A. Sirwardana, and D. R. MacFarlane, Synthesis, purification and characterization of ionic liquids, *Top. Curr. Chem.*, 290, 1–40, 2009.

13. S. Sowmiah, V. Srinivasadesikan, M.-C. Tseng, and Y.-H. Chu, On the chemical stabilities of ionic liquids, *Molecules*, 14, 3780–3813, 2009.

14. S. Zhang, X. Lu, Y. Zhang, Q. Zhou, J. Sun, L. Han, G. Yue, X. Liu, W. Cheng, and S. Li, Ionic liquids and relative process design, *Struct. Bond.*, 131, 143–191, 2009.

15. J. Dupont and P. A. Z. Suarez, Physico-chemical processes in imidazolium ionic liquids, *Phys. Chem. Chem. Phys.*, 8, 2441–2452, 2006.

16. A. S. Pensado, M. J. Comunas, and P. J. Fernandez, The pressure–viscosity coefficient of several ionic liquids, *Tribol. Lett.*, 31, 107–118, 2008.

17. M. H. Ghatee, F. Moosavi, A. R. Zolghadr, and R. Jahromi, Critical-point temperature of ionic liquids from surface tension at liquid-vapor equilibrium and the correlation with the interaction energy, *Ind. Eng. Chem. Res.*, 49, 12696–12701, 2010.

18. A. Marciniak, The solubility parameters of ionic liquids, *Int. J. Mol. Sci.*, 11, 1973–1990, 2010.

19. Polish Patent Application no P.389786.

20. W. Piekoszewski, W. Szczerek, and M. Tuszyński, The action of lubricants under extreme conditions in a modified four-ball tester, *Wear*, 249, 188–193, 2001.

13 Stability of Cutting Fluids Emulsions

Nadia G. Kandile, Ahmed M. Al-Sabagh,
Mahmoud R. Noor El-Din, and
David R. K. Harding

CONTENTS

This chapter discusses the role of tribology in metal cutting, the major features of cutting fluids, and the effect of their composition on their performance. It focuses on the action of cutting oils and the factors affecting the stability of emulsions formed from the cutting oils.

13.1 INTRODUCTION

13.1.1 TRIBOLOGY IN METAL CUTTING

Tribology is defined as the science and technology of interactive surfaces moving relative to each other. The science of tribology concentrates on contact physics and the mechanics of moving interfaces that generally involve energy dissipation. Tribological findings are primarily applicable to mechanical engineering and design where tribological interfaces are used to transmit, distribute, and/or convert energy. The contact between two materials and the rubbing of one surface with the other causes an inevitable process of wear. The contact conditions, enhancing the resistance of contact surfaces to wear, as well as optimizing the power transmitted by mechanical systems and the complex lubrication required, have become a specialized applied science and technical discipline that has seen a major growth in recent decades. Applied tribology is the field of research and application that encompasses a wide range of scientific fields. These include contact mechanics, kinematics, applied physics, surface topography, hydro- and thermodynamics, and many other engineering fields. Tribology deals with problems related to a great variety of physical and chemical processes and reactions that occur at tribological interfaces. Current demands for tribological advances to support improved productivity coincide with additional challenges posed by the increased utilization of engineered materials. Some of these materials are useful as tools and dies and can perform optimally due to their improved properties. Others are difficult-to-machine workpieces and create tribological problems during machining such as severe tool wear, high residual stresses in the machined surface, metallurgical and structural defects of machined surface, and many other problems. Metal cutting tribology is a branch of applied tribology that deals with a number of processes (mechanical, thermal, chemical, etc.) at the tool–chip, tool–workpiece, and chip–workpiece interfaces from a unified approach of energy transformation [1]. One should clearly realize that the major objective of metal cutting tribology is reduction in the energy spent in metal cutting. Other objectives include increased tool life, improved integrity of the machined surface, higher process efficiency, and stability. Metal cutting tribology can help with the selection of proper tool design, development and/or selection of proper tool materials including coatings, development and/or selection of proper cutting fluids, and optimization of the cutting process on the shop floor.

Metal cutting is a commonly used material removal process in the manufacturing industry. Cutting takes place by shearing action at a plane that is inclined relative to the machined surface by an angle known as shear plane angle. Previous research has shown that the energy consumed in the cutting operation is primarily due to plastic deformation and/or friction. This energy is dissipated in the form of thermal energy with 90% removed by the chips, 5% by the tool, and 5% by the workpiece [2]. Cutting fluids are commonly used in machining practice since these fluids take the heat away from the cutting zone, and reduce the temperature of the tool and increase tool life. Cutting fluids also provide lubrication at the chip–tool interface, make the chip flow smoother, and prevent buildup and edge formation. The long-term effect of used and disposed cutting fluids on the environment has raised much concern. The indiscriminate and extravagant use of cutting fluids in the manufacturing industry is being discouraged. Research has also shown potential health hazards to manufacturing workers who come in direct contact with cutting fluids [3,4]. Stringent rules and restrictions are being developed about the use and disposal of cutting fluids. This has increased the cost associated with the use of cutting fluid between 7% and 17% of the total manufacturing cost [5].

13.1.2 CUTTING FLUIDS

Cutting fluids or metalworking fluids (MWFs) are used in metal-forming and metal-cutting sectors as well as in the galvanic industry, where cooling, lubrication, and rust control are important at the tool–workpiece interfaces. They are also used in operations such as cutting, grinding, or rolling. MWFs help improve the life and function of the tool and account for up to 15% of the cost of the machining process. The metalworking industry classifies these fluids into three broad categories (Scheme 13.1) [6]:

- Oil-in-water (O/W) emulsions, also called "water-soluble oils" (although they are not soluble in water). These are obtained by mixing a mineral oil and an

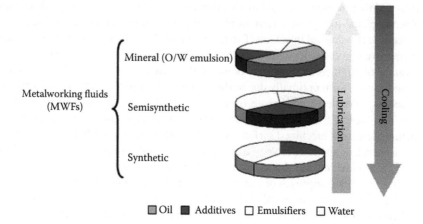

SCHEME 13.1 Classification of MWFs and their composition before being dispersed in water.

aqueous phase containing an emulsifier (surfactant). They may also contain additives to improve various lubricating properties. They are milky white with base oil concentrations of 1–10% w/w in water, and provide good lubrication.
- Synthetic fluids, also referred to as "chemical fluids." These do not contain mineral oil but are mixtures of several water-soluble compounds, such as emulsifiers, anticorrosion agents, or defoamers. They are clear or translucent solutions in water and often provide the best cooling performance among all MWFs.
- Semisynthetic fluids: These are a combination of O/W emulsions and synthetic fluids, and have properties common to the first two types.

The performance of MWF depends on their formulation. The emulsifier (surfactant) concentration plays a key role not only in lubrication property but also in properties such as emulsion stability, droplet size, and wettability that have a major impact on MWF performance. Biocides are added to MWFs in order to avoid proliferation of microorganisms [7,8].

MWFs increase the productivity and the quality of manufacturing operations by cooling and lubricating during metal-forming and metal-cutting processes [9].

Chemicals in cutting fluids pose potential health hazards to workers in polluted machining environments. After being used for some time, these fluids are degraded and eventually have to be disposed [10].

One potential way to avoid these environmental impacts is to discontinue or limit the use of MWFs, but a complete elimination of MWF function has proven impossible [11–14]. Dry/damp machining is being developed but it has negative environmental and economic impacts. Given that the machining of metals is essential to modern society and that MWF use is continuing to grow, it is necessary to design MWF systems with minimum impact on the environment.

13.1.2.1 Action of Cutting Fluids

When cutting fluids are applied, the existence of high contact pressure between chip and tool along the plastic part of the tool–chip contact length precludes any fluid access to the rake face. In spite of this, the theory that considers these fluids as boundary lubricants is the leading theory used to explain the marked influence of cutting fluids on the cutting process outputs (cutting force and temperature, surface finish and residual stresses in the machined surface, and tool wear) [15]. Despite a relatively large number of publications on cutting fluids, only very few discuss the role of a cutting fluid in the complex mechanics of the cutting process [16–22]. To account for cutting tool penetration to the rake face, four basic mechanisms have been suggested: access through the capillary network between chip and tool, access through voids connected with build-up edge formation, access into the gap created by tool vibration, and access due to propagation from the chip blackface through distorted lattice structure. However, no conclusive experimental evidences are available to support any of these mechanisms. It was observed that cutting fluids sometimes reduce the tool–chip contact length.

13.1.2.2 Types of Cutting Fluids

Currently, cutting fluids are classified into two main categories [23,24]:

1. Oil-based fluids
2. Chemical fluids

Oil-based fluids include straight oils, soluble oils, and Ag-based oils.

Straight oils, so called because they do not contain water, are basically petroleum, mineral, or ag-based oils. They may have additives designed to improve specific properties [25,26].

Soluble oils (also referred to as emulsions, emulsifiable oils, or water-soluble oils) are generally comprised of 60–90% petroleum or mineral oil, emulsifiers, and other additives [25,26,27]. A concentrate is mixed with water to form the metalworking fluid. When mixed, emulsifiers (a soap-like material) cause the oil to disperse in water, forming a stable "oil-in-water" emulsion [23,28].

Chemical fluids include synthetics and semisynthetics. Chemical cutting fluids, called synthetic or semisynthetic fluids, have been widely accepted since they were first introduced around 1945. They are stable, preformed emulsions that contain very little oil and mix easily with water. Chemical cutting fluids rely on chemical agents for lubrication and friction reduction [9].

Synthetic fluids contain no petroleum or mineral oil [22,28]. They were introduced in the late 1950s and generally consist of chemical lubricants and rust inhibitors dissolved in water. Like soluble oils, synthetics are provided as a concentrate that is mixed with water to form the metalworking fluid.

- Fluids vary in suitability for metalworking operations. For example, petroleum-based cutting oils are frequently used for drilling and tapping operations due to their excellent lubricity while water-miscible fluids provide the cooling properties required for most turning and grinding operations.
- Semisynthetics (also referred to as semichemical fluids) are essentially a hybrid of soluble oils and synthetics. The remaining portion of a semisynthetic concentrate consists mainly of emulsifiers and water. Wetting agents, corrosion inhibitors, and biocide additives are also present. Semisynthetics

are often referred to as chemical emulsions or preformed chemical emulsions since the concentrate already contains water and the emulsification of oil and water occurs during its production. The high emulsifier content of semisynthetics tends to keep the suspended oil globules small in size, decreasing the amount of light refracted by the fluid. Semisynthetics are normally translucent but can vary from almost transparent (having only a slight haze) to opaque. Most semisynthetics are also heat sensitive. Oil molecules in semisynthetics tend to gather around the cutting tool and provide more lubricity. As the solution cools, the molecules redisperse [24,25,29].

13.1.2.3 Alternatives to Cutting Fluids

In the pursuit of profit, safety, and convenience, a number of alternatives to traditional machining are currently under development. Dry machining has been around for as long as traditional machining, but it has seen a recent surge in interest as more people are realizing the true cost of cutting fluid management. Considering the high cost associated with the use of cutting fluids and projected escalating costs of disposal when stricter environmental laws are enforced, the choice seems obvious. Owing to the stricter environmental laws some alternatives have been sought to minimize or even avoid the use of cutting fluid in machining operations. Some of these alternatives are dry machining and machining with MQL. Other novel cutting fluids, such as liquid nitrogen, are also being explored for their unique properties. The following sections provide a few details about each of these technologies.

Dry machining, machining without the use of cutting fluids, has become a popular option for eliminating the problems associated with cutting fluid management. One of the greatest obstacles to acceptance of dry machining is the false belief that cutting fluids are needed to produce a high-quality finish. Studies have shown that with proper equipment and tooling, machining without fluids can produce a high-quality finish, and be less costly than machining with fluids [30]. MQL, also known as near-dry machining (NDM) or semidry machining, is another alternative to the traditional use of cutting fluids.

- Liquid nitrogen technology, one solution to the problem of cutting fluid management currently under development, is the use of liquid nitrogen as a coolant and lubricant. This technique is not the same as cryogenic machining, where the material to be cut is cooled to a very low temperature prior to the machining operation. Rather, the method currently under development uses liquid nitrogen to perform the cooling and lubricating job of the cutting fluid. Most of the part remains at ambient temperature while the flow of nitrogen is carefully directed to the point where it is needed. The small flow rate and low cost of liquid nitrogen make this technique a very attractive alternative [31].

This technique can be used on the equipment that has been designed for use with cutting fluids, and because nitrogen evaporates harmlessly into the air, there is no cutting fluid to dispose. If successful, this technique will provide an alternative to businesses that want to eliminate the use of traditional cutting fluids but cannot afford the capital expenditure required to purchase new dry-machining equipment. Reportedly, tool life and finish quality are also improved by this technique due to the low temperatures at the tool/part interface.

13.1.3 EMULSIFIER SYSTEMS

Mixed anionic:nonionic emulsifier systems are designed for petroleum- and bio-based MWFs to improve fluid lifetime, by providing emulsion stability under hard water conditions, a common cause of emulsion destabilization leading to MWF disposal.

With the evolution and modernization of metal mechanical industries, specific developments have required the need to formulate special lubricating fluids that meet the constantly changing market requirements. Cutting fluids are important elements in the industrial processes, used, for example, during metal-cutting or general metalworking operations. They are very complex mixtures that differ according to the type of operation implemented and the chemical nature of the metals handled in the industry [32]. The formulations are conceived from two basic, distinct fluids, typically water (W) and oil (O), and may therefore be classified as one of the two kinds: aqueous or nonaqueous. Fluids derived from O/W dispersions in the form of emulsions may also be prepared, and are commonly known as *soluble oils*. Unlike their denomination, these types of fluids are not water-soluble, but form emulsions in an aqueous environment, effectively dispersing oil within the continuous water phase [33].

Cutting fluids are very complex mixtures that differ according to the types of metalworking operation and the chemical nature of the workpiece metals [32]. Cutting oils are formulated from two basic distinct fluids: water (W) and oil (O). As a result, they may, therefore, be classified as one of two broad categories: aqueous or nonaqueous. Aqueous fluids include O/W dispersions or where oil is effectively dispersed in a continuous water phase [34].

These are basic mineral oils with specific additives to allow them to form O/W emulsions when added to water [35]. Cutting fluids also contain other additives such as anticorrosion, biocide, and antifoam agents [36,37]. The mineral oil is obtained from the distillation of petroleum, and its properties depend on the nature of crude oil. Mineral oils used in cutting fluid formulation can be grouped into two main classes: paraffinic and naphthenic.

The lubricating oils are largely used to enhance the finishing of metal surfaces, to reduce the deterioration of tools, and to protect them against corrosion. Biostable cutting fluids [38] possess many advantages with regard to the environment and the health of personnel who handle equipments and machinery. Manufacturers usually recommend as to which lubricating fluid is better suited for use and maintenance of a specific equipment. However, this does not imply total elimination of the problems related to longer life of equipments operating under heavy production conditions. Interruption of the operation of equipment due to the lubrication problem still occurs [39].

13.1.4 CUTTING LUBRICANTS

Some cutting lubricants are simply aqueous solutions of surfactants that have the desired extreme pressure (EP) properties. Such preparations containing no substantial proportion of oil are commonly used for metal-drawing lubricants, but do not appear to be as widely used for cutting, grinding, drilling, or machining [40]. A series of recent studies has contributed greatly to an understanding of the mechanisms by which cutting oils exert this effect [41,42]. The cutting energy was calculated from thermochemical energy input and the heat generated. Thus, the surface activity, that is, the molecular effects at the liquid–metal interface, enabled the metal to be ruptured with much greater energy efficiency. Similar experiments on the rolling of metals with and without aqueous surfactant solutions showed that the surface activity is also responsible for a much higher energy efficiency when the metal is deformed without being ruptured [43–45].

Soluble oils or more appropriately termed "emulsifiable oils" form milky emulsions when mixed with water. The most important characteristic of a soluble oil is that it emulsifies easily to form stable emulsions. Soluble oils have a rather complex composition. They usually contain two or more emulsifiers, coupling agent (CA), and stabilizing agents as well as additives to provide rust inhibition, lubricity, detergency, and resistance to bacterial attack [46–51]. The surfactant additive in "soluble oil" may act as an emulsifying agent, EP [52], and/or corrosion inhibitors. The relative concentrations of components are delicately balanced to ensure that the finished compositions will remain stable until used [53,54]. Some investigators reported that the sulfonates of calcium and sodium were included in the cutting fluid formulation as antiwear and EP additives. Misra and Skold reported on the effects of component molecular structure, pH, temperature, and concentration on the phase and aggregation behavior of aqueous formulations. The formulations comprise more than one polydisperse polyglycol-based compound and water-soluble polymer compounds. They investigated surface tension, turbidity, conductivity, and pH over a wide range of temperatures and concentrations [55]. Belluco and De Chiffre investigated the effect of new formulations of vegetable oils on surface integrity and part accuracy in reaming and tapping operations [56]. The performance of EP additives, including chlorinated paraffin, acid phosphate, and zinc dialkyl (or diaryl) dithiophosphate (ZDDP) was evaluated. Using the grinding test method, the index of material removal ability and the specimen hardness drop were evaluated. They also showed that the stability of cutting fluid was affected by the droplet size of emulsion [57].

Some investigators [58–61] reported that MQL cutting is a successful example of NDM operations, using actual milling operations. These studies showed equivalent to or sometimes superior cutting performance of MQL with a biodegradable synthetic ester lubricant relative to cutting with conventional coolant flood supply. The formulation of soluble oils has progressed from 1930s soluble oils, which were unstable after a few circulation cycles through the handling system. Today, soluble oils are stable to recirculation cycles in machine. Modern formulations use a package emulsifier system comprising new ionic surfactant soaps, oil-soluble sulfonates, and CAs [62].

13.1.5 LUBRICATING EMULSIONS

There are many components in a lubricating emulsion, each of which plays a role in the overall performance [63]. Most of these components are added in relatively small quantities, resulting in emulsions that are stable (Scheme 13.2).

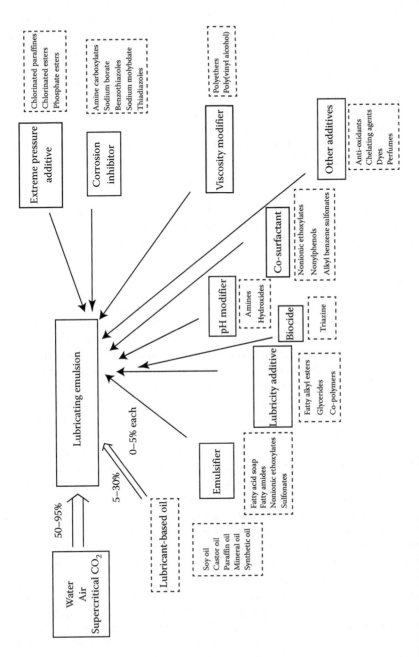

SCHEME 13.2 Lubricating emulsion composition.

There are some reports of air and supercritical carbon dioxide utilization technologies, for example, a method for lubricating a metal workpiece during a metalworking process includes delivering supercritical carbon dioxide to the workpiece during the metalworking process. The supercritical carbon dioxide acts as a lubricant, coolant, chip evacuator, and/or carrier for another lubricant or corrosion inhibitor [64,65]; however, emulsions are by far the most prevalent and relevant to the topics to be discussed.

The stability of O/W emulsions used for lubrication and the methods for studying stability are not different from those used in other applications. Zeta potential [66,67], surface tension, interfacial tension (IFT) [68], turbidity [69], oil-drop diameter [70], and simple physical observation of phase separation are often used to estimate emulsion stability. Early emulsion systems were prepared using medium viscosity mineral base oil [71], a sulfonate-type emulsifier, required EP, and antiwear additives. The most stable emulsions had pH readings on the higher end (8.8–9.6).

Surface tensions in the lower range of 33–42 mN m^{-1} gave more stable emulsions, which were in agreement with the study in [72]. A study using 17 different surfactants showed a correlation between lower surface tension and highly cutting oil performance.

13.1.5.1 Stable Emulsions

Stable emulsions are best formulated with emulsifiers or combinations of emulsifiers, which possess hydrophilic–lipophilic balance (HLB) values close to the required HLB of the oil phase. In this chapter, we have investigated the application of this established method for the development of multiple emulsions. This is of particular interest, since multiple emulsions are highly sensitive to variations in individual components because of the presence of two thermodynamically unstable interfaces. Multiple water-in-oil-in-water (W/O/W) emulsions are potential skin delivery systems for water-soluble active pharmaceutical ingredients because of their pronounced encapsulation properties.

The stable emulsions were prepared as follows: First, a suitable primary emulsion was developed based on the required HLB determination of the investigated oils. Second, and based on the required HLB, multiple W/O/W emulsions were developed using the most appropriate primary emulsion and 1% hydrophilic emulsifier blends in order to stabilize the second interface. In order to find the appropriate mixtures of hydrophilic emulsifiers, the required HLB for the primary W/O emulsion was determined using two different chemical classes of emulsifier blends, namely polyethoxylated ethers and polyethoxylated esters. The physicochemical parameters of the formulations were characterized by means of rheological measurements, droplet size, and creaming volume observations as well as by means of conductivity analysis. All methods were found to be appropriate for determining the required HLB with the exception of the rheological data. The primary emulsions tested required HLB values of 4.3–4.7 using paraffin as the oil phase and resulted in stable emulsions. Irrespective of the emulsifiers used, the finest droplets, lowest conductivity, and minimal creaming volume were obtained for multiple emulsions at the required HLB values between 15 and 15.5 using paraffin as the oil phase. Moreover, using a polyethoxylated ether instead of a polyethoxylated fatty acid ester resulted in more stable multiple emulsions.

13.1.5.2 Hydrophilic–Lipophilic Balance Values of the Emulsifiers

Owing to their distinct structure and properties, multiple emulsions are of particular interest for several drug-delivery approaches, including as carrier systems for the dermal application of pharmaceutical drugs [73–77]. Carrier systems are developed by empirical means. These empirical procedures are often extremely time consuming, since carrier systems require a well-defined ratio of ingredients [78]. It is, therefore, of great importance to apply the physicochemical properties of the selected constituents at an early stage of the formulation development process. Two of these major parameters are the HLB value of the emulsifiers used and, determined by empirical tests, the required HLB (rHLB) value for the oil phase investigated. Using these parameters allows reducing the number of experiments during the early formulation screening stage [79–82].

Griffin first established the HLB system to classify nonionic surfactants [79]. According to Griffin, lipophilic surfactants have low HLB values, and hydrophilic surfactants possess high HLB values.

Based on this system, W/O emulsions are obtained using surfactants with HLB values ranging between 3 and 8, whereas O/W emulsions are formed using surfactants with HLB values between 9 and 12. Stable emulsions are best formulated with emulsifiers or combinations of emulsifiers, which have HLB values close to the rHLB for the oil used in the formulation.

In order to determine the rHLB for an oil or oil phase, emulsions are produced with different ratios of emulsifier blends, representing various HLBs, and investigated for their separation property [83]. However, besides the HLB value, the chemical structure of the emulsifier influences the stability of emulsions. The main problem of stability, however, is the presence of two thermodynamically unstable interfaces, that is, the W/O interface of the primary emulsion and the O/W interface of multiple emulsions. Two different emulsifiers are, therefore, necessary for stabilization: one with a low HLB for the W/O interface and a second with a high HLB for the O/W interface. The effect of the nature and quantity of both emulsifiers on the properties of multiple emulsions has been discussed by several workers [78,84–87].

Selection of lipophilic surfactants and determining the rHLB for the primary W/O emulsion of multiple emulsions can be accomplished in the same way as for simple W/O emulsions. However, the determination of the rHLB for W/O/W multiple emulsions and the selection of the hydrophilic emulsifiers is more complex. It has been shown that both lipophilic and hydrophilic emulsifiers can adsorb on O/W interfaces [88]. The HLB value needed for emulsification of the oil phase to form O/W emulsion, when both hydrophilic and lipophilic emulsifiers are present, is the sum of the HLB values of all adsorbed surfactants. The amount of adsorbed lipophilic emulsifier on the O/W interface is, however, not defined and is dependent on its concentration in the formulation and the phase volume ratio [88]. It was discovered that the ideal HLB value for hydrophilic emulsifiers increases when the concentration of the lipophilic emulsifier is increased or when the amount of the hydrophilic emulsifier is reduced.

13.1.5.3 Surfactant Emulsifiers

Surfactant emulsifiers are widely used in MWFs to stabilize the oil and water components present in the formulations. They contain both a hydrophilic head group that has affinity for water, and a lipophilic tail group that is soluble in oil and oil-soluble components [89–91]. Such a fluid has the particular advantage that it combines the cooling property of water and the lubrication property of oil [92,93]. In general, a wide range of emulsifier properties and emulsifier types may be used to form and stabilize emulsions. The three broad categories of emulsifiers that can be considered are nonionic, anionic, and cationic. Cationic emulsifiers are not typically used in MWF formulations since they may be shipped out by the metal surfaces that are negatively charged at elevated pH (e.g., 8–10) of MWFs. Nonionic emulsifiers are electrically neutral and, therefore, have the advantage of being relatively unaffected by the hard water ion content of the MWF aqueous phase. While anionic emulsifiers are much more sensitive to hard water, they have the advantage of being more effective at reducing the O/W IFT and result in better wetting of the MWF on the metal surface. Anionic surfactants are cheaper than nonionic emulsifiers and also more amenable to waste treatment via emulsion breaking. Binary mixture of nonionic and anionic emulsifiers as a synergistic emulsifier system is also a common practice. In the design of such mixtures, it is necessary to establish a balance between anionic emulsifiers that minimize cost and increase waste treatability and nonionic emulsifiers that improve tolerance to hard water and salinity. Mixed emulsifier systems also increase the packing density of the emulsifier around the oil droplet, thus providing improved emulsion stability. Commercial cutting fluid concentrates generally contain a mixture of anionic and nonionic emulsifiers [94].

13.1.6 Factors Affecting Emulsion Stability

Water-in-oil (W/O) emulsions consist of water droplets in a continuous oil phase, and O/W emulsions consist of oil droplets in a water-continuous phase. Figures 13.1 and

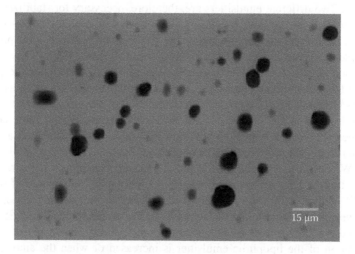

15 μm

FIGURE 13.1 Photomicrograph of an oil-in-water emulsion.

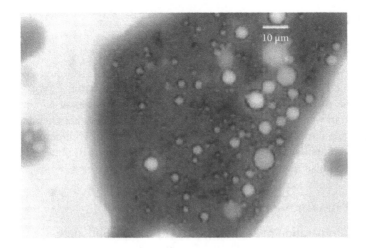

FIGURE 13.2 Photomicrograph of a water-in-oil-in-water emulsion.

13.2 show the two basic (O/W and W/O) types of emulsions. In the oil industry, W/O emulsions are more common (most produced oilfield emulsions are of this kind); therefore, the O/W emulsions are sometimes referred to as "reverse" emulsions.

From a practical standpoint, emulsion stability is the most critical parameter for analyzing oil–water emulsions. The higher the degree of emulsion stability, the more difficult the oil and water to be separated. Insight into the factors affecting stability will provide a greater understanding into the breaking, or deemulsification of emulsions. Under normal circumstances, viscous and heavy crude oils in which steam-assisted gravity drainage (SAGD) operations are economically viable, contain natural emulsifiers that will help to stabilize emulsions (this emulsion is kinetically stable).

Several factors affect the stability of reverse emulsions produced at oil wells:

1. Presence of surfactants that cause the oil to be dispersed in water as very small droplets
2. Shearing of the fluid during production, in which steam flooding causes the generation of smaller droplets, which increase the emulsion stability
3. Higher molecular weight organic compounds in the emulsion that increases the stability of the emulsion
4. Presence of fine solids such as clays, sand, or corrosion products injected during production increases the stability of the emulsion

Contrary to normal W/O emulsions, the temperature of crude oil does not play a significant role in stabilizing reverse emulsions. Minor viscosity change of the oil phase occurs when temperature is increased and thus increasing the fluid temperature does eliminate O/W emulsions.

Laboratory evaluation of crude oil emulsion stability involves agitation of the sample fluid in a reservoir under varying conditions.

13.2 EXPERIMENTAL

13.2.1 MATERIALS

Two types of paraffin base oils (viscosity grade (VG): 140–160 and 260–290 cSt) were obtained from Alameria Refining Company (Alexandria, Egypt). The properties of the paraffin oils are shown in Table 13.1. The water used in all experiments was tap water. The properties and the sources of the other chemicals used in this study are shown in Table 13.2. The chemicals were of commercial grade, and were used as supplied. A commercial cutting fluid (Dromus B) obtained from Shell Company was also investigated.

13.2.2 PREPARATION OF EMULSIFIERS

13.2.2.1 Preparation of Sodium Dodecyl Benzene Sulfonate

Dodecyl benzene sulfonic acid was neutralized by the addition of 10% aqueous NaOH solution dropwise with stirring until pH 7. The resulting sodium salt of dodecyl benzene sulfonic acid (SDBSA) was dissolved in isopropyl alcohol to precipitate any inorganic salt. The alcohol was distilled off and the anionic surfactant further purified overnight in a vacuum drying oven at 70°C [90].

TABLE 13.1

Some Properties of Paraffinic Base Oils Investigated

Test	Method	Result VG[a] = 140/160	VG = 260/290
Density at 15°C, g/cm³	ASTM D445	0.8723	0.8920
Appearance	Visually	Clear	Clear
Color	ASTM 1500	<1[b]	<1
Pour point, °C	ASTM 96	−3	0
Flash point (closed), °C	ASTM 93	196	211
Kinematic viscosity at 40°C, cSt	ASTM 445	29	68
Kinematic viscosity at 100°C, cSt	ASTM 445	5.3	7.9
Viscosity index	ASTM 2270	104	114
Total acid number (mg KOH/g)	ASTM 664	<0.05	<0.05
Percentage of aromatic content (CA%)[c]	Column chromatography	13	13
Percentage of paraffinic content (CP%)[d]	Column chromatography	63	63
Percentage of naphthenic content (CN%)[e]	Column chromatography	24	24

[a] VG: viscosity grade.

[b] Color <1 means the scale of color is defined by 16 glass standards of specified luminous transmittance and chromaticity, graduated in steps of 1.0 from 1.0 for the lightest color to 8.0 for the darkest.

[c] Aromatic content is divided into two catogeries: (i) benzene ring, e.g., benzopyrene etc. and (ii) nonbenzene ring, e.g., furane etc.

[d] Paraffinic content in different types of waxes, e.g., macro- and micro-crystalline waxes etc.

[e] Naphthenic content in cyclohexane, cyclopentane, etc.

TABLE 13.2
Chemicals Used in This Investigation

Materials and Abbreviations	Molecular Weight	Source	Abbreviation
Sodium hydroxide	40	WINLAB/Egypt	NaOH
Tween 85 (polyoxyethylene (e.o. = 10) sorbitan trioleate)	1905	WINLAB/Egypt	T
Oleic acid	282.46	Merck/Germany	OA
Lauryl alcohol (1-dodecanol)	186.33	Aldrich/Germany	LA
Monoethanolamine	61.08	Aldrich/Germany	MEA
Diethanolamine	105.14	Aldrich/Germany	DEA
Triethanolamine	140	Aldrich/Germany	TEA
Boric acid	61.83	Aldrich/Germany	BA
Isopropyl alcohol (2-propanol)	60.10	WINLAB/Egypt	PA
Xylene	106.17	Aldrich/Germany	X
Para toluene sulfonic acid	172.20	Aldrich/Germany	PTSA
Nonyl phenol ethoxylate	220.35	Aldrich/Germany	NPE9

13.2.2.2   Preparation of Sodium Oleate

Oleic acid (OA) was neutralized by the addition of 10% aqueous NaOH solution dropwise with stirring until pH 7. The resulting sodium salt was dissolved in isopropyl alcohol to remove any inorganic salt and the alcohol was distilled off. The anionic surfactant was further purified of traces of solvent by heating (70°C) under vacuum overnight [91].

13.2.2.3   Preparation of Oleic Diethanol Amide

In a 1-L single-neck flat-bottom flask fitted with a Dean-Stark trap and a condenser, 1 mol of OA was reacted with 1 mol of diethanolamine. The reaction ingredients were refluxed in 20 mL xylene solvent and 0.1 g p-toluene sulfonic acid catalyst with stirring at 140°C until the theoretical amount of water was collected. The obtained product of diethanolamide was dissolved in isopropanol and salted-out with supersaturated sodium chloride solution, the upper layer was taken and the isopropanol was distilled-off and the diethanolamide was obtained. The chemical equation is shown as follows [95]:

$$(HOCH_2CH_2)_2NH + C_8H_{17}CH=CHC_7H_{14}COOH \longrightarrow C_8H_{17}CH=CHC_7H_{14}CON(CH_2CH_2OH)_2$$

Diethanolamine   Oleic acid   Oleic diethanol amide

13.2.3   PREPARATION OF CORROSION INHIBITORS

13.2.3.1   Preparation of Boric Acid Diethanolamine

In a 1-L single-neck flat-bottom flask fitted with a Dean-Stark trap and a condenser, 1 mol boric acid and 1 mol diethanolamine in 20 mL of xylene and 0.1 g p-toluene sulfonic acid catalyst were refluxed with continuous stirring at 140°C, until the theoretical amount of water was collected. The solvent was then distilled off and the product was washed with hot water several times and dried [95]. The chemical reaction is shown as follows:

$$H_3BO_3 + (HOCH_2CH_2)_2NH \longrightarrow (HO)_2BN(CH_2CH_2OH)_2$$
Boric acid   Diethanolamine   Boric acid diethanolamine

13.2.3.2 Preparation of Dioleic Diethanolamine Diester

In a 1-L single-neck flat-bottom flask fitted with a Dean-Stark trap and a condenser, 2 mol of OA and 1 mol of diethanolamine in 20 mL of xylene solvent containing 0.1 g *p*-toluene sulfonic acid catalyst were refluxed with stirring at 140°C until the theoretical amount of water was collected. The product was dissolved in isopropanol and salted-out from a supersaturated sodium chloride solution. The upper layer was then separated, the isopropanol distilled off, and the diester product was isolated. The chemical reaction is shown as follows:

$$1(HOCH_2CH_2)_2NH + 2C_8H_{17}CH{=}CHC_7H_{14}COOH$$
Diethanolamine   Oleic acid

$$\downarrow$$

$$C_8H_{17}CH{=}CHC_7H_{14}COOCH_2CH_2NHCH_2CH_2OOCH_{14}C_7HC{=}CHC_8H_{17}$$
Diethanolamine diester

13.2.4 Cutting Oil Formulations

The formulation of a soluble cutting oil blend was made in three steps:

Step 1. the emulsifier package, which consists of a nonionic emulsifier (Tween 85) and an anionic emulsifier (SDBSA) in a 2:1 ratio, with two concentrations of the blend emulsifier used as 9% and 12%(w/w), were formulated by a blend of four components as follows.

Step 2.
  a. Lubricant: OA (4%)
  b. Biocide: triethanol amine (5%)
  c. Antirust: boric acid amide (2%)
  d. Stabilizer agent: lauryl alcohol (LA) (2%)

Step 3. Formulate the cutting fluid by blending the lubricant package with water to form the finished product, that is, the soluble cutting emulsion.

All emulsion formulations were aged for 12–15 h at approximately 25°C prior to stability evaluation [96].

13.2.5 EVALUATIONS

13.2.5.1 Stability of Oil Phase (Screening Test)

A stability test was conducted in accordance with the American Petroleum Institute standard test method (IP311). Approximately 75 mL of the test oil was placed into sample bottles. The bottles were capped and placed in an oven at 50°C for a period of 15–20 h. The bottles were then removed from the oven and immediately examined for any signs of turbidity, separation, or gelling. The blends that gave stable oil (with no gel formation and no separation) were selected for further investigation [91].

13.2.5.2 Emulsion Stability Test

This test was conducted according to the Institute of Petroleum (IP) standard test method IP 263 for emulsion stability on blend samples that gave the most stable soluble oil. Tests were conducted on 5% and 10% solutions of soluble oil in distilled water with 0.688 g/L of $CaCO_4 \cdot 2H_2O$ to give it hardness equivalent to 400 ppm in terms of $CaCO_3$. Emulsions with 5%, 10%, and 15% soluble oil were prepared as follows: 0.5, 1, and 1.5 mL of the formulated soluble oil were poured into 9.5, 8, and 8.5 mL of the prepared hard-water-graduated cylinder. The cylinder was then strongly shaken for 2 min and left for 24 h at room temperature. The cylinder was then examined for emulsion stability (oil or cream separation).

Soluble oil blends that exhibited remarkable emulsion stability [no separated oil/no cream after 24 h (mL oil/mL cream = 90)] were selected for further evaluation [91].

13.2.5.3 Iron Chip Corrosion Test (IP 125)

Corrosion tests with cast iron chips were carried out as follows. Four grams of cast iron chips, which had been washed with benzene, were immersed in the test fluid sample (4 mL) in a watch glass. The container was covered for 10 min and the fluid poured by tilting the watch glass. The rust-preventive effect was evaluated visually after 24 h. The amount of rust on the cast iron chips was as follows: no rust, 10; 1 case of rust, 9; 2 points of rust, 8; some points of rust, 7; many points of rust, 6; many points of rust and stains on the bottom of the watch glass, 5. A rating of 10 points corresponds to no appearance of rust. This method was based on the ASTM D 4627-92 [97–99].

13.2.5.4 Extreme Pressure and Antiwear Test

The standard test method (IP 239) was carried out in this study. The four-ball wear test (ASTM D-4172) determines the wear protection properties of a lubricant. Three metal balls are clamped together and covered with the test lubricant, while a rotating fourth ball is pressed against them in sliding contact. This contact typically produces a wear scar, which is measured and recorded. The smaller the average wear scar, the better the wear protection provided by the lubricant (Figure 13.3) [100].

13.2.5.5 Bacterial Resistance of Water Dilutable Metalworking Fluids "Rancidity Control"

Generally, microbial contamination occurs in an acidic medium rather than in an alkaline medium. The rise in pH values indicates better resistance of the fluid to

FIGURE 13.3   Four-ball machine consisting of a bearing ball.

microbial contamination. The pH of the O/W emulsions was determined at room temperature using a digital pH meter [92,101,102].

13.2.5.6   Biodegradability of Cutting Fluid Emulsion

This study aimed to investigate the ability of some microbial strains to utilize different constituent structures of the used emulsifiers of the emulsion samples as a sole source for carbon and energy. This was conducted in 100 mL batch flasks containing 30 mL basal salts medium (BSM) with initial pH 7 for bacterial and yeast strains and pH 6 for fungal strain. BSM was prepared according to the method reported before [92]. The incubation period was 10 days at 30°C in a shaking incubator (150 rpm). Bacterial and yeast growth were monitored by total colony-forming unit (TCFU) technique on tryptone glucose yeast extract medium (TGY) prepared according to Ref. [103] after 3 and 10 days of incubation. Fungal growth was monitored by total dry weight (D) technique [104] after 3 and 10 days of incubation. Table 13.10 represents the ratio of final/initial TCFU (I/I_o) for all tested bacterial and yeast strains and final/initial dry weight (D/D_o) for the tested fungal isolate [92].

13.2.5.7   Surface Tension and Surface Active Properties

At the critical micelle concentration (cmc), the value of most physical and chemical properties of the emulsifier solutions displays an abrupt change. The surface tension (γ) was measured at 303 K for different concentrations of the mixed emulsifiers. The values of γ were plotted against a natural log of the emulsifier concentration (ln C). The intercept of the two straight lines designates the cmc, where saturation in the surface-adsorbed layer takes place. Surface properties of the mixed emulsifier such as effectiveness (π_{cmc}), maximum surface excess (Γ_{max}), and minimum area per molecule (A_{min}) were calculated using Equations 13.1 through 13.5 [105]:

$$\pi_{cmc} = \gamma_o - \gamma_{cmc} \tag{13.1}$$

$$\Gamma_{max} = -10^{-7}\,[1/RT]\,[d\gamma/d\ln C]_T \tag{13.2}$$

$$A_{min} = 10^{16}/[\Gamma_{max} \cdot N_A] \tag{13.3}$$

$$\Delta G_{mic} = RT\ln cmc \tag{13.4}$$

$$\Delta G_{ads} = \Delta G_{mic} - [0.6022 \times \pi_{cmc} \times A_{min}] \tag{13.5}$$

where γ_o is the surface tension of pure water at the appropriate temperature, γ_{cmc} is the surface tension at cmc, Γ is the surface excess concentration in mol/dm^2, R is the universal gas constant ($R = 8.314$ J/(mol $\cdot$ K), T is the absolute temperature, ($t°C + 273$)K, γ is the surface or IFT mN/m, C is the concentration of the emulsifier (mol/L), A_{min} is the surface area per molecule of solute in square nanometers (nm^2/molecule), N_A is the Avogadro's number (6.023×10^{23} molecules/mol), ΔG_{mic} is the Gibbs free energy of micellization, and ΔG_{ads} is the Gibbs free energy of adsorption.

13.3   RESULTS AND DISCUSSION

13.3.1   MECHANISM

The primary functions of cutting fluids are to cool the tool and workpieces by transferring heat away from the cutting zone, and to lubricate the cutting zone so as to reduce the amount of heat generated. Secondary functions of cutting fluids include chip transport and corrosion inhibition. There are several proposed mechanisms to explain cutting fluid access to tool–chip interface (cutting zone). One of the most important mechanisms for a cutting fluid is the evaporation/condensation mechanism. The machine process may generate considerable heat that results in high enough temperature in the cutting zone to possibly vaporize the cutting fluid. It is unlikely that the cutting fluid can actually penetrate into the cutting zone during high-speed machining; however, in some cases, it may be possible for a small amount of cutting fluid to penetrate the cutting zone via such mechanisms as capillary action, vibration, or diffusion. When this is the case, the evaporation rate in different jet-impinging boiling regions can be estimated through a heat flux prediction. The amount of vapor will depend on the cutting zone temperature and the amount of fluid at the cutting zone. Also, most specialists proposed that cutting fluid accesses the cutting zone through the capillary network existing at the tool–chip interface. In this case, the cutting fluid can form a stronger lubrication film by physical and chemical adsorption on the capillary wall. Lubricating action of cutting fluid requires that the cutting fluid absolutely penetrates the capillaries of the cutting zone and that the penetration time should be less than capillary lifetime, that is, there is a difference between deviation of capillary lifetime and penetration time [106].

13.3.2 EVALUATION OF EMULSIFIER SYSTEMS FOR METAL WORKING FLUIDS

A series of trials were carried out to obtain the optimum stability of the emulsifier package. The package contained the following four components: surfactant package (SDBSA/Tween 85 as emulsifier mixture), OA, LA, and biocide. These components and the base paraffin oil were used in all the soluble oil formulations investigated. Different concentrations of these components were used to optimize the stability of emulsifier systems. The emulsions were evaluated for oil stability, emulsion stability, pH, and rust inhibition [106].

13.3.2.1 Formulated Oil Stability

Six different cutting oil formulations (C1–C6) were prepared by different combinations of surfactants with OA and LA. There were four emulsifier blends (E1, E2, E3, and E4) based on (NPE9 + SDBSA), (NPE9 + SOA), (T + SDBSA), and (T + SOA), respectively. These four blends and different percentages of OA and LA were used to form four emulsifier packages (P1, P2, P3, and P4). These four components and the base paraffin oil (160–190 cSt) were used to prepare six cutting oil formulations, namely C1–C6 as shown in Table 13.3. Table 13.3 shows two types of cutting oil formulations (C7 and C8) with the best emulsifier blend E3, OA, and LA for paraffinic oil (VG = 260–290 cSt). By inspection of the data in Table 13.3, it is clear that some cutting oil formulations (C1, C2, and C4) were unstable and produced oil or cream separation or gel formation after 24 h. The instability may be attributed to the low percentage of fatty acid and high percentage of LA. The gel formation may be the result of high concentration of the emulsifiers. The results indicated that the stability and performance of the soluble oil were related to the concentrations of the following five ingredients in the formulation: anionic surfactant, nonionic surfactant, OA, LA, and biocide. The optimum percentages of these ingredients were obtained after many sets of experiments. From Table 13.3, the following optimum concentrations of these ingredients were observed: surfactants (4%T + 2% SDBSA), OA 4%, and LA 2% maximum emulsion stability. In general, as the emulsifiers concentration increases, the emulsification power of the oil increased and oil flocculation decreased as the mole ratio of SDBSA and Tween 85 became 1:2. These results might be the indication of increasing emulsification power due to increasing repulsion between oil droplets when the ratio of SDBSA/Tween 85 system is 1:2. The electrical double layer would greatly increase repulsion due to the development of electrostatic forces that would be large enough to prevent coalescence. Finally, the emulsions prepared with the anionic/nonionic emulsifiers showed better wetting behavior than those prepared with a single emulsifier because of its stronger adsorption on the metal surface since the polar parts of the emulsifiers induce preferential

TABLE 13.3

Composition of Cutting Oil Formulations

Code	Emulsifier[a] Range, %	Oleic Acid (OA), Trial/Optimum Range, %	Lauryl Alcohol (LA), Trial/Optimum Range, %	Paraffin Oil, Trial/Optimum Range, %
C1	NPE9/SDBSA 2–6/0–4	2-12/4	1-10/2	80-90/88
C2	NPE9/SOA 2–6/0–4	2-12/4	1-10/2	80-90/88
C3	T/SDBSA 2–6/0–4	2-12/4	1-10/2	80-90/88
C4	T/SOA 2–6/0–4	2-12/4	1-10/2	80-90/88
C5	T/SOA 4/2	2-12/2–4	1-10/2	80-90/88–90
C6	T/SOA 4/2	2-12/4	1-10/0.5–2	80-90/88–89.5
C7	T/SDBSA 2–6/0–4	2-12/4	1-10/2	80-90/88
C8	T/SDBSA 4/2	2-12/4	1-10/2	80-90/88

[a] The abbreviations of the used emulsifiers were shown in Table 13.2.

orientation at the surface due to dipole interaction, which leads to lower spreading coefficients and higher works of adhesion. The work of adhesion has been recently related to the lubricating performance of O/W emulsions and, specifically, with the formation and the stability of a lubricant oil pool in the contact between rubbing surfaces [107].

13.3.2.2 Thermal Stability of the Soluble Oil Formulations

The thermal stability of cutting oil formula (C3) containing emulsifier package P3 for paraffinic oil (VG: 140–160 and 260–290 cSt) at 70°C is shown in Table 13.4. From the data obtained, it is clear that the oil formulation with paraffinic oil (VG = 260–290 cSt) gives the best result at high temperature [101].

13.3.2.3 Emulsion Stability of Soluble Oil Formulations

The emulsion stability of cutting oil formulation (C3) for paraffinic oil (VG = 260–290 cSt) after 24 h was studied. From the commercial package, it was found that the best concentration of cutting oil formulation (C3) was at 9% and 12% of E3 (Table 13.5). Emulsions with oil/water ratio (5/95), (10/90), and (15/85), using emulsifier package (P3), with 9% concentration of the oil phase E3 exhibited 96%, 96%, and 98% emulsion stability, respectively. The maximum emulsion stabilities of 97%,

TABLE 13.4

Thermal Stability of Cutting Oil Formulation (C3) Containing Emulsifier Package (P3) for Paraffinic Oil (140–160 and 260–290 cSt) at 70°C

Cutting Oil Formulation (C3)					
Emulsifier Package (P3)					
Emulsifier (E3) (4%T + 2% SDBSA) Optimum Concentration, %	OA	LA	Paraffin Oil	Thermal Stability, %	
				140–160 cSt	260–290 cSt
6	4	2	88	−ve[a]	+ve[b]
9				−ve	+ve
10				−ve	+ve
12				−ve	+ve

[a] −ve means that the emulsifiers packages are not thermally stable.
[b] +ve means that the emulsifiers packages are thermally stable.

99%, and 100% were obtained at 12% concentration of the oil phase E3. The data showed that cutting oil formulation (C3) at 12% concentration of E3 gave the best emulsion stability that was similar to the commercial package. The high stability of the cutting fluid emulsions was probably due to the presence of nonionic surfactant, which does not form a precipitate by the metallic ions in hard water. Its presence at the surface of the droplets reduces the electrostatic attraction and prevents coagulation. Moreover, the high stability of these emulsions in hard water could be due to steric repulsion between droplets [108].

13.3.3 SURFACE ACTIVE PROPERTIES

13.3.3.1 Surface Tension

Emulsifying agents act to lower surface tension, thereby allowing the formation of smaller O/W droplets with the same energy input. Emulsifiers also deter or retard droplet coalescence and flocculation by forming an interfacial film around the dispersed-phase droplets. The most common mechanisms of stabilizing emulsion droplets involve repulsive electrical or steric forces to retard coalescence. When two emulsion droplets with the same electric charge approach each other, a repulsive force is generated between the particles, thereby maintaining discrete and stable emulsion particles. Steric interactions occur when the droplets (henceforth referred to as "particles") have layers of emulsifier molecules at their water interface. Steric stabilization is especially important for nonionic emulsifiers because they take different orientations leading to "packing effects" that can prevent coalescence and flocculation [94].

The stability of an emulsion can be increased by making ionic emulsifiers with nonionic emulsifiers. Adsorption of ionic surfactants at the surface yields electrical repulsion between oil droplets. Mixed emulsifiers increase the solubilization

TABLE 13.5

Emulsion Stability of Cutting Oil Formulation Containing Emulsifier Package (P3) for Paraffinic Oil (260–290 cSt) after 24 h

Cutting Oil Formulation (C3)				Emulsion Stability					
Emulsifier Package (P3)				5/95[a]		10/90		15/85	
Emulsifier (E3)	OA	LA	Paraffin Oil	Oil Separated/Cream Separated[b]	Emulsion Efficiency[c], %	Oil Separated/Cream Separated	Emulsion Efficiency, %	Oil Separated/Cream Separated	Emulsion Efficiency, %
Optimum Range, %	4	2	88						
6				0/1	85	1/0	87	0/1	90
7				0/1	90	0/0	94	0/0	97
9				0/0	96	0/0	96	0/0	98
10				0/0	95	1/0	97	0/1	96
11				0/0	95	1/0	96	1/1	94
12				0/0	97	0/0	99	0/0	100
13				0/0	98	1/0	95	1/1	93
Commercial sample				0/0	96	0/0	97	0/0	99

a Ratio of cutting oil to water, for example, 5/95 means 5 mL of cutting oil added to 95 mL water.

b mL of the separated oil (0–5 for 5/95 oil/water ratio) for example, 0/0 means no oil and no cream are separated after 24 h.

c The percentage of oil separated in oil-in-water emulsion after a certain time.

TABLE 13.6

Surface Active and Thermodynamic Properties for the Prepared Emulsifier Blend (E3) (4% T + 2% SDBSA)

Emulsifier	cmc × 10⁴ (mol/L)	γ (mN/m)	Π (mN/m)	$\Gamma_{max} \times 10^{10}$ (mol/cm²)	$A_{min} = A_{cmc}$ (nm²)	ΔG_{mic} kJ/mol	ΔG_{ads} (kJ/mol)
SDBSA	1.122	38	34.3	1.16	142.84	−23	−24
Tween 85	6.562	44	28.3	1.68	98.89	−18	−21
Emulsifier blend (E3) (4% T + 2% SDBSA)	30	42	30.3	1.19	139.45	−14	−17

capacity of emulsifiers and may facilitate spontaneous emulsification when the oil phase is brought in contact with water. The presence of an anionic emulsifier in the emulsifier mixture gives a negative electric charge to the oil droplets. It was observed from Table 13.6 that the IFT and surface tension for the mixed emulsifier were less than that for the individual emulsifiers, indicating synergistic behavior of the mixed surfactant system.

13.3.3.2 Critical Micelle Concentration

Surface tension versus surfactant concentration plots indicates that each emulsifier is molecularly dispersed at low concentration, leading to a reduction in surface tension. This reduction in surface tension increases with increasing concentration of the emulsifier. At cmc, the emulsifier molecules form micelles, and are in equilibrium with the free emulsifier molecules. At the cmc, saturation of the surface-adsorbed layer takes place. The cmc of mixed surfactants is significantly lower than that of the pure emulsifiers (see Figure 13.4). This demonstrates that interactions between the two different emulsifier components in the mixed micelles were taking place [109,110]. The investigated polymeric mixed emulsifiers prefer adsorption over micellization as seen in Table 13.6, since their ΔG_{ads} was more negative than ΔG_{mic}. A larger negative ΔG_{ads} than ΔG_{mic} supports the emulsification process. It is an indication that the polymeric mixed emulsifier would like to adsorb on the interface rather than form micelles. The lower surface excess, Γ_{max}, and larger area per molecule, A_{min}, for the mixed emulsifier than for the individual emulsifiers produced greater emulsification efficiency for the mixed emulsifier. Hence, synergism between the two emulsifier molecules in the formation of mixed micelles was observed. This indicates stronger interaction between the two different emulsifier molecules than between the same emulsifier molecules.

13.3.3.3 Interfacial Tension

Oil and water do not mix due to the IFT between the two phases, typically around 30–35 mN/m. In order to compatibilize oil and water, emulsifying agents are added to lower the interfacial tension. More importantly, surfactants create an energy

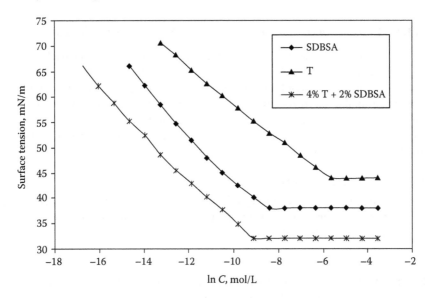

FIGURE 13.4 Surface tension: ln C isotherm for anionic, nonionic, and 1 anionic:2 non-ionic ratio of cutting fluid emulsifiers.

barrier between the oil droplets to prevent the droplets from coalescing [111]. The IFT plays a crucial role in emulsion stability. Low interfacial tension is necessary for emulsion formation, but it is not the only factor to obtain emulsion stability [112]. A low IFT causes rapid migration of surface active agents to the O/W interface [113]. The IFT drops significantly to around 16 mN/m for SDBSA and to 21 mN/m for Tween 85 (T), whereas for the mixed emulsifiers (4% T + 2% SDBSA), the IFT drops more to 11 mN/m. This was clear in Table 13.6.

13.3.3.4 Hydrophile Lipophile Balance

An emulsifier is a substance that stabilizes an emulsion by increasing its kinetic stability. The first task of an emulsifier is to reduce the IFT and the work required for the emulsification. The emulsifying efficiency of an emulsifier is related to the polarity of the molecule or the relation between the contributions of the polar hydrophilic head and the nonpolar lyophilic tail. A rule of thumb in emulsion technology is that the water-soluble emulsifiers tend to give O/W emulsions and oil-soluble emulsifiers give W/O emulsions. The HLB number helps the formulator decide as to what type of emulsifier is needed to emulsify a specific oil. For example, preparation of a W/O inverse emulsion will require emulsifiers with HLB values of 4–6. In contrast, preparing an O/W emulsion requires emulsifiers with HLB values of 8–14. It was determined that stable petroleum formulations required a narrow HLB range of 6–12 [114]. The required HLB for emulsifying paraffinic mineral oil in water is from 10 to 11 [115–117]. HLB for SDBSA is 11 [118,119] and that for Tween 85 is also 11 [120]. As a result, the HLB for the two blended emulsifiers is 11. The value of HLB is in the range of the oil/water emulsion system.

13.3.4 Evaluation of Emulsifier Systems in Metalworking Fluids

A series of trials were carried out to obtain the optimum emulsifier package with optimum stability. Each package contains four components: SDBSA, T, lubricant material (L), and CA. These four components and the base paraffin oil were used in all prepared soluble-oil formulations. The different percentages of these components were incorporated to optimize the stability of the emulsifier systems. The following properties of the emulsions were evaluated: oil stability, emulsion stability, pH, rust inhibition, antiwear, and biodegradability [121].

13.3.4.1 Thermal Stability

The thermal stability of the cutting oil with paraffinic oil (VG = 260–290 cSt) at 70°C is shown in Table 13.4. This formulation with paraffinic oil gave the best thermal stability result that was attributed to its high viscosity.

13.3.4.2 Emulsion Stability

The emulsifier package containing anionic emulsifier (SDBSA), nonionic emulsifier (T), lubricant material (L) and CA, and the base paraffin oil (VG = 260/290) was used in different cutting oil formulations as shown in Table 13.5. Some formulations resulted in oil or cream separation or gel formation after 24 h. It was observed that the stability and performance of the soluble oil were related to the concentrations of the four ingredients SDBSA, T, L, and CA in the formulation. The package with optimum stability was obtained after many trials and had the following composition: 1% SDBSA, 2% T in the emulsifier blend, 4% lubricant, and 2% CA.

13.3.4.3 Emulsion Stability of the Soluble Oil Formulations

The emulsion stability of the cutting oil formulation and the commercial package for paraffinic base oil (VG = 260–290 cSt) after 24 h was studied. From Table 13.5, it was found that the best composition of the cutting oil formulation was with 9% and 12% of the emulsifier blend. At oil/water ratios 5/95, 10/90, and 15/85, the emulsifier package with 9% emulsifier blend concentration exhibited 96%, 96%, and 98% emulsion stability, respectively. Whereas the maximum emulsion stability was 97%, 99%, and 100% for a 12% concentration of the emulsifier mixture. From the obtained data, it was found that the cutting oil formulation with 12% concentration of the emulsifier blend exhibited good performance properties as the commercial emulsifier package. Figure 13.5 compares unstable and stable emulsions obtained using different emulsion packages. The stable cutting fluid emulsion in Figure 13.5 contains 12% concentration of the emulsifier blend. Its stability is high due to the presence of the nonionic emulsifier Tween 85 [108]. Scan electron microscope (SEM) was used to rank the emulsion stability as a function of oil/water ratio. The stability of the emulsion decreases with oil/water ratio as follows: 15/85 > 10/90 > 5/95 (Figure 13.6).

These mixed emulsifier systems were found to induce long-term emulsion stability against coalescence via a synergistic "two-part" mechanism. According to such a mechanism, the role of the emulsifier was first to "delay" the recoalescence

| Broken emulsion for a bad formulation | Stable emulsion for a good formulation (9% and 12% concentration of C8) |

FIGURE 13.5 Comparison of unstable and stable emulsion obtained using different surfactant packages.

phenomenon, and second to induce further droplet breakup during emulsification by rapidly covering the new (naked) interface and reducing the IFT in order to provide long-term stability [122].

13.3.5 RUST CONTROL

Cutting fluids used on machine tools should inhibit rust formation. If not, the machine parts and the workpiece will be damaged. Freshly cut ferrous metals tend to rust rapidly since any protective coating is removed by the machining operation. Neat cutting oil prevents rust formation but does not cool as effectively as water. Water is the best and most economical coolant but causes the parts to rust. Cutting fluids must contain rust inhibitors to inhibit or prevent the rusting process [123]. Rust inhibitors also impart a protective film on cutting chips to prevent their corrosion and the formation of difficult-to-manage chunks or clinkers [102]. The ability to provide corrosion protection is important to MWFs. Emulsions were evaluated for rust formation on iron. Corrosion tests with cast iron chips showed a degree of 10, which means no appearance of rust. Three types of anticorrosion materials were used in this investigation, namely, boric acid diethanolamine (BDEA), dioleic diethanolamine diester (DODEA), and (BDEA + DODEA) blend as shown in Table 13.7. The corrosion test was performed for 3% of cutting oil/water V/V 5:95, 10:90, and 15:85. It was obvious from Figure 13.7 that a positive synergetic effect was attained when using a blend of two types of corrosion inhibitors (BDEA and DODEA) with 8% ratio. From the data obtained in Table 13.7 and Figures 13.6 through 13.8, it was found that the individual anticorrosion materials gave a poor result such as the formation of rust at all three cutting oil/water ratios. A positive synergetic effect was obtained when using a blend of two types of corrosion inhibitors (BDEA and DODEA) with 8% concentration, at 5:95, 10:90, and 15:85 cutting oil/water ratios and 9% and 12% emulsifier blend concentrations.

TABLE 13.7

Anticorrosion Properties of the Best Prepared Formulations of the Cutting Oils with Different Corrosion Inhibitors and Evaluating at (5/95, 10/90, and 15/85) Oil/Water Ratios after 24 h

Cutting Oil Formulation (C3)				Anticorrosion Properties						
Emulsifier Package (P3)		Paraffin Oil	Corrosion Inhibitor, 5%	5/95		10/90		15/85		pH
Emulsifier (E3)				No. of Pits	Staining Extent and Intensity	No. of Pits	Staining Extent and Intensity	No. of Pits	Staining Extent and Intensity	
OA	LA									
Optimum Concentration, %										
4	2	88								
9			BDEA[a]	0	0–1	1	2–1	2	2–3	8
12				1	1–0	1	1–2	3	3–2	
9			DODEA[b]	0	0–0	0	0–2	1	1–2	8.5
12				0	0–0	1	1–0	1	0–1	
9			2% BDEA + 3%	0	0–0	0	0–0	0	0–0	9
12			DODEA	0	0–0	0	0–0	0	0–1	
Commercial sample				0	0–0	0	0–0	0	0–0	

Note: Corrosion inhibition efficiency test was reported as: (1) number of pits and (2) staining extent and intensity. For example, (1) 0/0–0: no pits, no staining, and no intensity; (2) 0/1–1: no pits, low staining, and low intensity; (3) 1/3–2: low numbers of pits, high staining, and medium intensity.

[a] BDEA: boric acid diethanolamine.

[b] DODEA: dioleic diethanolamine diester.

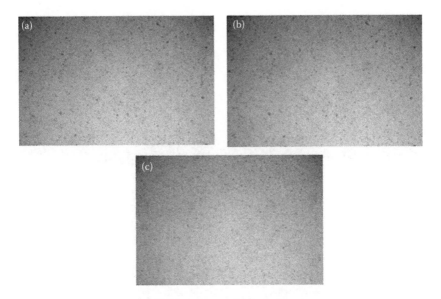

FIGURE 13.6 SEM micrograph for the (a) 5% emulsion (5/95 oil/water ratio), (b) 10% emulsion (10/90 oil/water ratio), and (c) 15% emulsion (15/85 oil/water ratio) for the emulsifier package with 12% concentration of the emulsifier blend.

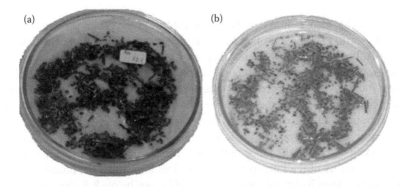

FIGURE 13.7 Difference between the use of (a) an individual corrosion inhibitor (rust appear) and (b) a blend between two types of corrosion inhibitors (no incidents of corrosion were observed).

13.3.6 RANCIDITY CONTROL (pH)

In the early days of the industrial revolution, the only cutting fluid was lard oil. Lard oil starts to spoil and gives off an offensive odor after a few days. This rancidity is caused by bacteria and other microscopic organisms that grow and multiply on lard. Modern cutting fluids are also susceptible to the same problem. As a result, most cutting fluids contain some type of biocide to control the growth of bacteria and fungi and make the fluid more resistant to rancidity. Microbial growth in fluids

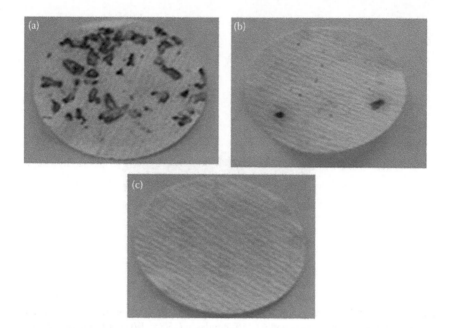

FIGURE 13.8 The filter paper (a) without antirust, (b) with the individual antirust, and (c) with blended antirust after 24 h.

can significantly reduce fluid life. Thus, the control of microbial growth is a must. Bacterial growth can be controlled by routine cleaning of the sump or with the use of biocides. Improper control of microbial growth will also alter the pH of the fluid. As the fluid becomes rancid or septic, the pH drops; in other words, the solution becomes more acidic. Acid produced by bacteria could dissolve metal chips and fines and also possibly cause the material to be a hazardous waste. Bactericides too high in concentration can be harmful to the human skin [123].

Rancidity is monitored by measuring pH. Ideally, the pH for water-based MWFs should be kept in the range of 8.6–9.3. The slightly alkaline range optimizes the cleaning ability of the fluid, prevents corrosion, minimizes the potential for dermatitis, and controls biological growth. If the pH drops below 8.5, the fluid loses its efficiency, attacks ferrous metals (rusting), and displays significant increase of biological activity. A pH of greater than 9.3 may cause dermatitis and corrosion of nonferrous metals. pH values too high or too low can prove hazardous to operators and also pose problems with waste water disposal. A good cutting fluid should be stable during storage and use. Most cutting fluids are now formulated with bactericides and other additives to control microbial growth, enhance fluid performance, and improve fluid stability [102,123].

The pH of the investigated formulations was in the neutral range (pH = 7) as shown in Table 13.8. On using the anticorrosion (rust control) blend (8% concentration of BDEA and DODEA), the pH was increased to a suitable pH (pH = 9) as shown in Table 13.7. Antirust additives can be considered as biocide. This alkalinity range between 8.5 and 9.0 was nearly close to the commercial sample.

TABLE 13.8
Stability of Oil/Water Emulsion and pH of Cutting Oil Formulations

Code	Emulsifier Range, %	Stability of Oil/Water Emulsion, %	pH
C1	NPE9/SDBSA 2-6/0-4	>45	7
C2	NPE9/SOA 2-6/0-4	>30	6.5
C3	T/SDBSA 2-6/0-4	60–70	7.5
C4	T/SOA 2-6/0-4	>40	6.5
C5	T/SOA 4/2	55–65	7.5
C6	T/SOA 4/2	50–60	7.5
C7	T/SDBSA 2-6/0-4	60–75	7.5
C8	T/SDBSA 4/2	75–92	7.5

Generally, biocides should be used in as low a concentration as possible. Owing to the availability of a variety of bacteria in MWFs, using a single biocide may control only some bacterial species while allowing others to proliferate. Thus, water-soluble formulations require biocides to keep bacterial populations from metabolizing corrosion inhibitors and surfactants, reducing product lubricity, and creating rancid odors. A wide variety of biocides are available because different microorganisms respond to different biocides and because biocide performance is sensitive to different MWF formulations. Five biocides lose activity over time and must be added regularly (and sometimes on an emergency basis) to maintain microbial populations at sufficiently low levels [124].

13.3.7 WEAR PROPERTIES OF FLUID LUBRICANTS

From the data on antiwear property in Table 13.9 and Figure 13.9, it was found that, generally, the emulsifier package 15/85 gave antiwear property better than emulsifier package 10/90. The emulsifier package "15/85 and 10/90" exhibited wear rates 0.2 and 0.6 g/s, respectively. In spite of the currently used cutting oil, commercial sample gave 1.1 g/s, the studied formulations of cutting oil gave results better than the commercially used cutting oil package.

13.3.8 BIODEGRADABILITY

Biodegradation refers to the set of processes by which organisms break down large molecules into smaller molecules. The ultimate goal of biodegradation is the conversion of

TABLE 13.9

Antiwear Property of the Best Prepared Formulations of the Cutting Oils and a Commercial Sample and Their Evaluation at 5/95 and 10/90 Oil/Water Ratio

Rotating Speed, rpm[a]	Wear Rate (g/s) × 10^{-6}		
	5/95	10/90	Commercial Sample
50	0.2	0.6	1.1
150	0.03	0.8	2.2
250	0.6	1.1	10.5
350	1.0	1.6	63.8
500	1.1	2.3	139.4

[a] rpm: Revolutions per minute.

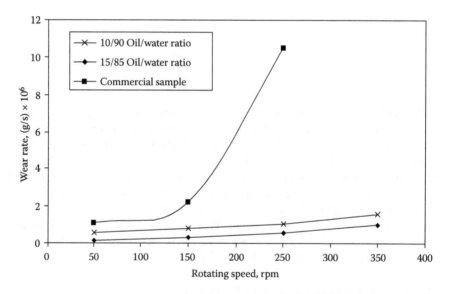

FIGURE 13.9 Rotating speed versus wear rate for some cutting emulsions at 10/90 and 15/85 oil/water ratio and the commercial one (test conditions: load 40–100 kg, temperature 70–100°C, speed 50–350 rpm, and time duration 10 s to 10 h).

organic molecules into carbon dioxide and energy. Biodegradable materials are susceptible to biochemical breakdown by the action of microorganisms. Hence, biodegradability is one of the most important properties of materials with regard to the environment [123]. In the biodegradability test, the new cutting fluid exhibited good degradation rates, passing the level for ready biodegradability. Hence, it was easily biodegradable under anaerobic conditions. The results listed in Table 13.10 and Figure 13.10 indicated the ability of the studied microbial strains to grow on the studied emulsion as a sole

TABLE 13.10

Growth of Microbial Strains with Different Substrate Media at Different Concentrations (5, 10, and 15 mg/L) of the Emulsion Sample from 12% Formulation (C8)

Gram +ve *Micrococcus luteus* RM1

Substrate Concentration (%)	I/I_o	
	3 day	10 day
5	6×10^3	1.1×10^2
10	1.1×10^4	1.2×10^5
15	5.7×10^3	1.7×10^4

Gram +ve *Bacillus sphaericus* HN1

Substrate Concentration (%)	I/I_o	
	3 day	10 day
5	1.8×10^3	1.2×10^5
10	1.6×10^4	1.3×10^5
15	2.3×10^4	4×10^6

Gram −ve *Pseudomonas aeruginosa* HR1

Substrate Concentration (%)	I/I_o	
	3 day	10 day
5	3.9×10^3	1×10^5
10	2.7×10^3	2.7×10^6
15	8.6×10^4	4.3×10^6

Candida parapsilosis NSh45

Substrate Concentration (%)	I/I_o	
	3 day	10 day
5	5.9×10^2	1×10^5
10	7.4×10^2	1.2×10^6
15	2×10^4	1.4×10^5

Aspergillus flavus Shf4

Substrate Concentration (%)	D/D_o	
	3 day	10 day
5	3.12	7
10	5	8.75
15	3.43	7.81

Substrate used:

Gram +ve RM1 and HN1: 1 g of tryptone + 0.5 g yeast extract + 1 g NaCl/100 mL bidistilled water.

Gram −ve HR1: 0.5 g of tryptone + 0.3 g yeast extract + 0.1 g glucose/100 mL bidistilled water.

Candida parapsilosis NSh45: 0.3 g yeast extract + 0.3 g malt extract + 1 g glucose + 0.5 g peptone/100 mL bidistilled water.

Aspergillus flavus Shf4: 30 g potato extract + 2 g glucose + 0.5 g yeast extract + 2 g agar/100 mL bidistilled water.

Blank at 5, 10, and 15% concentration of emulsion sample from 12% formulation

Gram+ve *Micrococcus luteus* RM1 at 5, 10, and
15% concentration of emulsion sample from 12% formulation

Gram−ve *Pseudomonas aeruginosa* HR1 at 5, 10, and
15% concentration of emulsion sample from 12% formulation

Candida parapsilosis NSH45 at 5, 10, and
15% concentration of emulsion sample from 12% formulation

Gram−ve *Pseudomonas aeruginosa* HR1 at 5, 10, and
15% concentration of emulsion sample from 12% formulation

Aspergillus flavus Shf4 at 5, 10, and
15% concentration of emulsion sample from 12% formulation

FIGURE 13.10 Growth of microbial strains on different concentrations of the emulsion sample from 12% formula.

source of carbon and energy, indicating the emulsion's biodegradability. From the ecological point of view, this cutting fluid was not aggressive to the environment and it can be easily treated and disposed off. Finally, one can conclude that the surface active properties, HLB, IFT, and synergistic effect of cutting fluid ingredients play a crucial and important role in the performance of cutting fluid emulsion.

13.4 CONCLUSION

Cutting fluids are used in metal forming and working for a variety of reasons such as improving tool life, reducing workpiece thermal deformation, improving surface finish, and flushing away chips from the cutting zone. Four cutting oil formulations based on four emulsifier packages were investigated. The performance of cutting oil was investigated in terms of oil stability, oil thermal stability, emulsion stability, corrosion inhibition, antiwear, and biodegradability.

From the investigation, it can be concluded that

1. The best cutting oil package consists of 1 SDBSA; 2 T with 9% and 12% concentration, 2% L, 4% CA, 8% concentration of BDEA and DODEA, and VG: 260–290 cSt paraffin oil.
2. Emulsion stability was best at 15/85 and 10/90 oil/water ratios.
3. The 15/85 oil/water ratio gave the best antiwear property.
4. The new cutting fluid was readily biodegradable.

The new cutting oil formulations performed better than the commercial sample.

ACKNOWLEDGMENT

The authors gratefully acknowledge the exploration laboratory of the Egyptian Petroleum Research Institute (EPRI), Egypt for technical support.

REFERENCES

1. V.P. Astakhov, *Tribology of Metal Cutting*, Elsevier, Amsterdam, 2006.
2. M.C. Shaw, The assessment of machinability. *Proceedings of the Conference on Machinability*, The Royal Commonwealth Society, London on 4–6 October 1965.
3. J. Fuchs, J. Burg, J.G. Hengstler, U. Bolm-Audorff, and F. Oes, DNA damage in mononuclear blood cells of metal workers exposed to N-nitrodiethanolamine in synthetic cutting fluids, *Mutat. Res.*, 342, 95–102, 1995.
4. S.M.A. Suliman, M.I. Abubakr, and E.F. Mighani, Microbial contamination of cutting fluids and associated hazards, *Tribol. Int.*, 30, 753–757, 1997.
5. F. Klocke and G. Eisenblatter, Dry cutting CIRP, *Ann. Manuf. Technol.*, 46, 519–527, 1997.
6. J.M. Benito, A. Cambiella, A. LoboF, G. Gutiérrez, J. Coca, and C. Pazosormulation, Characterization and treatment of metalworking oil-in-water emulsions, *Clean Technol. Environ. Policy*, 12, 31–41, 2010.
7. H. Bataller, S. Lamaallam, J. Lachaise, A. Graciaa, and C. Dicharry, Cutting fluid emulsions produced by dilution of a cutting fluid concentrate containing a cationic/nonionic surfactant mixture, *J. Mater. Process Technol.*, 152, 215–220, 2004.
8. E. Fernandez, J.M. Benito, C. Pazos, J. Coca, I. Ruiz, and G. Rıos, Regeneration of an oil-in-water emulsion after use in an industrial copper rolling process, *Colloids Surf. A*, 263, 363–369, 2005.
9. J. Childers, The chemistry of metalworking fluids. In *Metalworking Fluids*. J. P. Byers (Ed). pp. 165–190, Marcel Dekker, New York, 1994.
10. M. Greeley and N. Rajagopalan, Impact of environmental contaminants on machining properties of metalworking fluids, *Tribol. Int.*, 37, 327–332, 2004.
11. R.B. Aronson, Why dry machining?, *Manuf. Eng.*, 117, 33–36, 1995.
12. H. Popke, Th. Emmer, and J. Steffenhagen, Environmentally clean metal cutting processes machining, on the way to dry cutting. *Proc. Inst. Mech. Eng., Part B: J. Eng. Manuf.*, 213, 329–332, 1999.
13. P. Sreejith and B. Ngoi, Dry machining: Machining of the future, *Mater. Process. Technol.*, 101, 287–291, 2000.
14. T. Wakabayashi, H. Sato, and I. Inasaki, Turning using extremely small amounts of cutting fluids, *JSME Int. J. Series C—Mech. Syst., Mach. Elem. Manuf.*, 41(1), 143–148, 1998.

15. J.A. Bailey, Friction in metal machining—Mechanical aspect, *Wear*, 31, 243–253, 1975.
16. J.A. Williams. The action of lubricants in metal cutting, *J. Mech. Eng. Sci.*, 19, 202–212, 1977.
17. E.D. Doyle, J.G. Horne, and D. Tabor, Frictional interactions between chip and rake face in continuous chip formation, *Proc. Royal Soc. London*, A366, 173–183, 1977.
18. L. De Chiffre, Mechanics of metal cutting and cutting fluid action, *JSME Int. J. Mach. Tool Des. Res.*, 17, 225–234, 1977.
19. L. De Chiffre, Mechanical testing and selection of cutting fluids, *Lubr. Eng.*, 36, 33–39, 1980.
20. L. De Chiffre, Frequency analysis of surfaces machined using different lubricants, *ASLE Trans.*, 27, 220–226, 1984.
21. L. De Chiffre, What can we do about chip formation mechanics?, *CIRP Ann. Manuf. Technol.*, 34, 129–132, 1985.
22. L. De Chiffre, Function of cutting fluids in machining, *Lubr. Eng.*, 44, 514–518, 1988.
23. C.A. Sluhan, Selecting the right cutting and grinding fluids, *Tool. Prod.*, 40–50, May 1994.
24. K. Bienkowski, Coolants & lubricants—The truth, *Manuf. Eng.*, 89, 90–96, 1993.
25. R.B. Aronson, Machine tool 101: Part 6, Machine servers, *Manuf. Eng.*, 29, 47–52, 1994.
26. R.J. Tuholski, Don't forget the cutting fluid, *J. Ind. Technol.*, 9(4), 1–5, 1993.
27. Iowa Waste Reduction Center, *Cutting Fluid Management in Small Machine Shop Operations—First Edition*, University of Northern Iowa, Cedar Falls, Iowa, p.43, 1990.
28. J.R. Koelsch, Honing fluid performance, *Manuf. Eng.*, 1, 51–55, 1994.
29. E. Oberg, F.D. Jones, H.L. Horton, and H.H. Ryffel, *Machinery's Handbook*, Industrial Press Inc., New York, p. 2543, 1992.
30. R.K. Springbom (Ed.), *Cutting and Grinding Fluids: Selection and Application*, A.S.T.M.E., Dearborn, Michigan, pp. 102–104, 1967.
31. J. Rozzi, *Final Report: Low Cost Machining without Cutting Fluids*, National Center for Environmental Research, US EPA, University of Northern Iowa, USA, 2002.
32. J.D. Silliman (Ed.), *In Cutting and Grinding Fluids: Selection and Application*, SME, Dearborn, Michigan, Second Edition, pp. 119–135, 1992.
33. A. Menniti, K. Rajagopalan, T.A. Kramer, and M.M. Clark, An evaluation of the colloidal stability of metal working fluid, *J. Colloid Interface Sci.*, 284, 477–488, 2005.
34. A.J.S. Liem and D.R. Woods, Review of coalescence phenomena, *AIChE Symp. Ser.*, 70(144), 8–24, 1974.
35. B.O. Haglund and P. Enghag, Characterization of lubricants used in the metalworking industry by thermoanalytical methods, *Thermochim. Acta*, 283, 493–499, 1996.
36. H. Ashjian, T.J. Giacobbe, F.C. Loveless, C.R. Mackerer, N.J. Novick, and T.P. O'Brien, Bioresistant surfactants and cutting oil formulations, United States Patent 5,985,804, 1999.
37. M. Kobessho and K. Matsumoto, Metal working oil composition, United States Patent 5,908,816, 1999.
38. S.K. Misra and R.O. Sköld, Lubrication studies of aqueous mixtures of inversely soluble components, *Colloids Surf. A*, 170, 91–106, 2000.
39. M.A. Elbaradie, Cutting fluids: Part I. Characterisation. *J. Mater. Proc. Technol.*, 56, 786–797, 1996.
40. E.M. Trent, *Metal Cutting Coolants and Lubricants*, 2nd Edn, Chapter 10, pp. 221–241, Butterworths, London, 1984.
41. W.H. Oldacre, Cutting fluid and process of making the same, U.S. Patent No. 1,604,068, 1926.
42. W.M. Stocker and T. Hicks, *Metal Cutting: Today's Techniques for Engineers and Shop Personnel*, McGraw-Hill, New York, p. 186, 1980.
43. P. Walstra, Formation of emulsions, In: P. Becher (Ed.), *Encyclopedia of Emulsion Technology: Basic Theory*, Vol. 1, Marcel Dekker Inc., New York, pp. 57–128, 1983.

44. F.P. Bowden and F. Tabor, *Fraction and Lubrication of Solids*, Clarendon Press, Oxford, England, 1950.
45. D. Dowson, *History of Tribology*, Longmans Green, New York, p. 154, 1979.
46. M.C. Shaw, Metal-cutting process as a means of studying the properties of extreme-pressure lubricants, *Annals N.Y. Acad. Sci.*, 58, 962–978, 1951.
47. K.B. Gilbert and C. Singer, *History of Technology*, Vol. IV, Oxford University Press, London, p. 417, 1958.
48. K.K. Kaocarp Nippon, Water soluble metal working lubricants, *Jpn. Kokai, Tokyo, Koho, Jpn.*, 59, 227–991, 1983.
49. Y. Barakat, S.A. El-Kholy, A.M. Abdel-Fattah, and A.M. Zaatar, Some emulsifiable oil formulations for metal cutting fluids, *Egypt. J. Petrol.*, 5, 49–61, 1996.
50. L.M. Prince, *Microemulsions: Theory and Practice*, Academic Press, New York, p. 10, 1977.
51. G. Men, Y. Peng, L. Daosheng, L. Kejian, Y. Yi, W. Feng, H. Kai Fn, and J. Heng, Preparation of machine oil cutting fluid for band saw and evaluation of its performance, *Petrol. Sci. Technol.*, 10, 1251–1261, 2001.
52. A. Eckard, I. Riff, and J. Weaver, Formulation of soluble oils with synthetic and petroleum sulfonates, *Lubr. Eng.*, 53(6), 17–22, 1997.
53. M.J. Neal, *Tribology Handbook, Lubrication in Metal Working and Cutting*, Newness Butterworths, London, 1973.
54. G.W. Rowe, *Introduction of Principles of Metalworking*, St. Martin's Press, New York, pp. 265–269, 1965,
55. S.K. Misra and R.O. Skold, Phase and aggregational studies of some inversely soluble aqueous formulations, *Colloids Surf.*, A, 179, 111–124, 2001.
56. W. Belluco and L. De Chiffre, Surface integrity and part accuracy in reaming and tapping stainless steel with new vegetable based cutting oils, *Tribol. Int.*, 35, 865–870, 2002.
57. I.S. Morton, Development testing of metalworking lubricants, *Ind. Lubr. Tribol.*, 24, 163–169, 1972.
58. S.C. Cermak and T.A. Isbell, Synthesis and physical properties of caphea-oleic estorides and esters, *J. Am. Oil Chem. Soc.*, 81, 297–303, 2004.
59. S. Watanabe, Characteristic properties of water soluble cutting fluids additives derived from various esters, *Recent Res. Develop. Oil Chem.*, 5, 39–49, 2001.
60. T.L. Johnson and T.C. Nash, Primary amine alternatives to triethanolamine in metalworking fluid formulations, *Lubr. Eng.*, 42, 402–410, 1982.
61. S. Watanabe, New water soluble cutting fluids additives from phosphonic acids for aluminum materials, *Rivista Italia. Sostanze Grasse*, 82(2), 97–99, 2005.
62. E.L.H. Bastian, *Standard Handbook of Lubrication Engineering, American Society of Lubrication Engineers*, McGraw-Hill, New York, 1968.
63. M.D. Kenneth and B.K. Sharma, Emulsification of chemically modified vegetable oils for lubricant use, *J. Surf. Deterg.*, 14, 131–138, 2011.
64. F. Zhao, A. Clarens, A. Murphree, K. Hayes, and S.J. Skerlos, Structural aspects of surfactant selection for the design of vegetable oil semi-synthetic metalworking fluids, *Environ. Sci. Technol.*, 40, 7930–7937, 2006.
65. S.J. Skerlos, K.F. Hayes, and A.F. Clarens, U.S. Patent 2006/0247139, 2006.
66. L. Kong, J.K. Beattie, and R.J. Hunter, Electroacoustic study of concentrated oil-in-water emulsions, *J. Colloid Interface Sci.*, 238, 70–79, 2001.
67. A. Cambiella, J.M. Benito, C. Pazos, and J. Coca, Interfacial properties of oil-in-water emulsions designed to be used as metalworking fluids, *Colloids Surf. A*, 305, 112–119, 2007.
68. R. Nagarajan and E. Ruckenstein, Molecular theory of microemulsions, *Langmuir*, 16, 6400–6415, 2000.

69. J. Deluhery and N. Rajagopalan, A turbidimetric method for the rapid evaluation of MWF emulsion stability, *Colloids Surf. A*, 256, 145–149, 2005.

70. I. Roland, G. Piel, L. Delattre, and B. Evrard, Systematic characterization of oil-in-water emulsions for formulation design. *Int. J. Pharm.*, 263, 85–94, 2003.

71. V.A. Serov, A.I. Maksimova, and S.B. Dorfman, Emulsols with antiwear and extreme-pressure additives, *Chem. Technol. Fuels Oils*, 18, 34–35, 1982.

72. A.M. Al Sabagh, N.A. Maysour, N.M. Nasser, and M.R. Sorour, Some cutting oil formulations based on local prepared emulsifiers, Part I: Preparation of some emulsifiers based on local raw materials to stabilize cutting oil emulsions, *J. Dispersion Sci. Technol.*, 27, 239–250, 2006.

73. S. Fukushima, M. Nishida, and M. Nakano, Preparation of and drug release from W/O/W type double emulsions containing anticancer agents using an oily lymphographic agent as an oil phase, *Chem. Pharm. Bull.*, 35, 3375–3381, 1987.

74. L.A.M. Ferreira, M. Seiller, J.L. Grossiord, J.P. Marty, and J. Wepierre, Vehicle influence on *in vitro* release of metronidazole: Role of w/o/w multiple emulsion, *Int. J. Pharm.*, 109, 251–259, 1994.

75. A.J. Khopade and N.K. Jain, Multiple emulsions containing rifampicin, *Pharmazie*, 54, 915–919, 1999.

76. K. Lindenstruth and B.W. Muller, W/O/W multiple emulsions with diclofenac sodium, *Eur. J. Pharm. Biopharm.*, 58, 621–627, 2004.

77. T. Schmidts, D. Dobler, P. Schlupp, C. Nissing, H. Garn, and F. Runkel, Development of multiple W/O/W emulsions as dermal carrier system for oligonucleotides: Effect of additives on emulsion stability, *Int. J. Pharm.*, 398, 107–113, 2010.

78. T. Schmidts, D. Dobler, C. Nissing, and F. Runkel, Influence of hydrophilic surfactants on the properties of multiple W/O/W emulsions, *J. Colloid Interface Sci.*, 338, 184–192, 2009.

79. W.C. Griffin, Calculation of HLB values of non-ionic surfactants, *J. Soc. Cosmet. Chemists*, 5, 249–256, 1953.

80. P. Prinderre, P. Piccerelle, E. Cauture, G. Kalantzis, J.P. Reynier, and J. Joachim, Formulation and evaluation of o/w emulsions using experimental design, *Int. J. Pharm.*, 163, 73–79, 1998.

81. L.O. Orafidiya and F.A. Oladimeji, Determination of the required HLB values of some essential oils, *Int. J. Pharm.*, 237, 241–249, 2002.

82. R.C. Pasquali, M.P. Taurozzi, and C. Bregni, Some considerations about the hydrophilic–lipophilic balance system, *Int. J. Pharm.*, 356, 44–51, 2008.

83. W.C. Griffin, Classification of surface active agents by HLB, *J. Soc. Cosmet. Chemists*, 1, 311–326, 1949.

84. N. Jager-Lezer, I. Terrisse, F. Bruneau, S. Tokgoz, L. Ferreira, D. Clausse, M. Seiller, and J.L. Grossiord, Influence of lipophilic surfactant on the release kinetics of water-soluble molecules entrapped in a W/O/W multiple emulsion, *J. Control. Release*, 45, 1–13, 1997.

85. F. Tirnaksiz and O. Kalsin, A topical w/o/w multiple emulsion prepared with Tetronic 908 as a hydrophile surfactant: Formulation, characterization and release study, *J. Pharma. Pharm. Sci.*;8, 299–315, 2005.

86. J. Jiao and D.J. Burgess, Rheology and stability of water-in-oil-in-water multiple emulsions containing Span 83 and Tween 80, *AAPS Pharm. Sci.*, 5, 1–2, 2003.

87. S. Geiger, S. Tokgoz, A. Fructus, N. Jager-Lezer, M. Seiller, C. Lacombe, and J.L. Grossiord, Kinetics of swelling-breakdown of a W/O/W multiple emulsion: Possible mechanisms for the lipophilic surfactant effect, *J. Control. Release*, 52, 99–107, 1998.

88. S. Magdassi, M. Frenkel, N. Garti, and R. Kasan, Multiple emulsions II: HLB shift caused by emulsifier migration to external interface, *J. Colloid Interface Sci.*, 97, 374–379, 1984.

89. N.S. Canter, HLB: A new system for water-based metalworking fluids, *Tribol. Lubr. Technol.*, 64, 10–12, 2005.

90. A.M. Al-Sabagh, M.R. Noor El-Din, N.E. Maysour, and N.M. Nasser, Investigation of factors affecting cutting oil formulations prepared by emulsifiers based on linear alkyl benzene and oleic acid-maleic anhydride esters, *J. Dispersion Sci. Technol.*, 29, 740–747, 2008.

91. A.M. Al-Sabagh, N.E. Maysour, N.M. Nasser, and M.R. Sorour, Some cutting oil formulations based on local prepared emulsifiers. Part I: Preparation of some emulsifiers based on local raw materials to stabilize cutting oil emulsions, *J. Dispersion Sci. Technol.*, 27, 239–250, 2006.

92. D.N. Rao, R.R. Srikant, and Ch. S. Rao, Influence of emulsifier content on properties and durability of cutting fluids, *J. Braz. Soc. Mech. Sci. Eng.* 29, 396–400, 2007.

93. M.M. Clark and F. Fiessinger, Mixing and scale-up, In *Mixing in Coagulation and Flocculation*, pp. 282–308, American Water Works Association Research Foundation, Denver, CO, February, 1991.

94. J. Zimmerman, A. Clarens, K. Hayes, and S. Skerlos, Design of hard water stable emulsifier systems for petroleum and bio-based semi-synthetic metalworking fluids, *Environ. Sci. Technol.,* 37, 5278–5288, 2003.

95. A.I. Vogel, *A Textbook of Practical Organic Chemistry*, Third Ed., Wiley, New York, 1956.

96. A. Cambiella, J.M. Benito, C. Pazos, J. Coca, M. Ratoi, and H.A. Spikes, The effect of emulsifier concentration on the lubricating properties of oil-in-water emulsions, *Tribol. Lett.*, **22**, 53–65, 2006.

97. M. Sokovic and K. Mijanovic, Ecological aspects of the cutting fluids and their influence on quantifiable parameters of cutting processes, *J. Mater. Process. Technol.*, 109, 181–189, 2001.

98. S. Watanabe, T. Fujita, M. Sakamoto, T. Kuramochi, and Kawahara, Characteristic properties of cutting fluid additives derived from the adducts of diamines and acid chlorides, *J. Amer. Oil. Chem. Soc.*, 70, 9, 1993.

99. H. Tomoda, Y. Sugimoto, Y. Tania, and S. Watanabe, Characteristic properties of cutting fluid additives derived from the reaction products of hydroxyl fatty acids with some acid anhydrides, *J. Surf.. Deterg.*, 1, 533–537, 1998.

100. V.P. Astakhov, *Tribology and Interface Engineering Series, Tribology of Metal Cutting*, Elsevier Ltd, Oxford, UK, 2006.

101. K. Sutherland, Managing cutting fluids used in metal working, *Filtration Separation*, 45, 20–23, 2008.

102. Iowa Waste Reduction Center, *Cutting Fluid Management for Small Machining Operations: A Practical Pollution Prevention Guide*, 3rd ed., University of Northern Iowa, Iowa, USA, 2003.

103. H.J. Benson, *Microbiological Applications*, 6th ed., Wm. C. Brown Publishers, Dubuque, IA, 1994.

104. A. Willard and A. Ruth, *Mycologia*, 80, 855–858, 1988.

105. M.J. Rosen, *Surfactants and Interfacial Phenomena*, John Wiley & Sons, New York, Chichester–Brisbane–Toronto, 1978.

106. J. Liu, R. Han, and Y. Sun, Research on experiments and action mechanism with water vapor as coolant and lubricant in green cutting, *Int. J. Mach. Tools Manuf.*, 45, 687–694, 2005.

107. A. Cambiella, J.M. Benito, C. Pazos, and J. Coca, Interfacial properties of oil-in-water emulsions designed to be used as metalworking fluids, *Colloids Surf. A*, 305, 112–119, 2007.

108. E. Dickinson, C. Ritzoulis, and M.J.W. Povey, Stability of emulsions containing both sodium caseinate and Tween 20, *J. Colloid Interface Sci.*, 212, 466–473, 1999.

109. A.M.A. Omar, Micellization and adsorption of anionic/nonionic polymeric surfactants for metalwork fluid at different interfaces, *Ind. Lubr. Tribol.* 56, 171–176, 2004.

110. A.M. Al-Sabagh, M.A. Abdul-Raouf, and E. Abdel-Raheem, Surface activity and light scattering investigation for some novel aromatic polyester amine surfactants, *Colloids Surf. A*, 251,167–174, 2004.

111. N.S. Stamkulov, K.B. Mussabekov, S.B. Aidarova, and P.F. Luckham, Stabilisation of emulsions by using a combination of an oil-soluble ionic surfactant and water-soluble polyelectrolytes. I: Emulsion stabilisation and interfacial tension measurements, *Colloids Surf. A*, 335, 103–106, 2009.

112. A.M. Al-Sabagh, N.N. Zaki, and A.M. Badawi, Effect of binary surfactant mixtures on the stability of asphalt emulsions, *J. Chem. Technol. Biotechnol.*, 69, 350–356, 1997.

113. Y. Xu, J. Wu, T. Dabros, H. Hamza, S. Wang, M. Bidal, J. Venter, and T. Tran, Breaking water-in-bitumen emulsions using polyoxyalkylated DETA demulsifier, *Can. J. Chem. Eng.*, 82, 729–835, 2004.

114. M.J. Rosen, Wetting and its modification by surfactants, In: M.J. Rosen (Ed.), *Surfactants and Interfacial Phenomena*, John Wiley & Sons Inc., New York, pp. 174–199, 1978.

115. Th.F. Tadros, *Applied Surfactants: Principles and Applications*, Wiley-VCH, Weinheim, 2005.

116. R.C. Pasquali, M.P. Taurozzi, and C. Bregni, Some considerations about the hydrophilic–lipophilic balance system, *Int. J. Pharm.*, 356, 44–51, 2008.

117. P.M. Kruglyakov, *Hydrophile-Lipophile Balance of Surfactants and Solid Particles: Physicochemical Aspects and Applications*, Elsevier Science, Amsterdam, The Netherlands, 2000.

118. G. Pan, C. Jia, D. Zhao, C. You, H. Chen, and G. Jiang, Effect of cationic and anionic surfactants on the sorption and desorption of perfluorooctane sulfonate (PFOS) on natural sediments, *Environ. Pollut.*, 157,325–330, 2009.

119. X. Shen, Y. Sun, Z. Ma, P. Zhang, C. Zhang, and L. Zhu, Simultaneous elimination of dissolved and dispersed pollutants from cutting oil wastes using two aqueous phase extraction methods, *J. Hazard. Mater.*, 140, 187–193, 2007.

120. P. Bruheim, H. Bredholt, and K. Eimhjellen, Effects of surfactant mixtures, including Corexit 9527, on bacterial oxidation of acetate and alkanes in crude oil, *Appl. Environ. Technol.*, 65, 1658–1661, 1999.

121. M.A. El Baradie, Cutting fluids, Part I: Characterization, *J. Mater. Process. Technol.*, 56, 786–797, 1996.

122. R. Pichot, F. Spyropoulos, and I.T. Norton, Mixed-emulsifier-stabilised emulsions: Investigation of the effect of monoolein and hydrophilic silica particle mixtures on the stability against coalescence, *J. Colloid Interface Sci.*, 329, 284–291, 2009.

123. S. Oberwalleney and P. Sheng, *Framework for an Environmental-Based Cutting Fluid Planning in Machining Facilities*, University of California at Berkeley, Department of Mechanical Engineering, John Wiley & Sons, Hoboken, NJ, 1996.

124. J.P. Byers, *Metalworking Fluids*, 2nd Ed., Taylor & Francis, New York, 2006.

14 The Chemistry of Alkylcarbonates and Their Application as Friction Modifiers

Leslie R. Rudnick and Carlo Zecchini

CONTENTS

Dialkyl carbonates have been used in lubricant applications since the early 1940s. Dialkyl carbonates can be obtained by several synthesis routes. A wide variety of different dialkyl carbonates can be prepared to provide materials with required viscosity or degree of water solubility or biodegradability. Dialkyl carbonates posses properties and performance features desirable in synthetic lubricants. Since dialkyl carbonates are polar molecules, they can serve as friction modifiers in lubricants similar to esters.

14.1   INTRODUCTION

Dialkyl carbonates represent a class of functionalized synthetic fluids generally obtained by transesterification of dimethyl carbonate (DMC). The carbonate moiety represents the central core of either a symmetrical or asymmetrical compound depending on whether one or two alcohols are used. The physical properties and chemical reactivity of the dialkyl carbonates strongly depend on the size and shape of the alcohols. Alcohols can be chosen from a wide variety of natural and synthetic alcohols.

Dialkyl carbonates have been used in various applications, both as lubricant base fluids and as performance fluid components, due to their physical and chemical properties, and performance characteristics. They have low toxicity, are nonpolluting, and have favorable economics.

Organic carbonates can be used to manufacture many products, primarily polycarbonates and polyurethanes, but also pesticides and herbicides, pharmaceutical products, polyimide films, and electrolytic fluids for lithium batteries. These may also be used as a fuel additive, lubricant, and as a solvent.

Recently, the parent compound, DMC, has been reported to provide benefit as a gasoline additive, and diethyl carbonate (DEC) can be used in both gasoline and diesel fuel, and as a component in hydraulic fluids [1]. DMC and DEC have very good solubility in gasoline. Combustion analysis indicates that the ignition delay of an engine fueled with DMC-diesel blended fuel is longer, but combustion duration is much shorter, and the thermal efficiency is increased compared with that of a control. Further, if injection is also delayed, NO_x emissions can be reduced while particulate matter (PM) emissions are still reduced significantly [2]. DMC and DEC are used as oxygenated additives to reduce emissions of particulates and NO_x, O_3, and CO. The oxygen contents of DMC (53%) and DEC (40.6%) are much higher than the oxygen content of ethyl-*tert*-butyl ether/*tert*-amyl methyl ether (ETBE/TAME) (15.7%). DMC also has a blending octane value of 106.5, similar to ETBE, and low vapor pressure. Reid vapor pressure (RVP) and blending RVP of organic carbonates are significantly lower than those of methyl-*tert*-butyl ether (MTBE). The photochemical ozone creation potential (POCP) value for DMC is 2.5 compared to ETBE, which is 24.2. DMC has the lowest POCP of all the oxygenated volatile organic compounds (VOCs) [3].

Dialkyl carbonates are similar to conventional esters in terms of performance, with improved seal compatibility and absence of the formation of acidic components on thermo-oxidative and hydrolytic degradation, and they generally have lower toxicity than esters [1].

Disubstituted carbonates containing aromatic and aliphatic structural components have been proposed as lubricants for use with the newer refrigerants of the hydrofluorocarbon (HFC) type [4]. The aromatic portion of the structure provides good lubricity whereas the aliphatic portion provides stability and detergency. Lubricant oil compositions containing glycol ether carbonate are used in chain oils, automotive gear oils, lubricant oils for fibers, and metal rolling oils.

The quantity of dialkyl carbonates as base fluid in finished products ranges from 5% up to 30% in automotive products and from a few percent to 100% in industrial applications.

14.2   HISTORICAL PERSPECTIVE

Dialkyl carbonates have been used in lubricant applications since the early 1940s. In a 1941 patent, Fincke and Bartlett reported the use of a small amount of carbonates (1–5%), mostly from low-MW alcohols, to enhance the spreading and penetrability of mineral oil-based lubricants [5]. In 1945, Knutson and Graves reported on the use of a minor proportion of dialkyl carbonates in mineral oil-based formulations to improve the load carrying capacity of the lubricant. Alkyl groups employed ranged from ethyl to lauryl alcohol and also included the use of aromatic alcohols [6].

Dialkyl carbonates of C_{1-10} alcohols having low pour points and a high viscosity index were described by Mikeska and Eby in 1953 [7]. Bartlett described the use of polymers derived from monomers of carbonic ester as pour point depressants [8,9]. In 1956, Cottle, Knoth, and Young reported on the use of carbonates of C_{10-20} alcohols, which were obtained from oxosynthesis of branched olefins [10].

In 1982, Koch and Romano claimed lubricant formulations containing dialkyl carbonates that were obtained by transesterification of DMC with high-molecular-weight oxoalcohols. The resulting products exhibit improved characteristics and performance [11].

A 1989 patent claims lubricant formulations containing organic carbonates as improved lubricants for cold steel rolling [12]. Also, in a 1990 patent, Fisicaro and Gerbaz claimed lubricant formulations containing dialkyl carbonates obtained by transesterification of DMC with high-molecular-weight branched oxoalcohols. These alcohols were obtained by oxosynthesis of linear olefins, followed by cryogenic separation of the linear fraction [13].

The first commercial application of dialkyl carbonates as lubricant components was in 1987, when AgipPetroli introduced dialkyl carbonates as a new synthetic base fluid in the formulation of semisynthetic, gasoline engine oils [14]. AgipPetroli extended the application of dialkyl carbonates to the formulation of other types of automotive lubricants, diesel, and super high-performance diesel (SHPD) engine oils, gearboxes, and two-stroke/four-stroke oils. Dialkyl carbonates were also introduced in the field of industrial lubricants in the formulation of press molding, cold rolling, and metalworking fluids.

Alkylcarbonates (C_{14}–C_{15}) are also used in the metal industry as cleaning agents. The product detaches dust particles from metals after welding and has replaced silicone oil for this application. After pressing and molding a metallic material with the press mold, the press-molding oil is washed and removed with warm water. One very interesting characteristic is the possibility for the oil to be removed from the metals using a warm water shower [15].

A base oil containing a C_{14}–C_{15} dialkyl carbonate can be used to provide a lubricating fluid composition for dynamic pressure bearings. This composition is based on a fluid containing fine magnetic particles dispersed in a dialkyl carbonate oil containing an amine as antigelling agent. By adding an antigelling agent, a viscosity index improving agent, and a metal inactivating agent to the fluid, the characteristics, particularly viscosity, high-temperature gelation, and the like, can be markedly improved. To obtain good volatility, the base dialkyl carbonate oil can be blended with polyalpha olefins [16].

14.3 COMMERCIAL PRODUCTION

The first commercial production of carbonates suitable for application in the lubricant field was started in 1985 by Enichem Synthesis (Italy), with a capacity of 3000 tons/year. Starting raw materials were DMC and a mixture of synthetic oxoalcohols in the C_{14}–C_{16} range.

The process involves four main steps:

1. Transesterification of oxoalcohols using DMC
2. Recovery of azeotrope: methanol/DMC 70/30
3. Purification of dialkyl carbonates
4. Distillation of azeotrope methanol/DMC to recover these two products

China currently produces the largest quantity of DMC and other derivative carbonates. In China, there are at least 10–15 producers of DMC, most of them using the technology starting from CO_2 and ethylene/propylene oxide. Only one or two small-scale units are still producing using phosgene. Other major producers are UBE in Japan/Spain and Isochem/Framochem (still using phosgene technology) in Hungary.

Total production capacity in the world market is currently estimated at about 90 kt/year, nearly entirely produced in China. DMC captive capacity (produced for use by the manufacturer worldwide) is estimated to be around 170 kt/year, from Sabic (formerly GE Plastic) Japan/Spain from carbon monoxide/methanol (used to produce diphenylcarbonate for aromatic polycarbonate), from Asahi Taiwan from ethylene oxide, and from UBE Japan from methyl nitrite. Captive consumption in the Chinese market is estimated to be around 70 kt/year. Globally, worldwide production is around 330 kt/year.

DMC demand is growing rapidly worldwide and China alone accounts for 50% of worldwide consumption. Actually, the most important application is for coating with 55% used as solvent. Other applications include 25% for producing agrochemicals, 10% for batteries, 5% pharma, and 5% others (as for instance cosmetics, etc.). A strong and growing market is for lithium batteries, as witnessed by the number of recent patents in the area (more than 200 in the last few years). A hybrid lithium-ion battery system is made up of a graphite anode, a nonaqueous organic electrolyte that acts as an ionic path between the electrodes, and a transition metal oxide cathode. Lithium-ion battery electrolytes typically consist of a lithium salt and various additives dissolved in an organic solvent. The most popular solvents are mixtures of cyclic carbonates, for example, ethylene carbonate (EC), propylene carbonate (PC), and linear carbonates such as DMC and DEC [17].

14.4 CHEMISTRY

Dialkyl carbonates consist of a polar carbonate group with one carbon and three oxygen atoms in the same plane, and two alkyl chains bonded to two oxygen atoms. These alkyl groups are free to rotate about the axis of the carbon–oxygen bond as shown in Figure 14.1.

The chemical structure of the carbonate group and the pendant alkyl groups collectively determine the physical and chemical properties, and the tribological performance of dialkyl carbonates. Both the physical and chemical properties affect the performance of these materials when they are used as lubricant fluids. This includes the polarity of the dialkyl carbonate, which is affected by the ratio of carbon in the pendant alkyl groups to the carbonate functional group, and the reactivity of the pendant alkyl groups, which depends on the branching and degree of unsaturation present in these groups.

14.4.1 SYNTHESIS METHODS

Dialkyl carbonates are diesters of carbonic acid, and are derived from the condensation of carbonic acid with various alcohols as depicted in Figure 14.2.

Since carbonic acid has poor chemical stability under conditions necessary for the condensation of alcohols, this synthesis methodology is not practical. Furthermore, the half-ester formed in the first condensation step is also very unstable and cannot undergo the second condensation reaction necessary to prepare dialkyl carbonate. Fortunately, alternative methods are available for the synthesis of dialkyl carbonates.

14.4.2 CURRENT COMMERCIAL ROUTES

The most effective commercial route for the production of higher dialkyl carbonates is the transesterification of DMC, which is depicted in Figure 14.3. Consequently, the most important step of commercial routes is the synthesis of DMC.

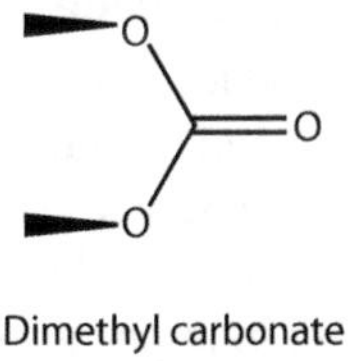

Dimethyl carbonate

FIGURE 14.1 An example of the structure of dialkyl carbonate.

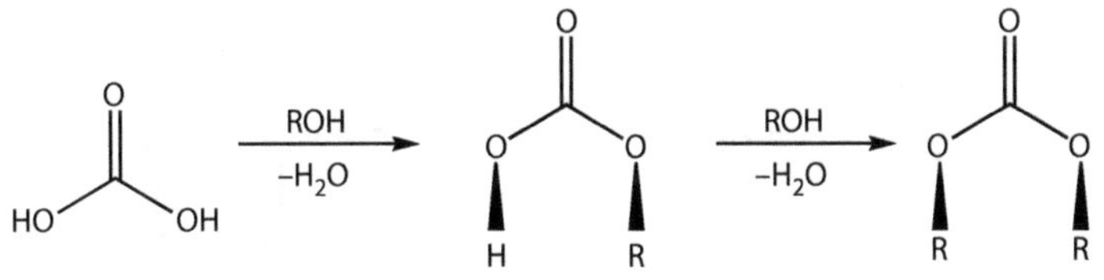

FIGURE 14.2 Example of synthesis of dialkyl carbonate.

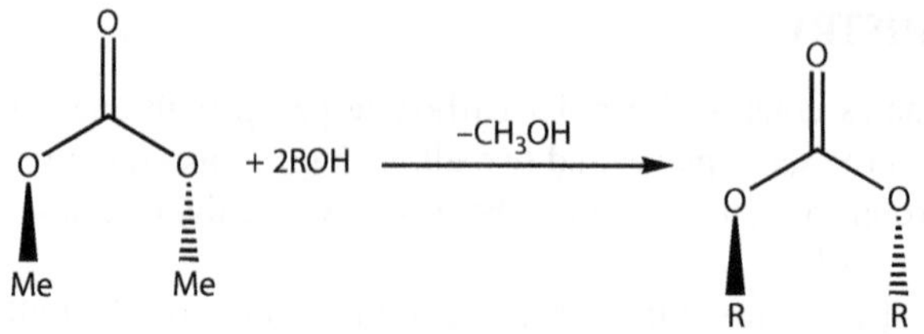

FIGURE 14.3 Synthesis of higher dialkyl carbonates by transesterification.

14.4.3 PRODUCTION OF DIMETHYL CARBONATE

There are three commercial routes for the production of DMC, all involving methanol as a methylating agent.

14.4.3.1 Synthesis from Phosgene

The application of phosgene for the synthesis of alkyl carbonates is depicted in Figure 14.4. This synthesis route has been widely applied in the past to industrial-scale production as the sole way to produce DMC. As in many transesterification syntheses, quantitative yields (>99%) are obtained by the reaction of the ester with excess alcohol. The excess alcohol is subsequently recovered by distillation and reused.

It is likely that environmental considerations and a generally more "green" approach to synthesis processes will replace this approach. The major problem with this process is in the handling and disposal of phosgene, chlorine compounds, the large quantity of NaCl produced, and the need to eliminate impurities of chlorine derivatives in the final product. These are deleterious impurities for solvent application of carbonates on metals, coating, in pharma, and in the production of lithium batteries. We have to take into consideration the fact that these three applications are the most important markets for DMC and that many skin and hair care preparations contain a combination of various dialkyl carbonates.

14.4.3.2 Synthesis from Carbon Monoxide

An important industrial route to produce DMC begins with methanol and carbon monoxide, in the presence of oxygen and a catalytic system based on Cu^+/Cu^{2+} [18].

The reaction takes place in two separate steps (Figure 14.5). The overall reaction of the two-step synthetic procedure is summarized in the reaction scheme of Figure 14.6.

This synthesis route is increasing in importance because it utilizes readily available starting materials. The process is relatively economical and is generally "green" when compared to reactions involving phosgene.

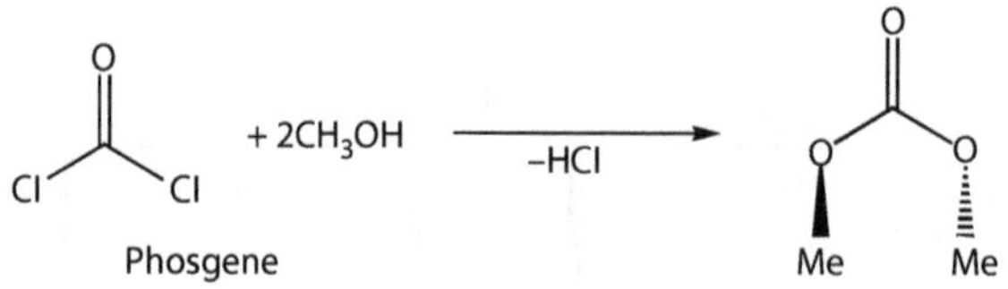

FIGURE 14.4 Synthesis of dialkyl carbonate from phosgene.

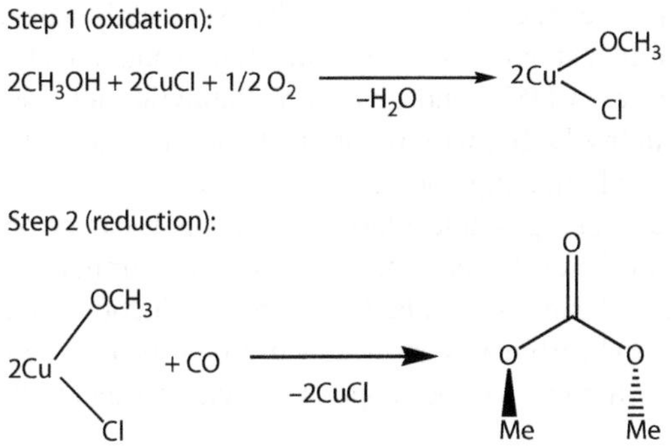

Step 1 (oxidation):

$$2CH_3OH + 2CuCl + 1/2\,O_2 \xrightarrow[-H_2O]{} 2Cu\begin{smallmatrix}OCH_3\\ \\Cl\end{smallmatrix}$$

Step 2 (reduction):

$$2Cu\begin{smallmatrix}OCH_3\\ \\Cl\end{smallmatrix} + CO \xrightarrow[-2CuCl]{} \text{[dimethyl carbonate]}$$

FIGURE 14.5   Two step synthesis of alkyl carbonates using copper catalyst.

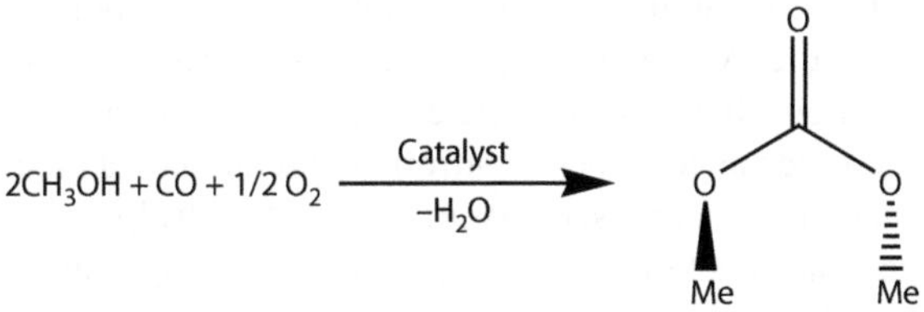

$$2CH_3OH + CO + 1/2\,O_2 \xrightarrow[-H_2O]{Catalyst} \text{[dimethyl carbonate]}$$

FIGURE 14.6   Overall reaction scheme of industrial synthesis of dialkyl carbonate.

14.4.3.3   DMC from Methyl Nitrite

Another way to produce DMC is by using NO as a redox reagent with methanol and CO, in the presence of a metal catalyst such as palladium. This reaction involves the formation of methyl nitrite from methanol, NO, and oxygen. Methyl nitrite is subsequently reacted with CO to give DMC and NO to recycle. The main hazard may be related to safety concerning the explosive limits. There are also some concerns about toxicity and handling of methyl nitrite [19].

14.4.3.4   Synthesis from Carbon Dioxide

By direct reaction of carbon dioxide and ethylene oxide, or propylene oxide, the corresponding cyclic five-member carbonates are easily obtained in high yield as shown in Figure 14.7.

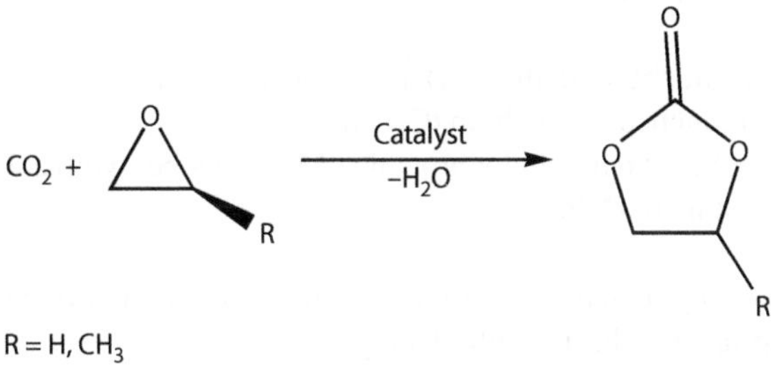

$$CO_2 + \text{[epoxide]} \xrightarrow[-H_2O]{Catalyst} \text{[cyclic carbonate]}$$

R = H, CH$_3$

FIGURE 14.7   Synthesis of dialkyl carbonate from carbon dioxide and oxides of ethylene and propylene.

In a second step (ester exchange process), the cyclic carbonate can be transesterified with methanol in a standard transesterification plant. This technology is based on a separation between DMC and glycols by catalytic and azeotropic separation to produce high-quality DMC, with by-products of ethylene and propylene glycols. This process requires high temperature and pressure.

Dialkyl carbonates of high-molecular-weight alcohols can be obtained directly by transesterification of the cyclic carbonates. This route eliminates the need for DMC as an intermediate and can result in better overall production economics.

Unsaturated dialkyl carbonates are widely used for the production of polycarbonates. However, these are beyond the scope of this chapter and will not be discussed.

14.5   PROPERTIES AND PERFORMANCE CHARACTERISTICS

The basic structure of the dialkyl carbonate molecule, the carbonic group, provides polarity to the molecule. As a lubricant, this functional group is responsible for several important functions. It provides the potential for interaction with the tribological surface, which increases its tendency to adsorb and remain on a metal surface much better than a nonpolar lubricant, such as mineral oils, polyalphaolefins and polyinternalolefins. The polar functional group also provides an opportunity to improve the solubility of polar additives in a formulated lubricant.

Most of the physical properties of carbonates such as viscosity, pour point, flash point, and fire point are functions of the size and shape of the alkyl groups. The isobaric vapor–liquid equilibria for alkyl carbonates and alcohols have been experimentally determined [20]. The isothermal vapor–liquid equilibria of dimethyl and diethyl carbonates with a variety of materials have been studied [21–25]. Thermal and oxidative stability are a function of the size and shape of the alkyl groups and also depend strongly on whether the alkyl groups are unsaturated. Generally, unsaturated alkyl groups will lower the thermal and oxidative stabilities of a molecular structure. Tables 14.1 and 14.2 show some of the physical property data for several low- and high-molecular-weight alkyl carbonates, respectively. From these data the effect of structure and molecular weight on viscosity and other properties can be observed.

The viscosity of dialkyl carbonates produced by transesterification generally range from <1 to 40 cSt at 100°C. Alcohols that impart this viscosity range to dialkyl carbonates are available with the following structures:

1. Natural, linear, short chain (C_{6-8}) and long chain (C_{12-18})
2. Synthetic, branched, long chain (C_{12-18})
3. Polyols, synthetic: trimethylol propane (TMP), neopentylglycol (NPG), and neopentaerythritol (NPE)

This viscosity range can be expanded by a suitable choice of alkyl group functionality and appropriate synthesis methodology.

Several current commercial products in addition to DMC include diisooctyl carbonate (DIOC), dibutyl carbonate (DBC), and DEC. The physical and performance properties of these materials can be found in Tables 14.3 through 14.5, respectively.

TABLE 14.1
Physical Properties of Low-Molecular-Weight Dialkyl Carbonates

Carbonate	d_4^{20a}	BP (°C)	MP (°C)	Flash Point (°C)[b]	n_D^{20c}	Viscosity at 25°C (cP)	CAS Registration Number
Dimethyl	1.073	90.2	4	14[c]	1.3687	0.664 (at 20°C)	616.38.6
Diethyl	0.976	125.8	−43	33[c]	1.3843	0.868 (at 15°C)	105.58.8
Di-n-propyl	0.941	166	−40	62[c]	1.4022	1.27	623.96.1
Di-isopropyl	0.920	147	−40	46[c]	1.3917	1.14	6482.34.4
Diallyl	0.994	97	—	—	1.4280	—	15022.08.9
Di-n-butyl	0.924	208	−40	92[c]	1.4099	1.72	542.52.9
Di-2-ethylhexyl	0.897	173	<−50	—	1.4352	4.4 cSt (40°C)	14858.73.2
Allyldiglycol	1.143	160	−4	177 (o)	1.4503	9.0	142.22.3
Ethylene[d]	1.322 (39°C)	248	39	145	1.4158	1.92 (40°C)	9649.1
Propylene[e]	1.207	242	−49	130	1.4189	2.53	108.32.7
Methyl-isopropyl	0.969	120	<−40	33[c]	1.3830	0.77	—
Methyl-n-propyl	0.983	131	<−40	37[c]	1.3962	0.84	—
Methyl-n-butyl	0.964	162	<−40	54[c]	1.3970	1.01	—

Note: The data in the table were taken from the material safety data sheet (MSDS) of each material.
[a] d_4^{20} Density at 20°C referred to the density of water at 4°C.
[b]   Data from closed cup (c) and open cup (u) test procedures.
[c] n_D^{20} Refractive index at 20°C of D line of sodium.
[d]   1,3-Dioxolan-2-one.
[e]   4-Methyl-1,3-doxolan-2-one.

A comparison of the thermal stability of these materials has been shown to be a function of the molecular weight of the dialkyl carbonate (personal communication with Carlo Zecchini of Polimerieuropa). These data show that deposit formation is related to the molecular weight of the dialkyl carbonate.

In general, a carbonate with a long chain has a good viscosity index, good hydrolytic stability, and should not give rise to acidity during an eventual thermooxidative degradation.

The most important features are the high lubricity, high biodegradability, excellent antiwear, and compatibility with mineral oils and synthetic fluids.

A higher-molecular-weight carbonate, C_{14}–C_{15} dialkyl carbonate, is reported to have lubricant application in automotive and industrial engines and equipment. C_{14}–C_{15} dialkyl carbonate can be employed in gasoline and diesel engines and in gearbox applications (personal communication with Carlo Zecchini of Polimerieuropa) [26].

Carbonates, depending on their structures, find application as lubricants in the fields of drilling fluids, oil-based muds (OBM), emulsion oil/water pseudo oil-based

TABLE 14.2

Physical Properties of High-Molecular-Weight Dialkyl Carbonates

Carbonate	Kinematic Viscosity, cSt, 100°C	Kinematic Viscosity, cSt, 40°C	Viscosity Index[a]	Pour Point (°C)	NOACK Volatility, %[b]
iso-C_{10}	2.5	9.4	86	−57	—
n-C_{12}	3.4	12.2	169	+5	—
iso-C_{13}	3.9	19.8	81	−42	—
oxo C_{12-15}	3.8	15.7	144	+5	15
oxo C_{12-15}/oxo i-C_{13} (1:1)	3.9	17.2	115	−21	16
oxo C_{14-15} branched	4.1	18	126	−36	12
iso-C_8/C_6 diol	4.7	24.5	110	−34	—
oxo C_{12-15}/C_4 diol, (9:1)	5.2	23.9	153	0	10.8
iso-C_{18}	11.0	180	—	−18	—
iso-C_8/TMP	12.4	151	62	−30	—

[a] Viscosity index ASTM D2270.
[b] NOACK volatility ASTM D5800.

muds (POBM), metalworking fluids, textile softeners, and plasticizers/release agents, and in engine oils.

The advantages of using C_{14}–C_{15} dialkyl carbonate include better engine protection, which results in longer engine life; extended drain intervals are reported as is better cold starting performance (personal communication with Carlo Zecchini of

TABLE 14.3

Physical Properties of Diisooctyl Carbonate

Viscosity, cSt 40°C	4.45
Viscosity, cSt 100°C	1.51
Pour point (°C)	−40
Density, 25°C (g/mL)	0.89
Flash point, °C (COC)[a]	140
Boiling point, °C (10 mm Hg)	173
Biodegradability, MITI mod.[b]	31% after 28 days
Hydrolysis, pH 10, 70°C, % after 7 days	0.13
Hydrolysis, pH 10, 70°C, % after 14 days	0.19
Water solubility (mg/L)	Negligible

Note: The data in the table were taken from the MSDS of each material. Personal communication with Carlo Zecchini of Polimerieuropa.
[a] Cleveland Open Cup method (ASTM D92).
[b] Ministry of International Trade and Industry (Japan).

TABLE 14.4
Physical Properties of Dibutyl Carbonate

Boiling point (°C)		207.5
Density, 20°C (g/mL)		0.92
n_D^{20} (C)[a]		1.4117
Molecular weight		174
CAS number		542-52-9
Viscosity, cP at 20°C		1.72
Dielectric constant at 25°C		1.0
Water solubility, % at 20°C		1.5
Hydrolysis at 70°C, pH 10, %	7 days	0.44
	14 days	0.92

Note: The data in the table were taken from the MSDS of
        each material.

[a] Index of refraction at 20°C.

TABLE 14.5
Physical Properties of Diethyl Carbonate

Boiling point (°C)	127
Density at 20°C	0.97
Molecular weight (g/mol)	118
Viscosity at 20°C (cP)	0.868
Dielectric constant at 25°C	2.82
Water solubility (%) at 20°C	Insoluble
Thermal stability	High

Note: The data in the table were taken from the MSDS
        of each material.

Polimerieuropa) [26]. Furthermore, this dialkyl carbonate can be used in low-smoke lubricating compositions for two-phase (two-stroke) engines [27].

In industrial applications, these fluids are used in both compressor and metalworking fluids as well as in the textile industry. These materials are reported to provide better lubricity performance resulting in lower maintenance. The C_{14}–C_{15} dialkyl carbonates are also reported to have high biodegradabililty (>65%, after 28 days) (personal communication with Carlo Zecchini of Polimerieuropa).

For metalworking in the aluminum industry, the use of C_{14}–C_{15} dialkyl carbonate results in less oil staining, less carbon residue, and a more uniform gloss sheet. This cleaner surface finishing results in higher productivity and the application of C_{14}–C_{15} dialkyl carbonate uses less oil. As a cleaning agent, the product detaches dust particles from metal after welding, and it has replaced silicone oils in this application. In this case, the formulation contains more than 90% of DAC.

Application of this material in the steel industry also results in a clean surface sheet and low oil consumption, resulting in overall high productivity. The fluids used form stable microemulsions.

Applications in the textile industry are (a) in the area of biodegradable softening agents for finishing without the need for silicone, and (b) as a lubricant of the reeds of a loom [28]. For this application, the concentration of C_{14}–C_{15} carbonate in oil formulation is about 90%.

In general, the longer-chain dialkyl carbonate provides a good viscosity index, good hydrolytic stability, and no acidic compounds are formed due to thermooxidative degradation. From a performance point of view, the C_{14}–C_{15} dialkyl carbonate provides high lubricity, high biodegradability, and excellent antiwear performance. C_{14}–C_{15} dialkyl carbonate is completely compatible with other synthetic base fluids as well as with mineral oil-based fluids [36]. In fact, the lubricant fluid used in steel cold rolling consists of C_{14}–C_{15} dialkyl carbonate formulated with base mineral oil at a concentration of 2.5% [26].

C_{14}–C_{15} dialkyl carbonate is a pure, odorless, colorless nonoily emollient that is stable to oxidation. Because of these features, it has also found application in the cosmetic industry, having good feel and spreadability and compatibility with other oils and pigments. Another application for C_{14}–C_{15} dialkyl carbonate can be as a cold resistance plasticizer for vulcanizable composition on chlorinated or nitrile rubbers [29]. C_{14}–C_{15} dialkyl carbonate is able to form stable water-in-oil and oil-in-water emulsions. Its biodegradability makes it environmentally friendly. The physical properties for C_{14}–C_{15} dialkyl carbonate are shown in Table 14.6. Oleochemical carbonates have been reviewed [30,31].

TABLE 14.6
Physical Properties of C_{14}–C_{15} Dialkyl Carbonate

Viscosity, cSt 40°C	15–21
Density (g/mL)	0.87–0.90
Pour point (°C)	–40
Volatility	5% at 242°C, 20% at 245°C, 40% at 250°C
Volatility, NOACK DIN 51581	16%
CAS number[a]	164907-75-9
ELINCS number[b]	401-240-2
INCI number[c]	C_{14-15} dialkyl carbonate
CTFA number[d]	2541
Water solubility at 25°C (mg/L)	<0.44

Note: The data in the table were taken from the MSDS of each material. Personal communication with Carlo Zecchini of Polimerieuropa.
[a] Chemical Abstracts Service (CAS).
[b] European List of Notified Chemical Substances (ELINCS).
[c] The International Nomenclature of Cosmetic Ingredients (INCI).
[d] Cosmetic, Toiletry, and Fragrance Association (CTFA).

C_{14}–C_{15} DAC is a viscous oil and does not have a crystalline structure. The melting point is considered to be around –40°C. At about –9°C, solid material starts precipitating from this material. Although C_{14}–C_{15} DAC has a higher viscosity index than PAO and mineral oil, it has a higher pour point and less low-temperature flowability than PAO.

Another application can be for unlocking screws/nuts and bolts due to the fact that the product has a strong ability to deeply penetrate as a wetting agent, which allows softening and disintegration of the structure of rust due to its high capillarity and low surface tension.

Moreover, it forms a thin highly protective film on metals, preventing rust and corrosion by maintaining the film for a long time, with lower consumption and lower maintenance. As an expeller it penetrates deeply, dissolves dirt and grease, and removes any type of moisture that could cause possible power dissipation.

Dialkyl carbonate also has good compatibility with hydrocarbon oils and ester oils. For metal surfaces subjected to very severe operating conditions, it is necessary to use products that form a stable lubricating film on the surface. The C_{14}–C_{15} dialkyl carbonate also has interesting extreme pressure properties, compared to other oils, without the use of additives based on phosphorus, sulfur, or chlorine (Table 14.7) [27,32].

Glycerol carbonate is also referred to as glycerol cyclic 1,2-carbonate and 4-hydroxy-1,3 dioxolane-2-one. It is a dialkyl carbonate that is formed by the reaction of two of the three hydroxyl groups of glycerol. Glycerol carbonate can be obtained by reaction between urea and glycerol with zinc salt as catalyst, or from glycerol and DMC in a catalyzed reaction [33,34]. The structure of glycerol carbonate shown in Figure 14.8 has a free hydroxy group that contributes to its polarity and other properties.

There is the potential to modify the viscosity and polarity of the glycerol carbonate moiety by reacting, for example, aliphatic or aromatic groups with the free hydroxyl group. For the aliphatic case, this would provide a long-chain structure that is similar to fatty acids or fatty alcohols, which are known to impart improved tribological properties to lubricants. The main applications for glycerol carbonate are

TABLE 14.7
Dialkyl Carbonate: Lubricant Performance

	Method	C_{14}–C_{15} DAC	Mineral Oil	Diesters	PAO
Viscosity, cSt, at 100°C	ASTM D445	4.00	4.28	3.65	3.73
Evaporation loss, % 200°C, 7 h		15	30.5	15.1	15.5
Viscosity index	ASTM D2270	124	95	144	115
Extreme pressure (maximum load) (kg/cm²)	ASTM D2596	5.5			2.5
Four ball wear test wear scar diameter (mm)	ASTM D4172	0.57	3.81	1.0	8.85

Note: The data in the table were taken from the MSDS of each material.

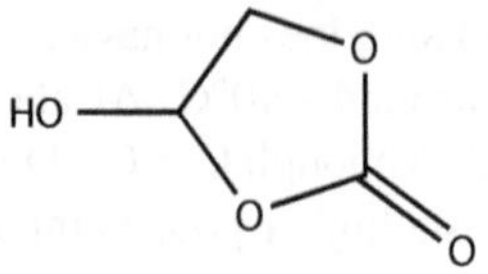

4-hydroxy-1,3 dioxolane-2-one

FIGURE 14.8 Structure of glycerol carbonate.

in pharma and cosmetic preparations as moisturizers in cosmetic creams and in hair and skin conditioners, and as detergent in laundry and cleaning compounds [35,36].

Although not generally considered in lubricant applications, the lubricity imparted by these materials contributes to overall performance [22,23]. In fact, the product has good adhesion to metallic surfaces and has good resistance to oxidation and hydrolysis. Glycerol carbonate, as well as other cyclic carbonates such as ethylene and PCs, after reaction with a hydrocarbyl amine, can be used as an additive. These materials, when used in a minor amount, can improve wear inhibition and impart friction modifier properties to a lubricating oil [37].

Glycerol carbonate (GLICA) is also used as a reactant for the synthesis of water-dispersible polyisocyanate resin, crosslinkable acrylic or epoxy acrylic coating, or thermosetting powder coating. Transesterification of glycerol carbonate by rapeseed methyl ester (RME), catalyzed by organotin catalysts, proved to be the best pathway for the synthesis of long-chain glycerol carbonate ester (GCE). These GCEs

TABLE 14.8
Properties of Glycerol Carbonate

Test	Method	Value
Viscosity, cP 100°C	ASTM D445	3
Viscosity, cP 38°C	ASTM D445	26
Density 20°C (g/mL)	ASTM D1298	1.4
Flash point (°C)	ASTM D93	212
Boiling point, °C (0.1 mmHg)		125–135
Molecular weight	Calculate	118
CAS number[a]	—	931-40-8
EINECS number[b]	—	213-235-0
Vapor pressure (kPa/20°C)	—	<0.0001
Pour point (°C)	ASTM D97	<–50
pH, 1 wt% in water		5.5
Biodegradability		Readily biodegradable
Water solubility, %		100

Note: The data in the table were taken from the MSDS of each material. Personal communication with Carlo Zecchini of Polimerieuropa.

[a] Chemical Abstracts Service (CAS).

[b] European Inventory of Existing Commercial Substances (EINECS).

have high thermal stability, as was shown by differential thermal analysis [38]. Characterization of the GCEs showed that they were excellent additives for metal machining lubricants, with lubricating performance comparable with that of trimethylpropane oleate [39,40].

GLICA is completely miscible with water, ethanol, methanol, acetone, and ethyl acetate. It is not miscible with hexane, MTBE, diethyl ether, and toluene. The physical properties of glycerol carbonate are summarized in Table 14.8.

14.5.1 Physical Properties

With only a few exceptions, all of the industrially important dialkyl carbonates are colorless liquids with specific gravity of less than 1.00. Dialkyl carbonates generally possess high boiling points and exhibit a pleasant odor.

Notably, the physical state depends on the type and molecular weight of alcohols: high-molecular-weight ($>C_{10}$) linear alcohols easily lead to solid products. In addition, low-molecular-weight carbonates with high symmetrical structure, for example, DMC and EC, exhibit melting points above 0°C.

Dialkyl carbonates are generally soluble in organic solvents, mostly in polar organic solvents such as alcohols, ethers, ketones, esters, and so on.

Calculations have been conducted to determine the pressure dependence of various properties of dialkyl carbonates (DMC and DEC) such as on density, isobaric thermal expansivity, and isothermal compressibility [1]. The dynamic viscosities and densities of dimethyl and diethyl carbonates have also been studied [41]. This work was done with potential application for the use of dialkyl carbonates in refrigeration systems, including automobile air conditioning systems. The viscometrics as a function of temperature are critical for this application since a vehicle will experience varying climates depending on the country of use and also in situations involving long-distance travel through different climatic conditions.

Cyclic five-member carbonates, ethylene, and PCs display good water solubility (100%, 100%, and 21.4%, respectively). Other low-molecular-weight carbonates, such as dimethyl and diethyl, have very limited water solubility (12.8% and 0%, respectively). High-molecular-weight carbonates are essentially insoluble in water. Lower dialkyl carbonates generally form azeotropic mixtures with water and/or organic solvents. This aspect increases the number of steps needed related to the production of dialkyl carbonates via transesterification.

Physical properties of the most important dialkyl carbonates are listed in Table 14.1 (low-molecular-weight carbonates) and in Table 14.2 (high-molecular-weight carbonates) and for other specific dialkyl carbonates in Tables 14.3 through 14.8.

14.5.2 Chemical Properties

Dialkyl carbonates also represent an important class of starting materials and intermediates in organic synthesis because of their ability to undergo the following reactions:

- To give urethanes, carbamates, and ammonia and amine ureas
- As alkylating agents, for amines, phenols, and acids

- As oxyalkylating agents, to produce five-membered cyclic carbonates
- Carbomethylation to give malonates, via Claisen condensation
- To give beta-ketoesters and alpha-cyanoesters from nitriles
- As carbonylating agents (DMC) for ketones, nitriles, hydroxy amides, and amino alcohols
- As methylating agents (DMC)
  - Quaternization of tertiary amines to produce quaternary ammonium salts
  - Transesterification of alcohols to give higher carbonates (oligo- and polycarbonates, cyclic, and aromatic carbonates). Relevant examples are diethyleneglycol
  - Bis-allylcarbonate (monomer for plastic lenses) and EC
  - Diphenyl carbonate via the new nonphosgene route. This method uses an intermediate dialkyl carbonate, usually DMC, as the source of carbonate functionality

Additional applications as reactant are

- Production of carbonate polyoxazolidines used in polyurethane coatings and sealants [42]
- DMC as foaming agent in polyurethane, through release of CO_2 as an alternative to chlorofluorocarbons to produce flexible low-density foam [43]

Apart from these synthetic uses, the structure of the carbonate group leads to other applications due to the fact that dialkyl carbonates undergo thermal and/or catalytic decomposition to alcohols, carbon dioxide, and olefins, and therefore, in principle, decompose without acid formation. The thermolysis of urethanes, in the presence of DMC, provides the possibility to produce isocyanates by a nonphosgene route [44]. Similarly, in the presence of water and/or hydrolyzing agents, hydrolysis can lead back to the formation of alcohols, carbonic acid, and, consequently, carbon dioxide. Dialkyl carbonates are generally stable to hydrolysis in acidic media, except under extreme conditions (e.g., very high temperature, high concentration). In this case, CO_2 is evolved. Dialkyl carbonates, however, hydrolyze at an appreciable rate in basic media where the carbonate readily undergoes alkaline hydrolysis to alcohol and carbonate ions, at the same rate as the corresponding esters. Since lubricant degradation (hydrocarbon oxidation) generally results in the formation of acidic species, this is a desirable property. In general, the hydrolytic stability of dialkyl carbonates increases with increasing molecular weight of the alkyl groups. The shape of the alkyl groups can also affect the rate of hydrolysis due to the steric effect of the alkyl group.

The properties that most directly affect the use of these materials as lubricant components include:

- Stability to hydrolysis in acidic media
- No acid formation on thermo-oxidation
- High biodegradability
- Low toxicity

- Moderate thermal stability
- Environmental compatibility, which reduces disposal costs and negative environmental impact
- High solvent power
- Friction reduction

Thermal stability is an important performance criterion for components of lubricant formulation. DBC, for example, is stable at 200°C for at least 20 h. Raising the temperature to 250°C, however, results in appreciable decomposition in 5–8 h.

Owing to this performance, dibutyl carbonate (DBC) has been studied for use in drilling fluids as base component of water in oil pseudo oil-based mud (W/O POMB) as a substitute for the current OBM [45]. In this case, it is possible to produce POMB with excellent rheological and other properties: plastic viscosity (PV), yield point (YP), gelling, stability, high temperature high pressure (HTHP) filtration control, low toxicity, and low environmental impact. DBC has been suggested for use in the cleaning of incrustations and of oily mud wastes from the casings before their cementation. For this application, it seems that DBC may have some problems due to its sharp smell and to the long contact time required for complete solubility of the incrustations. In fact, it is slightly longer than OBM (i.e., at normal temperature DBC needs 15 min versus OBM 5 min) but increasing the temperature reduces the time difference. The advantages obtained from using organic carbonates to this type of application are effectiveness in removing contaminants; simplification of the equipment used in this application; being able to operate in an open system as the emissions are produced; owing to their characteristics of biodegradability; low ozone depletion potential (ODP) and atoxicity; and they do not create any toxicological problems either for humans or for the environment.

Because of this, they can, therefore, also be used in offshore drilling activities, such as for example, in the washing of casings where, in practice, a substantial hydrolytic stability of the solvent is required. In any case, the nontoxicity of the degradation products is also a requirement [46,47].

14.5.3 PERFORMANCE CHARACTERISTICS

The performance of dialkyl carbonates in engine oil applications has been reported [14]. The main conclusions of this work can be summarized as follows.

14.5.3.1 Tribology

The presence of the carbonic group in dialkyl carbonates imparts lubricity properties to these products due to the interaction between the carbonic group and the metal surfaces. A comparison of linear-oscillating friction and wear test data for C_{14}–C_{15} dialkyl carbonate with other mineral and synthetic base fluids of similar viscosity at 100°C confirms the excellent lubricity properties of dialkyl carbonates [27].

14.5.3.2 Viscometrics

Viscometric data indicate that the percentage amount of dialkyl carbonates in a formulation necessary to meet the requirement of low-viscosity multigrade engine oils

(5W/XX, 10W/XX) is comparable with the percentage amount of other synthetic base fluids (PAO, esters, etc.) of the same viscosity.

14.5.3.3 Oxidation and Thermal Stability

Oxidation and thermal stability tests show an outstanding performance of dialkyl carbonates: the absence of acidic product formation at the end of the tests was an advantage. This property may affect the overall behavior of fully formulated lubricants in that corrosive wear phenomenon can be positively controlled.

14.5.3.4 Elastomer Compatibility

Elastomer compatibility, a critical performance feature and specification for synthetic base fluids, is generally satisfactory for dialkyl carbonates. Carbonates display good compatibility even with silicone elastomers, which are the most sensitive to base fluid polarity. In particular, semisynthetic multigrade engine oils formulated with dialkyl carbonates exhibit better performance in silicone rubber swelling test than similar formulations made with polyalphaolefins or polyol esters.

14.5.3.5 Wear Protection

The engine performance of multigrade oils containing dialkyl carbonates shows excellent antiwear properties. Dialkyl carbonates do not interfere with antiwear additives, as exhibited by other ester-type basestocks. Results have been obtained in a variety of standard engine tests.

14.5.3.6 Sludge Protection

Formulations containing dialkyl carbonates provide good sludge control, better than that exhibited by pure mineral-based formulations containing the same package at the same concentration. Results were obtained from VE and M102E Black Sludge tests. This performance can be related to the particular chemical structure of dialkyl carbonates, with a particularly strong polar group and two alkyl chains in the same molecule, resulting in a synergistic effect with traditional dispersant additives in the control of sludge.

14.6 TOXICOLOGY, BIODEGRADABILITY, AND HANDLING PROPERTIES

14.6.1 Toxicology

Dialkyl carbonates can be classified as nontoxic substances for humans and for the environment, according to the relevant tests and experiments that have been carried out to assess this aspect. Among others, the effects of contact, ingestion, and inhalation, in acute, subacute, and chronic conditions have been investigated indicating that dialkyl carbonates are not dangerous to humans.

As far as the environment is concerned, not all dialkyl carbonates are completely soluble in water. Dialkyl carbonates exhibit very low volatility, and they have to be removed from water and soil if any contamination occurs. They have a low degradation rate and the degradation does not form harmful products. Based on these considerations, dialkyl carbonates can be considered as nondangerous for the environment.

DMC exhibits good toxicological properties. The product is not an irritant to the eyes and skin, and has acute oral toxicity of $LD_{50} > 6000$ mg/kg (mouse). DEC exhibits good toxicological properties. The product is slightly irritating to the eyes and skin and has acute oral toxicity of $LD_{50} > 1570$ mg/kg (mouse). DBC exhibits good toxicological properties as well. Tests have shown that it is not an irritant to the skin or the eyes (rabbit test), and has an LD_{50} oral (rat) of >2000 mg/kg. C_{14}–C_{15} dialkyl carbonate has an LD_{50} oral (rat) of 10 mL/kg and an LD_{50} dermal (rat) of 5 mL/kg. Skin irritation, eye irritation, skin sensitization (rat), biological, ecotoxic, and subcutaneous effects are all reported as negative for carbonates (personal communication with Carlo Zecchini of Polimerieuropa).

14.6.2 BIODEGRADABILITY

Biodegradability test data (Table 14.9) show that, although for lower-molecular-weight dialkyl carbonates the polarity improves water solubility, which may indirectly improve biodegradability, the carbonate group does not ensure biodegradability.

DBC exhibited the following biodegradability performance using The Organization for Economic Co-operation and Development (OECD 301 C) test: COD (28 days) 68%, BOD (28 days) 72%.

The biodegradability of dialkyl carbonates strongly depends on the biodegradability of the alkyl groups present in the molecule. There are several structural features that can have an impact on the extent of biodegradability of various lubricant fluids. Some of these include the extent of branching of the hydrocarbon chains, the hydrocarbon chain length, the proportion of oxygen in the structure, and olefinic content. Polarity is affected in dialkyl carbonates by the chain length because there is already oxygen functionality due to the carbonate group. In addition, any oxygen present in the alkyl chains further increases the polarity of the fluid. This can result

TABLE 14.9
Biodegradability Test Results on Dialkyl Carbonates

Carbonate of	Biodegradability, % (28 days)
–CH_3OH	90
–oxo C_{12-15}/oxo i-C_{13}	75
–oxo C_{14-15} branched	65
–oxo i-C_{13}	60
–TMP oxo C_{12-15}/i-C_8	15
–i-C_{18}	Negligible
Propylene	94
Glycerol	Readily (>60%)
Diisooctyl	31
n-Butyl	68

in greater water solubility, which can impact the extent (and rate) of biodegradation since the test is conducted in aqueous media.

14.6.3 HANDLING

No special procedure and care are required for handling dialkyl carbonates. It is recommended that standard hygiene and safety procedures be followed. Dialkyl carbonates are stable at ambient temperature and atmospheric pressure. They do not require any special precautionary measure for storage, and can be safely stored in carbon steel tanks for long periods of time. One note of caution should be made about DMC. This dialkyl carbonate is flammable and, therefore, caution should be taken on the method of storage of this material.

In summary, dialkyl carbonates can be obtained by several synthesis routes. A wide variety of different dialylcarbonates can be prepared to provide materials with required viscosity or degree of water solubility or biodegradability. Dialkyl carbonates provide properties and performance features desirable in synthetic lubricants.

ACKNOWLEDGMENTS

The authors would like to acknowledge G. Fisicaro and G. Gerbaz for their original chapter on dialkyl carbonates in the second edition of the book *Synthetic Lubricants and High-Performance Functional Fluids* (Marcel Dekker) from which this chapter has been expanded, updated, and revised [48]. (The authors attempted via the Internet and through professional colleagues to contact the original authors without success to update their chapter.) LRR would also like to thank Carlo Zecchini, organic carbonates marketing consultant in Italy, Brajendra Sharma of USDA, Peoria, IL, USA, and Chris Klumph of SNPE NA, Princeton, NJ, USA for providing new data and helpful discussion for inclusion in this chapter.

REFERENCES

1. L. Lugo, V. Luna, J. Garcia, E. R. Lopez, M. J. P. Comunas, and J. Fernandez, Fluid Phase Equilibria, 2003.
2. X.-C. Lu, W.-G. Zhang, X.-Q. Qiao, and Z. Huang, Fuel design concept for improving the spray characteristics and emissions of diesel engines. *Proc. IMechE. Vol. 219 Part D: J. Automobile Eng.*, 547–557, 2005.
3. M. E. Jenkin and G. D. Hayman, Photochemical ozone creation potentials for oxygenated organic compounds: Sensitivity to variations in kinetic and mechanistic parameters. *Atmos. Environ.* 33, 1275–1293, 1999.
4. T. Takeno, K. Mizui, and K. Takahata, In Proceedings of the International Compressor Engineering Conference, Purdue, IN, USA, pp. 1045–1054; K. Mizui and Y. Furaya, US Patent 5,114,605, 1992, Mitsui Petrochemical Industries Ltd., Lubricant oil for refrigerators.
5. M. F. Fincke and J. H. Bartlett, Standard Oil, US Patent 2,263,265, 1941, Standard Oil Development Company, Art of improving lubricating oils.
6. A. T. Knutson and E. F. Graves, Lubrizol Corp., US Patent 2,387,999, 1945, Lubrication.
7. L. A. Mikeska and L. T. Eby, Standard Oil Development Company, US Patent 2,651,657, 1953, Synthetic lubricating oil.

8. J. H. Bartlett, Standard Oil Development Company, US Patent 2,673,185, 1954, Polymerized carbonate ester lubricating oil additives.
9. J. H. Bartlett, Esso Research and Engineering Company, US Patent 2,718,504, 1955, Oil solution of polymerized carbonate ester.
10. D. L. Cottle, F. Knoth, and D. W. Young, Esso Research and Engineering Company, US Patent 2,758,975, 1956, Synthetic lubricants.
11. P. Koch and U. Romano, Assoreni, Ital. Pat. 20264 A/82, 1982, Synthesis of higher alcohol carbonates and their use as synthetic lubricants.
12. E. Brandolese, Euron, Ital. Pat. 20191 A/89, 1989, Lubricating fluid for the coldrolling of steel.
13. G. Fisicaro and G.P. Gerbaz, AgipPetroli, Ital. Pat. 21812 A/90, 1990, Lubricant compositions for autotraction.
14. G. Fisicaro and S. Fattori, in Proceedings of the Conference on Synthetic Lubricants, Sopron, Hungary, 1989.
15. M. Hayashi, G. Minami, K. Matsunaga, T. Fujii, and T. Hayashi, US Patent 5727410, 1998, Matsushita Electronics Corporation, Press-molding oil and method for manufacturing press-molded products by using the same.
16. M. Hayakawa, T. Kobayashi, K. Shimizu, Y. Matsumura, M. Onoyama, and K. Nagano, US Patent 5629274, 1997, Sankyo Seiki Manufacturing Company, Ltd., Lubricating fluid composition for dynamic pressure bearing.
17. B. Scrosati, *Chim. Ind. (Milan)* 79, 463, 1997.
18. M. Massi, M. Mauri, U. Romano, and F. Rivetti, Dimethyl carbonate: A new building block for organic chemicals production. *Ing. Chim. Ital.*, 21(1–3), 1985.
19. T. Matsuraki and A. Nakamura, *Catal. Surv. Jpn.* 1(!), 77–88, 1997, DOI: 10.1023/A: 1019020812365.
20. H.-P. Luo, W.-D. Xiao, and K.-H. Zhu, Isobaric vapor–liquid equilibria of alkyl carbonates with alcohols fluid phase equilib. *Fluid Phase Equilibria*, 175, 91–105, 2000.
21. F. Comelli and R. Francesconi, Isothermal vapor-liquid equilibria measurements, excess molar enthalpies, and excess molar volumes of dimethyl carbonate + methanol + ethanol, and propan-1-ol at 313.15 K. *J. Chem. Eng. Data* 42, 705–709, 1997.
22. F. Comelli, R. Francesconi, and S. Otani, Isothermal vapor-liquid-equilibria of dimethyl carbonate plus diethyl carbonate in the range (313.15–353.15) K. *J. Chem. Eng. Data* 41, 534–536, 1996.
23. M. J. Cocero, F. Mato, I Carcia, J. C. Cobos, and H. V. Kehiaian, Thermodynamics of binary mixtures containing organic carbonates. 2. Isothermal vapor-liquid equilibria for diethyl carbonate + cyclohexane + benzene, or +tetrachloromethane. *J. Chem. Eng. Data* 34, 73–76, 1989.
24. M. J. Cocero, F. Mato, I Carcia, and J. C. Cobos, Thermodynamics of binary mixtures contains organic carbonates, 3. Isothermal vapor-liquid equilibria for diethyl carbonate + cyclohexane + benzene, or tetrachloromethane. *J. Chem. Eng. Data* 34, 443–445, 1989.
25. M. J. Cocero, I Carcia, J. A. Gonzalez, and J. C. Cobos, Thermodynamics of binary mixtures containing organic carbonates Part VI. Isothermal vapor-liquid equilibria for dimethyl carbonate + normal alkanes. *Fluid Phase Equilibr.* 68, 151–161, 1991.
26. E. Brandolese, EP Patent 0393749 A2—Agip Spa, 1990, Lubricant fluid for coldrolling of steel.
27. F. Berti and U. Rivetti, EP Patent 636681 A2, Euron, 1994, Low smoke lubricating composition for two phases engine.
28. Lamberti SpA, EP Patent 507973, 1991, Composition biodegradables utilized as softeners or lubricant in fabrics of textiles materials.
29. E. Traverso, F. Rivetti, and L. Spelta, US Patent 5580916, Enichem Synthesis 1996, Vulcanizable composition of nitrile rubbers.

30. M. Dierker, Oleochemical carbonates—An overview. *Lipid Technol.*, 16(8), 130–134, 2004.
31. J. A. Kenar, Current perspectives on oleochemical carbonates. *Inform* 15(9), 580–582, 2004.
32. P. Koch and U. Romano, EP Patent 0089709—Agip Spa/Anic, 1983, Synthesis of higher alcohols carbonates for synthetic lubricants.
33. O. Gomez et al., Synthesis of GLICA from glycerol and DMC, *Applied Catalysis*, September 2009.
34. J. B. Bell, V. A. Currier, and J. D. Malkemus, Jefferson Chemical Company, US Patent 2915529, 1959, Method for preparing glycerin carbonate.
35. J. Kahre, T. Loehl, H. Tesman, and H. Hensen, DE Patent 19756454, 1999, Henkel KGAA, Surface active compositions, especially cosmetics containing glycerol carbonate as emulsifier.
36. T. Loehl and A. Behler, DE Patent 19826327, 1999, Henkel KGAA: Cyclic carbonates for casing and drilling.
37. W. R. Ruhe, US Patent 0160198, 2010, Chevron Oronite, Friction modifiers derived from amines and cyclic carbonates.
38. Mouloungui, Structure and properties of GLICA esters. *Eur. Lipid Sci. Technol.* 103, 216–222, 2001.
39. C. Sylvain, Agrice card—Onidol Avenue George V—Paris cedex: Processing rapeseed methyl esters and glycerol for lubricant.
40. K. Mizui and Y. Furuya, EP Patent 0426153 A1, 1991. Mitsui Petrochemical Ind., Use of glycol ether carbonate as refrigeration lubricant.
41. Baylaucq, M. J. P. Comunas, C. Boned, A. Allal, and J. Fernandez, High pressure viscosity and density modeling of two polyethers and two dialkyl carbonates. *Fluid Phase Equilibria* 199, 249–263, 2002.
42. A. Greco, US Patent 5219979, 1992, Enichem synthesis, Polyoxazolidines with a structure containing carbonate groups.
43. D. Stefani, EP Patent 1092746 A1, 2001, Enichem Spa, Process for the preparation of low density flexible polyurethane foam.
44. F. Rivetti, The role of dimethylcarbonate in the replacement of hazardous chemicals. *Academie de Science, Chemistry* 3, 497–503, 2000.
45. H. Mueller, DE Patent 40 18 228, 1991. Henkel KGAA, Fluid drill oil treatment based on carbonic acid diesters.
46. C. Savastano, US Patent 5346615, 1994. Eniricerche S.p.A., Deasphalting and demetalating crude petroleum or its fractions.
47. F. Mizia and F. Rivetti, EP Patent 1083247, 2000. Enichem S.p.A., Use of organic carbonates as solvents for the washing of metal surfaces.
48. G. Fisicaro and G. Gerbaz, In: *Synthetic Lubricants and High-Performance Functional Fluids*, Second edition, L. R. Rudnick, and R. L. Shubkin (eds.), Marcel Dekker, New York, pp. 313–323, 1999.

15 Theoretical and Practical Treatments of Surface and Bulk Properties of Aqueous Mixed Surfactant Systems
Mixed Monolayers, Mixed Micelles Formation, and Synergism

Nabel A. Negm

CONTENTS

Theories for evaluating mixed surfactant systems are systematically reviewed, paying special attention to several points concerning the mixed surfactant systems, including interaction of counterions with the micelles, composition of mixed monolayers and mixed micelles, activity coefficients of the different surfactants in the mixed systems, and synergism observed in micellization as well as adsorbed film formation upon mixing. These theories discuss the effect of the different mixed systems with a variety of surfactants (charged surfactants, e.g., cationic, anionic, zwitterionic; and uncharged surfactants, e.g., nonionic surfactants). The studied mixed system is composed of the cationic surfactant dodecyl dimethyl phenyl ammonium bromide (C_{12}DMPAB) and nonionic surfactant decyl dimethyl phosphine oxide (C_{10}DMPO) surfactants denoted as 1 and 2, respectively. It is shown that the composition of mixed micelles (X_1) equilibrated with singly dispersed surfactant species in bulk solution phase and also the composition of adsorbed film phase (Z_2). Almost all combinations are discussed in terms of the respective interaction parameters, β^σ, β^m, where β^σ and β^m are the interaction parameters of the components in the adsorbed mixed monolayer and in mixed micelles, respectively. In addition, surface excess concentration (Γ_{max}), molecular area (A_{min}), and minimum surface Gibbs energy (G^s_{min}) for evaluation of synergism are discussed.

15.1 INTRODUCTION

In most practical applications, mixtures of surfactants, rather than individual surfactants, are used. In some cases, these are not deliberate mixtures such as those used in commercial surfactants. Even when designated by the name of an individual surfactant, for example, sodium lauryl sulfate, commercial surfactants are mixtures of surface active materials as a result of the nonhomogeneous raw materials used in their manufacture and/or the presence of unreacted raw materials and by-products. In other cases, mixed surfactants are obtained nondeliberately even if the used surfactant is ultrapure because it could be hydrolyzed by the aqueous medium (H_2O) to other contaminants. These contaminants could be much more surface active than the parent surfactant and have a higher adsorption tendency. This makes the system a mixed surfactant system.

In other cases, mixed surfactant systems have been of interest and are deliberately produced because they provide better performance characteristics in many technological applications, such as pharmaceutical, food, detergent, cosmetics, flotation, and enhanced oil recovery [1–3]. In fact, surfactants used in industrial applications are often deliberate mixtures because mixed surfactant systems offer a behavior different than that expected from the pure component solutions, which reinforces their interest from both theoretical and technological standpoints [1,2]. Furthermore, mixed micellar systems can serve as good models for the study of molecular interactions in complex supramolecular aggregates and also serve to mimic the behavior of biological systems and some of their functions in ion transport and drug delivery [1]. For all the above-noted reasons, the study of mixed surfactant systems, with a wide variety of experimental techniques and methods (e.g., surface tension measurements, interfacial tension measurements, conductivity, IR, H^1-NMR, SEM, or light scattering) has become increasingly important.

When two or more surfactants are present in water solution, mixed micelles are formed as a result of a complex balance of intermolecular forces [2]. Early theories involved in mixed micelles considered that the formation of mixtures of surfactants with the same homologous series was an ideal process. However, many authors have recently suggested this to be an oversimplification [2,4–8]. That is because even between similar molecules the interaction occurs due to geometrical or steric factors.

The thermodynamic treatments of mixed surfactant systems with two or more surfactants are so far developed for mixed micellization. The treatments are based on several models including the following: (1) the pseudophase separation model, (2) the mass action model, (3) phase separation model, and (4) the small system thermodynamics. A detailed description of these models will be given next.

15.1.1 Thermodynamic Models of Treating the Mixed Surfactants Systems

The pseudophase separation model and the mass action model have been frequently utilized to analyze and explain the characteristic aspects of the micellization process of pure surfactants [2–4]. The pseudophase separation model assumes that micelles are a separate phase formed when the concentration reaches the critical micelle concentration (cmc) of the surfactant. This concentration represents the saturation concentration of the surfactant at the interface.

The usual mass action model considers that a single micellar species and monomeric surfactants are in association–dissociation at equilibrium, and then mass action law can be applied [9].

Both models exhibit the same advantages for characterizing some aspects of micellization, but also have some limitations either in the bulk or at the interface of the surfactant solutions.

The phase model prevents the discussion of the micellar size, irrespective of their size, whereas the mass action model assumes monodispersity of micellar size, that is, assumes a fixed micellar size.

The multiple equilibrium model [11], as the extension of the mass action model, considers multiple equilibria or a stepwise mechanism of the micellization process to account for the polydispersity of micelle size, and hence the micelle size is considered to be gradually increasing in size. This polydispersity is also considered in the small system thermodynamics approach applied to micellar solutions. The phase model and the small system thermodynamics have been extended to multicomponent micelles. The thermodynamic approaches also deal with the problem of counterions binding to the micelles in ionic surfactants and in mixed systems containing ionic components.

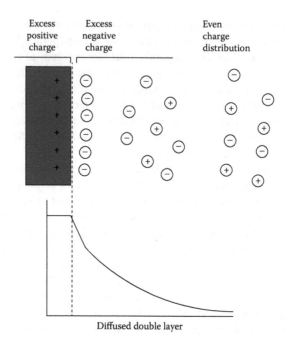

Excess positive charge	Excess negative charge		Even charge distribution

Diffused double layer

FIGURE 15.1 Counterions arranged around the positively charged micellar surface to form a Stern double layer, excess positive charge of the micellar surface, excess negative charges of negative counterions near the micelle surface, and even charge distribution of positive and negative counterions at the diffused double layer.

The topic of counterion binding to the micelles can be discussed as follows: the micellar aggregation of ionic surfactants is opposed by the head group's electrostatic repulsions, which are balanced, to a large extent, by the adsorption of counterions at the micellar surface. Generally, about 60–70% of the counterions are bound in the Stern layer at the micelle surface (Figure 15.1). Another point of view is that binding of counterions alone should not be taken into consideration, but the unbound counterions should be considered. This is due to the fact that unbound counterions are connected to the Stern layer with a decay layer from the Stern layer to the bulk of the solution. Also, there is a continuous arrangement of the counterions due to the association–dissociation arrangement of the counterions in the Stern layer. That arrangement is decayed with increasing distance away from the micellar surface. Thus, there is no definite distinction between the bound and unbound counterions.

In the pseudophase separation model, the effects of counterions are incorporated in the form of a charged pseudophase separation model because the electrically neutral micellar phase does not result in agreement between theoretical and experimental values for micelles with ionic surfactants.

15.1.1.1 Pseudophase Separation Models for Mixed Micellization of Nonionic Surfactants

The thermodynamic formulation of the mixed micellization process of nonionic surfactants using the phase separation model is simple due to the absence from complication due to the counterions bound to the micellar phase and the electrostatic effects. In this model, two boundaries are considered.

First, at the initial concentration of surfactant 1 and 2 (C) lower than the cmc (C_M) (i.e., $C \leq C_M$), the monomer concentrations of the two surfactants (C_1^m, C_2^m) are related to the (C) according to Equations 15.1 and 15.2:

$$C_1^m = y^C \tag{15.1}$$

$$C_2^m = (1 - y)C \tag{15.2}$$

where y is the initial mole fraction of surfactant 1 in the mixed surfactant system, C_1^m, C_2^m are the concentrations of surfactants 1 and 2 in the mixed micelles, respectively.

Second, at the monomeric surfactant concentrations higher than the cmc (i.e., $C \geq C_M$), the micellar mole fraction is expressed using Equation 15.3:

$$y = \frac{C - C_1^m}{C - C_1^m - C_2^m} \tag{15.3}$$

In this case, the solution is considered ideal due to the lack of counterions at the interface of the micelles formed, and accordingly, the surfactant concentrations in the micellar form are given using the individual cmc's of each surfactant (C_{M1}, C_{M2}) according to Equations 15.4 and 15.5, then rearranging these equations gives Equation 15.6:

$$C_1^m = xC_{M1} \tag{15.4}$$

$$C_2^m = (1 - x)C_{M2} \tag{15.5}$$

$$C_2^m = \left(1 - \frac{C_1^m}{C_{M1}}\right)C_{M2} \tag{15.6}$$

where x is the mole fraction of surfactant 1 in the mixed micelles formed.

Clint [10] made a simplification concerning (C_1^m, C_2^m, C, y) for the mixed system composed of two nonionic surfactants by considering the symbol ($\Delta = C_{M2} - C_{M1}$), the concentration of surfactant 1 in the mixed micellar phase to obtain Equation 15.7:

$$C_1^m = \frac{-(C - \Delta) + \left((C - \Delta)^2 + 4yC\Delta\right)^{1/2}}{2\left(C_{M2}/C_{M1} - 1\right)} \tag{15.7}$$

Also, the cmc of the ideal mixed system (C_M) can be calculated according to Clint's equation 15.8:

$$C_M = xC_{M1} + (1 - x)C_{M2} \tag{15.8}$$

Equation 15.8 was transferred to the summation form as represented in Equation 15.9 [11–13]:

$$C_M = \frac{1}{\sum (y_i / C_{Mi})} \tag{15.9}$$

Equations 15.6 through 15.9 are used to calculate the variables of the ideal mixed surfactant systems during the formation of the ideal mixed micelles.

To extend the equation of state for the nonideal mixed system, it is mandatory to introduce the activity coefficients (f_1, f_2) of the two surfactants in terms of the chemical potentials (μ_i^m, μ_i^b) of the micelles formed and the individual components, respectively. The chemical potential of the micelles formed is related to the chemical potential of the individual components and the activity coefficients according to Equation 15.10:

$$\mu_i^m = \mu_i^b + RT \ln x_i f_i \tag{15.10}$$

where R is the universal gas constant and T is the absolute temperature.

For binary systems of surfactants with nonideal behavior following the methodology of deriving the mixed system variables in Equations 10.6, 10.7, and 10.9 by using the measurable system values, Equations 15.11 through 15.14 are obtained [10]:

$$C_1^m = \frac{-(C - \Delta) + \left((C - \Delta)^2 + 4yC\Delta\right)^{1/2}}{2(f_2 C_{M2}/f_1 C_{M1} - 1)} \tag{15.11}$$

where $\Delta = f_2 C_{M2} - f_1 C_{M1}$; this term is analogous to the term $\Delta = C_{M2} - C_{M1}$ used in the ideal system after considering the activity coefficients of the two surfactants. These coefficients allow inclusion of the effect of the bound counterions and their influence on the dissociation of the amphiphiles.

$$C_2^m = \left(1 - \frac{C_1^m}{f_1 C_{M1}}\right) f_2 C_{M2} \tag{15.12}$$

$$C_M = \frac{1}{y/f_1 C_{M1} + (1 - y)/f_2 C_{M2}} \tag{15.13}$$

Also, the nonideal micellar composition can be formulated as Equation 15.14:

$$x = \frac{-(C - \Delta) + \left((C - \Delta)^2 + 4yC\Delta\right)^{1/2}}{2\Delta} \tag{15.14}$$

The above four Equations 15.11 through 15.14 are applicable for mixed micellar systems, including ionic and nonionic surfactants. The nonideality of the system is included in the activity coefficients. These coefficients are affected by the changes in the head group electrostatic repulsion energies, degree of counterions interaction, degree of surfactant ionization, and the steric interactions in the hydrophobic groups. Equations 15.11 through 15.14 were included in the study of Rubingh [14], who used these equations in addition to the regular solution approximations in order to express the micellar activity coefficients, f_1 and f_2, as given in Equations 15.15a and 15.15b [14]:

$$f_1 = \exp\beta(1 - x)^2 \tag{15.15a}$$

$$f_2 = \exp\beta x^2 \tag{15.15b}$$

where β is an interaction parameter between the components in the mixed micelles.

The treatment including this approximation has been applied to systems of nonideal binary mixed micelles containing ionic surfactants. It has been shown that a number of experimental data on C_M versus monomer composition can be well predicted by adjusting only the value of parameter β.

15.1.1.2 Pseudophase Separation Models for Mixed Micellization of Charged Surfactants

The main point in the study of charged mixed surfactant systems is the effect and extent of counterion binding to the micellar surface. The charged state of the mixed micelles containing ionic surfactants is described by the following quantities:

1. The micellar charge or surface potential
2. The effective degree of counterion binding to the micelles
3. The effect of added salts

The parameter β in Equations 15.15a and 15.15b is expressed in terms of the molecular interaction in the mixed micelles according to Equation 15.16 [14–16]:

$$\beta = \frac{N_A(\omega_{11} + \omega_{22} - \omega_{12})}{RT} \tag{15.16}$$

where ω_{11} is the energy of interaction between the surfactant 1 molecules, ω_{22} is the energy of interaction between the surfactant 2 molecules, ω_{12} is the energy of interactions between surfactant 1 and surfactant 2 molecules, and N_A is the Avogadro's number.

According to Equation 15.16, it is obvious that the total behavior of the mixed micellar system is dependent on the value of β. If the value of β determined with Equation 15.16 is negative, this implies a large interaction between the two different molecules (i.e., the system components), and the interaction between the two components is larger than that occurring between the identical components. A positive value of β is obtained for systems with large interaction between identical components.

Applying the condition of the micellization equilibrium ($\mu_i^m = \mu_i^b$) as represented in Equation 15.10, we obtain Equations 15.17a and 15.17b:

$$C_1^m = x_1 f_1 C_{M1} \tag{15.17a}$$

$$C_2^m = x_2 f_2 C_{M2} \tag{15.17b}$$

Expressing the values of f_1 and f_2 in Equations 15.15a and 15.15b from Equations 15.17a and 15.17b produces the equation of state of the mixed system of charged or ionic surfactant systems (Equations 15.18 and 15.19) as follows:

$$\frac{x^2 \ln\left(C_M y/C_{M1} x\right)}{(1-x)^2 \ln\left[C_M (1-y)/C_{M2}(1-x)\right]} = 1 \tag{15.18}$$

$$\beta = \frac{\ln\left(C_M y/C_{M1} x\right)}{(1-x)^2} \tag{15.19}$$

The molecular interaction parameters for the mixed monolayer formation by two different surfactants at an interface can be evaluated based on the application of the nonideal solution theory of thermodynamics regarding the mixed system at the interface [17–19] and provides Equation 15.20:

$$\frac{x^2 \ln\left(C_{12} y/C_1^o x\right)}{(1-x)^2 \ln\left[C_{12}(1-y)/C_2^o(1-x)\right]} = 1 \tag{15.20}$$

The interaction parameter and the activity coefficients of the two components at the interface can also be obtained using Equations 15.21 through 15.23 as follows:

$$\beta^s = \frac{\ln\left(C_{12} y/C_1^o x\right)}{(1-x)^2} \tag{15.21}$$

$$f_1 = \exp \beta^s (1-x)^2 \tag{15.22}$$

$$f_2 = \exp \beta^s x^2 \tag{15.23}$$

where $x, \beta^s, C_1^o, C_2^o, C_{12}$ are the mole fractions of surfactant 1 and the interaction parameters in the mixed monolayer, molar concentration of surfactant 1 and surfactant 2, and the molar concentration of the mixed monolayer, respectively.

15.1.2 PSEUDOPHASE SEPARATION MODELS FOR MIXED ADSORBED MONOLAYERS

By applying the approximation of Rosen [17,21] for the mixed monolayer at the interface, some points should be taken into consideration:

1. The two surfactants must be molecularly homogeneous and free from surface active impurities.
2. All solutions containing ionic surfactants should have the same total ionic strength, with a swamping excess of counterions.

3. The derivation of these equations is based on the assumption that the mixed micelle or monolayer can be considered to contain only surfactants; these structures are considered to contain no free water. This is reasonable when the surfactant molecules are so closely packed (e.g., at their maximum surface excess concentration in the monolayer) that all the water present can be considered to be bound to the hydrophilic head groups.

15.1.3 SYNERGISM IN SURFACE OR INTERFACIAL TENSION REDUCTION

Synergism with respect to the lowering of the concentration of surfactant mixture to produce a lower depression in the surface tension is called synergism in the surface (or interfacial) tension efficiency. From the relation upon which Equations 15.20 through 15.21 are based and from the definition of the synergism or antagonism of this type, it was shown mathematically that the conditions for synergism in surface tension reduction efficiency are [20]

In case of synergism (Equations 15.24 and 15.25):

$$\beta^\sigma < 0 \tag{15.24}$$

$$\left|\beta^\sigma\right| > \left|\ln(C_1^\circ / C_2^\circ)\right| \tag{15.25}$$

In case of antagonism or negative synergism (Equations 15.26 and 15.27):

$$\beta^\sigma > 0 \tag{15.26}$$

$$\left|\beta^\sigma\right| > \left|\ln(C_1^\circ / C_2^\circ)\right| \tag{15.27}$$

where β^σ is the interaction parameter of the components in the adsorbed mixed monolayer.

At the point of maximum synergism or maximum antagonism, that is, at the minimum or maximum aqueous phase total molar concentration, respectively, of mixed surfactants to produce a given surface tension, the mole fraction (α^*) of surfactant 1 in the solution phase equals its mole fraction at the interface and is given by Equation 15.28 [21]:

$$\alpha^* = \frac{\ln\left(C_1^\circ / C_2^\circ\right) + \beta^\sigma}{2\beta^\sigma} \tag{15.28}$$

The minimum or maximum aqueous phase total molar concentration of mixed surfactants in the system to produce a given surface tension is given by Equation 15.29:

$$C_{12,\min}^\circ = C_1^\circ \exp\left\{\beta^\sigma\left[\frac{\beta^\sigma - \ln\left(C_1^\circ / C_2^\circ\right)}{2\beta^\sigma}\right]^2\right\} \tag{15.29}$$

15.1.4 SYNERGISM IN MIXED MICELLE FORMATION IN AQUEOUS MEDIUM

Synergism in this respect is present when the C_M in the aqueous medium of any mixture of two surfactants is smaller than that of either individual surfactant. Antagonism in this respect is present when the C_M of the mixture is larger than the C_M of either surfactant in the mixture. From Equations 15.18 and 15.19 and the definition for this type of synergism or antagonism, the conditions for synergism or antagonism in this respect in a mixture containing two surfactants (in the absence of a second liquid phase) have been shown mathematically [22] to be

In case of synergism (Equations 15.30 and 15.31):

$$\beta^M < 0 \tag{15.30}$$

$$\left|\beta^M\right| > \left|\ln(C_{M1}/C_{M2})\right| \tag{15.31}$$

In case of antagonism or negative synergism (Equations 15.32 and 15.33):

$$\beta^M > 0 \tag{15.32}$$

$$\left|\beta^M\right| > \left|\ln(C_{M1}/C_{M2})\right| \tag{15.33}$$

At the point of maximum synergism or antagonism, that is, where the C_M of the system is at a minimum or maximum, respectively, the mole fraction α^* of surfactant 1 in the solution phase equals its mole fraction in the mixed micelle and is given by Equation 15.34:

$$\alpha^* = \frac{\ln(C_{M1}/C_{M2}) + \beta^M}{2\beta^M} \tag{15.34}$$

The minimum or maximum cmc value of the mixture is given by Equation 15.35:

$$C_{M,min} = C_{M1} \exp\left\{\beta^M \left[\frac{\beta^M - \ln(C_{M1}/C_{M2})}{2\beta^M}\right]^2\right\} \tag{15.35}$$

15.1.5 SYNERGISM IN SURFACE OR INTERFACIAL TENSION REDUCTION

The surface or interfacial tension reduction effectiveness exists when the mixture of two surfactants at the cmc reaches a lower surface or interfacial tension value compared to those obtained at their individual cmcs. The conditions for the synergism in surface tension reduction are obtained through the conditions represented in Equations 15.36 and 15.37:

$$\beta^\sigma - \beta^M < 0 \tag{15.36}$$

$$\left|\beta^{\sigma} - \beta^{M}\right| > \left|\ln\left(\frac{C_1^{0,\text{CMC}} C_{M2}}{C_2^{0,\text{CMC}} C_{M1}}\right)\right| \qquad (15.37)$$

In case of antagonism or negative synergism (Equations 15.37 and 15.38):

$$\beta^{\sigma} - \beta^{M} > 0 \qquad (15.38)$$

It is apparent from Equation 15.36 that synergism in surface tension reduction effectiveness can occur only when the attractive interaction between the two surfactants in the mixed monolayer at the interface is stronger than that in the mixed micelle in the solution phase. When the attraction between the two surfactants in the mixed micelles is stronger than in the mixed monolayer, it is possible for antagonism of this type to occur.

15.1.6 The Mass Action Model

The mass action model is a thermodynamic model for treating a mixed surfactant system simply and more fundamentally. In this model, the micellization process is considered as a reversible chemical reaction. The reactants are the surfactants and the bound counterions. Also, in this model, the standard free energy associated with micellization and dissociation of micelles formed is related to the standard state in which the components are dispersed at infinite dilution. This model is suitable because it considers that micelle formation is reversible. The degree of binding of surfactant molecules in the micellar form will depend on the aggregation number, binding of counterions, and the ratio of each surfactant in the initial solution and the micelles formed. The incorporation of the aggregation number in the equation of the state of this model addresses important issues such as the distribution of micellar shapes and sizes, the average aggregation number (considering the association–dissociation mechanism of the micelles formed), and the range of dispersity in mixed micelles.

In this model, the micelle is considered to be composed of g_i surfactant molecules of type i having a valence z_i and βg_s counterions. The chemical reaction representing the micelle formation is as follows (Equation 15.39) [23]:

$$g_1 A_1 + g_2 A_2 + \cdots + g_n A_n + \beta g_s \rightarrow \text{Micelle} \qquad (15.39)$$

The law of mass action is represented by Equation 15.40:

$$\frac{[X_m]}{[X_1]^{g_1} [X_2]^{g_2} \cdots [S]^{\beta g_s}} = \exp\left(-\frac{\Delta F^*}{RT}\right) \qquad (15.40)$$

where X_m is the mole fraction of micelles, X_1 is the mole fraction of surfactant (1), and S is the counterion mole fraction.

ΔF^* is the standard free energy associated with micelle formation from the dispersed amphiphiles at low concentration. ΔF^* depends on the shape of the micelles;

thus, it will include surface and volume contributions, in addition to dependence on the bound counterions.

In another approach to deal with the formed micelles as a phase formed from the main component of the surfactant molecules, which is the hydrophobic part, the contribution of the counterions is removed to produce the equilibrium equation 15.41:

$$\frac{[X_m]}{[X_1]^{g_1} [X_2]^{g_2} \dots [X_n]^{g_n}} = \exp\left(-\frac{\Delta F^\circ}{RT}\right) \qquad (15.41)$$

Equations 15.40 and 15.41 are comparable to each other and must yield the same mole fractions of the mixed micelles formed. Also, the values ΔF^* and ΔF° are different but related to each other.

Several studies deal with mixed surfactant systems, including cationic–cationic, cationic–nonionic, cationic gemini–cationic, and cationic gemini–nonionic mixed systems.

15.1.7 ANIONIC SURFACTANTS IN MIXED SURFACTANT SYSTEMS

The mixed micellar behavior of anionic surfactant sodium dodecyl sulfate (SDS) and cationic surfactant dodecyl ethyl dimethyl ammonium bromide (DDAB) at varying mole fractions of DDAB, that is, α_{DDAB} in aqueous solution of papain protein, has been investigated with the aid of spectroscopy and physicochemical measurements [24]. Thermodynamic parameters have been computed over the entire mole fraction range of DDAB. The Clint equation [10] and the regular approximation method have been used to investigate the interactions between mixed surfactants in the presence of papain protein. The two surfactants have similar tails; however, the charges on respective polar head groups are expected to have significant effects on their colloidal behavior. The cmc values of mixed surfactants have been estimated from fluorescence, conductivity, surface tension, and density measurements. The results show that lower and higher mole fractions of DDAB give negative deviations from ideality whereas intermediate mole fractions have a positive deviation. The aggregation number, N_{agg}, of mixed micelles has also been calculated. The turbidity or cloudiness at intermediate mole fractions demonstrates a reduction in the free monomer concentration due to neutralization of monomers of opposite charge. The effect of concentration of papain on mixed micellar behavior indicates that increasing the concentration of papain protein causes the critical aggregation concentration (cac) and cmc values to increase. The unfolding of papain protein polypeptide chain in the presence of a mixed surfactant has been observed [25].

The adsorption of sodium dodecyl sulfate, a polyethoxylated nonyl phenol, and well-defined mixtures thereof was measured on γ-alumina. A pseudophase separation model to describe mixed anionic–nonionic micelle formation, similar to the pseudophase separation model frequently used to describe mixed micelle formation, was developed. In this model, the regular solution theory was used to describe the anionic/nonionic surfactant interactions in the mixed micelle and a patch-wise adsorption model was used to describe surfactant adsorption on a heterogeneous solid surface [24].

The role of dipalmitoylphosphatic acid (DPPA) as a transfer promoter to enhance the Langmuir–Blodgett (LB) deposition of a dipalmitoyl phosphatidyl choline (DPPC) monolayer at air/liquid interfaces was investigated and the effects of Ca^{2+} ions in the subphase were discussed [26]. The miscibility of the two components at air/liquid interfaces was evaluated by surface pressure–area per molecule isotherms, thermodynamic analysis, and by direct observation using Brewster angle microscopy (BAM). Multilayer LB deposition behavior of the mixed DPPA/DPPC monolayers was then studied by transferring the monolayers onto hydrophilic glass plates at a surface pressure of 30 mN m^{-1}. The results showed that the two components, DPPA and DPPC, were miscible in the monolayer on subphases of both pure water and 0.2 mM $CaCl_2$ solution. However, an exception occurred at mole fractions of DPPA surfactant at $X_{DPPA} = 0.2$ and 0.5 at air/$CaCl_2$ solution interface, where a partially miscible monolayer with phase separation occurred. Negative deviations in the excess area analysis were found for the mixed monolayer system, indicating the existence of attractive interactions between DPPA and DPPC molecules in the monolayer. The monolayers were stable at the surface pressure of 30 mN m^{-1} for certain LB deposition as evaluated from the area relaxation behavior. It was found that the presence of Ca^{2+} ions had a stabilization effect for DPPA-rich monolayers, probably due to the association of negatively charged DPPA molecules with Ca^{2+} ions. Moreover, the Ca^{2+} ions may enhance the attachment of DPPA polar groups to the glass surface and also the interactions between DPPA polar groups in the LB multilayer structure. As a result, Y-type LB multilayer containing DPPC could be fabricated from the mixed DPPA/DPPC monolayers in the presence of Ca^{2+} ions [26].

Mixtures of sodium dioctyl sulfosuccinate (AOT) and sodium dodecyl sulfate (SDS) that were studied in water at 25°C by using surface tension, conductance, emf, and fluorescence emission methods exhibit synergism in the region where the mole fraction of AOT in the bulk solution (α_1) is less than 0.7 and ideality in the region where $\alpha_1 > 0.7$ [27]. The molal conductance versus concentration behavior of an aqueous solution of AOT is found to be different from that of other ionic surfactants with the exception of bile salts. Composition of the mixed micelle was evaluated and discussed using Rubingh's and Rodenas–Valiente–Villafruela (RVV) treatments [28]. The values of the counterion binding constant determined from the emf data show that the counterion binding behavior of the mixed micelle is controlled solely by AOT. The free energy for mixed micelle formation was calculated using Equation 15.42 [28]:

$$\Delta G_m^\circ = RT\left[(1 + \beta_c)\ln c_m + x_1 \ln \alpha_1 + x_2 \ln \alpha_2\right] \tag{15.42}$$

where α_1 and x_1 are the mole fraction and counterion binding of AOT, α_2 and x_2 are the mole fraction and counterion binding of SDS in the bulk solution, β_c is the interaction parameter between the two surfactant molecules, and ΔG_m° is the free energy of mixing.

The aggregation number determined by the fluorescence quenching method indicated that the number of AOT molecules remains constant and that of SDS decreases in the mixed micelle. Characteristics of the adsorption layer of the mixed surfactant system were also examined using the theoretical treatment of Rosen and Hua [28].

The effect of 2,2,2-trifluoroethanol (TFE) on micellar properties of Triton X-100 (TX-100) in aqueous solutions was investigated by cloud point (C_P), viscosity, surface tension, and fluorescence techniques [30–32]. The cmc values of the corresponding mixtures were obtained by the pyrene 1:3 ratio method and by surface tension data using the pendant drop technique. All the techniques provided about the same values for the cmc. Up to 0.83 M TFE increased the cmc by 30%. This small increase in the cmc is consistent with slight increase in the solubility of TX-100 in water. Fluorescence measurements indicate that the TFE decreased the aggregation number by about 30%. C_P decrease and intrinsic viscosity increase with TFE concentration are consistent with a preferential interaction of TFE with TX-100 micelles. TFE molecules form hydrophobic domains in the micellar palisade layer because of the hydrogen bond with the oxyethylene groups in TX-100. The intrinsic viscosity data are consistent with an increase in micelle hydrodynamic radius owing to the presence of TFE [29].

The mixing behavior of binary mixtures of the alkyl glucosides (C_nG)—octyl β-D-glucoside and decyl D-glucoside—in combination with sodium oleate (NaOl), and the amine oxide surfactants (AO)—N,N-dimethyl dodecyl amine oxide, N,N-bis(2-hydroxy ethyl) dodecyl amine oxide, and 3-lauramidopropyl-N,N-dimethylamine oxide—in combination with NaOl were investigated [30]. From the equilibrium surface tension measurements, the cmc data were obtained as functions of composition. Values of the cmc were analyzed according to both the regular solution model developed by Rubingh [14] for mixed micelles and Maeda's formulation [E] for ionic/nonionic mixed micelles. Two interaction parameters, β and $β_1$, were estimated from the regular solution model and Maeda's formulation, respectively. For NaOl/C_nG mixed systems, a decrease in the hydrocarbon chain length of C_nG resulted in a stronger interaction with NaOl, based on β and $β_1$ values. For NaOl/AO mixed systems, the bulkiness of the polar head groups of AO surfactants influenced the interaction between NaOl and AO. The dynamic surface tension measurements show that surface tension values of all surfactant solutions examined decreased with time. Time dependence of surface tension values for NaOl mixed systems was greatly influenced by the presence of NaOl rather than by the other component [30].

The mixture of the anionic O,O-bis(sodium-2-laurate)-p-benzenediol ($C_{11}p$PHCNa) and cationic alkanediyl-α,ω-bis(dimethyl dodecyl ammonium bromide) (C_{12}-2-E_x-C_{12} · 2Br) gemini surfactants has been investigated by surface tension and pyrene fluorescence [31]. The results show that the surface tension γ drops faster with total surfactant concentration C_T for $α_1 = 0.1$ or 0.3 than for $α_1 = 0.7$ or 0.9, where $α_1$ is the mole fraction of $C_{11}p$PHCNa in the bulk solution. The fast drop in γ for $α_1 < 0.5$ indicates strong adsorption at the air/water interface owing to the interaction between oppositely charged components, resulting in the formation of an adsorption double layer in the subsurface. The slow decrease in γ for $α_1 > 0.5$ is attributed to the preaggregation in the solution before the cmc.

The effect of NaBr on the aqueous two-phase regions of 0.10 mol kg^{-1} cetyl trimethyl ammonium bromide (CTAB)/sodium dodecyl sulfate (SDS)/H_2O mixed system at 45°C has been investigated [32]. The aqueous two-phase region with excess SDS or the aqueous two-phase region with excess CTAB move far from the equimolar line with the addition of NaBr. The effect of NaBr on shear viscosities of the systems at the two isotropic single-phase regions has also been studied. At each isotropic single-phase region, a single shear viscosity peak was observed for each m_{NaBr}. The addition of NaBr to the mixed CTAB/SDS/H$_2$O systems shifts the viscosity peaks away from the equimolar line, has stronger effect on the viscosity peak at the isotropic single-phase region with excess CTAB, and weaker effect on viscosity with excess SDS. Seven samples with 0.10 mol kg^{-1} CTAB/SDS/H$_2$O isotropic systems adjacent to the viscosity peaks were chosen to investigate the effect of NaBr on shear viscosity. The rheological behaviors as well as the shear viscosity values of the seven chosen systems all changed with the addition of NaBr. Salt-induced aqueous two-phase systems can be obtained with the addition of NaBr to the seven chosen systems. The common tendency of the seven chosen systems was the decrease of shear viscosities with the addition of NaBr and close to the phase boundaries, all fluids are thixotropic. These results indicate that the salt-induced aqueous two-phase systems originate from thixotropic isotropic systems [32].

Significant surface tension reduction (STR) of water was observed [33] for [Co(NH$_3$)$_6$](ClO$_4$)$_3$, [Co(en)$_3$](ClO$_4$)$_3$, [Co(bpy)$_3$](ClO$_4$), and [Co(phen)$_3$](ClO$_4$)$_3$ surfactants at a concentration range of 1.25–5.00 mM in the presence of sodium dodecyl sulfate (SDS) and sodium dodecyl benzene sulfonate (SBS). This suggests the formation of 1:1 and 1:2 association complexes. These complexes are {[complex]$^{3+}$(S$^-$)}$^{2+}$ and {[complex]$^{3+}$(S$^-$)}, where [complex]$^{3+}$ = [Co(NH$_3$)$_6$]$^{3+}$, [Co(en)$_3$]$^{3+}$, [Co(bpy)$_3$]$^{3+}$, or [Co(phen)$_3$]$^{3+}$, S$^-$ = DS$^-$ or BS$^-$. The effect of [Co(en)$_3$]$^{3+}$ on STR in SDS–water system is the largest due to a strong hydrophilic interaction between amino protons of [Co(en)$_3$]$^{3+}$ and sulfate oxygen atoms of DS$^-$. The effects of [Co(en)$_3$]$^{3+}$, [Co(bpy)$_3$]$^{3+}$, and [Co(phen)$_3$]$^{3+}$ on STR in SBS–water system are significant and almost the same. This means that the hydrophilic interaction between [Co(en)$_3$]$^{3+}$ and the sulfonate group is comparable to the hydrophobic interaction between [Co(bpy)$_3$]$^{3+}$ or [Co(phen)$_3$]$^{3+}$ and the phenyl group of BS$^-$. The Co(III) complexes of 1.25–5.0 mM are precipitated as {[complex]$^{3+}$(S$^-$)} at 0.029–0.173 mM of S$^-$. The precipitates {[Co(bpy)$_3$]$^{3+}$(S$^-$)$_3$} and {[Co(phen)$_3$]$^{3+}$(S$^-$)$_3$} can be dissolved at a molar ratio of [S$^-$]/[complex^{3+}] > 3.5 for SDS and >4.0 for SBS. This observation suggests that aggregated micelle [Co(bpy or phen)$_3$]$_2$(DS)$_7^-$ or [Co(bpy or phen)$_3$](BS)$_4^-$ is formed.

A study [34] describes the synthesis of three novel anionic gemini surfactants via a convenient and easily controllable method. The surfactants are (a) sodium-2,2-(6,6-(ethane-1,2-diylbis(azanediyl) bis(4-(octylamino)-1,3,5-triazine-6,2-diyl) bis(azanediyl) diethanesulfonate (C$_8$-2-C$_8$), (b) sodium-2,2-(6,6-(propane-1,3-diylbis (azanediyl) bis(4-(octylamino)-1,3,5-triazine-6,2-diyl) bis (azanediyl) diethane sulfonate (C$_8$-3-C$_8$), and (c) sodium-2,2-(6,6-(butane-1,4-diylbis (azanediyl) bis(4-(octylamino)-1,3,5-triazine-6,2-diyl) bis(azanediyl) diethane sulfonate (C$_8$-4-C$_8$) (Scheme 15.1). The interactions of these new anionic gemini surfactants with the conventional cationic cetyl trimethyl ammonium bromide (CTAB) are investigated in 0.1 mol L^{-1} NaCl aqueous solutions. The mixed systems are C$_8$-2-C$_8$/CTAB, C$_8$-3-C$_8$/CTAB, C$_8$-4-C$_8$/CTAB, and the mole factions (α_G) of C$_8$-n-C$_8$ (n = 2, 3, 4) were 0.5, 0.7, 0.9, respectively. These mixtures exhibit synergism in both surface tension reduction effectiveness and surface tension reduction efficiency. When α_G = 0.5, the three systems exhibit synergism in mixed micelle formation, whereas the other mole fractions do not show this synergism. The interactions in mixed adsorbed films are stronger than those involved in mixed micelles for all mixtures.

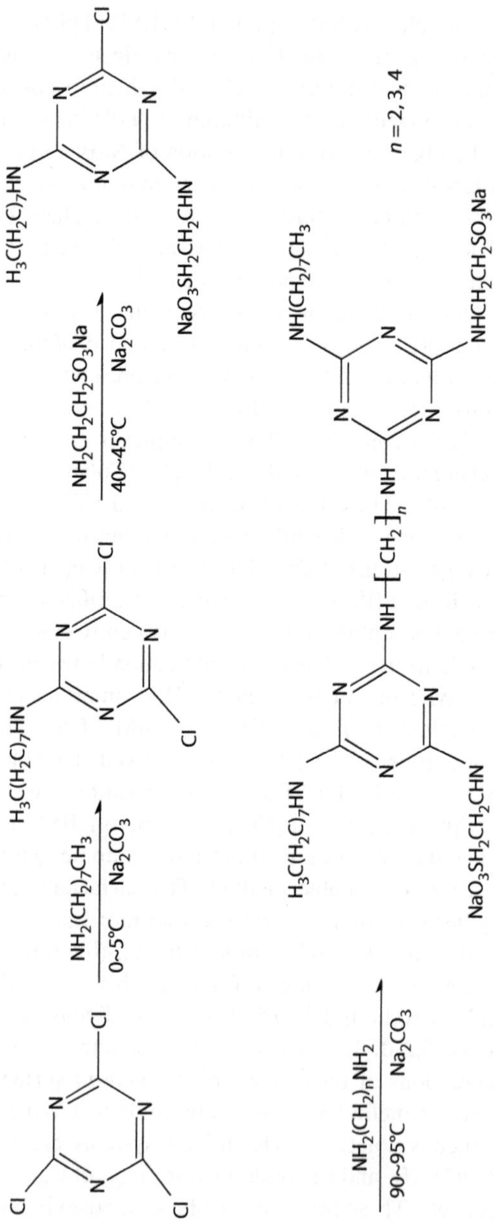

SCHEME 15.1 Surfactants used in study of Ref. [34].

15.1.8 CATIONIC SURFACTANTS IN MIXED SURFACTANT SYSTEMS

A compound of flavonol-based biosurfactant, C_8-substituted alkyl ammonium ethyl rutin (C_8AAER) for a potential pharmaceutical or agrochemical use, was prepared experimentally. The surface behavior of C_8AAER and its mixture with lecithin from soybean (LSB) had been studied [35]. C_8AAER, which has a pseudoamphoteric character, exhibits both liquid-condensed (LC) and liquid-expanded (LE) phase whereas LSB exhibits the form of LE phase only. The phase parameters of C_8AAER (including A_{limt}, π_{coll}) are observed to strongly depend on both the subphase temperature and pH, which regulate the degree of ionization. In addition, the observed positive deviation calculated from excess Gibbs free energies of the C_8AAER-LSB system suggests a repulsive interaction between C_8AAER and LSB at all mole fractions (X). Also, the interaction parameter is found to increase linearly with surface pressure, regardless of composition. Notably, the relationship of logarithmic activity coefficient versus X^2 reveals that the molecular interaction of C_8AAER-LSB can be adequately simulated using a simple regular mixture model. Importantly, lower C_s^{-1} values of this mixture relative to pure C_8AAER and LSB denote weak elasticity of mixed monolayers at mole fractions of 0.2 ~ 0.8. This indicates that direct addition of C_8AAER may exert an unfavorable influence on LSB membranes [35].

An extensive study of the foaming properties of a surfactant mixture consisting of the nonionic dodecyl dimethyl phosphine oxide (C_{12}DMPO) and the cationic dodecyl trimethyl ammonium bromide (C_{12}TAB) with molar mixing ratios of C_{12}DMPO:C_{12}TAB = 1:0, 5:1, 1:1, 1:5, 0:1 has been conducted above and below the cmc [36]. Foamability and foam stability were examined using the commercially available Foam Scan (Sparging), the standardized Ross-Miles (pouring), and a home-built winding (shaking) techniques. The focus, however, was on Foam Scan measurements since it allowed for evaluation of the foams liquid content. The foamability and foam stability of C_{12}TAB were found to be larger than that of C_{12}DMPO. The foamability continually increased with increasing C_{12}TAB content in the surfactant mixture, which reflects a reduction of the diffusion relaxation time (i.e., faster adsorption). Possible correlations were drawn between the foam properties and previously studied adsorption and foam film properties. Interestingly, the 1:1 mixture shows weak/negligible surfactant interactions but counter intuitively an increased foam stability compared to the single surfactant systems. However, at this ratio, charge neutralization occurs, which leads to the formation of a Newton Black Film, thus suggesting that the foam film type plays an important role in foam stability.

Micellar properties of binary mixtures of hexadecyl diethyl ethanol ammonium bromide surfactant with tetradecyl dimethyl ammonium, trimethyl ammonium, triphenyl phosphonium, diethyl ethanol ammonium, and pyridinium bromide surfactants have been characterized using conductometric and fluorescence techniques [37]. The cmc and the degree of counterion binding (δ) in the binary systems were determined from conductivity measurements. The results were analyzed using various existing theories to calculate micellar composition, activity coefficients, and the interaction parameter (β). Partial contribution of each surfactant, cmc_1, cmc_2, to the overall cmc value was also evaluated. Aggregation numbers and micropolarity of the mixed micelles were determined from fluorescence measurements. The results were discussed in terms of synergetic interaction in these systems on the basis of the head group–head group and tail–tail interactions and counterion binding.

The cmc values of cetyl pyridinium chloride (CPC) were determined in the presence of salicylate and benzoate ions in the less explored concentration region where the viscosity is Newtonian [38]. The cmc of CPC decreased from 9×10^{-4} to 7×10^{-7} and 3×10^{-6} mol kg^{-1} by adding about 0.3 mol kg^{-1} of salicylate and benzoate, respectively. The ortho hydroxyl group in salicylate thus has a remarkable influence on the micellization of CPC and the extent of this favorable effect is found to be about 3.5 kJ mol^{-1}. The Corrin–Harkins equation [35] was modified to explain the variation of cmc with electrolyte concentration in the presence of mixed counterions. The slope of this equation does not provide the value of the total counterion binding constant (δ), but gives information about the lower limit to the value of δ, which is found to be 0.66. Addition of salicylate and benzoate increases counterion binding to CPC micelles compared to that in the presence of chloride alone. An adsorption isotherm was derived to estimate the surface excess of CPC in the presence of mixed counterions.

The behavior of mixed cationic gemini–nonionic surfactant systems in solution and at interface has been studied by surface tension measurements. Cationic geminis used were alkanediyl-ω-bis(dimethylhexadecyl ammonium bromide) and nonionic surfactants were the Tweens (Tween 20, 40, 60, and 80). The cmc values of the mixtures are lower than the values of individual surfactants and decrease with the increase in stoichiometric mole fraction of Tweens in the solution. This shows non-ideal, synergistic mixing of the two components. Other parameters calculated were cmc at ideal mixing conditions (cmc$_{id}$), surface excess concentration, and minimum area per surfactant head group. Gibbs energy of micellization and adsorption, mole fraction of surfactants in mixed micelles and monolayers, and interaction parameters were also evaluated [39].

From plots of surface tension (γ) as a function of solution composition and total surfactant concentration, we determined the cmc, minimum surface tension at the cmc (γ_{cmc}), surface excess (Γ_{max}), and mean molecular surface area (A_{min}). On the basis of the regular solution theory, the compositions of the adsorbed film (Z) and micelles (X^M) were estimated, and then the interaction parameters in the micelles (β^M) and in the adsorbed film phase (β^σ) were calculated. For all mole fraction ratios, the results showed synergistically enhanced ability to form mixed micelles as well as surface tension reduction. Furthermore, β was calculated by considering nonrandom mixing and head group size effects. It was observed that, for both the plain air/aqueous interface and micellar systems, the nonideality decreased as the amount of electrolyte in the aqueous medium was increased. This was attributed to a decrease of the surface charge density caused by increasing the concentration of bromide ions.

The physicochemical properties of mixed surfactant aqueous solutions at various proportions of dodecyl trimethyl ammonium bromide (DTAB) and undecafluoro-n-pentyl decaoxy ethylene ether (C$_5$F$_{11}$EO$_{10}$) have been investigated at 25°C using surface tension and conductivity measurements [40]. The cmc and minimum surface tension have been experimentally estimated for different DTAB molar fractions. The micellar composition and mutual interaction parameters have been deduced on the basis of theoretical treatments proposed by Clint [10] and Rubingh [14]. The results show a significant synergistic effect at about equimolar $DTAB/C_5F_{11}EO_{10}$ system, probably due to the efficient electrostatic self-repulsion reduction between DTAB cationic head groups related to the presence of the nonionic surfactant in the mixed micelles.

Measurements of the surface tension of aqueous solutions were carried out at 20°C for mixtures of cetyl trimethyl ammonium bromide (CTAB) with short-chain alcohols such as methanol and ethanol, as well as for 1-hexadecyl pyridinium bromide (CPyB) with the same alcohols [41]. The concentration of CTAB and CPyB in aqueous solutions was in the range from 10^{-5} to 10^{-3} M. Methanol and ethanol concentrations were in the range from 0 to 21.1 mM and from 0 to 11.97 mM, respectively. Moreover, the surface tension of aqueous solution mixtures of cationic surfactants with propanol in the concentration range from 0 to 6.67 mM was also investigated. The resulting isotherms of surface tension were compared to those calculated from the Szyszkowski and Connors equations [41,42]. The constants in these equations were determined by the least squares method. It appeared that the values of the constants depend on the type of surfactant and alcohol. From comparison of the experimental versus theoretical isotherms for surface tension, it was possible, in first approximation, to describe the relationship between the surface tension of mixtures of aqueous solutions of cationic surfactants and short-chain alcohol as a function of alcohol molar fraction in the bulk phase using Szyszkowski and Corrin's equations [41,42]. Furthermore, changes in the surface tension of aqueous solutions of CTAB and CPyB with alcohol mixtures at each constant concentration of cationic surfactant were predicted by the Fainerman and Miller equation [48]. Based on the surface tension isotherms, Gibbs surface excess concentration of cationic surfactants and alcohols at water–air interface was determined, and in the case of alcohol, this concentration excess was recalculated for Guggenheim–Adam. The Guggenheim–Adam surface excess concentration was applied to the determination of the real concentration of alcohol in the mixed surface monolayer. The real concentration of cationic surfactant was assumed equal to Gibbs surface excess concentration [42,43].

The composition of the surface mixed monolayer was discussed with regard to the standard free energy of cationic surfactant and alcohol adsorption at water–air interface determined by different methods. The standard free energy of adsorption of "pure" cationic surfactants determined from the Langmuir and Aronson and Rosen equations [44] was compared to that deduced on the basis of C_p, and the surface tension of the cationic surfactant tail and tail–water interface tension.

Miscibility and interaction of decyl dimethyl phosphine oxide (DePO) with ammonium chloride (AC), hexyl ammonium chloride (HAC), and dodecyl ammonium chloride (DAC) in adsorbed films and micelles were studied by surface tension measurements [45]. Phase diagrams were drawn for the mixed adsorption, mixed micelle formation, and equilibrium between adsorbed films and micelles. Nonideal mixing of DAC and DePO was characterized by a negative excess Gibbs free energy and positive excess area of adsorption and negative excess Gibbs free energy of micelle formation. It is concluded that the interaction between DAC and DePO in adsorbed films and micelles is larger than that between the same surfactants alone due to two factors: (a) ion–dipole interactions between the head groups of DAC and DePO and (b) alkyl chain-alkyl chain interactions.

Aggregation and cloud point (C_p) behavior of lanthanide ions have been studied for mixtures of Triton X100 (TX100) with novel nonionic cyclophanic surfactants with varying lengths of polyoxyethylene and hydrophobic moieties (C_nE_m) based on the calixarene platform. The dynamic light scattering data reveal the contribution of the large size lamellar or stack-like mixed aggregates in C_nE_m-TX100 solutions. Aggregation and C_p behavior of TX100-C_nE_m mixed solutions are quite different from those of conventional nonionic surfactants. The effect of the hydrophobic substituent and polyoxyethylene chains length of C_nE_m on the C_p extraction of La(III), Gd(III), and Lu(III) in the mixed TX100-C_nE_m micellar solutions is discussed in relation to their aggregation and cloud point behavior. The result indicates that the cyclophanic structure of C_nE_m is the key reason for the formation of large lamellar-like aggregates with TX100, exhibiting the unusual C_p behavior [46].

The interaction in two mixtures of a nonionic surfactant AEO9 ($C_{12}H_{25}O(CH_2CH_2O)_9H$) with two ionic surfactants was investigated [47]. The two mixtures were AEO9/sodium dodecyl sulfate (SDS) and AEO9/cetyltrimethylammonium bromide (CTAB) at a 0.5 molar fraction of AEO9. The surface properties of the surfactants, cmc, effectiveness of surface tension reduction (γ_{cmc}), maximum surface excess concentration (Γ_{min}), and minimum area per molecule at the air/solution interface (A_{min}) were determined for both individual surfactants and their mixtures. Significant deviations from ideal behavior (attractive interactions) of the nonionic/ionic surfactant mixtures were observed. Mixtures of both AEO9/SDS and AEO9/CTAB exhibited synergism in surface tension reduction efficiency and mixed micelle formation, but neither exhibited synergism in surface tension reduction effectiveness.

The synergism and foaming behavior of a mixed surfactant system consisting of a nonionic surfactant (polyethoxylated alkyl ether C_nE_m) and a fatty acid soap (sodium oleate) were studied [48]. The micellar interaction parameter (the β-parameter) was determined from the cmc following the approach of Rubingh's regular solution theory [14]. For both the $C_{12}E_6$/sodium oleate and the $C_{14}E_6$/sodium oleate mixtures, the results indicate a fairly strong attractive interaction (negative β-values), which were in agreement with previous data reported for other nonionic/anionic surfactant systems. The characteristics of the foam produced from the surfactants were evaluated using a glass column equipped with a series of electrodes to measure the conductance of the foam, which enabled the water content of the foam to be determined. The total foam volume was almost the same for all concentrations and surfactants. The amount of liquid in the foam at 100 s after the air flow was turned-off decreased in the order NaOl > $C_{12}E_6$ > $C_{14}E_6$. Also, the mixtures had the same foam volumes as pure surfactants at the same concentration. However, both mixtures had higher concentrations of liquid in the foam when the mole fraction of the nonionic surfactant in the solution of the mixed surfactant system was greater than about >0.3.

The influence of adding hydrophobic and hydrophilic nonionic surfactants and a mixture of equal parts of the two on the rheological properties of a mixture of equal parts of microcrystalline cellulose and ibuprofen with water was assessed by capillary rheometry [49] at 5% and 25% concentration. The mixtures were also used to form pellets by extrusion/spheronization and their in vitro dissolution in simulated intestinal fluid was measured. As with previous rheological studies of these types of pastes, their flow behavior was non-Newtonian (shear thinning). Other rheological parameters were determined in terms of extensional flow and elastic parameters of recoverable shear and compliance. By comparison with previous studies with the model drug, it was determined that these properties were indicative of the ability of the formulations to produce satisfactory pellets. The investigations also identified that the concentration but not the type of surfactant determined the rheological properties of the wet mass. All the formulations produced round pellets with a narrow size distribution, whereby the median size increased with the concentration of the surfactant. The formulations with 25% level of each surfactant provided a rapid drug release (100% within 30 min), where the formulations with 5% or mixed surfactants provided lower drug release profiles.

The interfacial and micellization properties of various compositions, binary and ternary mixtures of gemini homologs surfactants, viz., tetramethylene-1,4-bis(N-hexadecyl-N,N-dimethyl ammonium bromide), pentamethylene-1,5-bis(N-hexadecyl-N,N-dimethyl ammonium bromide), and hexamethylene-1,6-bis(N-hexadecyl-N,N-dimethyl ammonium bromide), with various compositions were investigated at 25°C using conductometric and tensiometric methods [50]. The micellar and adsorption characteristics such as composition, activity coefficients, mutual interaction parameter, minimum area per molecule, free energy of micellization, and adsorption were evaluated and compared. The ideality/nonideality of the mixed micelles was tested using Clint [10], Rubingh [14], and Rubingh–Holland approaches [2,15]. The mixed systems were found to undergo synergistic interaction, more so in the mixed monolayer than in mixed micelle formation. The micellar and interfacial properties were found to depend on the nature of the surfactants that were used.

The micellization of the gemini surfactant pentamethylene-1,5-bis(tetradecyl dimethyl ammonium bromide) (14-5-14) has been investigated in water and water–organic mixed solvent media [51]. The organic solvents used were 1,4-dioxane (DO), dimethyl formamide (DMF), and ethylene glycol (EG). The conductivity in water and in mixed media (water + organic solvent) was measured at different temperatures. The data were used to estimate the cmc and degree of counterion dissociation (a) of the micelle. The study showed that the micellization tendency of the surfactant decreased in the presence of organic solvents. Also, the increase in the cmc values was comparatively low below 20% (v/v) of organic solvents showing the predominance of water character in the bulk phase at lower compositions of the organic solvents. Thermodynamic parameters were also obtained from the temperature dependence of the cmc values. The standard free energy of micellization was found to be negative in all the cases, and became less negative with increase of the cosolvent content.

The nature and strength of the interactions between the 1,3-bis(dimethyl hexadecyl) propane diammonium dibromide (16-3-16) gemini surfactant and a series of nonionic polyoxyethylene (20) sorbitan ester surfactants was investigated [52]. The sorbitan surfactants had laurate (Tween 20), stearate (Tween 60), or oleate (Tween 80) alkyl tails. The cmc values of the mixed gemini-Tween systems were determined using the dü-Nouy ring surface tension method, and the results have been analyzed using Clint [10], Rubingh [14], and Motomura [52] theories for mixed micellar systems. The results demonstrate a synergistic mixing behavior between the Tween surfactants and the gemini surfactant, where the strength of interaction was dependent on the chain length and saturation of the Tween alkyl tail.

Mixed micellization of dimeric cationic surfactants with monomeric cationic surfactants was investigated [53]. The dimmer surfactants were tetramethylene-1,4-bis(hexadecyl dimethyl ammonium bromide) (16-4-16) and hexamethylene-1,6-bis(hexadecyl dimethyl ammonium bromide) (16-6-16). The monomeric cationic surfactants were hexadecyl trimethyl ammonium bromide (CTAB), cetyl pyridinium bromide (CPB), cetyl pyridinium chloride (CPC), and tetradecyl trimethyl ammonium bromide (TTAB). The mixture was investigated using conductivity and steady-state fluorescence quenching techniques. The behavior of mixed systems, their compositions, and activities of the components were analyzed using Rubingh's regular solution theory [14]. The results indicate synergism in the binary mixtures. Ideal and experimental cmc's (i.e., cmc* and cmc, respectively) show nonideality, which is confirmed by β values and activity coefficients. The micelle aggregation numbers (N_{agg}) were evaluated using steady-state fluorescence quenching at a total concentration of 2 mM for CTAB/16-4-16 or 16-6-16 and 5 mM for TTAB/16-4-16 or 16-6-16 systems. The result indicates that the contribution of conventional surfactants was always more than that of the gemini surfactants. The micropolarity, dielectric constant, and binding constants (K_b) of the mixed systems were also evaluated from the ratios of the respective peak intensities (I_1/I_3) [53].

Mixed micellization and surface properties of binary mixtures of cationic gemini surfactant with conventional surfactants were determined [54]. The gemini surfactant was butanediyl-α,ω-bis(dimethyl cetyl ammonium bromide) (G4, 16-4-16). The conventional surfactants were cetyl pyridinium chloride (CPC), sodium bis(2-ethylhexyl) sulfosuccinate (AOT), and polyoxyethylene-10 cetyl ether (Brij-56). The investigations were conducted using conductometric and tensiometric methods. The data were analyzed using theoretical models of Rubingh [14], Rosen [17], Clint [10], and Maeda [35]. The analysis gave the interaction parameter, minimum area per molecule, surface excess, mixed micelle composition, free energies of micellization and adsorption, and activity coefficients. The activity coefficients and experimental cmc values were less than unity, indicating synergism in micelles as well as at the interface. Also, an expansion of the minimum area per molecule was observed in the binary systems, which was supported by low values of the packing parameter.

The properties of mixed monolayers composed of the cationic gemini surfactant ($[C_{18}H_{37}(CH_3)_2N^+-(CH_2)_3-N^+(CH_3)_2C_{18}H_{37}] \cdot 2Br^{-1}$, abbreviated as 18-3-18) and stearic acid (SA) at the air/water interface were investigated with a Langmuir–Blodgett (LB) balance [55]. Excess areas for different mixed monolayer compositions were obtained and used to evaluate the miscibility and nonideality of mixing. Because of the electrostatic attractive interactions between 18-3-18 and SA, the excess areas indicated negative deviations from ideal mixing. The compounds 18-3-18 and SA were miscible at the air/water interface, as was determined by atomic force microscopy (AFM) images of the LB films transferred onto mica substrates. The attenuated total reflectance (ATR) infrared spectra showed that SA in the mixed monolayers was ionized completely at X = 0.67 and formed a "cationic–anionic surfactant," with 18-3-18 and SA owing to the electrostatic interaction between the head groups.

The adsorption and micellization behavior of novel sugar-based gemini surfactants was investigated [56]. The surfactants were N,N-dialkyl-N,N-digluconamide ethylenediamine, Glu(n)-2-Glu(n), where n is the hydrocarbon chain length of 8, 10, and 12. The investigation involved static/dynamic surface tension, fluorescence, dynamic light scattering (DLS), and cryogenic transmission electron microscopy. The static surface tension of the aqueous Glu(n)-2-Glu(n) solutions measured at the cmc was significantly lower than that of the corresponding monomeric surfactants. This suggests that the gemini surfactants, newly synthesized in the current study, were able to form a closely packed monolayer film at the air/aqueous solution interface. The strong molecular association is supported by the remarkably (approximately 100–200 times) lower cmc of the gemini surfactants compared to the corresponding monomeric ones. Using a combination of fluorescence and DLS data, it was suggested that a structural transformation of the Glu(n)-2-Glu(n) micelles occur with increase in surfactant concentration. The measurements confirm the formation of worm-like micelles of Glu(12)-2-Glu(12) at a concentration above the cmc.

Pyrene fluorescence and Kraft temperature measurements have been carried out for various combinations of cationic gemini (m-2-m) with zwitterionic surfactants by varying the length of the hydrophobic tail over the whole mixing range [57]. The results have been evaluated using the regular solution theory. All the mixtures of cationic gemini + zwitterionic surfactants indicate the presence of synergistic interactions, which largely decrease at higher hydrophobicity of both components. A greater concentration of gemini component in the mixed micelles induces stronger synergism, which decreases with increase of the chain length of the hydrophobic tail of the gemini component. Kraft temperature measurements also indicate the presence of strong synergistic interactions, which decrease with increase in the length of the hydrophobic tail of both components.

Mixed micelle formation from a mixture of cationic gemini (12-s-12, s = 4, 6) and zwitterionic (N-dodecyl-N,N-dimethyl glycine, DDG) surfactants was investigated [58]. The study involved measuring the surface tension of aqueous solution as a function of total concentration at various pH, temperature of 25°C, and atmospheric pressure. The results were analyzed by applying the regular solution theory (RST) [14]. The analysis allows for the calculation of the excess Gibbs energy of micellization purely on the basis of thermodynamic equations. The synergistic interactions of all the investigated cationic gemini + zwitterionic surfactant mixtures were found to be dependent on the pH of the solution and the length of hydrophobic spacer of gemini surfactant. The resulting excess Gibbs free energy was negative for all the systems investigated.

15.2 EXPERIMENTAL

15.2.1 MATERIALS

Surfactants used in this study were dodecyl dimethyl phenyl ammonium bromide (C_{12}DMPAB) (Sigma Aldrich, 99%) (cationic surfactant) (Scheme 15.2) and decyl dimethyl phosphine oxide (C_{10}DMPO) (Sigma Aldrich, 99.9%) (nonionic surfactant) (Scheme 15.2). The surfactants were used as obtained without further purification. The water used was bidistilled water using Pyrex apparatus. The buffer solution was a standard solution obtained from Merck.

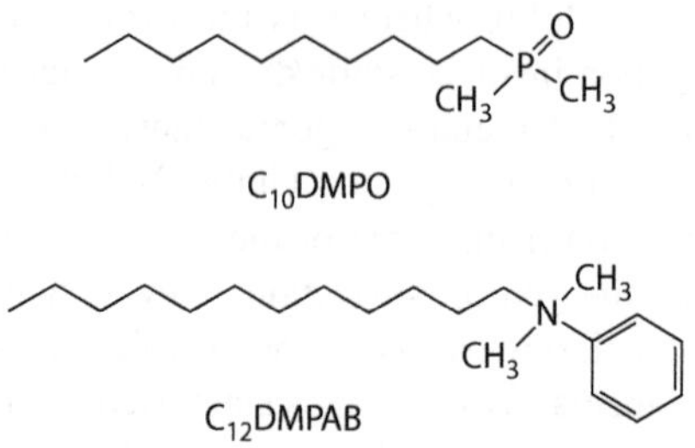

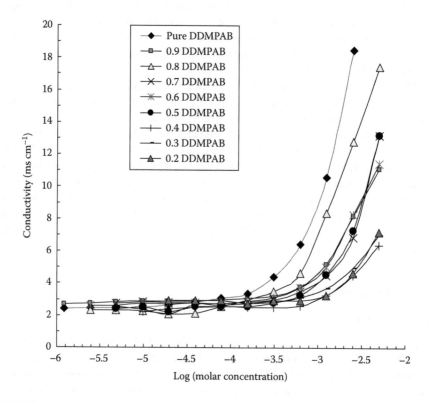

SCHEME 15.2   Surfactants under investigation.

15.2.2   SURFACE TENSION MEASUREMENTS

The surface tension (γ) was measured by the ring method using a K6 tensiometer (Krüss GmbH, Germany) at a temperature of 25°C [59]. The temperature (±0.5°C) was maintained by circulating thermostated water through a jacketed vessel containing the solution. The ring was cleaned by heating in alcoholic flame. The measurements were taken until constant surface tension values indicated that equilibrium had been reached. In all cases, the measurements were conducted for the single

FIGURE 15.2   Relation between total surfactant concentration and the conductivity of the solution at 25°C.

cationic surfactant and the different mixed systems in the presence of phosphate buffer to maintain the pH of the medium at 7 to ensure that the $C_{10}DMPO$ surfactant was in the nonionic form. The accuracy of γ measurements was within ± 0.5 mN m^{-1}.

15.2.3   CONDUCTIVITY MEASUREMENTS

Cyber Scan 320 conductivity meter (Eutech Instruments, France, sensitivity of 2 μS cm^{-1}, accuracy of 0.7%) was employed to perform the conductivity measurements at 25°C [60]. Equimolar stock solutions of the cationic and nonionic surfactants were prepared in double distilled water and then the desired mole fractions were obtained by mixing precalculated volumes of the stock solutions. The conductivity at each mole fraction was measured by successive addition of concentrated solution in pure water. A break in the conductivity versus total concentration curve signals the onset of the micellization process (Figure 15.2).

15.3   RESULTS AND DISCUSSION

15.3.1   BEHAVIORS AT THE SOLUTION–AIR INTERFACE

Figure 15.3 represents the relation between the concentration of the studied surfactant and its surface tension at 25°C. The surfactant molecules have the tendency to adsorb at the air–water interface in which the head groups are attached to the surface

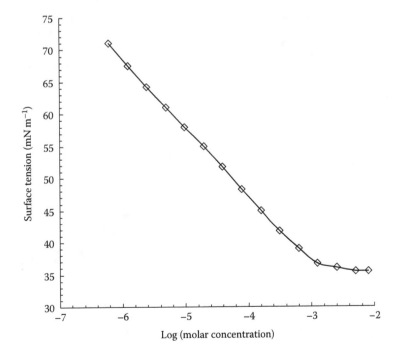

FIGURE 15.3   Surface tension versus log concentration of the pure cationic surfactant (dodecyl dimethyl phenyl ammonium bromide, $C_{12}DMPAB$) at 25°C.

by the attraction force. The hydrophobic chains are oriented in parallel position to decrease the interaction with the aqueous phase, which consequently decreases the surface tension values. As a result, increasing the concentration of the surfactants in the solution decreases the surface tension gradually, as well as the adsorbed amounts of the surfactant molecules at the interface.

Figures 15.4 through 15.5 represent the relation between the surface tension and log C of the mixed surfactant systems at various initial mole fractions of surfactant 1 at 25°C. The amount of adsorbed surfactant molecules can be calculated using Gibb's equation (Equation 15.43) [61]:

$$\Gamma_{max} = \left(\frac{1}{nRT}\right)\left(\frac{\partial \gamma}{\partial \ln C}\right) \tag{15.43}$$

The term $\partial \gamma / \partial \ln C$ is related to the migration of surfactant molecules from the bulk of the solution to the air–water interface and define the influence of concentration variation on surface tension variation. It also reflects the packing of surfactant molecules at the interface. Increase in the factor $\partial \gamma / \partial \ln C$ indicates the

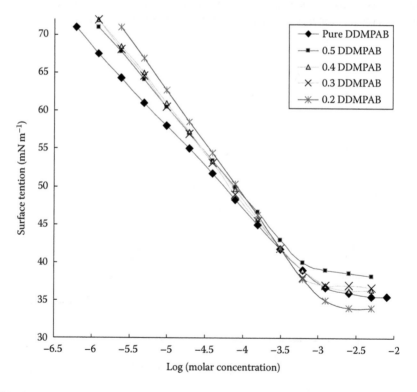

FIGURE 15.4 Surface tension versus log concentration of mixed cationic–nonionic surfactant solutions at 0.2, 0.3, 0.4, and 0.5 mole fractions of the cationic component (C_{12}DMPAB) at 25°C.

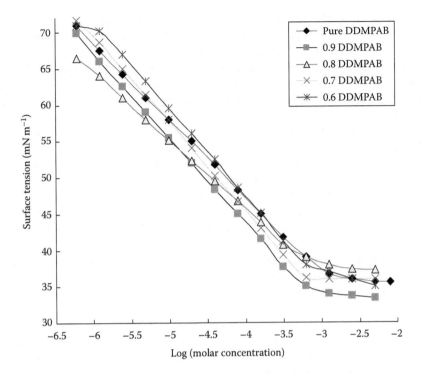

FIGURE 15.5 Surface tension versus log concentration of mixed cationic–nonionic surfactant solutions at 0.7, 0.8, and 0.9 mole fractions of the cationic component (C_{12}DMPAB) at 25°C.

increasing tendency of the surfactant molecules toward adsorption at the interface and vice versa. The integer number (n) represents the number of ionic species produced from the ionization of the surfactant molecules in the aqueous medium. In quaternary surfactant molecules, which have one positive charge and one negative counterion, n is equal to 1; while in the case of diquaternary or dianionic surfactants, n is equal to 2.

The minimum area that is occupied by the surfactant molecule at the solution–air interface can be calculated using Γ_{max} values according to Equation 15.44 [9,62]:

$$A_{min} = \frac{10^{20}}{N_A \Gamma_{max}} \qquad (15.44)$$

Γ_{max} is determined at surface saturation concentration (near the cmc) values. It is a measure of the effectiveness of surfactant adsorption, since it is the maximum value of surfactant adsorption at the interface. The adsorption effectiveness is closely related to the packing of surfactant molecules in the adsorbed film. The packing of the adsorbed film at the air–water interface is an important factor in surfactant applications such as foaming, wetting, detergency, and emulsification [63].

Data in Table 15.1 reveal that the maximum surface excess (Γ_{max}) of the cationic surfactant ($\alpha = 1$) is 0.923 mol cm^{-2}. The Γ_{max} values of the different mixed systems of C_{12}DMPAB–C_{10}DMPO are higher than that of the cationic surfactant (C_{12}DMPAB). Also, a proportional decrease of Γ_{max} values with varying mole fraction of C_{12}DMPAB component (α_1) is observed. Such behavior was observed in several recent studies of cationic (quaternary or gemini surfactants)–nonionic surfactant mixed systems [48–52]. Figure 15.6 shows the increase of the maximum surface excess at the interface with decreasing cationic surfactant mole fraction (α_1). It is clear that the maximum surface excess and the minimum surface area are inverse values to each other, which reflects the effect of packing the surfactant molecules at the interface in the area occupied by these molecules.

It is clear that the large head group of the cationic surfactant increases the area occupied by each molecule at the interface, which is responsible for the large A_{min} value of pure cationic surfactant C_{12}DMPAB. The decrease of cationic mole fraction (α_1) decreases the A_{min} values due to the following: (i) the population of C_{12}DMPAB component at the interface decreases, which decreased A_{min} values; and (ii) with the increase of C_{10}DMPO nonionic surfactant mole fraction ($1 - \alpha_1$), the interaction increases between the phosphonium group and the positively charged cationic head groups (Figure 15.6).

Several investigators suggest that branching near the charged head groups of cationic surfactant in cationic–nonionic surfactant mixtures increases the interaction between the different mixed molecules either in the mixed micelles or in the mixed monolayer formed. However, this effect is not seen in the specific case of a cationic polyethyleneoxide chain because the polyoxyethylene chains are able to form a complex with the counterions of the cationic surfactants.

In case of no interaction between the mixed components in the binary system (i.e., ideal mixing), the values of A_{ideal} in this case will be a sum of the

TABLE 15.1

Surface Parameters of the Individual and Mixed Surfactant Systems (α_1 Represents the Initial Mole Fraction of the C_{12}DMPAB Surfactant) at 25°C

α C_{12}DMPAB/ C_{10}DMPO	cmc$_{ideal}$ (mM)	cmc$_{mix}$ (mM)	γ_{cmc} mN m^{-1}	Γ_{max}/ mol m^{-2}	$A_{min(expt)}$/ Å^2	$A_{min(ideal)}$/ Å^2
0.0/1.0	2.040	2.040	35	1.568	105.9	105.9
0.2/0.8	1.220	0.832	35	1.188	139.8	120.7
0.3/0.7	1.016	0.741	38	1.179	140.8	128.1
0.4/0.6	0.870	0.655	37	1.084	153.2	135.5
0.5/0.5	0.761	0.609	40	1.075	154.4	142.9
0.6/0.4	0.677	0.568	38	1.041	159.5	150.3
0.7/0.3	0.609	0.535	36	1.039	159.8	161.7
0.8/0.2	0.553	0.498	39	1.037	160.2	165.1
0.9/0.1	0.507	0.481	35	1.017	163.3	172.5
1.0/0.0	0.468	0.468	37	0.923	179.9	179.9

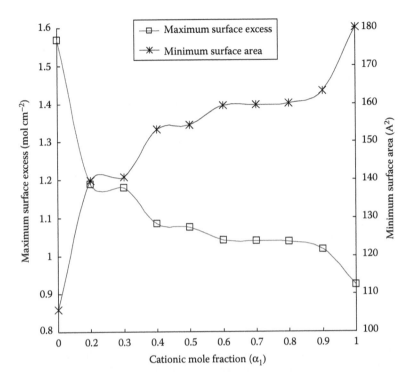

FIGURE 15.6 Variation of maximum surface excess and minimum surface area of the mixed system at different mole fractions of cationic component (C_{12}DMPAB) at 25°C.

contribution of different components depending on their initial mole fractions (α_i), see Equation 15.45:

$$A_{ideal}^{mix} = \alpha_1 A_{min}^1 + \alpha_2 A_{min}^2 \tag{15.45}$$

Deviation of $A_{min(ideal)}$ from experimental values ($A_{min(expt)}$) is due to the interaction between the mixed components. Comparison between $A_{min(ideal)}$ and $A_{min(expt)}$ values in Table 15.1 shows that $A_{min(expt)}$ values are larger than A_{ideal} values. This reveals the presence of repulsive interaction between the head groups in the mixed systems. The bulky head group of C_{12}DMPAB surfactant is responsible for the increase of $A_{min(expt)}$ values than the $A_{min(ideal)}$. These head groups are more readily accommodated at the convex micellar surface than the plain air–water interface. It is also clear that $A_{min(expt)}$ of the mixed systems at $\alpha_1 = 0.8$ and 0.9 are lower than the ideal. The lowering in $A_{min(expt)}$ values indicates a significant attractive interaction between the mixed components at the initial mole fractions mentioned (Figure 15.7). This observation was supported by the values of the interaction parameter in the interfacial mixed monolayer ($\beta°$). These values are highly negative, which indicates a positive synergism in this mole fraction region as will be seen later.

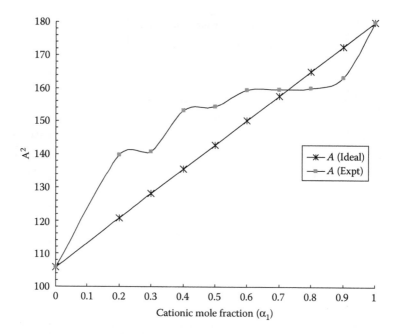

FIGURE 15.7 Ideal and experimental minimum surface areas of the mixed systems at different mole fractions of cationic component (C_{12}DMPAB) at 25°C.

15.3.2 BEHAVIOR OF MIXED MONOLAYER

The two fundamental characteristics of surfactants in solution are monolayer formation at the air–water interface and micelle formation in the bulk of solution. The mole fraction (X^σ_{ideal}) of component (C_{12}DMPAB) in the adsorbed monolayer can be calculated according to Equation 15.20.

The application of the nonideal solution theory of Rosen [20,28] using Equation 15.21 evaluates the molecular interaction parameters for the mixed monolayer of the mixed systems at the interface. The activity coefficients (f^σ_1, f^σ_2) of the surfactants in the mixed monolayer formed at the interface are related to β^σ according to Equations 15.22 and 15.23.

Solving Equations 15.20 through 15.23 yields the mixed monolayer parameters, including X^σ_1, X^σ_{ideal}, and β^σ. It is clear that less mole fractions X^σ_1 of C_{12}DMPAB are incorporated in the adsorbed monolayer at the interface than the ideally calculated (X^σ_{ideal}) values (Figure 15.8). This can be attributed to the interaction between the branched head groups, which makes their adsorption at the interface planer unfavorable. In addition, the interaction parameter of the adsorbed monolayer (β^σ) is always negative indicating the presence of some sort of attraction between the different head groups. This attraction may be due to the electrostatic interaction between the positively charged ammonium group (N^+) and the partially negative phosphonium group. At higher initial mole fractions of C_{12}DMPAB ($\alpha_1 = 0.9, 0.8, 0.7$), β^σ has large negative values, which indicates the presence of high attractive interaction between the different head groups. This result is in agreement with the lower A_{exp} values than

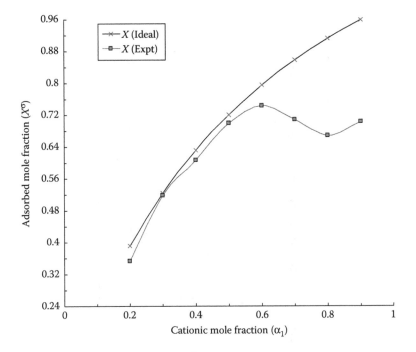

FIGURE 15.8 Variation of the mole fraction of the cationic surfactant in the adsorbed monolayer at different mole fractions of the cationic component ($C_{12}DMPAB$) at 25°C.

the expected ($A_{(min)ideal}$) at the mole fractions mentioned. The large negative values of β^σ show the existence of synergism in the mixed monolayer formation at the air–water interface.

The results of β^σ and X_1^σ are supported by the low values of activity coefficients of cationic surfactant (f_1^σ), which may be due to their low partitioning into the monolayer formed. The activity coefficients calculated with the Rubingh model [14] (Equations 15.22 and 15.23) of the regular solution theory are less than unity, indicating nonideal behavior and attractive interaction between surfactants in the mixed monolayer formed. The calculated data of the interaction parameters at the adsorbed mixed monolayer are summarized in Table 15.2.

15.3.3 MIXED MICELLES FORMATION

The cmc of pure or mixed surfactant systems formed from two or more components are determined by several techniques. The most commonly used and reproducible methods are surface tension and electrical conductivity.

If the system has two components, it is expected that two cmc values exist: the ideal value and nonideal value. The ideal value is expected when there is no interaction between different components. As described before, the value of cmc will differ depending on the type of interaction between different components.

TABLE 15.2

Interfacial and Micellar Interaction Parameters of C$_{12}$DMPAB Surfactant in the Mixed Surfactant Systems at 25°C

α C$_{12}$DMPAB/C$_{10}$DMPO	cmc$_{ideal}$ (mM)	X$^m_{ideal}$	X$^m_1$	X$^o_{ideal}$	X$^o_1$	β^m	β^o	f$_1$	f$_2$
0.0/1.0	2.040	—	—	—	—	—	—	—	—
0.2/0.8	1.220	0.521	0.512	0.391	0.354	−1.539	−0.548	0.694	0.669
0.3/0.7	1.016	0.514	0.593	0.524	0.518	−1.337	−0.864	0.801	0.625
0.4/0.6	0.870	0.593	0.656	0.632	0.608	−1.345	−0.177	0.853	0.560
0.5/0.5	0.761	0.656	0.718	0.720	0.698	−1.235	−0.265	0.906	0.529
0.6/0.4	0.677	0.718	0.774	0.794	0.742	−1.184	−0.606	0.941	0.492
0.7/0.3	0.609	0.774	0.828	0.857	0.709	−1.143	−2.166	0.967	0.457
0.8/0.2	0.553	0.828	0.870	0.911	0.668	−1.293	−4.872	0.978	0.376
0.9/0.1	0.507	0.870	0.929	0.959	0.702	−1.279	−5.644	0.994	0.332
1.0/0.0	0.468	—	—	—	—	—	—	—	—

The ideal cmc of the mixed system containing two or more surfactants can be calculated (by considering no interaction between the two components) according to Clint's equation (Equation 15.8) [10].

The experimental values of the cmc of $C_{12}DMPAB$-$C_{10}DMPO$ mixed system at different initial mole fractions extracted from the surface tension measurements are listed in Table 15.1. The data show a deviation from the ideal values calculated using Equation 15.8. Hence, Equations 15.8 and 15.9 differentiate between the ideal and experimental values of the cmc. The cmc deviation shows higher values of the cmc_{ideal} than the experimental values (Figure 15.9).

In the mixed cationic–nonionic surfactants under consideration, the cmc_{exp} is lower than the cmc of the nonionic surfactant ($C_{10}DMPO$) and is close to the cmc of the cationic component ($C_{12}DMPAB$) (Figure 15.9). The results show a synergistic effect in the micellar form due to lowering of the actual cmc values.

The data show that the cationic surfactant contributes to the micellar form more than $C_{10}DMPO$. Furthermore, the data indicate that the cationic surfactant molecules can easily partition into the mixed micelles formed. The cmc_{expt} values of the mixed systems at all the initial mole fractions studied are less than the ideal cmc values, thus showing negative deviation. This indicates synergism in the mixed micellar system. The two hydrophobic tails increase the hydrophobicity of the cationic surfactant molecules and hence the mixing is much favorable than that expected in the ideal state.

Maeda [35] stated that both head–head groups and chain–chain interactions were present in the mixed systems, which are neglected in the Rubingh approach [2,14,15].

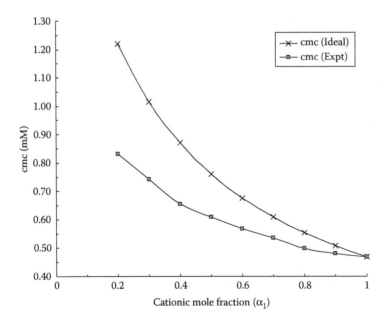

FIGURE 15.9 Ideal and experimental values of the critical micelle concentration of the mixed systems at different mole fractions of the cationic surfactant ($C_{12}DMPAB$) at 25°C.

The latter approach considers only head–head group interaction. By considering both types of interactions, the larger deviation between the experimental and ideal cmc values of ionic–nonionic mixed systems can be explained.

The ideal partitioning of the cationic surfactant in the mixed micellar form (X^m_{ideal}), which determines the value of the micellar mole fraction of C_{12}DMPAB component in the mixed micelles was calculated using Equation 15.14 (Table 15.2); whereas, the micellar mole fractions of the mixed micelles formed by two different surfactants (X^m_1) are determined using Equation 15.18.

It is clear from the data in Table 15.2 that X^m_1 is larger than X^m_{ideal}. This indicates the large contribution of the cationic surfactant to the micellar form than the non-ionic one. The difference between the ideal and nonideal micellar mole fractions indicates the presence of nonideality in the mixed micelles. It also explains why the cmc values of the mixed systems are near the cmc value of the cationic component C_{12}DMPAB.

Figure 15.10 shows the effect of the mole fraction of the cationic surfactant in solution on the ideal and the nonideal micellar mole fractions. The large difference between the two values proves the presence of negative synergism in the mixed micellar phase.

The synergism in the interaction between different components in the mixed micelles in terms of interaction parameter (β^m) can be numerically calculated using the Rubingh equation [14] (Equation 15.19). The β^m parameter is an indication of the

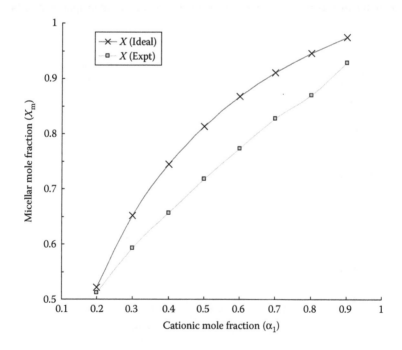

FIGURE 15.10 Ideal and experimental micellar mole fraction of the cationic component (C_{12}DMPAB) at different initial cationic mole fractions at 25°C.

degree of interaction between two different surfactants in the mixed micelles formed relative to that between the identical surfactant molecules under identical conditions. As stated before, zero value of β^m indicates ideal mixing with no change in the interaction between the different surfactant molecules. On the other hand, negative β^m values indicate attractive interaction (i.e., synergism) and positive β^m values indicate repulsive interaction (i.e., antagonism). The calculated values of β^m listed in Table 15.2 show that all mixed systems with different initial mole fractions of C_{12}DMPAB surfactant exhibit negative β^m values, that is, synergistic interaction.

The existence of synergism in mixtures containing surfactants depends not only on the strength of interaction between them but also on relevant properties of individual surfactants of the mixture [34]. The least was observed for the mixed systems containing 0.6 and 0.7 mole fractions of C_{12}DMPAB, whereas most synergism was observed for a 0.2 mole fraction of C_{12}DMPAB.

The activity coefficients, f^m, of the individual surfactants in the mixed micelles are related to the interaction parameter through Equations 15.22 and 15.23. The activity coefficient is a parameter which determines the ideality of mixing process. The deviation of f_1^m, f_2^m from unity shows the deviation of the system from ideality. On the basis of the activity coefficient values, the most ideally behaved systems relative to the cationic component, C_{12}DMPAB, are for f_1 values closest to unity. These are shown for cationic–nonionic systems having initial cationic mole fractions between 0.6 and 0.9 (Table 15.2).

15.3.4 ENERGETIC CHARACTERISTICS OF THE INDIVIDUAL AND MIXED SYSTEMS

The standard free energies of adsorption (ΔG_{ads}°) and micellization (ΔG_{mic}°) of the individual cationic and nonionic surfactant under investigation are calculated using thermodynamic and Gibbs Equations 15.46 and 15.47:

$$\Delta G_{mic}^\circ = -RT \ln \text{cmc} \tag{15.46}$$

$$\Delta G_{ads}^\circ = \Delta G_{mic}^\circ - 0.6023 \Pi_{cmc} A_{min} \tag{15.47}$$

where Π_{cmc} is the effectiveness and is defined as the difference between the surface tension of pure water and the surface tension of the surfactant solution at the cmc.

It is clear from the data in Table 15.3 that the standard free energies of adsorption and micellization are always negative, which indicates that the two processes are spontaneous. Moreover, the adsorption free energies are more negative than the micellization. This indicates the high tendency of the pure and mixed systems to be adsorbed at the interface than to micellize in the bulk. However, the surfactant molecules are in equilibrium between the two phases, either adsorbed monolayer or micelles.

The activity coefficients of the mixed systems at different initial mole fractions are used to calculate the excess free energy of mixing using Equation 15.48:

$$\Delta G_{ex} = RT \left[X_1 \ln f_1 + (1 - X_1 \ln f_2) \right] \tag{15.48}$$

TABLE 15.3
Energetic Parameters of the Individual and Mixed Surfactant Systems at Different Initial Mole Fractions of C_{12}DMPAB Surfactant at 25°C

α C_{12}DMPAB/ C_{10}DMPO	ΔG^o_{mic}, kJ mol^{-1}	ΔG^o_{ads}, kJ mol^{-1}	ΔG_{ex}, kJ mol^{-1}	G^s_{min}, J mol^{-1}
0.0/1.0	−31.1	—	—	22.3
0.2/0.8	−35.6	−66.7	−9.6	29.5
0.3/0.7	−36.1	−65.0	−8.1	32.2
0.4/0.6	−36.8	−69.1	−7.6	34.1
0.5/0.5	−37.1	−66.9	−6.3	37.2
0.6/0.4	−37.5	−70.1	−5.2	36.5
0.7/0.3	−37.8	−72.4	−4.1	34.6
0.8/0.2	−38.1	−70.0	−3.7	37.6
0.9/0.1	−38.3	−74.7	−2.1	34.4
1.0/0.0	−38.5	−76.4	—	40.1

The excess free energy of mixing determines the energetic difference between the energy of the pure micelles, which consisted of individual surfactant and the mixed micelles formed as a result of mixing the two components. The negative ΔG_{ex} values, presented in Table 15.3, suggest that the mixed micelles formed are more stable than the micelles of individual surfactants. The lowering in the ΔG_{ex} values indicates the stability of the formed mixed micelles. The introduction of a large mole fraction of the cationic surfactant in the micellar form of C_{12}DMPAB–C_{10}DMPO mixed system at high initial cationic mole fractions renders the micelle more stable.

For evaluation of synergism in mixing, another thermodynamic quantity, known as free energy of a surface at equilibrium, is used and calculated according to Equation 15.49:

$$G^S_{min} = A_{min} \gamma_{cmc} N_A \qquad (15.49)$$

G^S_{min} value not only contains the contribution of A_{min} but also γ_{cmc}, which affects mixed monolayer formation, and hence synergism. It may be defined as the work needed to make a surface area per mole or free energy change accompanied by transition from the bulk phase to the surface phase of solution. Lower the value of G^S_{min}, more thermodynamically stable the surface formed or more surface activity is attained, which is a measure of the evaluation of synergism. The resulting values are lower in magnitude with the addition of anionic surfactant (Table 15.3).

15.3.5 SYNERGISM

The existence of synergism in mixtures containing two surfactants has been shown to depend not only on the strength of the interaction between the surfactants (measured by the values of the β parameter) but also on the properties of the individual

TABLE 15.4

Synergistic Data on Mixed Surfactant Systems at Different Initial Mole Fractions of C_{12}DMPAB Surfactant at 25°C

α C_{12}DMPAB/ C_{10}DMPO	$\lvert\ln(C_1^\sigma/C_2^\sigma)\rvert$	$\lvert\ln(C_1^m/C_2^m)\rvert$	$\lvert\beta^m\rvert$	$\lvert\beta^\sigma\rvert$	$\beta^\sigma - \beta^m$	$\lvert\beta^\sigma - \beta^m\rvert$	$\lvert\ln(C_1^\sigma C_2^m/ C_2^\sigma C_1^m)\rvert$
0.2/0.8	0.94	1.47	1.54	0.55	0.99	0.99	0.53
0.3/0.7			1.34	0.86	0.87	0.87	
0.4/0.6			1.35	0.18	1.02	1.02	
0.5/0.5			1.24	0.27	0.97	0.97	
0.6/0.4			1.18	0.61	0.58	0.58	
0.7/0.3			1.14	2.17	−1.07	1.07	
0.8/0.2			1.30	4.87	−3.58	3.58	
0.9/0.1			1.28	5.64	−4.37	4.37	

surfactant components [21,64–67]. Thus, the conditions for synergism in surface tension reduction efficiency (when the total concentration of the mixed surfactant required to reduce the surface tension of the solvent to a given value is less than that of either individual surfactant) is shown in Equations 15.24 and 15.25.

It is apparent that C_{12}DMPAB–C_{10}DMPO mixtures exhibit synergism in surface tension reduction efficiency at the initial mole fractions of C_{12}DMPAB in the range of 0.7–0.9, since all β^σ values are negative and meet the second condition (Table 15.4).

Synergism in a mixed micelle formation exists when the cmc of the mixture is less than that of the individual surfactants of the mixture. The conditions for synergism to exist in the mixture are mentioned in Equations 15.30 and 15.31. It is apparent that the C_{12}DMPAB–C_{10}DMPO system exhibits synergism in mixed micelle formation when α(cationic) = 0.2, whereas the other surfactant mixtures do not show (Table 15.4) this synergism.

Synergism in surface tension reduction effectiveness (when the surface tension of the mixture at its cmc is lower than that of the individual surfactants at their respective cmc's) depends on the values of both β^σ and β^M in addition to the cmc's of the individual and mixed surfactant systems [67–70]. The condition for this type of synergism is given in Equations 15.36 and 15.37. This type of synergism was found in C_{12}DMPAB–C_{10}DMPO mixed systems at the initial cationic component mole fractions (α_{cationic}) of 0.7–0.9 (Table 15.4).

15.4 SUMMARY AND CONCLUSIONS

The surface active agents used in domestic and industrial applications are mixtures of several components with different surface activity properties. Hence, the behavior of mixed surfactants must be understood to deal with the properties of these mixtures. In the first section of this chapter, the theories describing surfactant mixtures were briefly reviewed. Several theories are used to deal with the surfactant mixtures: some theories considered ideal interaction between the different surfactants and others neglect ideal conditions. The main common feature between these two types of theories was consideration of the effect of the counterions surrounding the micelles. In the second section of the chapter, an investigation of the interaction between cationic surfactant (namely dodecyl diethyl anilinium bromide, $C_{12}DMPAB$) and nonionic surfactant (namely decyl dimethyl phosphineoxide, $C_{10}DMPO$) is discussed. The investigation showed the presence of synergism in the adsorbed mixed monolayer and also in the mixed micelles. The synergism of surface tension efficiency and surface tension effectiveness was found at certain initial mole fractions of the cationic surfactant component. The energetic treatment of the mixed surfactant systems at different mole fractions showed that the mixed monolayers and the mixed micelles are energetically stabilized.

ACKNOWLEDGMENT

I am greatly thankful to Professor Dr. Elshafie A. M. Gad for introducing and helping me to learn the field of mixed surfactants.

REFERENCES

1. S.D. Christian and J.F. Scamehorn (Eds.), *Solubilization in Surfactant Aggregates*, vol. 55, Marcel Dekker, New York, 1995.
2. P.M. Holland and D.N. Rubingh (Eds.), *Mixed Surfactant Systems*, Am. Chem. Soc., Washington, DC, 1992.
3. K. Ogino and M. Abe (Eds.), *Mixed Surfactant Systems*, Marcel Dekker, New York, 1993.
4. E. Junquera and E. Aicart, Mixed micellization of dodecyl ethyl dimethyl ammonium bromide and dodecyl trimethyl ammonium bromide in aqueous solution, *Langmuir*, 18, 9250–9258, 2002.
5. J.L. Lopez-Fontan, A. Gonzalez-Perez, J. Costa, J.M. Ruso, G. Prieto, P.C. Schulz, and F. Sarmiento, Thermodynamics of micellization of tetraethyl ammonium perfluorooctyl sulfonate in water, *J. Colloid Interface Sci.*, 297, 10–16, 2006.
6. P.C. Schulz, J.L. Rodriguez, R.M. Minardi, M.B. Sierra, and M.A. Morini, Are the mixtures of homologous surfactants ideal?, *J. Colloid Interface Sci.*, 303, 264–271, 2006.
7. G. Basu Ray, I. Chakraborty, S. Ghosh, S.P. Moulik, and R. Palepu, Physicochemical studies on the interfacial and bulk behaviors of sodium *n*-dodecanoyl sarcosinate (SDDS), *Langmuir*, 21, 10958–10964, 2005.
8. T. Chakraborty, S. Ghosh, and S.P. Moulik, The methods of determination of critical micellar concentrations of the amphiphilic systems in aqueous medium, *J. Phys. Chem. B*, 109, 14813–14821, 2005.
9. J.N. Israelachvilli, *Intermolecular and Surface Forces*, Academic Press, New York, 1992.
10. J.H. Clint, Micellization of mixed nonionic surface active agents, *J. Chem. Soc. Faraday Trans.*, 71, 1327–1334, 1975.
11. H. Lange and K.H. Beck, Zur Mizellbildung in Mischlösungen homologer und nichthomologer Tenside, *Kolloid. Z. Polym.*, 251, 424–431, 1973.
12. C. Treiner, The thermodynamics of micellar solubilization of neutral solutes in aqueous binary surfactant systems, *Chem. Soc. Rev.*, 23, 349–356, 1994.
13. K. Shinoda, T. Nakagawa, B. Tamamushi, and T. Isemura, In: *Colloidal Surfactants*, Academic Press, New York, 1963, Chapter 1.
14. D.N. Rubingh, Mixed micelle solution. In: *Solution Chemistry of Surfactants*, Vol. 1; K.L. Mittal (Ed.) Plenum Press, New York, 1979, pp. 337–354.

15. P.M. Holland and D.N. Rubingh, Nonideal multi-component mixed micelle model, *J. Phys. Chem.*, 87, 1984–1990, 1983.
16. A. Müller, The behaviour of surfactants in concentrated acids part IV. Investigation of the synergism of binary cationic/cationic mixtures in concentrated sulphuric acid, *Colloids Surfaces. A*, 57, 239–247, 1991.
17. M.J. Rosen, *Surfactants and Interfacial Phenomenon*, 2nd ed., p. 393, Wiley, New York, 1989.
18. J. Penfold, E. Staples, L. Thompson, and I. Tucker, The composition of non-ionic surfactant mixtures at the air/water interface as determined by neutron reflectivity, *Colloids Surfaces A*, 341, 127–132, 1995.
19. X.Y. Hua and M.J. Rosen, Synergism in binary mixtures of surfactants: I. Theoretical analysis, *J. Colloid Interface Sci.*, 90, 212–219, 1982.
20. X.Y. Hua and M.J. Rosen, Conditions for synergism in surface tension reduction effectiveness in binary mixtures of surfactants, *J. Colloid Interface Sci.*, 125, 730–732, 1988.
21. D.S. Murphy, Z.H. Zhu, X.Y. Yuan, and M.J. Rosen, Relationship of structure to properties of surfactants. Isomeric sulfated polyoxyethylene alcohols, *J. Am. Oil Chem. Soc.*, 67, 197–204, 1990.
22. M.J. Rosen and S.B. Sulthana, The interaction of alkyl glycosides with other surfactants, *J. Colloid Interface Sci.*, 239, 528–543, 2001.
23. R. Nagarajan, Molecular theory for mixed micelles, *Langmuir*, 1, 331–341, 1985.
24. J.J. Lopata, K.M. Werts, J.F. Scamehorn, J.H. Harwell, and B.P. Grady, Thermodynamics of mixed anionic/nonionic surfactant adsorption on alumina, *J. Colloid Interface Sci.*, 342, 415–426, 2010.
25. S.K. Mehta and G. Ram, Behavior of papain in mixed micelles of anionic–cationic surfactants having similar tails and dissimilar head groups, *J. Colloid Interface Sci.*, 344, 105–111, 2010.
26. Y.L. Lee, J.Y. Lin, and C.H. Chang, Thermodynamic characteristics and Langmuir–Blodgett deposition behavior of mixed DPPA/DPPC monolayers at air/liquid interfaces, *J. Colloid Interface Sci.*, 296, 647–654, 2006.
27. E. Rodenas, M. Valiente, and M.S. Villafruela, Different theoretical approaches for the study of the mixed tetraethylene glycol mono-*n*-dodecyl ether/hexadecyl trimethyl ammonium bromide micelles, *J. Phys. Chem. B.*, 103, 4549–4554, 1999.
28. M.J. Rosen and X.Y. Hua, Surface concentrations and molecular interactions in binary mixtures of surfactants, *J. Colloid. Interface Sci.*, 86, 164–172, 1982.
29. O.G. Singh and K. Ismail, Micellization behavior of mixtures of sodium dioctyl sulfosuccinate with sodium dodecyl sulfate in water, *J. Surf. Deterg.*, 11, 89–96, 2008.
30. K.R. Harris, P.J. Newitt, and Z.J. Derlacki, Alcohol tracer diffusion, density, NMR and FTIR studies of aqueous ethanol and 2,2,2-trifluoroethanol solutions at 25°C, *J. Chem. Soc. Faraday Trans.*, 94, 1963–1967, 1998.
31. D.P. Hong, M. Hoshino, R. Kuboi, and Y. Goto, Clustering of fluorine-substituted alcohols as a factor responsible for their marked effects on proteins and peptides, *J. Am. Chem. Soc.*, 121, 8427–8432, 1999.
32. S. Kuprin, A. Gräslund, A. Ehrenberg, and M.H.J. Koch, Nonideality of water-hexafluoropropanol mixtures as studied by X-ray small angle scattering, *Biochem. Biophys. Res. Commun.*, 217, 1151–1156, 1995.
33. F.G. Blanco, M.A. Elorza, C. Arias, B. Elorza, I.G. Escalonilla, C. Civera, and P.A.G. Gomez, Interactions of 2,2,2-trifluoroethanol with aqueous micelles of Triton X-100, *J. Colloid. Interface Sci.*, 330, 163–169, 2009.
34. S. Zhao, H. Zhu, X. Li, Z. Hu, and D. Cao, The interaction of novel anionic gemini surfactants with cetyl trimethyl ammonium bromide, *J. Colloid. Interface Sci.*, 350, 480–485, 2010.

35. H. Maeda, A simple thermodynamic analysis of the stability of ionic/nonionic mixed micelles, *J. Colloid. Interface Sci.*, 172, 98–103, 1995.
36. J. Zhao, J. Liu, and R. Jiang, Interaction between anionic and cationic gemini surfactants at air/water interface and in aqueous bulk solution, *Colloids Surfaces A*, 351, 141–146, 2009.
37. L.S. Hao, P. Hu, and Y.Q. Nan, Salt effect on the rheological properties of the aqueous mixed cationic and anionic surfactant systems, *Colloids Surfaces A*, 353, 187–195, 2010.
38. S. Sovanna, T. Suzuki, M. Kojima, S. Tachiyashiki, and M. Kita, Surface tension reduction (STR) in aqueous solutions of anionic surfactants with cobalt(III) complexes, *J. Colloid. Interface Sci.*, 332, 194–200, 2009.
39. Kabir-ud-Din, G. Sharma, and A.Z. Naqvi, Micellization and interfacial behavior of binary surfactant mixtures based on cationic geminis and nonionic Tweens, *Colloids Surfaces A*, 385, 63–71, 2011.
40. F. He, R. Li, and D. Wu, Monolayers of mixture of alkyl amino methyl rutin and lecithin at the air/water interface, *J. Colloid. Interface Sci.*, 349, 215–223, 2010.
41. K. Szymczyk, A. Zdziennicka, B. Janczuk, and W. Wojcik, Behaviour of cationic surfactants and short chain alcohols in mixed surface layers at water-air and polymer-water interfaces with regard to polymer wettability. I. Adsorption at water-air interface, *Colloids Surfaces A*, 265, 147–153, 2005.
42. S.E. Moore, M. Mohareb, S.A. Moore, and R.M. Palepu, Conductometric and fluorometric investigations on the mixed micellar systems of cationic surfactants in aqueous media, *J. Colloid. Interface Sci.*, 304, 491–496, 2006.
43. T. Mukhim, J. Dey, S. Das, and K. Ismail, Aggregation and adsorption behavior of cetyl pyridinium chloride in aqueous sodium salicylate and sodium benzoate solutions, *J. Colloid. Interface Sci.*, 350, 511–515, 2010.
44. M.L. Corrin and W.D. Harkins, *J. Am. Chem. Soc.* 69, 683–689, 1947.
45. S. Javadian, H. Gharibi, Z. Bromand, and B. Sohrabi, Electrolyte effect on mixed micelle and interfacial properties of binary mixtures of cationic and nonionic surfactants, *J. Colloid. Interface Sci.*, 318, 449–456, 2008.
46. H. Belarbi, D. Bendedouch, and F. Bouanani, Mixed micellization properties of nonionic fluorocarbon/cationic hydrocarbon surfactants, *J. Surf. Deterg.*, 13, 433–439, 2010.
47. E. Carey and C. Stubenrauch, Foaming properties of mixtures of a non-ionic (C_{12}DMPO) and an ionic surfactant (C_{12}TAB), *J. Colloid. Interface Sci.*, 346, 414–423, 2010.
48. V.B. Fainerman, S.A. Zholob, M.E. Leser, M. Michel, and R. Miller, Adsorption from mixed ionic surfactant/protein solutions-analysis of ion binding, *J. Phys. Chem.*, 108, 16780–16785, 2004.
49. K.A. Connors and K.A. Wright, Dependence of surface tension on composition of binary aqueous-organic solutions, *Anal. Chem.*, 61, 194–198, 1989.
50. J.M. Rosen and S. Aronson, Relationship of structure to properties in surfactants. 11. Surface and thermodynamic properties of *N*-dodecyl pyridinium bromide and chloride, *Colloids Surfaces A*, 3, 201–209, 1981.
51. A. Zdziennicka and B. Janczuk, Behavior of cationic surfactants and short chain alcohols in mixed surface layers at water–air and polymer–water interfaces with regard to polymer wettability. I. Adsorption at water-air interface, *J. Colloid. Interface Sci.*, 333, 374–383, 2010.
52. H. Iyota, K. Abe, N. Ikeda, K. Motomura, and M. Aratono, Nonideal mixing of alkyl ammonium chloride and decyl dimethyl phosphine oxide surfactants in adsorbed films and micelles, *J. Colloid. Interface Sci.*, 322, 287–293, 2008.
53. A. Mustafina, L. Zakharova, J. Elistratova, J. Kudryashova, S. Soloveva, A. Garusov, I. Antipin, and A. Konovalov, Solution behavior of mixed systems based on novel amphiphilic cyclophanes and Triton X100: Aggregation, cloud point phenomenon and cloud point extraction of lanthanide ions, *J. Colloid. Interface Sci.*, 346, 405–413, 2010.

54. Z. Zhiguo and Y. Hong, Interaction of nonionic surfactant AEO_9 with ionic surfactants, *J. Zhejiang Univ. Sci.*, 6, 597–601, 2005.
55. K. Theander and R.J. Pugh, Synergism and foaming properties in mixed nonionic/fatty acid soap surfactant systems, *J. Colloid. Interface Sci.*, 267, 9–17, 2003.
56. F. Podczeck, A. Maghetti, and M. Newton, The influence of non-ionic surfactants on the rheological properties of drug/microcrystalline cellulose/water mixtures and their use in the preparation and drug release performance of pellets prepared by extrusion/spheronization, *Eur. J. Pharmaceutical Sci.*, 37, 334–340, 2009.
57. K. Din, M.S. Sheikh, M.A. Mir, and A.A. Dar, Effect of spacer length on the micellization and interfacial behavior of mixed alkanediyl-α,ω-bis (dimethylcetyl ammonium bromide) gemini homologues, *J. Colloid Interface Sci.*, 344, 75–80, 2010.
58. K. Din, A. Koya, and Z.A. Khan, Conductometric studies of micellization of gemini surfactant pentamethylene-1,5-bis(tetradecyl dimethyl ammonium bromide) in water and water-organic solvent mixed media, *J. Colloid. Interface Sci.*, 342, 340–347, 2010.
59. J.R. Akbar, R.D. Eubry, D.G. Marangoni, and S.D. Wettig, Interactions between gemini and nonionic pharmaceutical surfactants, *Can. J. Chem.*, 88, 1262–1270, 2010.
60. N. Azum, A.Z. Naqvi, M. Akram, and K. Din, Studies of mixed micelle formation between cationic gemini and cationic conventional surfactants, *J. Colloid. Interface Sci.*, 328, 429–435, 2008.
61. K. Din, M.S. Sheikh, and A.A. Dar, Interaction of a cationic gemini surfactant with conventional surfactants in the mixed micelle and monolayer formation in aqueous medium, *J. Colloid. Interface Sci.*, 333, 605–612, 2009.
62. R. Li, Q. Chen, D. Zhang, H. Liu, and Y. Hu, Mixed monolayers of gemini surfactants and stearic acid at the air/water interface, *J. Colloid. Interface Sci.*, 327, 162–168, 2008.
63. K. Sakai, S. Umezawa, M. Tamura, Y. Takamatsu, K. Tsuchiya, K. Torigoe, T. Ohkubo et al., Adsorption and micellization behavior of novel gluconamide type gemini surfactants, *J. Colloid. Interface Sci.*, 318, 440–448, 2008.
64. M.S. Bakshi and K. Singh, Synergistic interactions in the mixed micelles of cationic gemini with zwitterionic surfactants: Fluorescence and Kraft temperature studies, *J. Colloid Interface Sci.*, 287, 288–297, 2005.
65. K. Singh and D.G. Marangoni, Synergistic interactions in the mixed micelles of cationic gemini with zwitterionic surfactants: The pH and spacer effect, *J. Colloid. Interface Sci.*, 315, 620–626, 2007.
66. N.A. Negm, I.A. Aiad, and S.M. Tawfik, Screening for potential antimicrobial activities of some cationic uracil biocides against wide spreading bacterial strains, *J. Surf. Deterg.*, 13, 503–511, 2010.
67. N.A. Negm and S.M. Tawfik, Studies of monolayer and mixed micelle formation of anionic and conventional nonionic surfactants in presence of adenosine-5-monophosphate, *J. Solution Chem.*, 41, 335–350, 2011.
68. N.A. Negm and M.F. Zaki, Synthesis and characterization of some amino acid derived schiff-bases bearing nonionic species as corrosion inhibitors for carbon steel in 2N HCl, *J. Disp. Sci. Technol.*, 30, 649–655, 2009.
69. N.A. Negm and S.A. Mahmoud, Synthesis, surface and thermodynamic properties of carboxymethyl-(di-2-ethanol)-ethyl alkanoate ammonium bromide amphoteric surfactants, *Egypt. J. Petrol.*, 14, 79–88, 2005.
70. D. Danino, Y. Talmon, and R. Zana, Alkanediyl, alpha, omega-bis(dimethyl ammonium bromide) surfactants (dimeric surfactants). 5. Aggregation and microstructure in aqueous solutions, *Langmuir*, 11, 1448–1456, 1995.

16 Tribological Properties of Aqueous Solutions of Silicone Polyethers

Marian W. Sulek and Malgorzata Zieba

CONTENTS

The tribological properties of aqueous solutions of silicone polyethers and pure polyethers are evaluated. Three compounds, whose molecules differ in the siloxane chain length and the number of ethylene glycol and propylene glycol units in the polyether chain, are investigated. The lubricating properties of the solutions are determined using a four-ball tribometer (T-02 tester). The measurements are carried out under constant loads of 2, 3, 4, and 5 kN. Antiseizure characteristics (scuffing load, seizure load, limiting pressure of seizure) are also investigated by carrying out the test with the load increasing with time. The effect of concentration and chemical structure of silicone polyether on tribological properties is discussed. These model lubricants exhibit exceptionally beneficial tribological properties. Studies on their application to highly loaded friction pairs will be carried out in the future.

16.1 INTRODUCTION

Water can be used in specific applications as a lubricant base. It is cheap, readily available, nonflammable, and also environmentally friendly. Owing to its high heat capacity and conduction, water exhibits very good cooling properties. Water also has negative tribological properties such as a corrosive effect on metals, low viscosity, and poor lubricity. The latter can be improved by incorporating appropriate additives in the lubricant composition. The additives should be safe to humans and the environment during production, application, and utilization, and they should also be water-soluble. Satisfactory results were obtained after application of the following additives: alkylpolyglucosides, sodium lauryl sulfate and its ethoxylated derivative, ethoxylates of castor oil, fatty alcohols and amines, fatty acid methyl esters from rapeseed oil, and sorbitan esters [1–5]. The effects of the structure and concentration of these compounds in lubricants on resistance to motion, wear, and antiseizure properties were analyzed.

Silicone oils (poly(dimethylsiloxanes), PDMS) have a number of interesting properties for application in lubricants [6–10]. Their major disadvantage is, however, their insolubility in water. Attempts have thus been made to apply silicone polyethers as additives to modify the lubricity of water.

16.2 APPLICATION OF SILICONES AND THEIR DERIVATIVES AS LUBRICANTS

Numerous applications of silicones result from their atypical properties. They can be used in a wide temperature range, practically from –50°C to 250°C. They are stable to oxidation and resistant to aqueous solutions of acids, bases, and salts [6–7]. They are physiologically neutral and can be used in biomedical and pharmaceutical materials and as components of cosmetics and household chemical products [6–8]. They have a hydrophobizing property and form stable films on solid surfaces. Under friction conditions, the silicone film undergoes rupture and squeezes out of the friction area. This possibility increases in metal–metal contacts but decreases in metal–plastic and plastic–plastic contacts. This negative property is a result of the hydrophobic properties of silicone molecules [9–10].

Silicone oils also have some disadvantages. They do not blend well with other oils. They also exhibit low solubility of performance-enhancing additives. Silicone oils have poor antiwear properties [9–10]. The disadvantages of silicone oils can be eliminated by the application of appropriate additives: lubricity modifiers (e.g., lithium stearate, molybdenum disulfide, mica, carbon black), thickeners (e.g., colloidal silica, cellulose derivatives), and inactive fillers (e.g., chalk) [1,9–10].

This chapter deals with the model lubricants based on water- and silicone-based surface active compounds. Silicone derivatives with hydrophobic siloxane and poly(propylene glycol) chains and a hydrophilic poly(ethylene glycol) chain have been synthesized and investigated as additives modifying water lubricity. The comb-like silicone polyether derivatives have amphiphilic properties (Figure 16.1) [6–8].

Silicone polyethers are nonionic surfactants and can form micelles in solutions [6–8]. Compared with organic surface active compounds, they exhibit higher surface activity [1,8,11–20]. For example, the surface tension (σ) values for silicone polyethers are in the range of 20 mN/m, whereas the corresponding values for aqueous solutions of ethoxylated cetyl alcohols are about 40 mN/m [1].

Surface activity and solubility of surfactants in hydrophobic and hydrophilic media can be modified by varying the chain length of ethylene oxide (PEG) or propylene

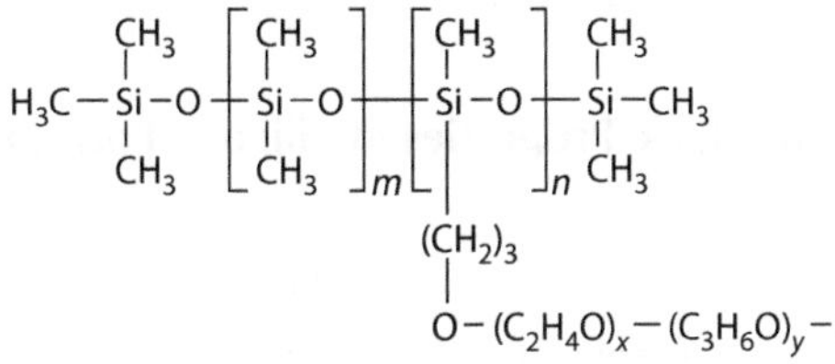

$$H_3C-\underset{\underset{CH_3}{|}}{\overset{\overset{CH_3}{|}}{Si}}-O\left[\underset{\underset{CH_3}{|}}{\overset{\overset{CH_3}{|}}{Si}}-O\right]_m\left[\underset{\underset{(CH_2)_3}{|}}{\overset{\overset{CH_3}{|}}{Si}}-O\right]_n\underset{\underset{CH_3}{|}}{\overset{\overset{CH_3}{|}}{Si}}-CH_3$$

$$O-(C_2H_4O)_x-(C_3H_6O)_y-H$$

FIGURE 16.1   Structure of comb-like silicone polyether.

oxide (PPG). An increase in the number of PEG groups results in increased water solubility whereas an increase in the number of PPG groups leads to increased solubility in nonpolar media [6–8].

16.3   EXPERIMENTAL SECTION

16.3.1   MATERIALS

The physicochemical and tribological properties of aqueous solutions of silicone polyethers depend considerably on the polysiloxane backbone chain length and on the number of ethylene glycol and propylene glycol units on the branch chain. Three silicone materials (Figure 16.1) produced by Evonik Industries were used in this study. Their siloxane chain lengths (m, n) were different and their ethoxylation (x) and propoxylation (y) degrees were also different. Structural features and basic properties of silicone polyethers are shown in Table 16.1.

The evaluation of tribological properties was carried out for water and aqueous solutions of silicone polyethers at weight concentrations of 1%, 4%, 10%, and 100%. The water used to prepare solutions was distilled twice.

Bearing balls 0.5″ in diameter made of 100Cr6 bearing steel (surface roughness $R_a = 0.32$ µm, hardness 60 ÷ 65HRC) were used. Before the tests, all elements of the friction pairs were carefully chemically cleaned. The cleaning process was conducted in an ultrasonic cleaner and consisted of four stages: preliminary washing in extraction naphtha, washing in acetone (5 min), washing in ethyl alcohol (5 min), and, finally, washing the samples in distilled water.

16.3.2   METHODS

A T-02 four-ball tribotester produced at the Institute for Sustainable Technologies in Radom (Poland) was used for evaluation. Its structure and methodology have been described elsewhere [21]. The schematic of the arrangement of balls in the T-02 four-ball tribometer is shown in Figure 16.2.

The tribometer (T-02) used in the tests has the following technical parameters:

- Rotational speed of the spindle: 0–1800 rpm
- Contact load: 0–7200 N
- Rate of load increase: 409 N/s

TABLE 16.1

Structural Features and Basic Properties of Silicone Polyethers Used in the Tests

Property	Silicone Polyether		
	PEG/PPG-14/0	PEG/PPG-14/4	PEG/PPG-20/20
Siloxane chain length (m, n)	$m = 13$	$m = 18$	$m = 32$
	$n = 5$	$n = 5$	$n = 6$
Ethoxylation degree (x)	$x = 14$	$x = 14$	$x = 20$
Propoxylation degree (y)	$y = 0$	$y = 4$	$y = 20$
Cloud point of aqueous solutions of silicones (concentration—4 wt%)	>85°C	67°C	50°C
Surface activity in aqueous solutions	Capable of micelle formation	Capable of micelle formation	Capable of micelle formation
Solubility in media	Water	Water	Water
	Ethyl alcohol	Ethyl alcohol	Ethyl alcohol
	1,2-Propylene glycol	1,2-Propylene glycol	Isopropyl myristate
Trade name	Abil B 8843	Abil B 8851	Abil B 8863
Notations in figures	A	B	C

FIGURE 16.2 Schematics of the arrangement of balls in the T-02 four-ball tribometer.

The device used had microprocessor software which made it possible to conduct tests according to standard testing procedures.

The following types of tests were carried out:

- Tests at a constant load allowing for determination of resistance to motion variations with time for individual lubricant compositions. A constant preset load ranging from 2 to 5 kN was used. The duration of the tests was 15 min and the rotational speed was 200 rpm. The coefficient of friction (μ) was calculated from the measured frictional torque (M_T) as follows:

$$\mu = K\frac{M_T}{P},\tag{16.1}$$

where
  μ = coefficient of friction (–)
  M_T = frictional torque (N · m)
  P = load (N)
  K = a dimension coefficient equal to 222.47 m^{-1} taking into account the distribution of forces present in the tribosystem

- Test with linearly increasing load allows for the determination of the quantities characterizing the scuffing process: scuffing load (P_t), seizure load (P_{oz}), and limiting pressure of seizure (p_{oz}). They are determined on the basis of the analysis of the dependence of frictional torque (M_T) as a function of linearly increasing load (P) (Figure 16.3) and appropriate calculations (Equation 16.2):
  a. Scuffing load (P_t)—pressure after which a boundary layer rupture takes place resulting in a sharp increase in frictional torque. High P_t values indicate that the lubricant components are capable of forming a stable lubricant film. Scuffing load is expressed in newtons.
  b. Seizure load (P_{oz})—at which the frictional torque increases above 10 N · m. It is treated as a load that causes seizure of the tribosystem in a four-ball tester and its immobilization. High P_{oz} values indicate that the lubricant is capable of preventing seizure. It is expressed in newtons.

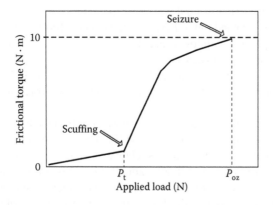

FIGURE 16.3 Typical dependence of frictional torque (M_T) on linearly increasing load.

c. Limiting pressure of seizure (p_{oz}) is a unit load expressed in N/mm^2
which describes the pressure present in the tribosystem at seizure load
(P_{oz}). It is calculated from the following equation:

$$p_{oz} = 0.52\frac{P_{oz}}{d^2},\qquad(16.2)$$

where
p_{oz} = limiting pressure of seizure (N/mm^2)
P_{oz} = seizure load (N)
d = wear scar diameter measured after the test (mm)
0.52 = coefficient taking into account the distribution of forces in the
tribosystem

The experimental conditions were as follows: rate of load increase 409 N/s, load
from 0 N to 7200 kN, constant rotational speed of the spindle, 500 rpm.

The measure of wear in the tests at both constant and increasing loads was a mean
value of wear scar diameters of the balls (d) parallel and perpendicular to the direction of friction. A polar reflection microscope produced by PZO-Warszawa (Poland)
was used to measure wear scar diameter. Mean d values from three independent
measurements were calculated and plotted.

16.4   RESULTS AND DISCUSSION

The physicochemical studies [1,11] indicate that silicone polyether molecules
undergo adsorption on the steel surface and form a surface phase. Under friction
conditions, it can transform into a lubricant film reducing resistance to motion and
wear. To confirm this thesis, the values of friction coefficient (μ) and wear scar diameter of the ball (d) as well as the parameters characterizing seizure (P_t, P_{oz}, p_{oz}) were
determined.

16.4.1   FOUR-BALL TESTS AT CONSTANT LOAD ($P = 2000$–5000 N)

The coefficient of friction (μ) and wear scar diameter of the ball (d) were obtained
after the tests were carried out at loads of 2000, 3000, 4000, and 5000 N. Aqueous
solutions of compounds with low concentrations (1%, 4%, 10%) and 100% silicones
were used as lubricating substances.

The dependence of the coefficient of friction on load (2, 3, 4, 5 kN) for solutions
with various concentrations of silicone polyethers (1%, 4%, 10%, and 100%) is shown
in Figure 16.4.

It was possible to carry out tests with pure water only at the lowest load (2 kN). At
higher loads, water caused seizure to occur (Figure 16.4). A significant decrease of
μ at 2 kN was obtained for 1% solutions relative to water. An increase in the concentration of silicone did not result in a meaningful decrease in the coefficient of friction. However, contrary to expectations, an increase in load resulted in a significant
decrease of μ by as much as 50%. Pure silicone ethers (100% solutions) exhibited
the lowest resistance to motion which was practically independent of the chemical

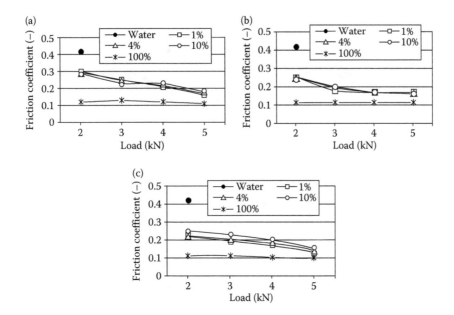

FIGURE 16.4 Dependence of mean friction coefficient on load for water and aqueous solutions of silicone polyethers: (a) PEG/PPG-14/0, (b) PEG/PPG-14/4, and (c) PEG/PPG-20/20. T-02 four-ball tester, constant load (2000–5000 N), rotational speed 200 rpm, test duration 900 s, room temperature.

structure of the compound and the applied load (*P*). With increasing load, the differences between the μ values at low-concentration solutions (1%, 4%, 10%) and for 100% solutions decreased. However, for the highest load (5 kN), the difference was less than the measuring error.

Wear scar diameters of the balls were measured after each test and the results are given in Figure 16.5.

The effect of silicone ethers on reducing wear scar diameters relative to water (Figure 16.5) was particularly significant at the lowest load (2 kN) at which the *d* value decreased over 2.5-fold. The differences in the *d* values between solutions and neat silicones were not as large as with μ (Figure 16.4). Also, the effect of the chemical structure of silicones on *d* was minor. An increase in wear is expected with increasing load. It should be mentioned, however, that the maximum wear scar diameter value at the highest load (5 kN) was comparable with the value for pure water at the lowest load (2 kN).

16.4.2 FOUR-BALL TEST AT INCREASING LOAD (*P* = 0–7200 N)

Tests with increasing load were also performed. They were carried out according to the procedure described in Section 16.3.2. The data from this test were used for the determination of scuffing load (P_t) and seizure load (P_{oz}). Aqueous solutions of silicone polyethers at concentrations of 1%, 4%, and 10% and pure silicone polyethers were tested.

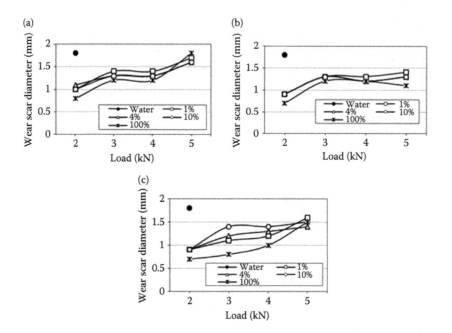

FIGURE 16.5 Dependence of wear scar diameter on load for water and aqueous solutions of silicone polyethers: (a) PEG/PPG-14/0, (b) PEG/PPG-14/4, and (c) PEG/PPG-20/20. T-02 four-ball tester, constant load (2000–5000 N), rotational speed 200 rpm, test duration 900 s, room temperature.

Typical plots of the dependences of friction torque (M_T) as a function of linearly increasing load for water and 1% aqueous solutions of silicone polyethers are shown in Figure 16.6.

In the case of the PEG/PPG-14/0 solution (Figure 16.6), frictional torque initially increases until about 2 kN and then stays constant until 3.8 kN. Further increase in

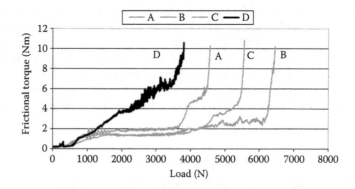

FIGURE 16.6 Dependences of frictional torque as a function of linearly increasing load for water (D) and 1% aqueous solutions of silicone polyethers: (A) PEG/PPG-14/0, (B) PEG/PPG-14/4, and (C) PEG/PPG-20/20. T-024-ball tester, load: 0–7200 N, rate of load increase 409 N/s, rotational speed 500 rpm.

load resulted in a sudden increase in frictional torque. Analogous dependence of load versus frictional torque was obtained for aqueous solutions of other silicone compounds. The sudden increase in frictional torque occurred at different loads for each of the compounds.

Plots of frictional torque (M_T) as a function of linearly increasing load for water and three neat silicone polyethers are presented in Figure 16.7.

No sudden increase in frictional torque, which would allow for determination of a scuffing load (P_t) value, was observed for the pure silicones (PEG/PPG-14/0 and PEG/PPG-14/4) (Figure 16.7). Therefore, the maximum load value for the test (7200 N) was assumed.

The scuffing load values (Figure 16.8) for 1%, 4%, and 10% concentrations of aqueous silicone polyethers seem to be relatively low. This results from unexpectedly high P_t values for pure compounds relative to water (200 N). Even in the case of the lowest concentrations, an increase in the scuffing load value (even fourfold) relative to the base is significant. In the case of pure PEG/PPG-14/0 and PEG/PPG-14/4 compounds, no scuffing was observed ($P_t = 7200$ N).

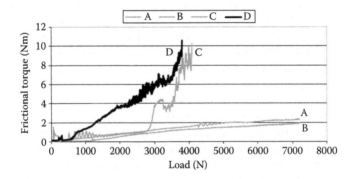

FIGURE 16.7 Dependences of frictional torque as a function of linearly increasing load for water (D) and neat silicone polyethers: (A) PEG/PPG-14/0, (B) PEG/PPG-14/4, and (C) PEG/PPG-20/20. T-02 four-ball tester, load: 0–7200 N, rate of load increase 409 N/s, rotational speed 500 rpm, room temperature.

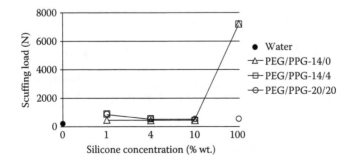

FIGURE 16.8 Dependence of scuffing load on concentration of silicone polyethers in water obtained from load versus friction data such as those shown in Figures 16.6 and 16.7.

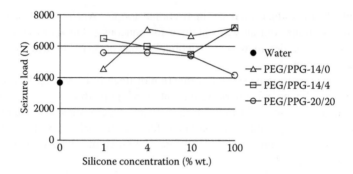

FIGURE 16.9 Dependence of seizure load on the concentration of silicone polyethers in water. T-02 tester.

The dependence of seizure load on the concentration of aqueous solutions of silicone polyethers and pure compounds is shown in Figure 16.9.

Seizure load (Figure 16.9) takes the value of 3700 N for water. A significant increase in P_{oz} of up to even 6500 N was observed at 1% concentration of silicone polyethers (PEG/PPG-14/4). A further concentration increase results in a decrease in the measured quantity and the P_{oz} values are the lowest in the presence of 10% solutions. No seizure of the system was observed in the case of pure compounds with the exception of PEG/PPG-20/20. Therefore, the highest possible load of 7200 was assumed as the P_{oz} value.

The dependence of limiting pressure of seizure on the concentration of silicone polyethers in water is shown in Figure 16.10.

The value of limiting pressure of seizure (Figure 16.10) for water as a lubricant base is 200 N/mm². Even a 1% addition of PEG/PPG-20/20 results in an increase in p_{oz} of up to sevenfold relative to water. The p_{oz} values decrease by nearly 30% for 1%, 4%, and 10% PEG/PPG-20/20 solutions. In the case of the PEG/PPG-14/0 and PEG/PPG-14/4 compounds, the values of limiting pressure of seizure remain at a relatively stable level of about $p_{oz} = 1100$ and $p_{oz} = 1150$ N/mm², respectively. In the case of pure silicones, the quantity studied took the values of below 1000 N/mm². Still, the values exceeded the ones obtained for water from 2 to almost 5 times.

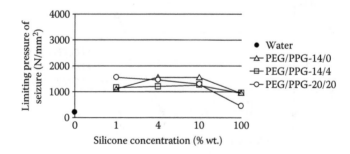

FIGURE 16.10 Dependence of limiting pressure of seizure on the concentration of silicone polyethers in water. T-02 tester.

TABLE 16.2

Mean P_t, P_{oz}, p_{oz} Values Obtained after Tests with Linearly Increasing Load for Water, Aqueous Solutions of Silicone Polyethers, and Pure Silicone Polyethers

Compound	Concentration (%)	P_t (N)	P_{oz} (N)	p_{oz} (N/mm²)	d (mm)
Water	0	200	3700	200	3.1
PEG/PPG-14/0	1	450	4600	1100	1.5
	4	450	7100	1550	1.5
	10	400	6700	1550	1.5
	100	7200	7200	950	2.0
PEG/PPG-14/4	1	850	6500	1150	1.7
	4	550	6000	1200	1.6
	10	450	5500	1250	1.5
	100	7200	7200	950	2.0
PEG/PPG-20/20	1	850	5600	1550	1.4
	4	550	5600	1450	1.4
	10	550	5400	1300	1.5
	100	550	4200	450	2.2

Note: T-02 tester, load 0–7200 N.

The results of the tests carried out at linearly increasing load for water, aqueous solutions of silicone polyethers, and pure silicone polyethers are given in Table 16.2.

16.5 SUMMARY

Silicone surfactants with an alkyl hydrophobic chain have been used as additives for modifying the lubricating properties of water [1–7]. The properties of silicones and silicone polyethers, particularly their biocompatibility, indicated that it may be possible to apply them to biological systems and to friction components in machines used in the cosmetics and pharmaceutical industries [8,9,13–21]. Another reason was their high surface activity characterized by low surface tension and wettability of steel surfaces [8,9,13–21].

The aim of this work was to investigate the tribological characteristics of aqueous silicone polyether solutions and neat silicone polyethers at high loads. The coefficient of friction (μ) and wear scar diameter (d) were determined on a four-ball tribometer at various constant loads (2, 3, 4, and 5 kN). The results showed that even at the highest load (5 kN), the system did not undergo seizure in the presence of silicone polyethers, whereas seizure occurred in water at 3 kN. The effect of silicone polyethers on resistance to motion was particularly significant. Contrary to expectations, the coefficient of friction decreased or remained constant with increasing load. Low values of μ were observed for neat silicone polyethers, which were over 4 times lower at 5 kN than that of water at 2 kN. Wear was also relatively low. At the load of 2 kN where no seizure was observed in water, the wear scar diameter for silicone derivatives was over 2.5 times lower than that for water.

High load carrying capacity of silicone polyethers was also verified using tests with increasing load at a rate of 409 N/s. Antiseizure properties were determined by analyzing changes in frictional torque as a function of load. As follows from the dependences obtained, the changes for the silicone solutions with the best antiseizure properties differed from typical correlations (Figure 16.3). They are characterized by a relatively large range of loads at which the torque value depends to a small degree on load and takes relatively low values (about $2 \, N \cdot m$). In the case of the PEG/PPG-14/0 and PEG/PPG-14/4 neat silicone polyethers, one can observe a low, practically linear, increase in torque as a function of load (Figure 16.7). Such beneficial antiseizure properties have not been observed for surfactants containing lubricants studied so far.

The results presented indicate that it is possible to use silicone polyethers as additives for modifying the lubricity of water. Also, it is observed that the neat silicone polyethers can be used as lubricant bases. Their tribological benefits at extremely high loads have been confirmed. Studies of other silicone polyethers will continue. Also, research aimed at explaining their lubrication mechanism will also be conducted.

REFERENCES

1. M.W. Sulek, *Aqueous Solutions of Surfactants in Materials Science of Tribological Systems*. Scientific Publishing House of Institute for Sustainable Technologies, Radom, 2009, (In Polish).
2. M.W. Sulek and T. Wasilewski, Influence of critical micelle concentration (CMC) on tribological properties of aqueous solutions of alkyl polyglucosides, *Tribol. Trans.*, 52, 12–20, 2009.
3. M.W. Sulek, T. Wasilewski, and M. Zieba, Tribological and physical-chemical properties aqueous solutions of cationic surfactants, *Ind. Lubric. Tribol.*, 62, 279–284, 2010.
4. M.W. Sulek and T. Wasilewski, Tribological properties of aqueous solutions of alkyl polyglucosides, *Wear*, 26, 193–204, 2006.
5. M.W. Sulek, T. Wasilewski, and K. J. Kurzydlowski, The effect of concentration on lubricating properties of aqueous solutions of sodium lauryl sulfate and ethoxylated sodium lauryl sulfate, *Trib. Lett.*, 40, 337–345, 2010.
6. R. Hill, *Silicone Surfactants*. Marcel Dekker, New York, 1999.
7. H. Maciejewski and B. Marciniec, Silicone polyethers. Synthesis, properties and applications, *Pol. J. Chem. Technol.*, 82, 138–141, 2006, (In Polish).
8. N. Nagatani, K. Fukuda, and T. Suzuki, Interfacial behaviour of mixed systems of glycerylether-modified silicone and polyoxyethylene-modified silicone, *J. Colloid Interface Sci.*, 234, 337–343, 2001.
9. G. Capporiccio, P.M. Cann, and H.A. Spikes, Additives for fluorosilicone oils at high temperature, *Wear*, 193, 261–268, 1996.
10. S.C. Ni, P.L. Kuo, and J.F. Lin, Antiwear performance of polysiloxane-containing copolymers at oil/metal interface under extreme pressure, *Wear*, 253, 862–868, 2002.
11. M.W. Sulek and M. Zieba, Silicone polyethers as effective additives modifying water lubricity, *Tribologia*, 226, 247–256, 2009, (In Polish).
12. C. Rodriguez, H. Uddin, K. Watanabe, H. Furukawa, A. Harashima, and H. Kunieda, Self-organization, phase behaviour, and microstructure of poly(oxyethylene) poly (dimethylsiloxane) surfactants in nonpolar oil, *J. Phys. Chem.*, 106, 22–29, 2002.

13. I. Baquerizo, M.A. Riuz, J.A. Holgado, M.A. Cabrerizo, and V. Gallado, Measurement of dynamic surface tension to determine critical micellar concentration in lipophilic silicone surfactants, *Il Farmaco*, 55, 583–589, 2000.
14. A. Wang, L. Jiang, G. Mao, and Y. Liu, Direct force measurement of comb silicone surfactants in alcoholic media by atomic force microscopy, *J. Colloid Interface Sci.*, 242, 337–345, 2001.
15. S.S. Soni, N.V. Sastry, V.K. Aswai, and P.S. Goyal, Micellar structure of silicone surfactants in water from surface activity, SANS and viscosity studies, *J. Phys. Chem.*, 106, 2606–2617, 2002.
16. Y. Lin and P. Alexandridis, Cosolvent effects on micellization of an amphiphilic siloxane graft copolymer in aqueous solutions, *Langmuir*, 18, 4220–4231, 2002.
17. R.M. Hill, M. He, Z. Lin, H.T. Davis, and L.E. Scriven, Lyotropic liquid crystal phase behaviour of polymeric siloxane surfactants, *Langmuir*, 9, 2789–2798, 1993.
18. M. He, R.M. Hill, Z. Lin, L.E. Scriven, and H.T. Davis, Phase behaviour and microstructure of polyoxyethylene trisiloxane surfactants in aqueous solution, *J. Phys. Chem.*, 97, 8820–8834, 1993.
19. T. Iwanaga and H. Kunieda, Effect of salts or polyols on the cloud point and liquid-crystalline structures of polyoxyethylene-modified silicone, *J. Colloid Interface Sci.*, 227, 349–355, 2000.
20. Z. Nemeth, G. Racz, and K. Koczo, Foam control by silicone polyethers—Mechanisms of cloud point antifoaming, *J. Colloid Interface Sci.*, 207, 386–394, 1998.
21. W. Piekoszewski, M. Szczerek, and W. Tuszynski, The action of lubricants under extreme pressure conditions in a modified four-ball tester, *Wear*, 240, 188–193, 2001.

Part IV

Advanced Tribological Concepts

17 Surfactants for Electric Charge and Evaporation Control in Fluid Bearing Motor Oil

T. E. Karis

CONTENTS

Fluid dynamic bearing spindle motors are now pervasive in magnetic recording disk drives. Two areas of continuous improvement for these motor bearings are to minimize the motor voltage and electric charge buildup in the oil and to minimize oil evaporation from the meniscus between the rotor and the stator. A small voltage across the motor bearing causes a high electric field across the submicron gap between the magnetic recording slider and the disk. Charge control additives were incorporated into typical fluid bearing motor base oil. Dielectric spectroscopy was used to measure the electrical conductivity and permittivity. The relative permittivity was found to increase linearly with the specific conductivity. The motor bearing voltage and discharge current were measured on running motors built with the experimental oils. An equivalent electric circuit model is proposed to relate the voltage and discharge current to the specific conductivity of the oil. Throughout the lifetime of a disk drive, the oil loss due to evaporation can be no more than a few milligrams. A surfactant monolayer on the surface of the oil should decrease the evaporation rate by increasing the activation energy barrier for the oil molecule to transition from the liquid to the vapor phase. Oil surface tension was measured for a diester hydrocarbon oil containing a wide range of oil-compatible surfactants, including polyalkylene glycols and fluorohydrocarbon esters. The most promising surfactants for oil evaporation control were found to be perfluoropolyether diol or tetraol esters with C6 or C8 hydrocarbon acid.

17.1 INTRODUCTION

The demand for lower vibration and longer service life has propelled the evolution of magnetic recording hard disk drive motors from ball bearings to fluid bearings [1–3]. Fluid bearings incorporate a self-pressurized oil film, <10 μm thick, between the rotor and the stator. Internal pressure is provided by a groove pattern on the shaft. The thermophysical properties of fluid bearing oils and formulation with antioxidant and corrosion inhibitor are described in Ref. [4]. This chapter focuses on materials for charge control and surfactants to reduce the oil evaporation rate.

Hydrocarbon oil employed in fluid bearings has intrinsically low conductivity. Shear flow of low-conductivity liquids produces bulk electric charge separation [5,6], and an electric potential develops across the oil film in the bearing. Since the magnetic recording slider flies within several nanometers of the magnetic recording disk surface, the electric charge and potential of the motor bearing should be kept within specified limits to avoid arcing and lubricant transfer to the slider [7].

Early formulation of conductivity additives for fluid bearing motor oils with conductivity improver was intended to match the 10–100 MΩ resistance of steel ball bearing spindle motors [8–11]. The goal was to provide sufficient oil conductivity to dissipate static charge generated by air shear over the spinning disks and tribocharge from slider–disk contacts [12,13]. However, fluid bearing motors built with this highly conductive oil formulation also had an excessively high discharge current.

Later oil formulations were developed with improved dielectric properties for charge control [14]. This chapter presents the oil dielectric properties that were measured for a variety of formulations with different types of conductivity improver additives. The motor running voltage and peak discharge current are shown for motors built with these formulations. A series RC equivalent circuit model is derived to relate the dielectric properties of the oil with motor voltage and peak discharge current. The influence of motor voltage and electric energy stored in the bearing on the likelihood and severity of arc discharge between the magnetic recording slider and the disk is discussed.

In addition to being thermally stable and having sufficient electrical conductivity with low electric charge separation, the oil evaporation rate must also be low enough so that there is no appreciable evaporation over the 5–7 year lifetime of the motor while running at its maximum internal operating temperature of about 80°C. The oil temperature–viscosity coefficient must be as low as possible so that the motor can start at cold temperature and maintain adequate stiffness at the operating temperature. The viscosity should be low to minimize viscous power dissipation. However, oil vapor pressure generally increases with decreasing viscosity at a given temperature. Various base oils with a wide range of composition, structure, and molecular weight were investigated with the goal of providing guidelines for selection of an oil composition with the lowest possible viscosity and vapor pressure. The relationship between oil vapor pressure and viscosity is explored in Karis and Nagaraj [15]. If there is a way to lower the viscosity of an oil without increasing the vapor pressure, it seems that it can only be done by increasing the flow-activation entropy.

Alternatively, the evaporation rate of a low-viscosity oil may be inhibited by a surfactant layer at the air–oil interface, similar to the effect of a fatty acid monolayer on the evaporation rate of water [16–19]. The surfactant decreases the evaporation rate by imposing an additional energy barrier for a molecule to transition from the liquid to the vapor phase. A molecule that is a surfactant is usually amphiphilic. For example, a molecule that is a surfactant in water (e.g., soap) contains a hydrophobic part and a hydrophilic part. The hydrophilic part of the surfactant molecule is water soluble, whereas the hydrophobic part of the surfactant molecule is insoluble and preferentially accumulates at the air–water interface. Similarly, a molecule that is a surfactant in oil contains an oleophobic part and an oleophilic part. The oleophilic part of the surfactant molecule is oil soluble, while the oleophobic part of the surfactant is insoluble. The oleophobic part preferentially accumulates at the oil–air interface. The degree of oleophobic/oleophilic phase separation is a function of the structure and composition of the surfactant, which governs its solubility in the oil. Specifically, if the oleophobic part of the surfactant is too soluble in the oil, the surface excess of surfactant at the oil–air interface is too small, and there is no effect on evaporation. On the other hand, if the oleophilic part of the surfactant is too insoluble in the oil, micelles are formed; in the extreme case, a separate liquid phase is formed. Thus, the oleophobicity/oleophilicity of the surfactant must be appropriately balanced.

A hydrocarbon surfactant that has a polar part and a nonpolar part, such as a polyalkylene glycol with a polar ether block and a nonpolar alkane block [20], can act as a surfactant in the moderately polar hydrocarbon ester oils typically used in the disk drive fluid bearings. Another surfactant that is suitable for use in fluid bearing motor oils comprises hydrocarbon blocks that are soluble in the oil and fluorocarbon blocks that are insoluble in the oil. The surfactant comprising a hydrocarbon block and a fluorocarbon block is referred to as a fluorohydrocarbon surfactant. A properly balanced fluorohydrocarbon ester surfactant comprises a fluorocarbon block that resides

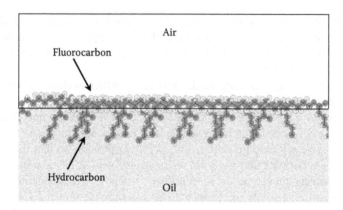

FIGURE 17.1 A schematic illustration showing the arrangement of fluorosurfactant at the oil–air interface.

primarily at the oil–air interface, and a hydrocarbon block, which resides within the hydrocarbon ester oil (Figure 17.1). The fluorocarbon surface layer between the oil and the air inhibits mass transfer of the oil molecules into the vapor phase by increasing the vaporization energy. Another type of surfactant is referred to as an antifoam agent. An antifoam surfactant displaces a foam-producing surfactant layer with an antifoam surfactant layer, which is unfavorable to foam stability [21]. Antifoam agents may also provide a surface layer that is capable of inhibiting oil evaporation.

Surface tension measured with an ester oil containing a wide range of surfactants along with the relative oil evaporation rates expected from a rate model shows their potential use for oil loss reduction [22].

17.2 MATERIALS

17.2.1 Surfactants for Charge Control

The single-component base oil for the charge control part of this work, referred to as NPG base oil, is neopentyl glycol dicaprate (decanoic acid,1,1′-(2,2-dimethyl-1,3-propanediyl) ester, CAS 27841-06-1). Trace amounts of butylated hydroxy toluene (BHT) stabilizer were present in some of the base oil lots. The following formulations were prepared and tested. Oil S contains approximately 0.5 wt% of a commercial conductivity additive (Stadis 450, Octel Starreon L.L.C., 8375 S. Willow St., Littleton, CO 80124). Stadis 450 is widely employed as the conductivity additive in jet airplane engine fuel [23,24]. From our chemical analysis, the active ingredients in Stadis 450 appear to be a mixture of polyamine and a polysulfone with dodecyl-benzene sulfonic acid. Stadis 450 was not analyzed for metallic elements such as Ba. Oil V contains 2 wt% of an oligomeric antioxidant (50 wt% reaction mixture of dioctyl diphenyl amine and phenyl naphthyl amine in a tetraester oil carrier, R.T. Vanderbilt, Vanlube 9317). Oligomeric antioxidants comprising aromatic amines are employed for high-temperature thermal stability of turbine oils [25]. The composition of Stadis-450 and the oligomer additives is further discussed in Ref. [26].

Several other additives that are likely to improve both the oil thermal stability and the electrical conductivity were incorporated in the NPG base oil for evaluation. Irganox L57 is octylated/butylated diphenyl amine (Ciba Specialty Chemicals). BHT is butylated hydroxy toluene, and BTA is 1H-benzotriazole; both are commercial reagent grades. OAN is the emeraldine form of oligomeric polyaniline with dodecyl benzene sulfonic acid (unsubstituted phenyl ring and a methyl group attached to the nitrogen). For example, see Ref. [27]. Polyaniline is insoluble in the base oil whereas oligoaniline is sparingly soluble in the base oil.

Oil A contained a para alkyl ester-substituted BHT (resembling Ciba Specialty Chemicals Irganox 1076 or L135), and a secondary aromatic amine resembling dioctyl diphenyl amine (also similar to Ciba Specialty Chemicals Irganox L57). Oil B is Oil A added with alkylated BTA. Oil C is Oil B with more of the aromatic amine.

17.2.2 SURFACTANTS FOR EVAPORATION CONTROL

The base oil for the surfactant studies was di(n-butyl) sebacate (DBS) [4]. This oil was selected because it is similar in structure and polarity to the actual fluid dynamic bearing oils, but it has a higher vapor pressure due to its lower molecular weight. The higher vapor pressure facilitates measuring the evaporation rate of the experimental oil formulations. The surfactants incorporated in the oil for surface tension reduction are listed in Table 17.1.

The Ztetraol and Zdol fluoroesters were custom synthesized in our laboratory. The chemical structures are shown in Table 17.2. Zdol and Ztetraol alone are completely insoluble in the fluid bearing motor oil. The solubility of the perfluoropolyether ester is adjusted by the hydrocarbon chain length (e.g., C6 or C8) and one or two hydrocarbon chains per end (Zdol or Ztetraol). These fluorosurfactants were synthesized by reacting low-molecular-weight Ztetraol 1000 or Zdol fraction with molecular weight 1000 with hexanoyl choride or octanoyl chloride (Sigma-Aldrich). Initially, 5 g of each perfluoropolyether was dissolved in 5 g of 3M Novec™ HFE-7100 solvent (methoxy-nonafluorobutane). The acyl chlorides were added with 20% molar excess to ensure complete conversion of the hydroxyl groups, 0.5 g at a time to avoid foaming, and mixed on a Vortex mixer between each addition. The solvent was stripped in a small Kugelrohr (short-path vacuum distillation apparatus). The distillation conditions were 60°C at 17 to 34 kPa vacuum, then 110°C at 98 kPa vacuum for 1 h. The composition and purity of the product was verified by Fourier transform infrared spectroscopy.

The PFD-HCA and PFP-HCA fluoroester surfactants were custom synthesized by Exfluor Corp., Round Rock, Texas. The distribution of the fluorosurfactants at the oil–air interface is schematically illustrated in Figure 17.1.

The Novec™ FC-4430 is a commercial nonionic polymeric fluorosurfactant that is based on perfluorobutane sulfonate and was provided by 3M Corp. The DuPont Zonyl 8857A and FSO-100 fluorosurfactants were selected for having good compatibility with the ester base oil.

Foam control agents, for example, alkyl polyacrylates and low-molecular-weight silicone oil, have been developed for use in lubricating oils [28]. These surfactants are also desirable for the control of aerosol mist in the fluid bearing cavity. Commercial

TABLE 17.1

Surfactants Tested for Evaporation Suppression by Surface Tension Reduction of Fluid Dynamic Bearing Motor Oil

Additive	Designation	wt%	Surface Tension, γ (mN/m)	$\Delta\gamma$ (mN/m)	Theoretical Relative Evaporation Rate (%)
Ztetraol ester C8 acid	Ztetraol B	0.5	22.64	10.29	19
Ztetraol ester C6 acid	Ztetraol C	0.8	21.07	11.86	15
Zdol ester C8 acid	Zdol B	0.7	21.27	11.66	15
Zdol ester C6 acid	Zdol C	0.5	23.23	9.70	21
Perfluoro-1H,1H,8H,8H-octanediol-di(n-hexanoate)	PFD-HCA6	0.7	32.34	0.59	91
Pentafluorophenyl hexanoate	PFP-HCA6	0.6	32.24	0.69	90
3M Novec 4430	Novec 4430	1.5	26.66	6.27	37
DuPont Zonyl FSO-100	Zonyl FSO-100	0.7	30.18	2.74	64
DuPont Zonyl 8857A	Zonyl 8857A	0.6	30.87	2.06	72
Ivanhoe FCA-1960	FCA-1960	0.7	31.65	1.27	82
Ivanhoe FCA-1910	FCA-1910	0.7	25.97	6.96	33
Brij 35	Brij 35	0.5	32.34	0.59	91
Brij 52	Brij 52	0.6	32.44	0.49	92
Brij 56	Brij 56	0.5	32.63	0.29	95
Brij 58	Brij 58	0.5	32.83	0.10	98
Brij 76	Brij 76	0.5	32.54	0.39	94
Brij 78	Brij 78	0.5	32.54	0.39	94
Polyamide	Sylvagel 6100	0.6	31.85	1.08	84

Note: The surface tension of the model base oil di(n-butyl) sebacate alone was 32.9 mN/m. The surface tension was measured at 25°C and the theoretical relative evaporation rate was calculated at 85°C.

foam control agents FCA-1960 and FCA-1910 provided by Ivanhoe Industries, Inc., Mundelein, Illinois were soluble in the fluid bearing motor oil.

Nonionic surfactant diblock copolymers (e.g., polyalkylene glycols) may be used to alter the droplet formation, hydrodynamic properties, and evaporation of the oil to provide evaporation and mist control. The diblock copolymer surfactant has two blocks: a nonpolar alkane and a polar polyether. In this case, the polar ether block is more soluble in the hydrocarbon ester base oil, whereas the nonpolar block, which is less soluble in the base oil separates out along the oil–air interface. This arrangement decreases the surface tension of the base oil. The structure and composition of the Brij polyaklyene glycols are given in Table 17.3. Variation of the surface activity is provided by the block lengths and the ratio of alkanol to polyethylene oxide.

Sylvagel 6100 is a polyamide gellant for use in low-polarity organic liquids. At low concentration, the Sylvagel is expected to also act as a surfactant, with negligible

TABLE 17.2
Molecular Structures of the Fluorosurfactants That Were Synthesized for Evaporation Suppression in Hydrocarbon Ester Oils

Name	Structure
Zdol ester	$R-CH_2CF_2O-[(CF_2CF_2O)_p(CF_2O)_q]_x-CF_2CH_2-R$

$$R=CH_3(CH_2)_y\overset{\displaystyle O}{\overset{\displaystyle \|}{C}}-O$$

Ztetraol ester	$R-CH_2\overset{R}{\underset{|}{C}HCH_2OCH_2CF_2O-[(CF_2CF_2O)_p(CF_2O)_q]_x-CF_2CH_2OCH_2\overset{R}{\underset{|}{C}HCH_2-R}$

$$R=CH_3(CH_2)_y\overset{\displaystyle O}{\overset{\displaystyle \|}{C}}-O$$

PFD-HCA6 ester	$CH_3(CH_2)_4\overset{\displaystyle O}{\overset{\displaystyle \|}{C}}-OCH_2(CF_2)_6CH_2O-\overset{\displaystyle O}{\overset{\displaystyle \|}{C}}-(CH_2)_4-CH_3$

PFP-HCA6 ester	

TABLE 17.3
Definition and Model Formula of the Brij Polyalkylene Glycol Samples
$$C_xH_{2x+1}(OCH_2CH_2)_yOH$$

Reagent	x	y
Brij 78	18	20
Brij 76	18	10
Brij 58	16	20
Brij 56	16	10
Brij 52	16	2
Brij 35	12	1

increase in the bulk viscosity of the oil. Sylvagel 6100 was provided by Arizona Chemical, LLC, Jacksonville, Florida.

17.2.3 Sample Preparation

Conductivity agents or surfactants were mixed into the base oil with a combination of stirring, sonication, and mild heating with a heat gun. Stirring was either by shaking or with a Vortex mixer. The nonconductive oligoaniline leukobase (colorless) was mixed with a stoichiometric amount of dodecylbenzene sulfonic acid, which formed the emeraldine (green-tinted) OAN. Even the low-molecular-weight OAN was not completely soluble in the ester base oil. The insoluble portion was filtered out after mixing (Pall Gelman Laboratory Acrodisc CR 25 mm Syringe filter with 1 μm PTFE membrane, P/N 4226 (VWR Scientific, 28143-928)).

Oil V with 2 wt% Vanlube 9317 in NPG base oil was prepared in our laboratory. Oil V Lot 1 and Oil V Lot 2 were two separate batches of 2% Vanlube 9317 in NPG base oil prepared by an oil vendor.

Fluid bearing disk drive spindle motors were fabricated with some of the experimental oils by a vendor.

17.3 APPARATUS AND PROCEDURES

17.3.1 Oil Dielectric Properties

Oil dielectric properties (relative permittivity and the loss factor) were measured using the TA Instruments Dielectric Analyzer (DEA) model 2970 with ceramic single surface sensors (1 mm spacing) at 50°C for 1 h in nitrogen. Each sensor was individually calibrated. Incoming sensors that deviated significantly were discarded. A thin film of the oil was smeared over the sensors with a wooden stick to cover the electrodes. Nominally, the DEA measurements were done at isothermal temperatures of 50°C and 80°C. At each temperature, six sinusoidal oscillation frequency sweeps were performed between 0.1 and 10,000 Hz, with 5 points per decade.

When DEA measurements were done at several different temperatures, frequency–temperature superposition was employed to extend the frequency range of the measurements by shifting along the frequency axis, relative to 50°C data. The frequency–temperature superposition procedure is illustrated with rheological and dielectric data for disk lubricants in Ref. [29].

The oil specific conductivity was usually derived from the loss factor data, some of which is shown in Figure 17.2a, by fitting with the limiting low-frequency form of the loss factor equation

$$\varepsilon'' = \frac{\sigma}{2\pi\varepsilon_0 f} \tag{17.1}$$

where ε'' is the dielectric loss factor, σ is the specific conductivity in S/m, ε_0 is the absolute permittivity of free space in F/m, and f is the excitation frequency in Hz. In some cases, single point values of conductivity were also calculated from the loss

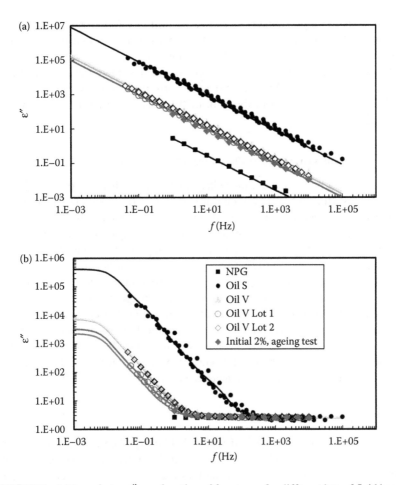

FIGURE 17.2 (a) Loss factor ε'' as a function of frequency for different lots of fluid bearing oils. (b) Relative permittivity ε'' as a function of frequency comparing different lots of fluid bearing oils. The smooth curves are fit to the discrete relaxation time series used to derive dc relative permittivity. The initial 2% aging test oil is another sample of Oil V prepared in our laboratory before aging. Reference temperature 50°C.

factor measured at 1 Hz, which is nearly the same as the specific conductivity calculated from the fit to the data over the whole range of frequency.

The relative permittivity provides a measure of the electric charge stored in the oil. Typical values of the relative permittivity, ε'', as a function of frequency are shown in Figure 17.2b. The relative permittivity data were fit to a discrete relaxation time series model:

$$\varepsilon' = \varepsilon_u + \sum_i \frac{\varepsilon_{r,i} - \varepsilon_u}{1 + \left(2\pi\tau_i f\right)^2} \qquad (17.2)$$

Here, ε_u is the high-frequency relative permittivity (often called the dielectric constant), and is about 2 for nonpolar hydrocarbons such as n-dodecane, $\varepsilon_{r,i}$ is the ith relaxation magnitude, and τ_i is the ith relaxation time.

The dc relative permittivity is the point where the smooth curves intersect the vertical axis in Figure 17.2b. The dc relative permittivity $\varepsilon'(0)$ is important because the electrical charge stored in the bulk of the fluid bearing oil is proportional to $\varepsilon'(0) = \sum_i \varepsilon_{r,i}$.

17.3.2 MOTOR VOLTAGE AND PEAK DISCHARGE CURRENT

Test rigs were set up to measure the voltage and peak discharge current on disk drive spindle motors while they were spinning at 10,000 or 15,000 rpm. The stator (stationary center part of the bearing) was mounted onto a conducting metal base. Electrical contact to the spinning rotor (rotating outer part of the bearing) was made with a conductive fiber brush or a small wire. Measurements were performed with a high impedance electrometer (Keithly model 617). The negative (ground) terminal of the electrometer was connected to the stator. For peak discharge current measurement, the electrometer was set to current sensing mode. The analog voltage output from the electrometer was acquired by a personal computer with an analog-to-digital converter card. Typical plots of the discharge current transient are shown in Figure 17.3. The highest current during the initial discharge transient when contact was first made between the stator and the hub is referred to as the peak discharge current. Motor voltage, v_0, and peak discharge current, i_p, are the average values of measurements on at least five motors.

17.3.3 SURFACE TENSION

The surface tension was measured with a Wilhelmy plate tensiometer. The plate was fabricated from a stainless-steel sheet with 1 cm perimeter. The surface of the plate was uniformly roughened by bead blasting. For each measurement, the plate was suspended above a pan of oil with the pan on an analytical balance accurate to within ±0.1 mg. The change in the reading of the oil pan weight was recorded when the plate first made contact with the surface of the oil. The surface tension (mN/m) = pull force (g) × 980. Note: mN/m = milli-Newton/meter, which is equivalent to the cgs unit of dyne/cm. Between measurements, the plate was cleaned with acetone, which is a good solvent for both the oil and the surfactants. The surface tension of the base oil and the base oil containing surfactants is given in Table 17.1.

17.4 RESULTS

17.4.1 DIELECTRIC SPECTROSCOPY

The dielectric loss factor and relative permittivity for the prototype oils and initial production batches are shown in Figure 17.2. There are orders of magnitude difference between the NPG base oil, the Oil V containing 2% Vanlube 9317, and Oil S (containing Stadis-450). The specific conductivity, relative permittivity at 1 Hz, $\varepsilon'(1\ \text{Hz})$, and dc relative permittivity, $\varepsilon'(0)$, for these oils are listed in Table 17.4.

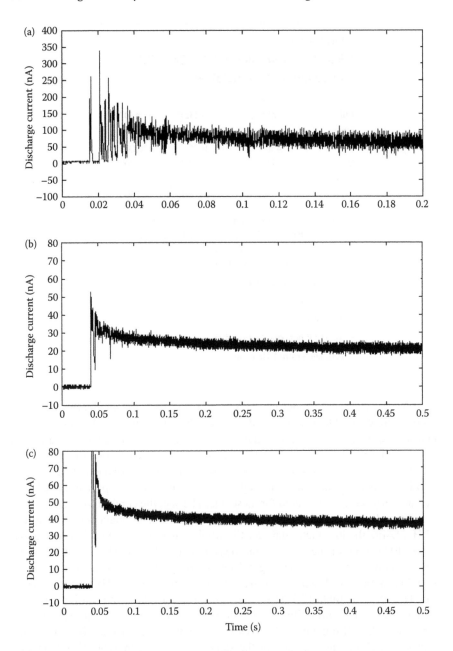

FIGURE 17.3 Typical discharge current measured on 15 k rpm fluid dynamic motor bearings. (a) NPG base oil, (b) original Oil V (2 wt% Vanlube 9317 in NPG base oil), and (c) vendor Oil C. (Courtesy of T-C. Fu, Hitachi Global Storage Technologies, Inc., San Jose, California.)

TABLE 17.4

Dielectric Properties of NPG Base Oil and Vendor Oil Formulations

Oil	σ (nS/m)	$\varepsilon'(1\ Hz)^a$	$\varepsilon'(0)^{b,a}$
NPG	0.1[c]	2[d]	2.5[e]
S	450	1000[b]	420,000
V	11.0	9.7	7800
V Lot 1	5.5	5.9	3300
V Lot 2	8.5	8.4	7300

Note: The oil specific conductivity from a regression fit to loss factor vs. frequency is σ, relative permittivity at 1 Hz is $\varepsilon'(1\ Hz)$, and dc relative permittivity is $\varepsilon'(0)$. Measurement temperature is 50°C.

[a] From relaxation time fit.

[b] Sum of relaxation strengths.

[c] Typical value.

[d] Value for nonpolar hydrocarbon *n*-dodecane.

[e] Estimated from 1 Hz point.

Variations in electrical properties between oil lots that are nominally the same, Original Oil V, and Oil V Lots 1 and 2 are attributed to uncontrolled low levels of impurities in the base oil. These impurities were present in such small amounts that they could not be identified by NMR spectroscopy measurements on the oils.

The dielectric properties of oils containing other types of conductivity agents are listed in Table 17.5. The low-frequency region of the dielectric spectrum needed to calculate $\varepsilon'(0)$ was not measured for these oils. The specific conductivity and relative permittivity listed in Table 17.5 are estimated from the DEA data points measured at 1 Hz. Using the loss factor data for the whole range of frequency between about 0.1 and 10,000 Hz, or the single point measured at 1 Hz, provides nearly the same value for conductivity. For the original Oil V, using all the loss factor data, gives 11.0 nS/m in Table 17.4, whereas using only the point value at 1 Hz for the same oil gives 11.5 nS/m in Table 17.5. The same holds true for relative permittivity. For the original Oil V, the permittivity from the three-relaxation time curve fit to the relative permittivity data (Equation 17.2) gives 9.7 in Table 17.4, whereas the single low-frequency (1 Hz) point estimate for the same oil gives 9.2 in Table 17.5. However, due to the large increase in ε' with decreasing frequency below 1 Hz (Figure 17.2b) ε' measured at 1 Hz cannot be used to estimate $\varepsilon'(0)$.

Among the various additives that were tested, the OAN was the most interesting. The filtered oil containing this additive was greenish tinted due to the fluorescence of the conjugated polymer [30,31]. However, even after filtration, there was still a tendency for sedimentation. When the oil containing OAN was held at 100°C for

TABLE 17.5
Dielectric Properties of NPG Base Oil with Different Types of Additives

Additive(s)	Concentration (wt%)	σ (nS/m)	ε′(1 Hz)
Vanlube 9317 (Oil V)	2	11.5	9.2
Irganox L57	1	0.2	2.8
BHT	0.7	0.2	3.1
OAN	0.1	2.0	3.1
	0.5	6.0	3.0
BTA	0.05	0.7	2.3
	0.50	0.7	2.4
A	0.1–1	0.6	2.7
B		0.7	3.0
C		0.7	3.0

Note: Single point values at 1 Hz, 50°C.

several hundred hours, sedimentation and conductivity loss were observed. From NMR spectroscopy, it appeared that the oligoaniline had chemically reacted with the diester base oil, which formed a gel.

All of the additives provided some increase in specific conductivity while the relative permittivity remained at least two orders of magnitude lower than that of Oil S.

The dielectric properties of Vanlube 9317 in NPG base oil with additive concentration between 0.5 and 4.0 wt% are shown in Tables 17.6 and 17.7. The samples in

TABLE 17.6
Dielectric Properties of NPG Base Oil with Different Amounts of Vanlube 9317 Showing the Results from Two Different Batches Prepared by a Vendor

Batch Code	Vanlube 9317 (wt%)	σ (nS/m)	ε′(1 Hz)
Amber	0.5	1.8	3.6
	1.0	2.5	3.7
	2.0	3.0	4.2
	4.0	6.5	6.1
Red	0.5	1.6	3
	1.0	2.9	4.2
	2.0	3.6	4.5
	4.0	4.5	4.8
	2.0 + 0.5% alkylated BTA	4.8	7.0

Note: Single point values measured at 1 Hz, 50°C.

TABLE 17.7
Dielectric Properties of NPG Base Oil with Different Amounts of Vanlube 9317 Prepared in Our Laboratory

Vanlube 9317 (wt%)	σ (nS/m)	ε'(1 Hz)
0.5	3.0	4.1
1.0	2.6	3.6
2.0	3.9	4.4
4.0	5.8	5.1

Note: Single point values measured at 1 Hz, 50°C.

Table 17.6 were prepared by a vendor, and the samples in Table 17.7 were prepared in our lab from a different lot of NPG base oil. With a 4× increase in concentration, the specific conductivity increased by 3 to 4×, and the relative permittivity at 1 Hz increased by less than 1.7×.

17.4.2 MOTOR VOLTAGE AND PEAK DISCHARGE CURRENT

Typical bearing current versus time from the initial electrical contact measured on running motors are shown in Figure 17.3. There are three main components of the current: (1) initial transient(s) that decay within 200–500 ms, (2) steady-state (run) current, and (3) a high-frequency noise component. The high-frequency noise is probably from the sliding contact between the conductive brush and the spinning motor hub. The steady-state current is caused by streaming charge separation in the oil shear flow [32]. The initial transient current flow is the dissipation of capacitive (electronic or ionic) charge separation. The peak of the initial transient current flow, or the peak discharge current, i_p, and the steady-state running voltage, v_0, are listed in Table 17.8.

Since v_0 is measured by a high input impedance electrometer, it is the voltage that corresponds to the fully charged state of the bearing (capacitor). The voltage on a running motor is between −400 and +140 mV. With respect to the stator as ground, the running voltage polarity was negative except for the motors built with Oil V, in which case the polarity was positive. The difference in polarity is probably because the charge transport mechanism for Vanlube 9317 is electronic hopping whereas the transport mechanism for the other conductivity additives is ionic. Even with the high-conductivity Oil S, the motor voltage is about the same as with the pure base oil, indicating that the motor voltage is controlled by properties other than oil conductivity and permittivity alone.

In contrast to motor voltage, the peak discharge current exhibits a clear dependence on additive formulation. By far, the highest peak discharge current is with the Oil S formulation, which has the highest conductivity and dc permittivity. There appears to be a relationship between the peak discharge current and the specific conductivity of the oil, which will be explored in Section 17.5.2.

TABLE 17.8

Electrical Properties of Fluid Bearing Motors Built with Experimental Oil Formulations

Motor rpm	Oil	v_0 (mV)	i_p (nA)	$\lvert i_p/v_0 \rvert$ (μS)	$\lvert v_0 \varepsilon'(0)\varepsilon_0 \rvert$ (C/m)
15 k	NPG	−280	338	1.2	6.2×10^{-12}
10 k	S	−270	1264	4.7	1.0×10^{-6}
15 k		−230	1378	6.0	8.5×10^{-7}
	V	+140	36	0.26	9.7×10^{-9}

Additives in NPG Base Oil, 15 k rpm Motors

Additive(s)	Concentration (wt%)	v_0 (mV)	i_p (nA)	$\lvert i_p/v_0 \rvert$ (μS)
Irganox L57	1	−220	43	0.20
BHT	0.7	−100	32	0.32
OAN	0.1	−150	34	0.23
	0.5	−400	112	0.28
BTA	0.05	−190	41	0.22
	0.50	−280	55	0.2
A	0.1 to 1	−111	215	1.95
B		−80	18	0.23
C		−330	81	0.25

Note: Shown are the average values from at least five motors built with each oil formulation. Motor voltage is v_0; peak discharge current is i_p. The magnitude of the motor peak discharge conductance is $\lvert i_p/v_0 \rvert$ and $\lvert v_0 \varepsilon'(0)\varepsilon_0 \rvert$ is the specific charge in the oil. ε_0 is the absolute permittivity of free space in F/m. F is farad = C/volt, and C is coulomb. Low-frequency data was not measured for all of the additives.

17.4.3 Surface Tension

The surface tension of the DBS base oil and the oil containing the amount of each surfactant listed in Table 17.1 is shown in Figure 17.4a. The surfactant effect on the theoretical reduction of the oil evaporation rate is calculated in Section 17.5.4.

17.5 DISCUSSION

17.5.1 RC Equivalent Circuit Model

A model is useful for the interpretation and understanding of the relationship between the oil dielectric properties and the peak discharge current. An RC equivalent circuit, which is a reasonable first step toward modeling the electrical discharge phenomenon, is shown in Figure 17.5. The motor self-charges to a voltage v_0, which is present until there is a short circuit between the rotor and the stator, as in the

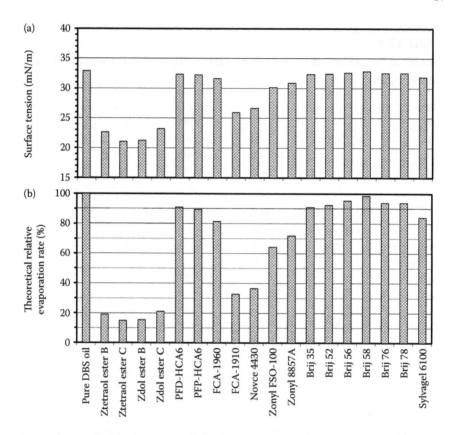

FIGURE 17.4 (a) Surface tension of the DBS base oil and the base oil containing surfactants and (b) the theoretical evaporation rate as a percent of the evaporation rate of the base oil calculated from the surface tension reduction by the surfactant.

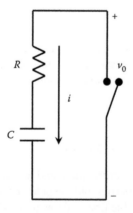

FIGURE 17.5 Series RC equivalent circuit model for the motor discharge transient. The switch is suddenly closed at time $t = 0$ with initial voltage $v = v_0$.

peak discharge current measurement. The transient current following the switch closure is

$$i(t) = \frac{v_0}{R} e^{-t/RC} \qquad (17.3)$$

where R is the spindle resistance in ohms, C is the spindle capacitance in Farads, and t is time. The total electric charge and energy in the fluid bearing oil is proportional to dc relative permittivity, $\varepsilon'(0)$ according to

$$\text{Electric charge} = v_0 C \propto v_0 \varepsilon'(0)\varepsilon_0 \qquad (17.4)$$

and

$$\text{Electric energy} = \frac{1}{2} v_0^2 C \propto \frac{1}{2} v_0^2 \varepsilon'(0)\varepsilon_0, \qquad (17.5)$$

where $\varepsilon_0 \approx 8.85 \times 10^{-12}$ F/m is the absolute permittivity of free space. The peak discharge current, i_p, is proportional to the oil specific conductivity, σ:

$$\text{Peak discharge current} = i_p = \frac{v_0}{R} \propto v_0 \sigma. \qquad (17.6)$$

From the RC model, it follows that the magnitude (absolute value) of the peak discharge current ratio to the motor voltage, $|i_p/v_0|$, is proportional to oil specific conductivity. This means that the charge stored in the bulk of the oil should discharge more quickly through oil that has a higher conductivity. Hence, a log–log plot of $|i_p/v_0|$ versus σ is ideally a straight line with slope 1.

17.5.2 PEAK DISCHARGE CURRENT

The peak discharge current is compared with the dielectric properties of the oil. From the RC model, the absolute value of the ratio of the peak discharge current to the motor voltage, $|i_p/v_0|$, is proportional to the oil conductivity. The quantity $|i_p/v_0|$ is referred to as the peak discharge conductance. The peak discharge conductance for motors built with various oil formulations is listed in Table 17.8. A log–log plot of the peak discharge conductance versus oil conductivity is shown in Figure 17.6. The proportionality expected from the RC model is shown by the straight line, which was fitted through the Oil S data point. Deviation from the proportional response increases with decreasing specific conductivity. The peak discharge conductance increased at low conductivity for the NPG base oil, whereas it leveled off for the other oils with conductivity below Oil V (except for Oil A, which had an unusually high peak discharge current). Some of the variation in $|i_p/v_0|$ could be from differences between individual motors within the tolerances of the manufacturing process.

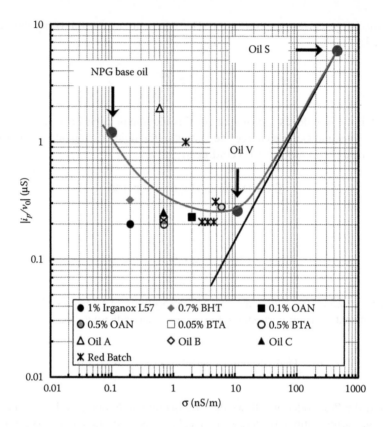

FIGURE 17.6 Motor peak discharge conductance versus oil specific conductivity for 15 k rpm fluid dynamic bearing motors built with various oil formulations. The smooth curve is drawn through the data points. The line is from a fit of the RC model to the Oil S data point.

The smooth curve sketched in Figure 17.6 shows what seems to be a tendency for the peak discharge conductance to have a minimum near 8 nS/m. This minimum is formed by the transition from ionic charge separation at high conductivity to electrostatic charge separation at low oil conductivity. For specific conductivities higher than about 10 nS/m, the peak discharge conductance is controlled by the ionic charge storage and dissipation through the bulk of the oil, which follows the RC model. The deviation from the RC model increases as the conductivity decreases below 10 nS/m, and the peak discharge current even increased for the very-low-conductivity NPG base oil. This type of deviation suggests that there is another charge/discharge mechanism, which increases in importance relative to that described by the RC model, in the limit of low conductivity. It may be that an electrostatic charge develops on the surfaces of the bearing with low-conductivity oils. Dissipation of the electrostatic charge is then mostly through the external circuit and does not rely on transport through the bulk of the oil. Hence, the peak discharge current does not decrease with decreasing conductivity when the oil conductivity is very low. Overall, the smooth curve with the minimum in Figure 17.6 shows that there is a minimum in the peak discharge conductance for oils with a specific conductance between 2 and 20 nS/m.

Combining the RC model (Equation 17.6) with the bearing geometry provides an equation for the peak discharge conductance for the higher-conductivity oils:

$$\left| \frac{i_p}{v_0} \right| = \left(\frac{A}{d} \right) \sigma, \tag{17.7}$$

where the geometry factor A/d is the ratio of the area A to the plate separation d of the equivalent parallel plate capacitor. From the RC model fit to the Oil S motor data point (solid line in Figure 17.6), $A/d = 13.3$ m. The geometry factor of the equivalent parallel plate capacitor will differ somewhat from the actual fluid bearing. The actual fluid bearing has cylindrical and parallel disk, or conical, regions, which provide radial and axial stiffness [33]. The equivalent geometry factor A/d is employed to calculate the fluid bearing resistance $R = 1/(\sigma(A/d))$, and the capacitance $C = \varepsilon'(0)$ $\varepsilon_0(A/d)$. For the fluid bearing motors built with type V oils and the oil properties in Table 17.4, $6.8 < R < 13.7$ MΩ and $0.38 < C < 0.94$ µF.

17.5.3 ELECTRIC CHARGE

A plot of the specific charge in fluid bearing $\left| v_0 \varepsilon'(0)\varepsilon_0 \right|$ versus specific conductivity σ is shown in Figure 17.7a. The power law slope of the regression line fit to the data is 1.4. This power law slope is consistent with a volume to surface area mapping. The specific charge in the bearing is approximately given by $\left| v_0 \varepsilon'(0)\varepsilon_0 \right| \propto \sigma^{3/2}$. The power law slope of 3/2 implies that bulk charge carriers accumulate at the bearing surfaces (rotor and stator).

Although they are generally considered to be independent from one another, the relative permittivity increased nearly linearly with specific conductivity. The relative permittivity for separately prepared samples with different concentrations of Vanlube 9317 in NPG base oil is plotted as a function of specific conductivity in Figure 17.8. The samples of different concentrations that are labeled as aged were held in an oven at 100°C for up to 1400 h, which slightly increased conductivity and permittivity. From the linear regression fit in Figure 17.8, $\varepsilon'(1 \text{ Hz}) = 2.5 + 0.5\sigma$, with σ in nS/m.

17.5.3.1 Slider Disk Contact

In the event of an asperity contact between the magnetic recording slider and the disk, the bearing could discharge through this tiny contact area causing damage to the recording head [34]. The asperity contact resistance is $R_c \approx 2/(\sigma_c(A_c/l_c))$, where σ_c is the specific conductivity of the carbon overcoat, A_c is the asperity contact area, and l_c is the length of the contact. The factor of 2 accounts for the resistance of the slider overcoat, which is assumed to have the same thickness and conductivity as the disk overcoat. With typical values of $\sigma_c \approx 10^{-10}$ S/m, $A_c \approx 10^{-12}$ m^2, and $l_c \approx 5 \times 10^{-9}$ m, then $R_c \approx 10^{17}$ Ω. Consequently, electric current flow through an asperity contact is expected to be negligibly small.

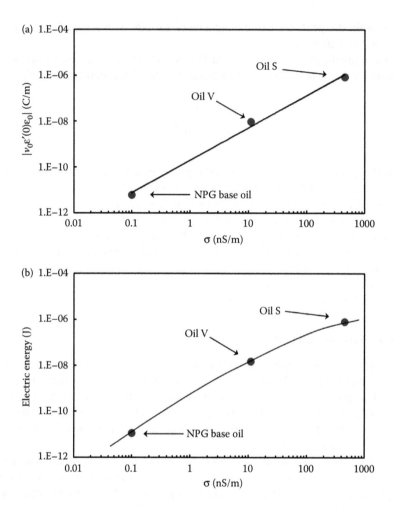

FIGURE 17.7 (a) Specific electric charge in the bearing oil (Equation 17.4) and (b) total electric energy stored in the fluid bearing (Equation 17.5) versus oil specific conductivity for 15 k rpm fluid bearing motors.

17.5.3.2 Corona Discharge

Damaged magnetic recording disk surfaces occasionally exhibit isolated defect sites, which may result from local melting and vaporization. Magnetic recording sliders sometimes show pitting removal of ceramic occlusions as particles from the ceramic slider body, usually near the trailing edges and corners. These particles may be found embedded near scratches in the disk overcoat and magnetic layers. Most of these defects are traced back to particulate contamination or damage during magnetic recording disk or disk drive manufacturing and assembly. Fluid bearing motor voltage and discharge are considered as possible contributing factors in the formation of such defects.

Defects could be created by an electric arc discharge [35] across the air gap between the magnetic recording disk and slider. At some critical electric field,

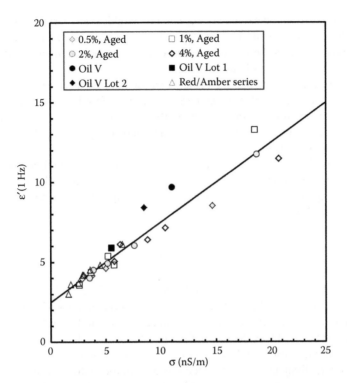

FIGURE 17.8   Relative permittivity ε' at 1 Hz plotted as a function of specific conductivity σ for various samples of NPG base oil containing Vanlube 9317. Concentration is wt%. The initial values for the aged samples are given in Table 17.7. The composition of the Red/Amber series is given in Table 17.6.

emitted electrons start to ionize the material that is being ejected from the cathode, which forms highly conductive plasma. The resulting electric arc, which can be as short lived as a nanosecond, forms molten debris and craters on the cathode [35,36].

Nitrogenated carbon overcoats on magnetic recording disks typically contain 10–15 at% nitrogen. The critical electric field for similar nitrogenated carbon films is 12×10^6 to 17×10^6 V/m, for electric arc discharge between the film and a steel ball in a vacuum [36]. For a magnetic recording slider flying with 10 nm gap between the slider and the disk, the range of critical voltage would then be 120–170 mV. That is within the range of motor running voltages measured for motors built with the various oil formulations listed in Table 17.8.

The gap between the slider and the disk is maintained by an air bearing, which generates pressure that provides the air bearing stiffness. The gap is less than the mean free path of air [37]. Since the gap is less than the mean free path of air, ions ejected from the cathode surface will more often impact on the anode surface before colliding with another gas molecule. The disk carbon overcoat is covered with about 1 nm of perfluoropolyether lubricant. Both the disk and slider surfaces contain adsorbed water and organic and inorganic compounds from the environment [38]. Since there is a wide range of materials on the surfaces, the work function to ionize material from the cathode is expected to exhibit some variation. Overall, the critical electric field should be most sensitive to the type of disk lubricant and the atomic composition of the overcoat because those are the dominant constituents of the surfaces at low relative humidity [39].

It is interesting to note that the critical electric field for discharge increases with decreasing conductivity of nitrogenated carbon, probably because there are fewer electrons available for emission to initiate the arc [36]. Also, the site of the erosion could be either on the slider or the disk surface, depending on the polarity of voltage across the motor bearing.

Conditions for arc discharge between the disk and slider could potentially exist for most of the oil formulations listed in Table 17.8. If the motor bearing fully discharges through the arc, the maximum adiabatic temperature rise ΔT in a volume of surface material V is

$$\Delta T = \frac{1}{2} \frac{v_0^2 C}{C_p V} \tag{17.8}$$

where $v_0^2 C/2$ is the electric energy in the bearing and C_p is the heat capacity of the surface material. The volume of surface material is the amount of cathode (disk or slider) overcoat and underlying layers, which participate in the discharge event. This is the volume that is ablated to a pit in the case of large ΔT. The actual ΔT will be less than the maximum estimated by Equation 17.8 if the discharge event does not last long enough to dissipate all of the electric charge in the motor bearing.

The maximum temperature increase by conversion of all the electric energy stored in the motor bearing to thermal energy according to Equation 17.8 is estimated as follows. The total electric energy in the fluid bearing, $v_0^2 C/2 = v_0^2 \varepsilon'(0)\varepsilon_0(A/d)/2$, is shown plotted as a function of specific conductivity in Figure 17.7b. In the limit of low σ, $v_0^2 C/2$ approaches proportionality to σ^2, and in the limit of high σ, $v_0^2 C/2$ approaches direct proportionality to σ. As with the two regions seen in the dependence of $|i_p/v_0|$ on σ, this change in the functional dependence on σ may indicate a transition from mostly electrostatic surface charge in the limit of low conductivity to bulk charge in the high conductivity limit.

A large ablation pit in the magnetic recording disk or slider is 1 μm long × 1 μm wide × 0.2 μm deep. The volume of this pit is used to estimate the volume V in Equation 17.8. The volumetric heat capacity of solids is typically $1.2 < C_p < 4.5$ MJ/m³K [40]. This estimate uses mid-range $C_p = 2.85$ MJ/m³K. For the Original Oil V, with $v_0^2 C/2 \approx 10^{-8}$ J from Figure 17.7b, the temperature rise $\Delta T \approx 175,000°$C, which is more than sufficient to vaporize material from the surface. In practice, the discharge is limited by the high relative velocity between the disk and slider and the finite characteristic time τ for discharge of the motor bearing. The motor bearing R and C calculated from the dielectric property measurements on the bulk oil (Section 17.5.2) provide $2.6 < \tau = RC < 13$ s. The discharge current charts measured on running motors in Figure 17.3 indicate a much shorter $\tau \approx 0.02$ s for decay of the initial transient. Thus, it is important to minimize both the motor bearing capacitance (dc relative permittivity of the oil) and the motor voltage in addition to providing sufficient conductivity.

17.5.4   SURFACE TENSION AND EVAPORATION RATE

As mentioned in the introduction section, the decrease in the surface tension by a surfactant layer can, in principle, be employed to estimate the increase in the activation energy barrier for an oil molecule to transition from the liquid to the vapor phase. The additional energy barrier is approximated by the energy to form a hole in the surfactant monolayer with the cross-sectional area of an oil molecule. The hole area is estimated from πa^2 where the equivalent circular radius of an oil molecule is $a \approx 0.5$ nm. The difference between the surface tension of the pure base oil and the surface tension of the oil containing the surfactant is $\Delta\gamma$. The energy to form the hole for an oil molecule to evaporate through the surfactant monolayer is then $\pi a^2 \Delta\gamma$. According to the rate model for evaporation [18], the hole energy adds to the vaporization "activation" energy. The effect of the surfactant on the evaporation rate is then approximately given by

$$\text{Relative evaporation rate } (\%) = 100 \times e^{-\left(\frac{\pi a^2 \Delta\gamma}{kT}\right)} \tag{17.9}$$

where k is the Boltzmann constant and T is the absolute temperature. The theoretical relative evaporation rate is listed in the last column of Table 17.1, and plotted in Figure 17.4. Fluorosurfactants derived from the perfluoropolyethers Zdol and Ztetraol are predicted to provide most suppression of oil evaporation. The FCA-1910 and Novec 4430 should also be considered for further evaluation. To validate the model prediction, motor oils containing the most promising surfactants should be tested in motors operating for extended periods of time at the use temperature.

Several factors that were not included in the model may become significant in the running motor. Measurements on the quiescent liquid at ambient temperature are insufficient to accurately predict the performance at the use temperature because the surface tension reduction (evaporation suppression) is temperature dependent through the temperature dependence of surfactant solubility. Surface shear flow in the meniscus between the rotor and stator was not included in the derivation of Equation 17.9. For example, surface shear flow may form dislocations or patches of surfactant and bare liquid.

17.6   SUMMARY AND CONCLUSIONS

Fluid bearing motor oil formulations with a variety of different conductivity improvers were prepared. The dielectric properties of the oil formulations were characterized by dielectric spectroscopy. Voltage and peak discharge current were measured on running motors built with the characterized oils. A series RC equivalent circuit model was developed to relate the motor voltage and peak discharge current with oil specific conductivity. There was a minimum in the peak discharge conductance near a specific conductivity of 8 nS/m. This minimum is attributed to a transition from primarily electrostatic surface charge with very-low-conductivity oils to bulk charge separation in the higher-conductivity oils. The electric charge separation increased with specific conductivity to the 1.4 power. Overall, the relative permittivity measured at 1 Hz increased linearly with specific conductivity.

In the magnetic recording disk drive, the electrical resistance of asperity contacts is so large that only a negligible charge is expected to flow by this pathway. However, the electric field across the air bearing of low-flying sliders over the range of observed motor running voltage is close to the critical field for electric arc discharge from nitrogenated carbon overcoats. A sufficiently energetic arc discharge could lead to pitting and debris generation. The best conductivity improver for charge control in fluid bearing motors provides sufficient conductivity to dissipate electrostatic charge with the least amount of electric energy stored in the fluid bearing.

High conductivity is the primary criterion for static charge dissipation in fuels and oils. Since the hydrocarbon is nonpolar, hydrocarbon conductivity is most often provided by an additive that solubilizes metal ions, and hence provides ionic conductivity, such as the mixed polymers with dodecylbenzene sulfonic acid of oil S. However, this method of charge transport incorporates physical charge separation because of the ionic charge transport mechanism. The ionic charge separation, in conjunction with high ionic mobility leads to a high discharge current. An alternative mechanism for conductivity in a nonpolar medium is by hopping conduction through electron donor–acceptor interactions [41]. In this case, there is no transport of metal ions or organic anions. Therefore, the oils V, with the nonionic polyaromatic oligomeric amine additive provide significantly higher conductivity than the pure base oil by hopping conduction. Oils V have a much lower dc permittivity than oil S while increasing conductivity enough to dissipate the static charge. Thus, the oils V are preferable for magnetic recording disk drive motor bearing.

A wide variety of surfactants suitable for use in hydrocarbon diester fluid dynamic bearing oils were selected and the surface tension was measured in a model fluid bearing oil di(n-butyl) sebacate. The corresponding reduction in the oil evaporation rate through a quiescent surfactant monolayer was estimated with a first-order rate model. The most promising surfactants for limiting the evaporation rate from the fluid bearing oil meniscus are those based on perfluoropolyether diol diester or tetraol tetraester with n-hexanoic or n-octanoic acid. These compounds provide the greatest differential solubility of the oleophobic block (perfluoropolyether) and the oleophilic block (hydrocarbon) in the oil. The hydrocarbon ester block is soluble and resides in the oil. The perfluoropolyether block is insoluble in the oil and resides at the oil–air interface. Since it provides the greatest decrease in surface tension, the perfluoropolyether forms the highest-pressure monolayer (as inferred from surface tension reduction). Consequently, it is energetically more unfavorable for an oil molecule to traverse through the perfluoropolyether monolayer in the process of vaporization.

The model does not account for shearing of the surfactant monolayer. It is possible that shear could form dislocations or free volume in the monolayer. Such defects provide evaporation sites from the bulk oil through the monolayer. However, some evaporation reduction should still be provided by the intact regions of the monolayer.

ACKNOWLEDGMENTS

Key technical contributions and guidance were provided by R. Kroeker of HGST San Jose Development Laboratory Advanced Mechanical Integration, and T-C. Fu and S. Y. Wong of HGST San Jose Development Laboratory Drive Mechanical

Development. Chemical and dielectric analyses were provided by J. M. Burns, M. D. Carter, and C. Hignite in the HGST San Jose Materials Analysis Laboratory. Thanks are due to O. Melroy and B. Marchon at the HGST San Jose Research Center for encouragement and discussions throughout this work. The author is grateful for the assistance of S. Saida and T. Watanabe at Nidec Corporation for preparation of the commercial oil lots and the specially built motors. A. Afzali in IBM Yorktown Research provided the OAN.

REFERENCES

1. S. Deeyiengyang and K. Ono, Analysis of ball-bearing vibrations of hard disk spindles, *J. Inf. Stor. Proc. Syst.*, *3*, 89–99, 2001.
2. R. Wood, Future hard disk drive systems, *J. Magn. Magn. Mater.*, *321*, 555–561, 2009.
3. C-P. R. Ku, J-Y. Juang, X. Sun, L. Huang, and F-Y. Huang, High frequency radial mode vibration in hard disk drive, *IEEE T. Magn.*, *47*, 1893–1898, 2011.
4. T.E. Karis, Lubricants for the disk drive industry, in: *Synthetics, Mineral Oils, and Bio-Based Fluids*, 2nd Edition, L. Rudnick (Ed.), CRC Press, Boca Raton, 2012.
5. J. Gavis and I. Koszman, Development of charge in low conductivity liquids flowing past surfaces: A theory of the phenomenon in tubes, *J. Colloid Interf. Sci.*, *16*, 375–391, 1961.
6. W-H. Tang, H-W. Yang, S-L. Yang, and C. Wu, Electrostatic discharge of aviation oil and damage protection, *Adv. Mat. Res.*, *393–395*, 1193–1197, 2012.
7. B-K. Tan, B. Liu, Y. Ma, M. Zhang, and S-F. Ling, Effect of electrostatic force on slider-lubricant interaction, *IEEE T. Magn.*, *43*, 2241–2243, 2007.
8. A.F. Diaz, R.D. Johnson, T.E. Karis, H.S. Nagaraj, and M.T. Nguyen, US Patent 5,641,841, Conductive lubricant for magnetic disk drives, 1997.
9. A.F. Diaz, R.D. Johnson, T.E. Karis, H.S. Nagaraj, and M.T. Nguyen, US Patent 5,744,431, Conductive lubricant for magnetic disk drives, 1998.
10. A.F. Diaz, R.D. Johnson, T.E. Karis, H.S. Nagaraj, and M.T. Nguyen, US Patent 5,886,854, Conductive lubricant for magnetic disk drives, 1999.
11. R.U. Khan, S. Murthy, N.S. Parsoneault, and H.L. Leuthold, US Patent 5,940,246, Disc drive hydro bearing lubricant with electrically conductive non-metallic additive, 1999.
12. Z. Feng, C. Shih, V. Gubbi, and F. Poon, A study of tribo-charge/emission at the head-disk interface, *J. Appl. Phys.*, *85*, 5615–5617, 1999.
13. M. Miyatake, F. Kugiya, and N. Kodama, Analysis of discharge mechanism in HDD, *IEEE T. Device Mat. Re.*, *11*, 323–327, 2011.
14. J.M. Burns, T-C. Fu, A.K. Hanlon, C. Hignite, T.E. Karis, R.M. Kroeker, and S.Y. Wong, US Patent 7,212,376, Disk drive system with hydrodynamic bearing lubricant having charge-control additive comprising dioctyldiphenylamine and/or oligomer thereof, 2007.
15. T.E. Karis and H.S. Nagaraj, Evaporation and flow properties of several hydrocarbon oils, *Tribol. T.*, *43*, 758–766, 2000.
16. R.J. Archer, and V.K. La Mer, The rate of evaporation of water through fatty acid mono-layers, *J. Phys. Chem.*, *59*, 200–208, 1955.
17. E.L. Foulds and R.G. Dressler, Performance of monolayer blends of odd and even carbon chain alcohols in water evaporation suppression, *Ind. Eng. Chem. Prod. Res. Dev.*, *7*, 75–79, 1968.
18. G.T. Barnes, The effects of monolayers on the evaporation of liquids, *Adv. Colloid Interfac.*, *25*, 89–200, 1986.
19. T.D. Tang, M.T. Pauken, S.M. Jeter, and S.I. Abdel-Khalik, On the use of monolayers to reduce evaporation from stationary water pools, *J. Heat Transf.*, *115*, 209–214, 1993.
20. A.Imhof, and D.J. Pine, Stability of nonaqueous emulsions, *J. Colloid Interf. Sci.*, *192*, 368–374, 1997.

21. A.W. Adamson, *Physical Chemistry of Surfaces*, John Wiley & Sons, New York, 1976.
22. F. Hendriks, T.E. Karis, and K. Shida, US Patent 7,998,913, Flow modifiers for improved magnetic recording device, 2011.
23. H.L. Walmsley, The avoidance of electrostatic hazards in the petroleum industry: Special issue, *J. Electrostat.*, *27*, 1992.
24. Wikipedia jet. [online]. http://en.wikipedia.org/wiki/Jet_fuel.
25. P. Bartle and Chr. Volkl, Thermo-oxidative stability of high-temperature stability polyol ester jet engine oils—A comparison of test methods, *J. Synth. Lubrication*, *17*, 179–189, 2000.
26. T.E. Karis, Lubricant additives for magnetic recording disk drives, in: *Lubricant Additives: Chemistry and Applications*, L. Rudnick (ed.), pp. 467–511, Marcel Dekker, New York, 2003.
27. J.P. Sadighi, R.A. Singer, and S.L. Buchwald, Palladium-catalyzed synthesis of mono-disperse, controlled-length, and functionalized oligoanilines, *J. Am. Chem. Soc.*, *120*, 4960–4976, 1998.
28. I.C. Callaghan, Antifoams for nonaqueous systems in the oil industry, In: *Defoaming: Theory and Industrial Applications*, P.R. Garrett (Ed.), Surfactant Science Series, *45*, pp. 119–150, Marcel Dekker, New York 1993.
29. T.E. Karis, B. Marchon, M.D. Carter, P.R. Fitzpatrick, and J.P. Oberhauser, Humidity effects in magnetic recording, *IEEE T. Magn.*, *41*, 593–598, 2005.
30. W.S. Huang and A.G. MacDiarmid, Optical properties of polyaniline, *Polymer*, *34*, 1833–1845, 1993.
31. S.F.S. Draman, R. Daik, and A. Musa, Synthesis and fluorescence spectroscopy of sulphonic acid-doped polyaniline when exposed to oxygen gas, *Int. J. Chem. Biol. Eng.*, *2*, 112–119, 2009.
32. Y. Zelu, T. Paillat, G. Morin, C. Perrier, and M. Saravolac, Study on flow electrification hazards with ester oils, *2011 IEEE International Conference on Dielectric Liquids (ICDL)*, Trondheim, June 26–30, 2011, pp. 1–4, DOI 10.1109/ICDL.2011.6015404.
33. Q.D. Zhang, S.X. Chen, and Z.J. Liu, Design of a hybrid fluid bearing system for HDD spindles, *IEEE T. Magn.*, *35*, 821–826, 1999.
34. A. Wallash and H. Zhu, Electrostatic discharge (ESD) breakdown between a recording head and a disk with an asperity, *IEEE T. Magn.*, *42*, 2492–2494, 2006.
35. O. Groning, O.M. Kuttel, E. Schaller, P. Groning, and L. Schlapbach, Vacuum arc discharges preceding high electron field emission from carbon films, *Appl. Phys. Lett.*, *69*, 476–478, 1996.
36. N.M.J. Conway, B. Godet, and B. Equer, Nitrogen containing hydrogenated amorphous carbon prepared by integrated distributed electron cyclotron resonance for large area field emission displays, *Mat. Res. Soc. Symp. Proc.*, *621*, Q1.6.1–Q1.6.6, 2000.
37. S-C. Kang, R.M. Crone, and M. Jhon, A new molecular gas lubrication theory suitable for head-disk interface modeling, *J. Appl. Phys.*, *85*, 5594–5596, 1999.
38. T.E. Karis and U.V. Nayak, Liquid nanodroplets on thin film magnetic recording disks, *Tribol. T.*, *47*, 103–110, 2004.
39. T.E. Karis, Water adsorption on thin film media, *J. Colloid Interf. Sci.*, *225*, 196–203, 2000.
40. Wikipedia heat capacity. [online]. http://en.wikipedia.org/wiki/Volumetric_heat_capacity
41. L.A. Bronshtein, Y.N. Shekhter, and V.M. Shkol'nikov, Mechanism of electrical conduction in lubricating oils (review), *Chem. Tech. Fuels Oils*, *15*, 350–355, 1979.

18 Adsorption of Surfactants on Hematite Used as Weighting Material and the Effects on the Tribological Properties of Water-Based Drilling Fluids

F. Quintero and J. M. González

CONTENTS

Natural hematite (Fe_2O_3) has been employed as an alternative weighting material to barite (Ba_2SO_4) for drilling fluids in several Venezuelan fields. Hematite has shown some physicochemical advantages over barite: a higher specific gravity and solubility in acid media and lower attrition rate. However, the most challenging issue related to hematite field applications has been to reduce its high wear potential. Previous research, whose results were validated on field tests with drilling fluids formulated with hematite, showed that the abrasion and erosion rates produced by hematite decreased with decrease of particle size and also depend on the morphology and angularity of the particle relative to barite. The objective of this work was to study the adsorption of surfactant molecules on hematite particles and its influence over the wear behavior of hematite particles on water-based drilling fluids. The ultimate goal was of developing new lubricant and wear-reducing additives to improve the performance of water-based drilling fluids. The results showed that the surfactant system studied adsorbs on hematite particles and this adsorption was influenced by the pH of the solution. Additionally, the results obtained in the abrasive test suggest that this property of the surfactant could be related to the wear reduction of hematite in this system. But it was necessary to perform additional laboratory tests to find a clear correlation between these parameters.

18.1 INTRODUCTION

The process of drilling wells requires the use of heavy equipment and tools that conform the bottom hole assembly (BHA), commonly used to improve geo-steering performance of the drill pipe and to acquire information from the subsurface. Since this assembly is attached near the drill bit, it usually suffers progressive wear associated with continuous rotation against the rock and through its internal components caused by the erosive action of drilling fluids on its external surface [1]. In the oil industry, there are drilling fluids of different nature, classified according to the external fluid phase as water-based, oil-based, and pneumatic or gas-based systems. Basically, drilling fluids are composed of a base fluid (water or oil), a weighting material (Ba_2SO_4, Fe_2O_3, or $CaCO_3$) and other additives to control rheological properties [2–5], fluid losses [6–8], and shale inhibition [9–12], among others. The type of drilling fluid selected for a drilling operation depends on the formation being drilled, the depth, the mechanical resistance, and the wellbore's pressure. Regardless of the type of drilling fluid, its main functions are to maintain hole integrity, transport the rock cuttings from the bottom hole to the surface, control formation pressure, and cool and lubricate the BHA.

The lubricity function is very important due to the existence of frictional forces during all stages of well construction [13] (drilling, completion, and maintenance); sources of frictional forces include pipe resistance to rotation (torque) and the raising and lowering movement (drag) inside the well in contact with either the wellbore (metal-to-rock) or the casing (metal-to-metal). These forces can be minimized by increasing the lubricity of the circulating working fluid. This can be achieved using lubricant additives, generally available as film producing liquids or solid beads, powders, or fibers. Liquid additives include glycols, oils, esters, fatty acid esters, surfactants [14], and polymer-based lubricants. Solid additives such as graphite, calcium carbonate flakes, glass, and plastic beads are used. Excessive torque and drag can cause unacceptable loss of power, making oil well operations less efficient, especially in high-angle and extended-reach wells [15–18].

In oil well construction and maintenance processes, particularly during drilling, all the equipment and fluid systems present different tribological phenomena and related problems. The dominant wear modes include impact wear, abrasion, and slurry erosion [19], and when not controlled or predicted, they can cause catastrophic failures of the equipment and the wellbore, with the ultimate loss of the hole. Abrasion and erosion wears are caused mainly by the content of solid particles used as weighting material in the drilling fluid formulation.

For many years, barite (Ba_2SO_4) has been used as the primary weighting material for oil-based drilling fluids. The purpose of the weighting material is to provide a specified density to the fluid system in order to control the formation pressure during the drilling process. However, the world reserve of barite is decreasing in quality and quantity whereas international demand for fluid densifiers is increasing every year, associated with the increase of worldwide drilling activity [1]. Owing to this consideration, Petróleos de Venezuela (PDVSA), the primary Venezuelan oil industry, has initiated a project to develop an alternative to replace imported barite by hematite produced in Venezuela. However, the principal barrier that has limited the massive use of hematite as a weighting material in drilling fluid systems has been its erosive and abrasive wear effect over metallic and nonmetallic components of the fluid circulation system (valves, pumps, pipes) and downhole tools (directional tools, motors, turbines) of the drilling rig. This effect worsens when drilling fluids are operated at high densities and high pumping rates. A previous investigation showed that the erosion rate produced by hematite decreases with decreasing particle size and also depends on the morphology and angularity of the particle relative to barite [1].

The objective of this work was to study the adsorption of surfactant molecules on hematite particles and how it affects the wear behavior of particles on water-based drilling fluids. The aim of this investigation is to develop new lubricants and wear-reducing additives for water-based drilling fluids with the following characteristics:

highly effective in reducing friction and wear at low concentration, compatible with other drilling fluid additives, able to support drilling conditions, low toxicity, and environmentally safe. It is believed that surfactants could accomplish this task.

18.2 ADSORPTION

The adsorption phenomenon is called the accumulation of one or more components on a surface. The substance that adsorbs is called an adsorbate and the material onto which it is adsorbed is called the adsorbent. The reverse process of adsorption is desorption [20].

The following are some commonly observed phenomenon of adsorption:

1. Adsorption is highly selective. The amount adsorbed depends largely on the nature of the adsorbent and surface pretreatment to which it has been subjected as well as the nature of the adsorbate. Increasing the surface area of the adsorbent and concentration of adsorbate will increase the amount adsorbed.
2. The kinetics of adsorption increase when the temperature increases but decrease with increasing amount adsorbent.
3. It is a spontaneous process, that is, ΔG (Gibbs free energy) is negative. It is generally associated with an increase in the order of the adsorbate molecules, which means that ΔS is negative, which, according to the equation $\Delta G = \Delta H - T\Delta S$, will usually have an exothermic process. The change in enthalpy when 1 mol of adsorbate is adsorbed in the appropriate amount of adsorbent is called the adsorption enthalpy.

Physical adsorption (physisorption) is associated with a higher adsorption enthalpy of $-40\,\text{kJ mol}^{-1}$ whereas values less than $-80\,\text{kJ mol}^{-1}$ are typical of chemical adsorption or chemisorption (considering that adsorption is generally an exothermic reaction). Physisorption enthalpy is comparable with the enthalpy of condensation, whereas the enthalpy of chemisorption is comparable with the enthalpy of chemical reactions.

Physisorption occurs most frequently, whereas chemisorption occurs only when the adsorbent and the adsorbate ionic covalent bond is formed between them. In general, the physisorption process can be reversed easily; on the contrary, chemisorption is difficult to reverse and usually occurs more slowly than physisorption.

The number of adsorbed layers in physisorption can vary in thickness from monolayers to multilayer of molecules because Van der Waals forces may extend from one layer of molecules to another. In contrast, chemisorption cannot, by itself, give rise to a multilayer thickness, due to the specificity of the link between the adsorbent and the adsorbate. However, it is possible to form subsequent layers of various molecules physically adsorbed on the chemisorbed first layer.

At constant temperature, the amount adsorbed increases on the surface of the adsorbent with the adsorbate concentration in solution. The relationship between the amount adsorbed on the surface (X) and the concentration (C) in the solution at equilibrium is called the adsorption isotherm. Only at very low concentrations

X is proportional to C only at a very dilute solution concentration. Generally, the adsorbed amount (X) increases disproportionally to the indicated solution concentration. This is mostly due to the gradual saturation of the surface. The adsorption isotherm can be represented by an equation of the form:

$$KC^n = \frac{X}{m} \Rightarrow KC^n = S \qquad (18.1)$$

where S (μg g^{-1}, mg mg^{-1}, μmol g^{-1}, mol g^{-1}) is the ratio of the amount adsorbed by unit amount of adsorbent, C (ppm, ppb, μmol L^{-1}, mol L^{-1}) is the equilibrium concentration of the solute, and K (μg L^{-n} g^{-1} mL^n, μg^{1-n} g^{-1} L^n, μmol^{1-n} g^{-1} L^n, μmol^{1-n} kg^{-1} L^n) and $n < 1$ are constants for the system at the evaluated temperature, where the constant n is usually less than unity. The most documented isotherms are Langmuir, Freundlich, and Brunauer–Emmett–Teller (BET) [21].

18.2.1 LANGMUIR ISOTHERM

Langmuir developed a model to predict the degree of adsorption of a gas on a surface as a function of fluid pressure [22]. This model assumes that

1. The adsorbate forms a monolayer on the surface.
2. All surface sites are equivalent.
3. No interaction occurs between adsorbed particles.
4. The adsorbed molecules do not move on the surface of the adsorbent.

The adsorption process can be represented by a chemical equation. If the adsorbate is a gas, we can write the equilibrium as

$$A(g) + S \xleftrightarrow{K} AS \qquad (18.2)$$

where A is the adsorbate gas, S is an unoccupied site on the surface, and AS represents a molecule at an occupied site or adsorbed on the surface. The equilibrium constant can be expressed as

$$K = \frac{[AS]}{[A] \cdot [S]} \qquad (18.3)$$

where [A] denotes the concentration of A per unit volume (mol L^{-1}), while the other two terms [S] and [AS] are two analogs or surface concentrations expressed per unit area (mol cm^{-2}). The principle of chemical equilibrium holds with these terms. The complete form of the Langmuir isotherm considers (Equation 18.3) in terms of surface coverage, θ, which is defined as the fraction of the adsorption sites to which a solute molecule has become attached or fraction of the occupied sites. An expression for the fraction of the surface with unoccupied sites is therefore $(1 - \theta)$. Given these definitions, and for [A] = C, we can rewrite the term [AS]/[S] in terms of θ and Equation 19.3 becomes

$$K = \frac{\theta}{(1 - \theta)C} \tag{18.4}$$

18.2.2 Brunauer–Emmett–Teller Isotherm

This model assumes that the gases that adsorb on the first layer or monolayer are molecules forming a multilayer and are able to generate two cases:

1. The first layer is chemisorbed and the others are physisorbed
2. All layers forming on top of the monolayer are physisorbed

The BET isotherm can be expressed through the following equation:

$$\frac{p}{v(p_0 - p)} = \frac{1}{v_m c} + \frac{(c - 1)p}{c v_m p_0} \tag{18.5}$$

where p (e.g., atm, MPa) and p_0 (e.g., atm, MPa) are the equilibrium and the saturation pressure of adsorbates at the temperature of adsorption, c is a constant related to the adsorption heat and it has something to do with the probability of adsorbing or desorbing, v_m (e.g., cm^3 g^{-1}) is the volume of gas required to form a monolayer on a unit gram of the sample, and v (e.g., L g^{-1}) is the volume of gas adsorbed per unit mass of solid [22,23].

18.2.3 Freundlich Isotherm

The Freundlich model takes into account the interaction between molecules adsorbed on different sites; it was one of the first models to relate the concentration of adsorbate on the surface to the concentration in the solution [22–24]. This isotherm is governed by the following equation:

$$m = k c^{\frac{1}{n}} \tag{18.6}$$

where m is the mass in grams of adsorbed per gram of adsorbent, c (mol L^{-1}) is the concentration of the substance adsorbed in solution, and k and n are constants.

18.2.4 Adsorption of Surfactants on Solid Surfaces

Surfactant adsorption is a process of transfer of surfactant molecules from bulk solution phase to the surface/interface [25,26]. The adsorption of surfactants at the solid–liquid interface plays an important role in many technological and industrial applications such as detergency [27], mineral flotation [28,29], dispersion of solids

[30], and oil recovery [31,32]. Significant progress has been made in understanding the adsorption/desorption behavior of single surfactants on solids. Many studies cover the solid/liquid interfacial behavior of surfactants because it is important for the theoretical understanding and practical application of surfactants [33,34]. The adsorption of surfactants at a solid–liquid interface is strongly influenced by several factors [24]:

1. The nature and structure of the groups in the solid surface such as the density and charge of adsorption sites on the heterogeneous surface
2. The molecular structure of surfactant hydrophilic head (ionic, nonionic) and hydrophobic group (length of the aliphatic chain or aromatic)
3. The characteristics of the aqueous solution (pH, temperature, presence of electrolytes, etc.)

All these factors determine the mechanism by which adsorption occurs and its efficiency.

The mechanisms that promote the adsorption of surfactant in solution in a liquid–solid or liquid–liquid interface include [24]

1. *Ion exchange:* the displacement of ions on the surface by surfactant ions. This applies, for example, to the adsorption of quaternary ammonium ions that displace hydrogen ions on the surface in the process of corrosion protection.
2. *Hydrogen bonding:* the process of polar bond between the hydrogen atom in surfactants and a negatively charged atom (O, N) on the surface. Hydrogen bonds between surfactants and the surface of mineral species have been proposed for a number of systems, particularly involving surfactants containing hydroxyl groups, phenolic, carboxylic, and amine.
3. *Ion pairing:* the adsorption of ionic surfactants in unoccupied charged sites. An example is the adsorption of cationic or amphoteric surfactants on negatively charged sites.
4. *London—Van der Waals adsorption forces:* These forces occur between substrates and nonpolar molecules, and also the forces of cohesion of fluids, often called dispersion forces. The frequency of oscillation of the electrons, which is responsible for these forces, is linked to the refractive index of the medium.
5. *Hydrophobic adsorption interactions lateral:* Occurs when the packing of surfactant molecules at the interface causes an interaction between the tail lipophilic neighboring molecules, allowing the molecules to migrate away from the aqueous environment.
6. *Adsorption due to polarization of π electrons:* This occurs when there is an attraction between an aromatic nucleus and positive site on the surface of the substrate.

In conclusion, the interaction or mutual attraction between adsorbate and adsorbent, which is responsible for adsorption, depends on the nature of forces involved; this can be chemical (covalent or electrovalent), called chemisorption, or physical forces (Van der Waals), known as physical adsorption or physisorption. This last is controlled by the pH of the solution with the adsorption of inorganic ions.

18.2.4.1 Adsorption and Wettability Changes on Solid Surfaces

Wettability is defined as the displacement of a fluid (gas or liquid) from a surface by another fluid. The phenomena of wettability are very important in many industries, for example, the adsorption of surfactants onto a solid is responsible for changes in the wettability of the solid surface [35]. The solid surface can be hydrophilic (high energy) or hydrophobic (low energy), and can also have charged groups, (positive or negative site); in addition, the conditions of the surfactant and surface can be changed by varying the pH or temperature.

The characterization of the adsorption of surfactants at a solid–liquid interface is more difficult than at the air–water interface for two reasons: first, the solid–liquid interfacial tension cannot be measured directly and, second, the relation between solid–liquid interfacial tension and the concentration of surface excess is not simple [35].

The interfacial tension between a solid and a liquid can be estimated by assessing the compatibility between the solid surface and water, which could be modified by the presence of surfactants. The equilibrium contact angle is a measure of the wettability of the system.

The equilibrium contact angle can be considered as a thermodynamic quantity because it is related to the free energy of solids. This relationship is expressed as Young's equation [36], which is

$$\gamma_{sv} = \gamma_{sl} + \gamma_{lv} \cos\theta \tag{18.7}$$

where γ_{sv} (e.g., mN m^{-1}, J m^2) is the surface tension of the solid, γ_{sl} (e.g., mN m^{-1}, J m^2) is the interfacial tension corresponding to the solid–liquid interface, γ_{lv} (e.g., mN m^{-1}, J m^2) is the surface tension of the liquid, and θ (degree $^\circ$) is the contact angle between the solid and the measurement liquid. The work of adhesion between solid–liquid is given by

$$W = \gamma_{lv} + \gamma_{sv} - \gamma_{sl} \tag{18.8}$$

Combining Equations 18.7 and 18.8 yields

$$W = \gamma_{lv}(1 + \cos\theta) \tag{18.9}$$

Equation 18.7 can be rearranged as follows:

$$\gamma_{sv} - \gamma_{sl} = \gamma_{lv} \cos\theta \tag{18.10}$$

Equation 18.10 establishes a relationship between the contact angle and wetting as follows:

1. Complete wetting: $\gamma_{sv} - \gamma_{sl} = \gamma_{lv} \rightarrow \cos\theta = 1 \rightarrow \theta = 0^\circ$
2. Partial wetting: $\gamma_{sv} - \gamma_{sl} \prec \gamma_{lv} \rightarrow \cos\theta \prec 1 \rightarrow \theta \succ 0^\circ$

The problem of Equation 18.7 is that the solid–liquid interfacial tension γ_{sl} is not directly measurable. If the surfactant in water is not volatile, we can assume that it is not adsorbed at the solid–gas interface. The solid surface tension γ_{sv} can be taken as independent of the nature of the liquid drop, and particularly with respect to the surfactant concentration. Under these experimental conditions where the surface energy of the solid is kept constant, variations in adhesion tension directly reflect what happens at the solid–liquid interface. In practice, to promote the wettability of the solid surface, the surfactant is added to minimize the solid–liquid interfacial tension [35]. The mathematical expression (18.7) presents two unknowns γ_{sl} and γ_{sv} and, to solve it, requires a second equation:

$$d\gamma_{ls} = -\Gamma_{1sl}d\mu_1 - \Gamma_{2sl}d\mu_2 \qquad (18.11)$$

The terms $\Gamma_{1sl}d\mu_1$ and $\Gamma_{2sl}d\mu_2$ in Equation 18.11 are obtained through a separate experiment where the surface concentration of excess surfactant is determined. This method of determining the surfactant concentration on the solid–liquid interface is called the method of exhaustion. According to this procedure, a known amount of solid is added to a known volume of an aqueous solution of surfactant at a known concentration. After a specific contact time, called the adsorption equilibrium time, the suspension is separated and the concentration of surfactant in the supernatant solution is measured. The difference between initial and final surfactant concentrations multiplied by the volume of liquid is the amount of surfactant adsorbed on the solid [35].

18.2.4.2 Models for Surfactant Adsorption on a Solid Surface

In general, the adsorption isotherms are interpreted by rechange of the slope of the plot of surface excess versus concentration. This allows us to divide the isotherm into regions, based on potential conformation of adsorbed surfactant in each region [37,38]. In more recent studies, the isotherm data are often associated with information that allows us to accurately determine the nature of adsorption. The surface charge, zeta potential, the counterion concentration, pH of the solution, and the conductivity of the solution have been contrasted with surface excess.

18.2.4.2.1 Two-Step Model

On a linear scale, adsorption isotherms show two plateau-type regions, as well as an increase in surface excess near the critical micelle concentration (cmc), as shown in Figure 18.1 [37,38].

The adsorption isotherm can be divided into four regions and the method of adsorption suggests the following: In Region I, the surfactant is adsorbed through electrostatic interactions with the substrate used. The surface excess is determined mainly by the surface charge. Adsorption is low, and thus the interactions between adsorbed surfactant molecules are negligible. In Region II, the substrate surface charge has been neutralized. However, the activity of surfactant in solution is not large enough to lead to any form of aggregation at the interface and still the surfactant is adsorbed as monomers. Abrupt increase in the rate of adsorption around the hemimicelles denotes the start of Region III. In this region, the concentration of surfactant in solution is

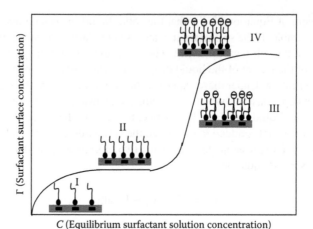

FIGURE 18.1 General shape of the adsorption isotherm in the two-step model. (Adapted from R. Atkin and V. Craig, *Adv. Colloid Interface Sci.*, 103, 219–304, 2003.)

enough to lead to hydrophobic interaction between surfactant monomers. The monomers are electrostatically adsorbed in Region II, and are thought to act as anchors (or nucleation sites) for the formation of hemimicelles. In Region III, the hemimicelle structure is not necessarily fully formed. In Region IV above the cmc, hemimicelles have fully formed along with high levels of surface coverage [37].

18.2.4.2.2 Four-Step Model

The analysis of the two-step model explains the common features of adsorption isotherms and is the only evaluation method available. Somasundaran and Fuerstenau proposed the guidance model of four steps for the interpretation of surfactant adsorption isotherms in a log–log plot [37]. The main advantage of using a log–log plot is that it amplifies the characteristics of the isotherm at low surface excess values.

The general form of isotherms plotted in log–log, and the morphology of adsorbed structures associated with each region of the four-step model is represented schematically in Figure 18.2. Region I of Somasundaran and Fuerstenau shows that the adsorption of surfactant monomers on to the substrate is through electrostatic forces, where the hydrophilic groups are in contact with the surface. Region II involves strong lateral interaction between the adsorbed monomers, causing the formation of primary aggregates and creating a surface hydrophobic zone. In the four-step model, this type of aggregate is known as a hemimicelle. Increases in the surface excess in Region III are thought to result from growth of the structures formed in Region II, without any increase in the number of surface aggregates. The presence of head groups facing into the solution renders the surface hydrophilic once more. The transition between Regions II and III is thought to be due to the neutralization of the surface charge, that is, the transition from Region II to III corresponds to the isoelectric point of the solid; therefore, adsorption in Region III occurs through the growth of existing aggregates rather than the formation of new aggregates due to

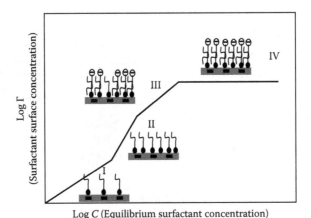

Log *C* (Equilibrium surfactant concentration)

FIGURE 18.2 General shape of the adsorption isotherm in the four-step model. (Adapted from P. Somasundaran and S. Krishnakumar, *Colloids Surfaces A.*, 123–124, 491–513, 1997.)

lack of positive adsorption sites. Finally, in Region IV, the morphology of the surface is assumed as a fully formed bilayer [37–39].

18.2.4.2.3 Adsorption of Ionic Surfactant on Polar Surfaces

Materials possessing charged surfaces include almost all the inorganic oxides and salts of technological importance (silica, alumina, titania, hematite, etc.) and silver halides, latex polymers containing ionic co-monomers, many natural surfaces such as proteins, and cellulose. It is very important to understand the interactions of such surfaces with surfactants in order to optimize their effects in applications such as paint and pigment dispersions, papermaking, textiles, pharmaceuticals, and oil industry (e.g., drilling fluid formulation and enhanced oil recovery) [34,40].

Owing to the number of interactions in systems containing charged surfaces and ionic surfactants, it is very important to closely control all of the variables in the system. As the adsorption proceeds, the dominant mechanism may go from ion exchange through ion binding to dispersion or hydrophobic interaction.

Adsorption isotherms of charged surfactants on oppositely charged surfaces generally show three well-defined regions of adsorption in which the rates vary because of changes in the mechanism of adsorption (represented schematically in Figure 18.3). One interpretation of such adsorption involves three consecutive mechanisms. In the first stage, Region I, adsorption occurs as a result of ion exchange in which closely associated "native" counterions are displaced by surfactant molecules. In this (Region I) stage, the surface charge or surface potential may remain essentially unchanged. In Region II, adsorption continues, and the ion pairing of surfactant molecules with surface charges may become important, resulting in a net decrease (isoelectric point). It is often found that in Region II the rate of adsorption will increase significantly. The observed increase may be due to the cooperative effects of electrostatic attraction and lateral interaction among adjacent hydrophobic groups of adsorbed surfactants as the packing density increases.

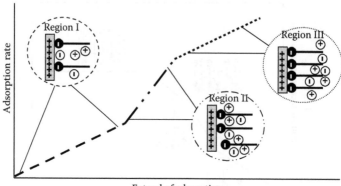

Extend of adsorption

FIGURE 18.3 Proposed mechanisms to explain various rates of ionic surfactant adsorption as a function of surface coverage and mode of adsorption: Region I—ion exchange; Region II—ion pairing; Region III—charge neutralization. (Adapted from M. Drew, *Surfactant Science and Technology*, Third edition, Chapter 10, pp. 329–349, John Wiley & Sons, New Jersey, 2006.)

In Region III, hydrophobic interactions between adjacent surfactant tails can predominate, often leading to the formation of aggregate structures or hemimicelles already postulated. If the hydrophobic interaction between surfactant tails is weak (because of short or bulky hydrocarbon chain) or if electrostatic repulsion between head groups cannot be overcome (because of the presence of more than one charge of the same sign or low ionic strength), the enhanced adsorption rate of Region II may not occur and hemimicelle formation may be absent. An additional result of the onset of dispersion-force-dominated adsorption may be the occurrence of charge reversal as adsorption proceeds [40].

18.2.4.2.4 Adsorption of Nonionic Surfactants

Recently, the adsorption of nonionic ethoxylated surfactants has attracted much attention due to their potential application in process detergency, cosmetics, enhanced oil recovery, and so on. The adsorption of nonionic surfactants differs from ionic surfactants mostly because of the absence of electrostatic interaction.

Nonionic ethoxylated alcohols exhibit strong adsorption on silica, but not on other minerals such as alumina. Since hydrogen bonding is relatively weak in comparison with electrostatic and chemical bonding, the nature of the aqueous solution at the solid–liquid interface will be of particular importance for the adsorption of nonionic surfactants. The lack of adsorption, for example, on certain minerals such as alumina is speculated to be due to the fact that the surfactant molecules are unable to disrupt the rigid water layer surrounding the substrate [39,40].

18.2.4.2.5 Adsorption of Surfactant Mixtures

Surfactants are generally used as mixtures in order to accomplish different purposes. A typical feature of the adsorption of ionic/nonionic surfactant mixtures and oppositely charged ionic surfactant mixtures is synergistic interaction at the interface as well as in solution. The adsorption of nonionic ethoxylated alcohols on alumina is negligible, but it could be enhanced by several orders of magnitude by coadsorption with an anionic surfactant. Similarly, anionic surfactants do not adsorb on negatively charged silica, but substantial adsorption could be achieved by coadsorption with a nonionic ethoxylated alcohol, which strongly adsorbs by itself on the surface. Enhanced surface activity of mixtures of nonionic and anionic surfactants has been demonstrated for several systems [39].

18.3 EXPERIMENTAL DETAILS

18.3.1 MATERIALS

18.3.1.1 Surfactants

Commercial anionic and nonionic surfactants were used as received without further purification. They were obtained from Petroleum and Petrochemical Service C.A. (PPS), Valencia, Venezuela. The anionic surfactant was a phosphate ester (trilaureth-4 phosphate) [41] and the nonionic surfactant was an ethoxylated lauryl alcohol with a degree of ethoxylation of four. The surfactants mixture contains 20% w/w ester phosphate and 80% w/w ethoxylated alcohol.

18.3.1.2 pH Modifier

Sodium hydroxide was used as a pH modifier (NaOH, 99% purity) and was obtained from Akzo Nobel, Germany. It was used to adjust the pH between 4, 7, and 10 unit.

18.3.1.3 Solids (Weighting Material)

Hematite was obtained from Micronizados Caribe C.A., Sucre, Venezuela. It was used as received. The hematite used in different experiments had the following particle size distribution (Figure 18.4): 90% of the particles had a diameter of less than 43.8 μm ($D_{0.9}$) and 50% of the particles had a diameter of less than 13.9 μm ($D_{0.5}$).

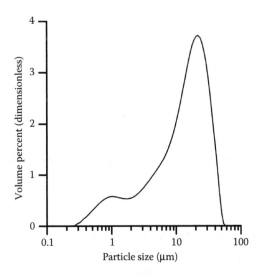

FIGURE 18.4   Hematite particle size distribution used in this study.

18.3.1.4 API Reference Bentonite

Powdered premium-grade sodium cation montmorillonite according to ISO specification 10416:2010 from the American Petroleum Institute (API). Bentonite is primarily used to increase the viscosity and carrying capacity of water-based drilling
fluids. It significantly reduces fluid loss and in combination with polymers forms
a thin tough filter-cake cleaning and a smaller concentration of low-gravity solids
suspended in the drilling fluid. The bentonite used was obtained from US Bentonite
Processing, Inc.

18.3.2 METHODS

18.3.2.1 Surface Tension Measurements

DCAT 11 instrument from Dataphysics was used to determine static surface tension
with the Wilhelmy's method at $25 \pm 1°C$. The results are presented as an arithmetic
mean of three independent measurements. The solutions were prepared using triply
distilled water.

18.3.2.2 Adsorption Isotherm Determination

Adsorption isotherms were determined by the batch equilibrium adsorption method.
In a typical procedure, 7.2 g of hematite was added to 80 mL of aqueous surfactant
solution at a specific surfactant concentration, while stirring for 1 min. Preliminary
experiments indicated that within the first 12 h, almost complete adsorption had
occurred, but it was decided to wait for 24 h of contact time to ensure complete
surfactant adsorption on the hematite particles. All experiments were carried out at
a controlled temperature of $25 \pm 1°C$.

   To determine maximum surfactant adsorption on hematite, a surface tension
measurement method was used [34]. The amount of surfactant adsorbed was calculated from the difference in surfactant concentration before and after adsorption at
a constant surface tension value of the average of three replicate measurements. In
order to measure the surface tension after adsorption, the hematite dispersion was
separated by a filtration process. The amount of solute adsorbed is expressed as surfactant mass/solid mass and was determined as follows:

$$A = (C_i - C_{eq}) \frac{V}{W} \qquad (18.12)$$

where C_i and C_{eq} are the initial and equilibrium liquid-phase concentrations of surfactant solution $(g\,L^{-1})$, respectively; V is the volume of the surfactant solution (L),
W is the mass of the dry adsorbent (g), and A (g surfactant/g solid) is the amount of
surfactant adsorbed per solid mass.

18.3.2.3 Measurement of Powder Contact Angle

Powder contact angle was measured using a dynamic contact angle meter (model
DCAT 11, from Dataphysics). The measurement consists of quantification of the
mass of the liquid penetrating the studied porous material. In a typical experiment,

a sample of the material was fastened to a pull rod connected to a sensor enabling measurement of the mass. The pull rod was placed above the vessel containing the measurement liquid. The surfaces of the sample and liquid should be parallel to each other. The vessel with the liquid was then raised until the liquid touched the sample. From this moment on, the measurement time and the sample mass gain were recorded. The measurement was concluded after a specified time had elapsed or the sample mass had stopped increasing.

When a liquid penetrates a single capillary of radius r, the distance of flow l in time t is given by the modified Washburn's equation [42–45].

$$l^2 = \frac{C\gamma\cos\theta}{2\eta}t \qquad (18.13)$$

where l (cm) is the distance of flow in time t (s), γ (mN m^{-1}) is the surface tension of the liquid, θ (°grades) is the advancing contact angle, η (mPas) is the viscosity of the liquid, and C (cm^5) is a constant to estimate the tortuous path of the capillaries.

The modified Washburn's equation is also used as dependence between wetting liquid mass and time. The relation between liquid mass and height in the column is linear as given by

$$w_{imb}^2 = \frac{C\rho^2\gamma\cos\theta}{2\eta}t \qquad (18.14)$$

where w_{imb} (g) is the mass in time t (s), γ (mN m^{-1}) is the surface tension of the liquid, θ (°grades) is the advancing contact angle, η (mPas) is the viscosity of the liquid, ρ (g cm^{-3}) is the density of the liquid, and C (cm^5) is a constant to estimate the tortuous path of the capillaries.

18.3.2.4 Zeta Potential Measurements

The zeta potential of hematite was measured on a Zetasizer Nano Z (Malvern, England). The particle size analyzer Mastersizer Hydro 2000G (Malvern, England) was used to measure hematite particle size distribution in aqueous media. The following procedure was used for sample preparation: 1.8 g hematite particles were mixed in a glass vessel with an aqueous surfactant solution at 1% w/v until reaching a volume of 20 mL. The pH of solutions was then adjusted as desired using NaOH, while stirring for 1 min. The mixture was kept at rest for 24 h at a controlled temperature of 25°C to ensure complete surfactant adsorption on the hematite particles and also to allow the sedimentation process. Then, 1 mL of supernatant was removed from the glass vessel with a syringe and used to fill the Zetasizer cell for zeta potential measurement.

18.3.2.5 Dispersion Stability Measurements

Different suspensions were prepared to evaluate the effect of surfactant on the dispersion stability of hematite particles in water. Particle samples (Fe$_2$O$_3$) of 1.8 g were mixed in the Turbiscan Lab® cell with a volume of 20 mL of aqueous surfactant solution at 1% w/v or distilled water according to the case. The suspensions were agitated at 3000 rpm for 1 min at 25°C. The dispersion stability was measured using Turbiscan Lab® Expert (Formulaction, France) in scan mode, by following the sedimentation of the particles over 2 h (one scan per 10 min).

18.3.2.6 Abrasiveness Test

Weighting materials used in drilling fluids can vary considerably in their relative abrasivity [1]. The laboratory method designed and used to measure and evaluate relative abrasiveness has been described before [46]. Briefly, the test is performed using a standard test blade attached to a high-speed mixer. The blade is used to mix the base drilling fluid containing 1 g mL^{-1} of the weighting material at a speed of 11,000 rpm. The mass loss of the blade after a time (min) of mixing is used to calculate the abrasiveness of the weighting material in milligrams per minute (mg min^{-1}).

The following procedure was used: (1) Prepare a base suspension by adding 15.0 g API reference bentonite to 350 mL distilled water in a container while stirring at 5000 rpm on the base suspension mixer. (2) Pour 300 mL of the base suspension into the mixing container. (3) For the surfactant evaluation, add 3.5 g of the surfactant mixture to the base suspension. (4) Add 300 g of the weighting materials into the base suspension while stirring at 5000 rpm for 20 min. (4) Adjust the pH using NaOH, only for surfactant evaluation. (5) Immediately prior to use, clean the blade by washing with detergent and a small brush, rinse thoroughly and dry. (5) Weight a freshly cleaned and dried abrasion test blade to the nearest 0.1 mg, then install the blade into the mixer. (6) Mix the base drilling fluid (base suspension, surfactant, and solids) at a speed of 11,000 rpm. Register the mass loss of the blade after mixing every 10 min for 60 min.

18.4   RESULTS AND DISCUSSION

18.4.1   SURFACTANT IN AQUEOUS SOLUTION

A fundamental property of a surfactant in aqueous solution is its capacity for self-association. The formation of a more or less dense monolayer of surfactant at an interface is the first manifestation of the tendency to associate. When the surfactant concentration increases in the aqueous phase, there is rapid saturation of the interfacial area, and consequently, the number of dissolved molecules tends to increase. At a certain concentration called cmc, surfactant association produces structures called micelles. Micelles are often spherical and contain dozens of molecules oriented so that the nonpolar part of the surfactant is removed from the aqueous environment [24].

Figure 18.5 shows the surface tension versus surfactant concentration at pH 4, 7, and 10 units. It is noted that the cmc values in aqueous solution at pH 7 and 10 are lower than at pH 4. A possible explanation is that the functional group of the surfactant (phosphate) molecule at pH 4 is partially ionized, which could affect the micellization process due to electrostatic repulsion. It is important to point out that the surfactant used is a nonionic/ionic mixture that can show unexpected behavior in an aqueous solution. Table 18.1 presents the cmc values at different pH.

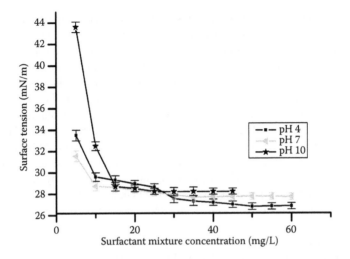

FIGURE 18.5 Adsorption isotherms (liquid–gas) for the surfactant mixture used at different pH (25°C).

TABLE 18.1

Values of Surface Tension and cmc at Different pH

Surfactant Description	γ_o at cmc$_{25°C}$ (± 0.1 mN m^{-1})			cmc (ppm, 25°C)		
	pH 4	pH 7	pH 10	pH 4	pH 7	pH 10
Surfactant mixture	26.8	27.7	28.2	50	35	25

When the pH is near 7 or 10, a higher deprotonation of the surfactant functional group occurs due to the addition of sodium hydroxide (NaOH). The lowering of the cmc in this case is mainly due to decrease in the thickness of the ionic environment around ionic head groups. The presence of an additional electrolyte (Na$^+$) decreases the electrical repulsion between the head groups in the micelle [24]. In addition, the surfactant mixture contains free phosphoric acid residues, which are neutralized to phosphate salts in the presence of sodium hydroxide in the solution. These salts could also reduce the surfactant cmc at high pH.

18.4.2 ADSORPTION OF SURFACTANT ON HEMATITE PARTICLES

18.4.2.1 Effect of pH on Adsorption Isotherms

The surfactant adsorption on the hematite particles was assessed through the classical method of exhaustion of surfactant. This method involves plotting the amount of surfactant adsorbed per unit mass of solid as a function of the surfactant concentration at equilibrium. This graphical representation or isotherm can be analyzed using the two-step model [34–35].

Usually, the changes in pH in aqueous media can generate significant changes in the adsorption process of ionic surfactants on solid surfaces. As the pH of the aqueous solution is reduced, the solid surface becomes more positive (below point zero charged) due to the adsorption of protons from the bulk solution onto the sites. As a result, there will be an increase in the adsorption of anionic surfactants.

Figure 18.6 shows the adsorption isotherm of the surfactant/hematite system at pH 4, 7, and 10. The adsorption curves at different pH are depicted as four regions. For clarity, the regions will be described at each pH. The first region at pH 4 (from 8 to 20 mg L^{-1}) presents a very low surfactant adsorption, which could be attributed to phosphate adsorption ($H_2PO_4^-$ ion is the major species in the solutions according to the dissociation constants of H_3PO_4) onto hematite; this occurs by an ion exchange between phosphate and hydroxyl groups on the hematite surface, leaving few or no positively charged sites to generate electrostatic interaction between the surfactant and the solid surface.

In case of Region I at pH 7 (from 6 to 9 mg L^{-1}) and 10 (from 8 to 15 mg L^{-1}), a higher surfactant adsorption is observed than at pH 4, the highest being at pH 10. In the pH range of 7–10, phosphoric acid deprotonation does not affect surfactant adsorption. The major species in solution at pH 7 is $H_2PO_4^-$ (80%) and at pH 10 are $H_2PO_4^-/HPO_4^{-2}$ (50/50%). Therefore, ion exchange between negatively charged phosphate species and negative charge on the surface does not occur due to electrostatic repulsion. Under these surface conditions, the nonionic surfactant molecules are more likely to adsorb onto the hematite particles. Based on these observations, hydrogen bonding is proposed to be the initial driving force for adsorption of nonionic surfactant molecules at pH 7 and 10.

When surfactant concentration is increased, the surfactant adsorption slightly increases at pH 4, but the transition to Region II is not clear for this system. However, at pH 7 and 10, a clear transition to Region II is observed. The isotherms show

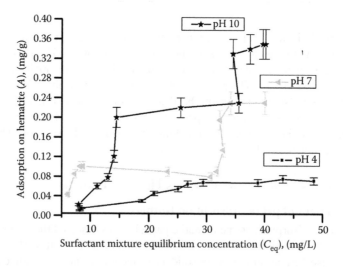

FIGURE 18.6 Adsorption isotherms (liquid–solid) for the surfactant mixture used at different pH (25°C).

plateaus at pH 7 and 10 (Figure 18.6). Again, the amount of surfactant adsorbed onto the particles is higher at pH 10 than at pH 4 or 7. It is proposed that the nonionic surfactant adsorbs onto the negatively charged surface, occupying all the available adsorption sites. This creates a hydrophobic environment on the surface particle. Once this occurs, the anionic surfactant adsorbs through hydrophobic chain–chain interaction.

Likewise, the adsorption isotherm at pH 4 does not show a noticeable transition to Region III, and the adsorption isotherm remains the same for surfactant equilibrium concentration higher than 30 mg L^{-1}. But at pH 7 and 10, Region III is clearly observed by an abrupt increase in surfactant adsorption just after the first plateau. At pH 7, Region III occurs from 30 to 35 mg L^{-1}, while at pH 10, it is observed at approximately 35 mg L^{-1}. In this region, surface aggregates (hemimicelles) are formed. For the system studied, it is believed that these aggregates are generated due to lateral interaction between hydrocarbon chains of the anionic and nonionic surfactants.

All three systems reach a second plateau, which corresponds to Region IV. This indicates the maximum surfactant amount adsorbed onto the solid particles. The highest amount of surfactant adsorbed per unit solid mass occurs at pH 10 and is approximately 0.35 ± 0.03 mg g^{-1}.

It should be noted that the exact shape of an isotherm will depend on various factors such as surfactant type, surfactant structure, electrolyte concentration, pH, and presence of cosurfactant and alcohol [47].

18.4.2.2 Effect of Surfactant Adsorption on Contact Angle

Surfactant molecules contain both hydrophilic and hydrophobic parts. They can adsorb to a significant extent even at a very low concentration. They can also form aggregates in solution and at the solid/liquid interface by hydrophobic interactions above a certain concentration. Surfactant adsorption on solid surface can lead to changes in a variety of interfacial phenomena such as wettability. Direct measurement of contact angle directly on fine or micrometer-size particles was conducted using the modified Washburn's method [48].

Figure 18.7 shows the penetration rate of the aqueous surfactant solution into a hematite solid bed. Sometimes the penetration rate is not proportional to particle wettability due to the complex effects of viscosity, surface tension, density, and bed porosity [49]. For the hematite particles studied, the penetration rate was greater for water than surfactant solutions (1% w/v) at all evaluated pH (Table 18.2). The result could lead to misinterpretation of the experiment in terms of effect on wettability. However, it is possible to calculate the advancing contact angle using the penetration rate data in conjunction with Washburn's equation (Equation 18.12). This procedure provides the contact angle values that are directly related not only with particle wettability but also with surfactant adsorption.

The aqueous surfactant solutions at pH 7 and 10 were able to wet significantly the hematite particles. These solutions gave contact angles of $32 \pm 2°$ and $25 \pm 2°$, respectively. On the other hand, the solution at pH 4 tends to wet the hematite particles less. This result is corroborated by the value of contact angle for this solution, which was $46 \pm 2°$ (Table 18.2). These results are in agreement with the observation of the adsorption isotherms, where the aqueous surfactant solution at pH 4 showed

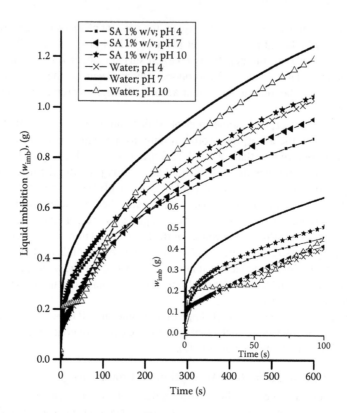

FIGURE 18.7 Penetration of aqueous surfactant solutions into a hematite bed.

TABLE 18.2

Contact Angles at Different pH in Aqueous Solution with or without Surfactant onto Hematite, at 25°C

Measure	Without Surfactant			With Surfactant 1% w/v		
	pH 4	pH 7	pH 10	pH 4	pH 7	pH 10
Contact angle (θ) $\pm 2°$	64	57	50	46	32	25

a lower adsorption than those of higher pH. In summary, the results indicate that surfactant adsorption makes hematite particles more hydrophilic, that is, the contact angles with aqueous surfactant solutions are lower than the solutions without surfactant, under the experimental conditions studied.

18.4.2.3 Effect of Surfactant Adsorption on Particle Zeta Potential

Figure 18.8 shows the variation of zeta potential of hematite particles in water as a function of pH. The results indicate that the isoelectric point or point of zero charge is at a pH of 6.3 [50].

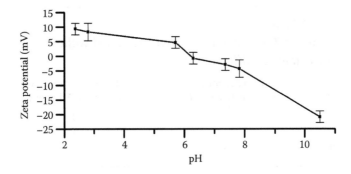

FIGURE 18.8   Zeta potential of hematite particles in water at different pH and 25°C.

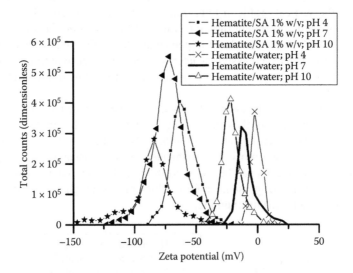

FIGURE 18.9   Hematite zeta potential in water, with and without surfactant, at different pH.

The zeta potential variations in aqueous media with and without surfactant at different pH are presented in Figure 18.9. It has been found that the surfactant mixture adsorbs onto the solid surface, generating a higher negative charge, and an increase in the average zeta potential of hematite particles in the presence of surfactant at different pH values is observed. The results can be summarized as follows: from +7 to −60 mV, from −3 to −70 mV, and from −18 to −87 mV, at 4, 7, and 10 pH, respectively (Table 18.3). These results suggest that the hydrophilic groups of the surfactant molecules adsorbed on the surface of the particles point toward the aqueous solution, which was the explanation offered when the adsorption isotherm was analyzed.

18.4.2.4   Effect of Surfactant on Dispersion Stability

Dispersion stability measurements and qualitative information about aggregation and sedimentation were obtained using an optical vertical scanner (Turbiscan Lab® Expert from Formulaction, France). The instrument combines transmitted and

TABLE 18.3

Hematite Zeta Potential in Water, with and without Surfactant, at Different pH

pH	Zeta Potential (mV) ±5 mV without Surfactant	Zeta Potential (mV) ±5 mV with Surfactant 1% w/v
4	7	−60
7	−3	−75
10	−18	−87

backscattered light measurement with a vertical scanning of the system studied. As a result the vertical profile of light transmitted and backscattered through each sample is obtained at different time intervals and translated into a horizontal axis. The experimental data is correlated in percentage to the light flux of two reference standards constituted by a polystyrene latex suspension (absence of transmission and maximum backscattering) and silicon oil (maximum transmission and absence of backscattering) [51].

Figure 18.10 presents the average backscattering percentage (ΔBS) as a function of time for the top region of the cell, the length analyzed was 10 mm for all samples, measured from the water–air interface to the bulk solution. In this zone is clearly observed the clarification process experimented by solid dispersion due to particle sedimentation. The results can be analyzed in the following manner: higher values of ΔBS indicate better dispersion stability, which means a high number of particles suspended in the aqueous medium [52].

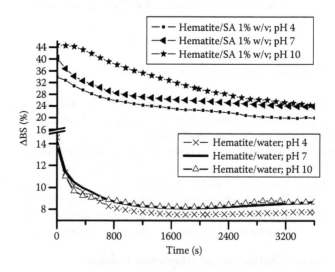

FIGURE 18.10 Effect of pH and surfactant on hematite particle dispersion stability at 25°C. Higher values of ΔBS indicate better dispersion stability, which means a high number of particles suspended in the aqueous media.

The hematite dispersion stability in pure water at different pH values (Figure 18.10) is defined by surface species in equilibrium with water molecules. The environments in which hematite dissolves have been subdivided according to the dominant Fe species in solutions in different pH ranges. The hematite dispersion stability is lower at pH values of about 4 when $Fe(OH)_3$ is the dominant ion and the stability is slightly higher for pH values of 7 and 10, where the dominant Fe species are $Fe(OH)^{+2}$ and $Fe(OH)_4^-$, respectively [53,54].

Surfactant addition considerably increases the hematite dispersion stability in water (Figure 18.10) under the test conditions. It is proposed that the nonionic surfactant of the system adsorbs onto the negatively charged surface creating a hydrophobic environment on the particle. This is then followed by adsorption of the anionic surfactant, through hydrophobic chain–chain interactions. This second adsorption is responsible for hematite particles acquiring a higher negative charge (Figure 18.9) and for strong electrostatic repulsion, which produces a better dispersion stability in water [52,55].

18.4.3 Effect of Surfactant on the Tribological Properties: Abrasiveness Test

Abrasive wear is the loss of material due to friction or impact by hard particles or hard protuberances that are forced against and move along a solid surface. Abrasive wear is commonly classified according to the type of contact and the contact environment. The type of contact determines the mode of abrasive wear. The two modes of abrasive wear are two-body and three-body abrasive wear. Two-body abrasive wear occurs when the abrasive particles or rough edges are fixed to a second body sliding over another with removal of material. In three-body abrasive wear, particles are free to roll, so as not to remove material from the first body at all times of contact [56].

The abrasiveness test is performed using a standard test blade attached to a high-speed mixer used to mix a base drilling fluid containing the weighting material. The mass loss of the blade is used to calculate the abrasive index (AI) of the weighting material in milligrams per minute (mg min^{-1}). A lower index indicates less abrasiveness by the solid particles. The abrasiveness test of weighting materials can be classified as three-body abrasive wear. It is proposed to use surfactants that adsorb effectively on the hematite particles in order to reduce the wear rate by hematite when used as weighting materials in drilling fluids formulation. In the present research, a nonionic/anionic surfactant mixture at a concentration of 1% w/v (relative to water) at different pH, was evaluated for its effect on wear properties.

Figure 18.11 shows the effect of the surfactant at different pH on abrasive wear. The slurries with surfactant (Table 18.4) at pH 4, 7, and 10 presented AIs at equilibrium of 4.2, 1.7, and 1.7 ± 0.1 mg min^{-1}, respectively. The slurry without surfactant (A) showed an AI at equilibrium of 3.7 ± 0.1 mg min^{-1}. The abrasiveness test results suggest that surfactant adsorption onto hematite particles could be related to wear reduction, that is, the higher the adsorption the lesser the abrasive wear. The surfactant addition in slurry (C, D) at pH 7 and 10 generate a reduction of about 54% in the AI in comparison with slurry (A). Slurry (B) at pH 4 reported an AI at equilibrium similar to the slurry without surfactant (A). This result can be correlated with the lower amount of surfactant adsorbed on hematite particles at pH 4 (Figure 18.6). It is

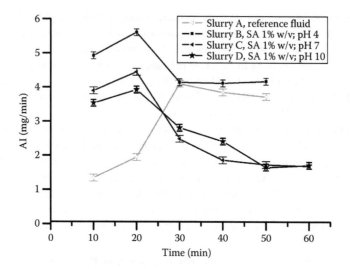

FIGURE 18.11   Effect of surfactant and pH on abrasive wear.

TABLE 18.4
Slurry Formulations Studied in Abrasion Test

Additives	Slurry Formulations			
	A (Reference Fluid)	**B, pH 4**	**C, pH 7**	**D, pH 10**
Base suspension	300	300	300	300
Surfactant mixture (nonionic/ anionic) (g)	—	3.5	3.5	3.5
Hematite (g)	300	300	300	300
Sodium hydroxide (NaOH) (g)		0.2	0.4	0.6

believed that the surfactant adsorbed not only on the metal surfaces but also on the hematite particles generates a highly packed and densely ordered film, which minimizes the impact of hematite particles on the test blade due to electrostatic repulsion between the negative charged surfaces, thus, leading to a lower mass loss.

18.5   SUMMARY

The adsorption of nonionic/anionic mixed surfactant system on hematite particles was investigated. Adsorption was influenced by the pH of the solution. The maximum surfactant adsorption per unit of hematite mass occurred in aqueous solution at pH 10, and was approximately 0.35 ± 0.03 mg g^{-1}. It is proposed that the nonionic surfactant adsorbs first onto the negatively charged surface, occupying all the available adsorption sites. This creates a hydrophobic environment on the particle, which allows the adsorption of the anionic surfactant through hydrophobic chain–chain interactions. Evidences for surfactant adsorption are particles show higher dispersion stability in water; particles become more hydrophilic; and particle zeta potential increases to larger negative values in the presence of mixed surfactant.

The nonionic–anionic surfactant mixture at a concentration of 1% w/v (with respect to water content), and different pH was evaluated for wear properties. The results of an abrasiveness test suggest that surfactant adsorption onto hematite particles could be related to wear reduction. However, additional investigations will be required in order to establish a clear correlation between wear and these parameters.

ACKNOWLEDGMENT

This research was sponsored by PDVSA Intevep S.A., Venezuela under project 0877-PRIE.

REFERENCES

1. G. Quercia, R. Belisario, and R. Rengifo, Reduction of erosion rate by particle size distribution (PSD) modification of hematite as weighting agent for oil based drilling fluids, *Wear*, 266, 1229–1236, 2009.
2. S. Baba Hamed and M. Belhadri, Rheological properties of biopolymers drilling fluids, *J. Petroleum Sci. Eng.*, 67, 84–90, 2009.
3. M. Dolz, J. Jiménez, M.J. Hernández, J. Delegido, and A. Casanovas, Flow and thixotropy of non-contaminating oil drilling fluids formulated with bentonite and sodium carboxymethyl cellulose, *J. Petroleum Sci. Eng.*, 57, 294–302, 2007.
4. V. Mahto and V.P. Sharma, Rheological study of a water based oil well drilling fluid, *J. Petroleum Sci. Eng.*, 45, 123–128, 2004.
5. R. Caenn and G.V. Chillingar, Drilling fluids: State of art, *J. Petroleum Sci. Eng.*, 14, 221–230, 1996.
6. V.C. Kelessidis, C. Papanicolaou, and A. Foscolos, Application of Greek lignite as an additive for controlling rheological and filtration properties of water bentonite suspensions at high temperatures: A review, *Int. J. Coal Geol.*, 77, 394–400, 2009.
7. H. Dehghanpour and E. Kuru, Effect of viscoelasticity on the filtration loss characteristics of aqueous polymer solutions, *J. Petroleum Sci. Eng.*, 76, 12–20, 2011.
8. T. Hamida, E. Kuru, and M. Pickard, Filtration loss characteristics of aqueous waxy hull-less barley (WHB) Solutions, *J. Petroleum Sci. Eng.*, 72, 33–41, 2010.
9. Z. Hanyi, Q. Zhengsong, H. Weian, and C. Jie, Shale inhibitive properties of polyether diamine in water-based drilling fluid, *J. Petroleum Sci. Eng.*, 78, 510–515, 2011.
10. Y. Qu, X. Lai, L. Zou, and Y. Su, Polyoxyalkyleneamine as shale inhibitor in water-based drilling fluids, *Appl. Clay Sci.*, 44, 265–268, 2009.
11. M. Khodja, J.P. Canselier, F. Bergaya, K. Fourar, M. Khodja, N. Cohaut, and A. Benmounah, Shale problems and water-based drilling fluid optimisation in the Hassi Messaoud Algerian oil field, *Appl. Clay Sci.*, 49, 383–393, 2010.
12. J. Guo, J. Yan, W. Fan, and H. Zhang, Applications of strongly inhibitive silicate-based drilling fluids in troublesome shale formations in Sudan, *J. Petroleum Sci. Eng.*, 50, 195–203, 2006.
13. R. Samuel, Friction factors: What are they for torque, drag, vibration, bottom hole assembly and transient surge/swab analyses?, *J. Petroleum Sci. Eng.*, 73, 258–266, 2010.

14. J.M. González, F. Quintero, R.L. Márquez, S.D. Rosales, and G. Quercia, Formulation effects on the lubricity of O/W emulsions used as oil well working fluids, in: *Surfactants in Tribology*, Volume 2, G. Biresaw and K.L. Mittal (Eds.), pp. 241–265, CRC Press, Taylor & Francis Group, Boca Raton, FL, 2011.

15. W.E. Foxenberg, S.A. Ali, T.P. Long, and J. Vian, Field experience shows that new lubricant reduces friction and improves formation compatibility and environmental impact, Paper SPE 112483 in SPE International Symposium and Exhibition on Formation Damage Control Proceedings, Lafayette, LA, 2008.

16. M.S. Aston, P.J. Hearn, and G. McGhee, Techniques for solving torque and drag problems in today's drilling environment, Paper 48939 in SPE Annual Technical Conference and Exhibition Proceedings, New Orleans, LA, 1998.

17. P. Skalle, K.R. Backe, S.K. Lyomov, L. Kilaas, A.D. Dyrli, and J. Sveen, Microbeads as lubricant in drilling muds using a modified lubricity tester, Paper SPE 14797 in IADC/SPE Drilling Conference Proceedings, Houston, TX, 1999.

18. J.J. Truhan, R. Menon, and P.J. Blau, The evaluation of various cladding materials for down-hole drilling applications using the pin-on-disk test, *Wear*, 259, 1308–1313, 2005.

19. J.D. Kercheville, A.A. Hinds, and W.R. Clements, Comparison of environmentally acceptable materials with diesel oil for drilling mud lubricity and spotting fluid formulations, Paper SPE 14797 in IADC/SPE Drilling Conference Proceedings, Dallas, TX, 1986.

20. I. Levine. *Physicochemical*, Fifth edition, Vol 1. McGraw Hill, España, 2005.

21. D. Mohapatra, D. Mishra, and K.H. Park, A laboratory scale study on arsenic(V) removal from aqueous medium using calcined bauxite ore, *J. Environ. Sci.*, 20, 683–689, 2008.

22. P. Ghosh, *Colloid and Interface Science*, Third edition, Chapter 6, pp. 222–225, PHI Learning Pvt. Ltd, New, Delhi, 2009.

23. D.A. McQuarrie and J.D Simon, *Physical Chemistry. A Molecular Approach.* University Science Books, New York, 1997.

24. M.J. Rosen, *Surfactants and Interfacial Phenomena*, John Wiley & Sons, New York, 1978.

25. L. Zhang and P. Somasundaran, Adsorption of mixtures of nonionic sugar-based surfactants with other surfactants at solid/liquid interfaces I. Adsorption of n-dodecyl-β-D maltoside with anionic sodium dodecyl sulfate on alumina, *J. Colloid Interface Sci.*, 302, 20–24, 2006.

26. J. Shibata and D.W. Fuerstenau, Flocculation and flotation characteristics of fine hematite with sodium oleate, *Int. J. Miner. Process.*, 72, 25–32, 2003.

27. J.C. López-Montilla, M.A. James, O.D. Crisalle, and D.O. Shah, Surfactants and protocols to induce spontaneous emulsification and enhance Detergency, *J. Surfactants Deterg.*, 8(1), 45–53, 2005.

28. Y. Hu, G. Qiu, and J.D. Miller, Hydrodynamic interactions between particles in aggregation and flotation, *Int. J. Miner. Process.*, 70, 157–170, 2003.

29. M. Clifton, T. Nguyen, and R. Frost, Effect of ionic surfactants on bauxite residues suspensions viscosity, *J. Colloid Interface Sci.*, 307, 572–577, 2007.

30. A. Dimirkou, A. Ioannou, and M. Doula, Preparation, characterization and sorption properties for phosphates of hematite, bentonite and bentonite-hematite systems, *Adv. Colloid Interface Sci.*, 97, 37–61, 2002.

31. S.K. Goel, Selecting the optimal linear alcohol ethoxylate for enhanced oily soil removal, *J. Surfactants Deterg.*, 1(2), 45–53, 1998.

32. M. El-Batanoney, Th. Abdel-Moghny, and M. Ramzi. The effect of mixed surfactants on enhancing oil recovery, *J. Surfactants Deterg.*, 2(2), 45–53, 1999.

33. V.M. Starov, Surfactant solutions and porous substrates: Spreading and imbibition, *Adv. Colloid Interface Sci.*, 111, 3–27, 2004.

34. M.A. Muherei, Equilibrium adsorption isotherms of anionic, nonionic surfactants and their mixtures to shale and sandstone, *Modern App. Sci.*, 3(2), 158–167, 2008.

35. G. Broze (Ed), *Handbook of Detergents—Part A, Properties, Surfactant Science Series Vol. 82*, Chapter 3, pp. 47–97, Marcel Dekker, New York, 1999.
36. M. Żenkiewicz, Methods for the calculation of surface free energy of solids, *J. Achievements. Mater. Manuf. Eng.*, 24, 1–9, 2007.
37. R. Atkin and V. Craig, Mechanism of cationic surfactant adsorption at the solid–aqueous interface, *Adv. Colloid Interface Sci.*, 103, 219–304, 2003.
38. F. Aixing, Adsorption of alkyltrimethylammonium bromides on negatively charged alumina, *Langmuir*, 13, 506–510, 1997.
39. P. Somasundaran and S. Krishnakumar, Adsorption of surfactants and polymers at the solid-liquid interface, *Colloids Surfaces A.*, 123–124, 491–513, 1997.
40. M. Drew, *Surfactant Science and Technology*, Third edition, Chapter 10, pp. 329–349, John Wiley & Sons, New Jersey, 2006.
41. D.J. Tracy and R.L. Reierson, Commercial synthesis of monoalkyl phosphates, *J. Surfactants Deterg.*, 5, 169–172, 2002.
42. B.X. Wang and W.P. Yu, Fluid phase equilibrium and its properties for unsaturated wet porous media, *Fluid Phase Equilib.*, 75, 197–212, 1992.
43. H.G. Bruil and I.J. van Aartsen, The determination of contact angles of aqueous surfactant solutions on powders, *Colloid Polym. Sci.*, 252, 32–38, 1974.
44. L. Carino, Kinetics of wetting of a compacted powder by aqueous solutions of surface active agents, *Colloid Polym. Sci.*, 254, 108–113, 1976.
45. L. Labajos-Broncano, M.L. González-Martín, and J.M. Bruque, On the evaluation of the surface free energy of porous and powdered solids from imbibition experiments: Equivalence between height–time and weight–time techniques, *J. Colloid Interface Sci.*, 262, 171–178, 2003.
46. ISO specification 10416:2010, *Recommended Practice for Laboratory Testing of Drilling Fluids*, Recommended, Washington, DC, Seventh Edition, February 2010.
47. W. Lv, B. Bazin, D. Ma, Q. Liu, D. Han, and K. Wu, Static and dynamic adsorption of anionic and amphoteric surfactants in the presence of alkali, *J. Petroleum Sci. Eng.*, 77, 209–218, 2011.
48. R.J. Crawford and D.E. Mainwaring, The influence of surfactant adsorption on the surface characterization of Australian coals, *Fuel.*, 80, 313–320, 2001.
49. T. Dang-Vu and J. Hupka, Characterization of porous materials by capillary rise method, *Physicochem. Probl. Miner. Process.*, 39, 47–65, 2005.
50. B. Bai, N.P. Hankins, M.J. Hey, and S.W. Kingman, In situ mechanistic study of SDS adsorption on hematite for optimized froth flotation, *Ind. Eng. Chem. Res.*, 43, 5326–5338, 2004.
51. S. Comba and R. Sethi, Stabilization of highly concentrated suspensions of iron nanoparticles using shear-thinning gels of xanthan gum, *Water Res.*, 43, 3717–3726, 2009.
52. J.M. González, F. Quintero, J.E. Arellano, R.L. Márquez, C. Sanchéz, and D. Pernia, Effects of interactions between solids and surfactants on the tribological properties of water-based drilling fluids, *Colloids Surfaces A.*, 391, 216–223, 2011.
53. G. Faure, *Principles and Applications of Inorganic Chemistry*, First edition, Chapter 14, pp. 270–274, Macmillan Pub. Co., CollierMacmillan, New York, 1991.
54. L. Zhang, L. Wan, N. Chang, J. Liu, C. Duan, Q. Zhou, X. Li, and X. Wang, Removal of phosphate from water by activated carbon fiber loaded with lanthanum oxide, *J. Hazard. Mater.*, 190, 848–855, 2011.
55. L. Zeng, X. Li, and J. Liu, Adsorptive removal of phosphate from aqueous solutions using iron oxide tailings, *Water Res.*, 38, 1318–1326, 2004.
56. G. Stachowiak and A.W. Batchelor, *Engineering Tribology*, Third Edition, Chapter 11, pp. 503–503, Butterwort-Heinemann, Burlington (U.S.A.), Oxford (UK), 2001.

19 Moving Contact Line Problem in Electrowetting

Relevance to Tribological Phenomenon

Ya-Pu Zhao and Ying Wang

CONTENTS

Electrowetting (EW) or electrowetting-on-dielectric (EWOD) has been widely used as a tool for the manipulation of microfluidics in microelectromechanical systems, and there have been rapid developments in the last two decades. In EW, when an external voltage is applied, the contact line of the droplet on solid surface moves until the droplet reaches a new equilibrium. The moving contact line (MCL) phenomenon in EW is a matter of the solid–liquid interface, in which tribology works. This chapter focuses on the MCL problem for EW or EWOD, in which the "Huh–Scriven paradox" is also valid. As a matter of fact, the MCL problem has remained an issue of controversy and debate for more than 40 years since the famous paper by Huh and Scriven in 1971. The difficulty stems partly from the fact that classical hydrodynamic equations coupled with the conventional no-slip boundary condition predict a singularity for the stress that results in a nonphysical logarithmically singular energy dissipation rate at the triple contact line.

After a concise review of the classical EW fundamental theories, we pay special attention to the nanoscale EW, electro-elasto-capillarity and EW on curved surfaces, which are currently being investigated in our group both experimentally and numerically. Precursor film (PF) for planar EW and precursor chain (PC) for EW in an interior corner, together with molecular kinetic theory analysis, are discussed in depth. Both PF and PC are the effective mechanisms to eliminate the nonphysical singular stress distribution and the logarithmically singular energy dissipation rate for the MCL problem under electric field.

19.1 INTRODUCTION

Along with the decreasing system size, properties of microelectromechanical systems (MEMS) and nanoelectromechanical systems (NEMS) will become different from that of macroscopic systems. Especially, the large surface-area-to-volume ratio raises serious adhesive and frictional problems for their operation, dictating that surface properties are of paramount importance in these miniaturized systems.

The manipulation of a liquid droplet in MEMS and NEMS has been developed rapidly in recent decades, owing to its wide potential use in various fields like biomedicine. Electrowetting (EW) or electrowetting-on-dielectric (EWOD), which alters wettability of a liquid droplet on a solid substrate by introducing a voltage, can be used as one useful method to manipulate individual droplets. Since individual droplets can be handled independently and more easily, which is better than conventional microfluidic devices that are based on continuous flow, EW or EWOD in MEMS and NEMS could provide more chances to application and open wider vistas. In EW, when an external voltage is applied, the contact line of the droplet on the solid surface moves until the droplet reaches a new equilibrium. This moving contact line (MCL) phenomenon in EW is a matter of solid–liquid interface, which is relevant to tribology. A good understanding of the MCL problems in EW will help in better design of MEMS and NEMS based on the EW principle.

Modern EW or EWOD was developed from electrocapillarity, which was first studied and described in detail by Gabriel Lippmann [1], who won the Nobel Prize for physics in 1908 for his method of reproducing colors photographically based on the phenomenon of interference. He carried out experiments to study the electrocapillarity of mercury in contact with electrolyte solutions, finding that the capillary depression of mercury could be influenced by the applied voltage between mercury and the electrolyte solutions. The explanation of this phenomenon was that the induced residual charge altered the solid–liquid interfacial tension γ_{sl}, and the Lippmann equations were put forward:

$$\sigma = -\frac{\partial \gamma_{sl}}{\partial V}, \tag{19.1}$$

$$c = \frac{\partial \sigma}{\partial V}, \tag{19.2}$$

where σ is the surface charge density, V is the value of the applied voltage, and c is the capacitance per unit area. Lippmann's PhD thesis, presented to Sorbonne University on July 24, 1875, was on electrocapillarity.

Following Lippmann's study, subsequent researches [2,3] concentrated on this voltage-influenced phenomenon, from electrocapillarity to EW, for more than a century. In early EW studies, the droplet was in direct contact with the electrode surface. This made it hard to put it into application because the electrolytic decomposition of water took place when the applied voltage exceeded a few hundred millivolts [4]. This problem was solved by placing a thin insulating layer on the substrate to separate it from the liquid droplet [5,6], which is called EWOD. In this situation, the applied voltage can be increased to hundreds of volts, and the variation in wettability is reversible over a very large range of contact angles, and thus made it possible to put it into real application. Since EWOD is superior to EW in application, current studies almost concentrate on EWOD rather than EW. Actually, the concepts of EWOD and EW are currently without strict distinction, and EWOD is often called as EW for short, except in special cases.

EWOD has attracted much attention, and relevant investigations have developed fast in recent decades. Lab-on-a-chip (LOC), which has been used in biomedical [7–10] and chemical [11–13] devices, is one of the most popular applications of EWOD. Such EWOD-based devices can be used to manipulate aqueous droplets ranging from nanoliters to microliters in volume [14]. Droplet motions such as dispensing, mixing, merging, splitting, and transport can be realized without the use of conventional pumps, valves, or channels. Transport of droplets is rapid and repeatable, and was demonstrated with well over 100,000 cycles of transfer for a single droplet [14]. Owing to its high level of integration and operational flexibility, LOC devices are promising for use in biomedical instruments, and considerable progress has been made in such applications as sample collection and preparation, DNA analysis and repair, protein recognition, and cell sorting [15]. Microlens is another example of EWOD-controlled application, where the drop is used as an optical lens. Compared with solid lenses, liquid lenses are more flexible with adjustable curvature and focal length, which can be controlled via EWOD. Berge and Peseux [16] pioneered this work by designing an optical system with variable focal length controlled merely by an external voltage. Their results showed that their optical systems were with high quality, speed, and reversibility, demonstrating the possibility of fabricating cheap electrically controllable lenses. More investigations [17–21] on microlens were carried out following their study, including an attempt to develop better conductive liquid droplet and to use liquid lens arrays, intending to optimize the devices and expand their applications. Besides, EWOD can also be used in electronic display technology (such as portable devices). "Droplet-on-a-wristband" for electric signal and droplet connections, which was put forward recently by Fan and coworkers [22], is a promising use of EWOD. Recently, the concept of electronic paper (e-paper) was proposed, and considerable attentions have been paid to such a substrate [23–25] owing to its advantages of flexibility, versatility, and low cost. Electronic devices can be readily fabricated on paper substrate, as was demonstrated by Kim and Steckl [26] who investigated several types of paper. E-paper provides more opportunities to improve the existing electronic display devices and develop new ones.

Scientifically speaking, EW or EWOD can be included into the broader MCL problem, in which the "Huh–Scriven paradox" is also valid. As a matter of fact, the MCL problem has remained an issue of controversy and debate for more than 40 years, ever since the famous paper by Huh and Scriven [27] in 1971. The difficulty stems partly from the fact that classical hydrodynamic equations coupled with conventional no-slip boundary condition predict a singularity for the stress that results in a nonphysical logarithmically singular energy dissipation rate at the triple contact line (TCL).

In this chapter, we will give a brief overview of the fundamentals of EWOD, in conjunction with our group's work. Before delving into a detailed discussion, EWOD-relevant characteristic time and length scales will be presented in Section 19.2. In

Section 19.3, EWOD theories, including the basic and extended equations, droplet actuation principles, and precursor films (PFs), will be introduced. The corresponding experiments and molecular dynamics (MD) simulations will be discussed subsequently in Section 19.4 before the conclusion section.

Note that EWOD is actually a very complex phenomenon, and we do not aim to cover all of the issues in this chapter. We will focus on the MCL problem in EW. PF for planar EW and precursor chain (PC) for EW in an interior corner, together with molecular kinetic theory (MKT) analysis, are discussed in depth. Both PF and PC are effective mechanisms to eliminate nonphysical singular stress distribution and the logarithmically singular energy dissipation rate for the MCL problem under an electric field. We hope that this chapter will help in better understanding of EWOD phenomenon and inspire new investigations and more applications, particularly for the MCL problem in EW.

19.2   CHARACTERISTIC TIME AND LENGTH SCALES RELEVANT TO EWOD PHENOMENON

EWOD-based devices have been used in various fields and applications, as was discussed in the preceding section. Along with the decreasing system size, micro- and nanosystems will perform differently from macroscopic systems. The forces that dominate in macroscopic systems may become unimportant in micro- and nanosystems, whereas others that can be neglected in macrosystems become significant. Surface tension effect is one such example that dominates at micro- and nanoscales due to increase in the surface-to-volume ratio with decreasing size. A clear understanding of relevant characteristic time and length scales would assist the study of the surface effects and help to design new micro- and nanosystems.

19.2.1   Characteristic Timescales

The period of a free droplet in free oscillation T_{LR} was given by Lord Rayleigh in 1879 [28], which is called Lord Rayleigh's period:

$$T_{LR} = \frac{\pi}{4}\sqrt{\frac{\rho D^3}{\gamma_{lv}}}, \tag{19.3}$$

where γ_{lv} is the liquid–vapor surface tension, ρ is the mass density of the droplet, and D is the droplet diameter. In conventional EWOD experiment, where $D \approx 2 \times 10^{-3}$ m, $\gamma_{lv} = 72 \times 10^{-3}$ N/m (at room temperature 25°C), and $\rho \approx 10^3$ kg/m^3, then Lord Rayleigh's period $T_{LR} \approx 8.3 \times 10^{-3}$ s.

Another often used characteristic time is the capillary characteristic time T_c, which is based on the mass of the droplet and is defined as follows:

$$T_c = \sqrt{\frac{m}{\gamma_{lv}}}, \tag{19.4}$$

where m is the droplet mass. Combining Equation 19.3 with Equation 19.4, the relationship between the two characteristic timescales can be obtained: $T_c = \sqrt{(8/3\pi)}T_{LR} \approx 0.92T$.

Taking the viscosity of the liquid into account, viscous characteristic time T_{vis} can be derived from the capillary number Ca (the definition of Ca is $Ca = (\eta v/\gamma_{lv}) = (\eta l/\gamma_{lv}t)$, where v is the characteristic velocity and t is the characteristic time), and its expression is [29]

$$T_{vis} \sim \frac{\eta l}{\gamma_{lv}} \tag{19.5}$$

where η is the viscosity of the liquid droplet and l is the characteristic length.

The magnitudes of some characteristic times relevant to EWOD are

1. The Maxwell relaxation time of bulk water $t \sim 10^{-12}$ s [30,31].
2. The characteristic timescale for dipolar reorientation in an uncharged nanotube is in the range of nanoseconds [32].
3. The response time of EWOD-based display devices at millimeter level is about 10 ms.

19.2.2 Characteristic Length Scales

There are many characteristic length scales related to surface effects, and it is not possible to list all of them. Only those relevant to EWOD are given here.

1. *EW number.* The Lippmann–Young (L–Y) equation is the basis for EWOD:

$$\cos\theta = \cos\theta_0 + \frac{\varepsilon V^2}{2d\gamma_{lv}}, \tag{19.6}$$

where θ_0 is the contact angle described by the Young equation $\cos\theta_0 = (\gamma_{sv} - \gamma_{sl}/\gamma_{lv})$, that is, contact angle without voltage. ε is the dielectric constant, d is the thickness of the dielectric film, γ_{lv} is the liquid–vapor surface tension, and V is the applied voltage. The last term of the L–Y equation is a dimensionless number, which is called the EW number:

$$Ew = \frac{\varepsilon V^2}{2\gamma_{lv}d}. \tag{19.7}$$

It represents the ratio of electrostatic energy to interfacial energy. The larger the EW number, the bigger is the change in contact angle due to applied voltage.
2. *The characteristic time of line tension effect.* The Young equation, which is the basis of wetting, can be regarded as a special case of the L–Y equation, that is, it describes EW without applied voltage. It does not take the three-phase

molecular interactions at the contact line into account. Considering the line tension effect, the Young equation should be modified for planar surface [33]:

$$\cos\theta = \cos\theta_0 - \frac{\tau}{\gamma_{lv}R} \tag{19.8}$$

where R is the radius of the contact line and τ is the line tension. The line tension can be positive or negative, depending on the properties of the droplet and the substrate material. It can be expressed approximately as [33]

$$\tau \approx 4\delta\sqrt{\gamma_{sv}\gamma_{lv}}\,\cot\theta_0, \tag{19.9}$$

in which δ stands for the average distance between the liquid and the solid molecules. The order of magnitude of the line tension is in the range from ~10^{-11} N to ~10^{-6} N [34]. From Equation 19.8, a characteristic length l_B can be derived:

$$l_B = \frac{|\tau|}{\gamma_{lv}}. \tag{19.10}$$

It measures the strength of the line tension relative to liquid–vapor surface tension. From the above equation, it can be inferred that the line tension effect should be considered only when the droplet size is below the micrometer level.

3. *Debye screening length (or Debye shielding length)*. When a particle that carries $+q$ charge is deposited in a continuous medium, the electrons in the medium will be attracted to form an electron sphere, outside of which charges are screened. A Debye sphere is such a volume in which there is a sphere of influence inside the volume while the influence vanishes outside. The Debye length is the radius of such a sphere [35,36], which is defined as

$$\lambda_D = \sqrt{\frac{\varepsilon\varepsilon_0 k_B T}{e^2 N_A \sum_i z_i^2 M_i}}, \tag{19.11}$$

where ε_0 is the permittivity of the medium (e.g., vacuum), k_B is the Boltzmann's constant, T is the absolute temperature, e is the charge of an electron, M_i is the molar concentration, and z_i is the valency. Debye screening length is an important characteristic length scale in EW. In conventional macroscopic EW, the diameter of the droplet is of the order of millimeters, which is much larger than the Debye screening length, that is, $D \gg \lambda_D$, so that the liquid is insulating. However, for EW at nanoscale, the droplet size D is much smaller than the Debye screening length [37], that is, $D \ll \lambda_D$. Thus, at nanoscale, the droplet becomes a conductor. Actually, in most macroscopic EW experiments, the droplet liquid is a conducting solution (e.g., 0.1 mol/L KCl solution) instead of just pure water.

19.3   ELECTROWETTING THEORIES

Classical EWOD theories were initially based on the ideal model, such as the system is macroscopic, the surface is planar and smooth, and the liquid droplet is a perfect conductor. Owing to its great application potential, EWOD has drawn much attention from various fields. With the development of EWOD, more intensive studies are conducted, concentrating on different aspects. For instance, the real surface configurations (such as curvature, roughness, etc.), EWOD under high voltage and below micrometer level are taken into consideration. Contact angle hysteresis is an accompanying phenomenon that always affects EWOD as well as wetting. It can be caused by many factors, and will affect droplet actuation by influencing the minimum actuation voltage. Temperature-induced effect should also be considered, which is related to Marangoni convection, because the introduction of voltage may cause uneven heat distribution in the droplet and at the liquid–vapor interface. At a microscopic level, many phenomena and properties may be different from those at macroscopic level. The properties of the molecular PF will be discussed in this section. At the end of this section, compared with classical EWOD, another form called spontaneous EW will be discussed.

19.3.1   CLASSICAL ELECTROWETTING THEORY: THE LIPPMANN–YOUNG EQUATION

The L–Y equation, which describes the relationship between contact angle and applied voltage, is the basis of EWOD:

$$\cos\theta = \cos\theta_0 + \frac{\varepsilon V^2}{2d\gamma_{lv}}. \tag{19.6}$$

A typical curve from an EWOD experiment is shown in Figure 19.1. It can be seen that the L–Y equation correctly predicts the experimental results under low voltage. However, it fails under high voltage. Contact angle saturation under high voltage will be discussed in detail in Section 19.3.2.2.

The L–Y equation can be derived from different approaches. Here, we will give three main approaches, which have also been discussed by Mugele [4] and Berthier [15]: thermodynamic approach, energy minimization approach, and electromechanical approach.

19.3.1.1   Thermodynamic Approach

The droplet directly contacted the metal surface in Lippmann's original work. In this case, when voltage is introduced, all of the counterions are assumed to be located at a fixed distance d_H from the surface based on the Helmholtz model, and an electrical double layer (EDL) forms at the solid–liquid interface. As was mentioned above, modern EW is EWOD by introducing a thin dielectric film to separate the droplet and the metal. The thickness of the EDL d_H (of the order of a few nanometers) is far less than that of the dielectric film, so that the effect of the EDL can be neglected. Taking the dielectric film as a component of the solid–liquid interface, the reduction of the effective interfacial tension γ_{sl}^{eff} is

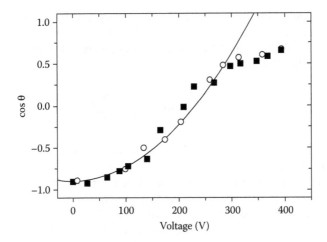

FIGURE 19.1 A typical curve of cosine of the contact angle (θ) versus applied voltage in electrowetting-on-dielectric (EWOD). The points are experimental results whereas the solid line is the theoretical prediction following the Lippmann–Young equation 19.6. (From F. Mugele and J. C. Baret, Electrowetting: From basics to applications, *J. Phys.: Condens. Matter*, 17, 705, 2005. With permission.)

$$d\gamma_{sl}^{eff} = -\sigma_{sl}dV. \qquad (19.12)$$

When a voltage is introduced, the electric energy is mainly stored in the dielectric film, very little of which is available to change the solid–liquid interfacial tension. σ_{sl} is the surface charge density at the solid–liquid interface and is given by

$$\sigma_{sl} = \frac{\varepsilon_0\varepsilon_d}{d}V. \qquad (19.13)$$

From Equation 19.12 and 19.13, the effective interfacial tension can be obtained:

$$\gamma_{sl}^{eff} = \gamma_{sl} - \frac{\varepsilon_0\varepsilon_d}{2d}V^2. \qquad (19.14)$$

Using the Young equation, by substituting γ_{sl}^{eff} for γ_{sl}, the L–Y equation can be derived.

19.3.1.2 Energy Minimization Approach

Consider the droplet, the dielectric film, the metal counter electrode, and the voltage source as a thermodynamic system. When a voltage is applied, the droplet will spread until its free energy F reaches a minimum value:

$$dF = \gamma_{sl}dA - \gamma_{sv}dA + \gamma_{lv}dA\cos\theta + dU - dW_B = 0 \qquad (19.15)$$

where γ_{sl}, γ_{sv}, and γ_{lv} are the solid–liquid, solid–vapor, and liquid–vapor interfacial energy, respectively. U is the electric energy, W_B is the work that the voltage source performs, and A is the droplet base area (see Figure 19.2). Dividing both sides of Equation 19.15 by dA gives

$$\gamma_{sl} - \gamma_{sv} + \gamma_{lv} \cos\theta + \frac{dU}{dA} - \frac{dW_B}{dA} = 0. \tag{19.16}$$

In order to get the contact angle θ, we need the values of dU/dA and dW_B/dA. According to electromagnetics, U is defined by the formula $dU = (1/2)\varepsilon E^2 d\Omega$, where E is the electric field intensity and Ω is the volume filled with electric field. The predominant effect in EWOD process is the distribution of the electric field, which is very complicated near the three-phase contact line. For simplicity but without loss of generality, the area with electric field singularity is supposed to be much smaller compared with the uniform field, so that the singularity within the vicinity of the contact line can be neglected and the electric field is taken as a uniform field. The electrostatic energy per unit area under the droplet base is

$$\frac{U}{A} = \int_0^d \frac{1}{2}\varepsilon_0\varepsilon_d E^2 dz. \tag{19.17}$$

Hence, the increase of electrostatic energy upon an infinitesimal increment of the droplet base area can be given by

$$\frac{dU}{dA} = \frac{1}{2}\varepsilon_0\varepsilon_d E^2 d = \frac{1}{2}\frac{\varepsilon_0\varepsilon_d}{d}V^2. \tag{19.18}$$

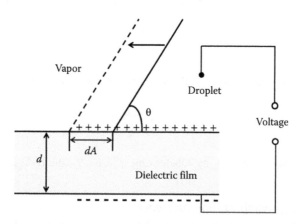

FIGURE 19.2 Schematic of the variation of contact angle with applied voltage. When the voltage increases, the droplet base area increases with an infinitesimal value dA to reach a new equilibrium. (With permission from H.J.J. Verheijen and M.W.J. Prins, Reversible electrowetting and trapping of charge model and experiments, *Langmuir*, 15, 6616. Copyright 1999, American Chemical Society.)

In the droplet spreading process, the work per unit area dW_B/dA done by the voltage source to redistribute the charge can be written as

$$\frac{dW_B}{dA} = V\sigma_{sl} = \frac{\varepsilon_0\varepsilon_d}{d}V^2. \tag{19.19}$$

Substitution of Equations 19.18 and 19.19 in Equation 19.16 gives the L–Y equation 19.6.

19.3.1.3 Electromechanical Approach

The droplet, which is deposited symmetrically on a planar thin dielectric solid, is assumed to be perfectly conductive and is surrounded by an immiscible, perfectly insulating fluid (for instance, vapor). Consider the right-half plane of the droplet (see Figure 19.3). The bottom electrode is grounded and an external potential of V is applied, so that $\Phi = V$ on S_{12} (the liquid–fluid interface) and S_{13} (the solid–liquid interface), $\Phi = 0$ on S_e, where Φ is the electrostatic potential and $E = -\nabla\Phi$. Within the surrounding fluid and dielectric field, the electrostatic potential satisfies the Laplace equation $\nabla^2\Phi = 0$. The electric field E is perpendicular to the surface of the droplet on the surrounding fluid side, and vanishes in the conducting droplet. Neglecting the osmotic contribution, the electrostatic force acting on the droplet surface is

$$F_d = \int_{S_{12}+S_{13}} T\cdot n dS, \tag{19.20}$$

where T is the Maxwell stress tensor:

$$T = -\frac{1}{2}\varepsilon E^2 I + \varepsilon E \otimes E, \tag{19.21}$$

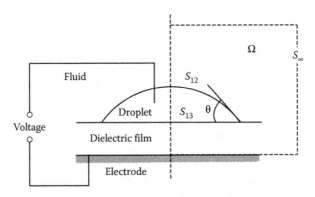

FIGURE 19.3 Schematic of interface domain to be studied in electrowetting (EW). θ is the contact angle, S_{12} is the liquid–fluid interface, S_{13} is the liquid–dielectric film interface, and S_∞ is the surface which is at an infinitely remote distance from the droplet.

where I is the second-order isotropic tensor and $\otimes$ is the tensor product. There is no tangential electric field on the surface of the conducting droplet on the fluid side, so that $E = E_n$ and $E = E \cdot n$. The electrostatic force in Equation 19.20 becomes

$$F_d = \int_{S_{12}+S_{13}} \frac{1}{2} \varepsilon E^2 n dS. \tag{19.22}$$

The solution to Equation 19.22 requires calculation of the electric field (or charge) distribution along the droplet surface. The normal stress on the dielectric film surface, which is caused by the electrostatic force at the solid–liquid interface, can be balanced by elastic stress. Therefore, what we are concerned most with is the electrostatic force at the liquid–fluid interface. As the three-phase contact line is approached, both the charge density and the electric field increase sharply owing to sharp-edge effects. Hence, the field within the vicinity of the contact line will mainly contribute to the electrostatic force at the liquid–fluid interface.

Taking the edge region of the droplet as an infinite planar wedge, Vallet et al. [38] and Kang [39] analyzed the electrostatic field within the vicinity of the edge region by using Schwarz–Christoffel conformal mapping:

$$Z = \int_{i\pi}^{w} \left(e^{w'} + 1\right)^{\beta} dw' + i\pi, \tag{19.23}$$

where the parameter β is defined as $\beta = 1 - \theta/\pi$. Equation 19.23 transforms the plane $Z = x + iy$ to the plane $w = u + iv$, and the transformed coordinates u and v are scaled by d/π and V/π (see Figure 19.4). For simplicity, the electric permittivities of the surrounding fluid and the dielectric film are assumed to be the same, so that the electric field is uniform in the transformed plane w and $E = V/\pi$. Then, the electric field in the original plane can be obtained:

$$E = E_n = \frac{V/\pi}{d/\pi} \frac{1}{dZ/dw} = \frac{V}{d} \frac{1}{|e^u - 1|^\beta}. \tag{19.24}$$

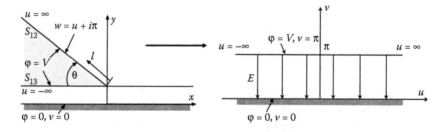

FIGURE 19.4 Schwarz–Christoffel transformation of the droplet edge region. The Z plane ($Z = x + iy$, x and y are the complex coordinates of the original plane) is transformed into the w plane ($w = u + iv$, u and v are the complex coordinates of the transformed plane). (Left figure: With permission from K.H. Kang, How electrostatic fields change contact angle in electrowetting, *Langmuir*, 18, 10318. Copyright 2002, American Chemical Society.)

The surface charge density ($\sigma = \varepsilon E$) that is related to the field is

$$\frac{\sigma}{\sigma_0} = \begin{cases} \dfrac{1}{(e^u - 1)^\beta}, & \text{on } S_{12} \\[3mm] \dfrac{1}{(1 - e^u)^\beta}, & \text{on } S_{13} \end{cases} \tag{19.25}$$

where $\sigma_0 = \varepsilon V/d$ is the charge density at the solid–liquid interface far from the contact line. The Maxwell stress tensor, which is calculated by Equation 19.21, can be expressed in the n-, t-, and z-axis system (n-axis is aligned with the normal liquid–fluid surface) using the following equation:

$$[T] = \varepsilon \begin{bmatrix} \dfrac{1}{2}E_n^2 & 0 & 0 \\[3mm] 0 & -\dfrac{1}{2}E_n^2 & 0 \\[3mm] 0 & 0 & -\dfrac{1}{2}E_n^2 \end{bmatrix}. \tag{19.26}$$

On the wedge surface, where $w = u + i\pi$, the mapping equation becomes

$$\frac{dZ}{dw} = \frac{dZ}{du} = \frac{dx + idy}{du} = \frac{dx}{du} + i\frac{dy}{du}. \tag{19.27}$$

Then, the distance from the wedge apex l can be obtained:

$$\frac{dl}{du} = \frac{d}{\pi}\frac{\sqrt{(dx)^2 + (dy)^2}}{du} = \frac{d}{\pi}|\frac{dZ}{dw}| = \frac{d}{\pi}|e^u - 1|^\beta, \tag{19.28}$$

$$l = \frac{d}{\pi}\int_0^u |e^{u'} - 1|^\beta \, du'. \tag{19.29}$$

For a droplet with a small contact angle, that is, $\theta \to 0$ and $\beta = 1 - \theta/\pi \to 1$, we can obtain the relationship between the electric field and the distance [40]:

$$E_n \sim \frac{1}{|l|^{1/2}}. \tag{19.30}$$

The field singularity occurs near the wedge region, especially at the wedge tip. In fact, from Equation 19.24, it can be seen that the field is always singular near the wedge tip (where $u = 0$) for the contact angle ranging from 0 to π. Studies [38–41] show that this singularity effect occurs in a very confined region with a length scale of magnitude d. Since the charge density and the Maxwell stress are both related to the electric field, both will be singular within the region of $O(d)$. The electrostatic stress will therefore be confined within the vicinity of the edge region. With Equations 19.25 and 19.28, the net electrostatic force acting on the upper side of the wedge can be calculated:

$$F_e = \frac{\varepsilon V^2}{2\pi d} \int_0^\infty \frac{1}{(e^u - 1)^\beta} du = \frac{\varepsilon V^2}{2d} \operatorname{cosec} \theta. \tag{19.31}$$

The horizontal and vertical components of this force are

$$F_{ex} = \frac{\varepsilon V^2}{2d}, \quad F_{ey} = \frac{\varepsilon V^2}{2d} \cot \theta. \tag{19.32}$$

Divided by the liquid–vapor surface tension γ_{lv}, the horizontal part is exactly the EW number in the L–Y equation. Actually, the L–Y equation can be obtained from the balance of forces in the horizontal direction at the contact line.

The thermodynamic and electromechanical approaches discussed in Sections 19.3.1.1 and 19.3.1.3, respectively, represent two different interpretations of the EWOD phenomenon. While one regards the reduction of the contact angle as originating from the change of the solid–liquid interfacial tension, the other considers it as the Maxwell tensor that "drags" the droplet to spread out. It is still dubious which mechanism is the reason, though the two interpretations seem to be related in a sense. The possible interpretations to EW are summarized in Figure 19.5. More investigations are required to clarify the phenomenon. However, regardless of which interpretation is responsible for the phenomenon, the L–Y equation always works and is the basis of EW.

19.3.2 EXTENDED LIPPMANN–YOUNG EQUATIONS

19.3.2.1 Electrowetting on Rough Surfaces

Real surfaces are always rough with topography patterns [41,42], such as the configuration of lotus leaves. While the L–Y equation is for EWOD on ideal smooth surfaces, new models need to be established to describe the contact angle variation on rough surfaces.

The Cassie and Baxter method for a composite surface was used to extend the classical EW equations to microstructured surfaces [43]. For simplicity, the rough surface is represented by hemispherically topped cylindrical asperities (see Figure 19.6). By using the energy minimization approach, an extended equation is derived:

$$\cos \theta_c = f_1 \left(\cos \theta_0 + \frac{1}{2} \frac{\varepsilon_0 \varepsilon_d V^2}{d \gamma_{lv}} \right) - f_2, \tag{19.33}$$

where θ_c is the contact angle on rough surface, θ_0 is the contact angle described by the Young equation, $f_1 = dA_{sl1}/dA_{sl}$ and $f_2 = dA_{sl2}/dA_{sl}$ are the surface roughness factors shown in Figure 19.6. The contact area on the rough surface is divided into two parts at the composite interface: one part is for the solid–liquid interface A_{sl1} while

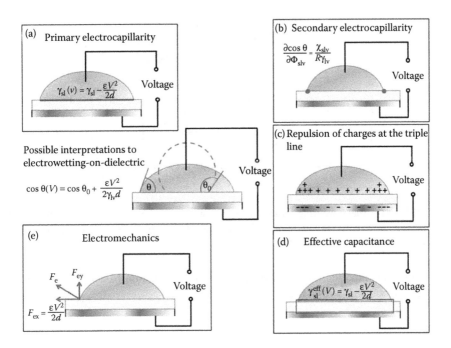

FIGURE 19.5 Possible interpretations (a) through (e) to the reduction of the contact angle (CA) as a result of the applied voltage: primary electrocapillarity, secondary electrocapillarity, repulsion of like-charges at the triple line, effective capacitance and electromechanics. (a) Primary electrocapillarity means that the solid–liquid interfacial energy changes by applying a voltage, that is, $\gamma_{sl}(V) = \gamma_{sl} - (\varepsilon V^2/2d)$, thus leading to the reduction of CA. (b) Secondary electrocapillarity indicates that the applied voltage causes polarization of the triple line, leading to change in the line tension and as thereby in CA, that is, $\partial \cos \theta/\partial \Phi_{slv} = \chi_{slv}/R\gamma_{lv}$, where χ_{slv} is the line density of the electric charges and Φ_{slv} is the electrostatic potential on the triple line. (c) Repulsion of like charges at the triple line shows that it is repulsion that may cause the contact line moves and droplet spreads. (d) Effective capacitance shows that the dielectric film is taken as part of the solid–liquid interface. The effective solid-surface energy changes as the voltage is applied, that is, $\gamma_{sl}^{eff}(V) = \gamma_{sl} - (\varepsilon V^2/2d)$. The dielectric film works as a capacitor and almost all of the electric energy is stored in it. (e) Electromechanics illustrates that the reduction of CA is a result of Maxwell stress tension or the electrostatic force. (Y. Wang and Y.P. Zhao, Electrowetting on curved surfaces, *Soft Matter*, 8, 2599, 2012. By permission of The Royal Society of Chemistry.)

the other is for the liquid–air interface A_{sl2}. Using the dimensionless EW number $Ew = \varepsilon_0 \varepsilon_d V^2/2d\gamma_{lv}$, Equation 19.33 can be rewritten as

$$\cos \theta_c = f_1 \left(\cos \theta_0 + Ew \right) - f_2. \tag{19.34}$$

Equation 19.33 or Equation 19.34 is the extended L–Y equation based on the Cassie–Baxter model. Without applied voltage, that is, $V = 0$, Equation 19.34 reduces to the Cassie–Baxter equation.

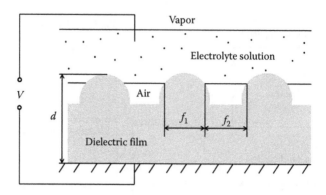

FIGURE 19.6 Electrowetting on an optimized rough surface with hemispherically topped cylindrical asperities. The air remains in the grooves. The roughness factor in the solid–liquid area is $R_1 \cdot f_1 = dA_{sl1}/dA_{sl}$ and $f_2 = dA_{sl2}/dA_{sl}$.

Using the same method, another extended L–Y equation based on the Wenzel model can be established [44]:

$$\cos\theta_c = R_1\left(\cos\theta_0 + Ew\right) \tag{19.35}$$

where $R_1 = A_{sl1(actual)}/A_{sl1(apparent)}$ is the roughness factor [45], which is defined as the ratio of the actual-to-apparent (geometric) solid–liquid contact areas.

It was found that during the EW process on rough surfaces, the contact angle might display a step change [46,47] (see Figure 19.7). This is because wetting on rough surfaces is initially in the Cassie–Baxter state; but after a threshold value of

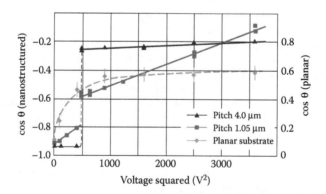

FIGURE 19.7 Cosine of the contact angle as a function of the applied voltage squared for molten salt on nanostructured and planar substrates. The one line (the dotted one) is the variation of $\cos\theta$ on a planar substrate. The other two lines (solid ones) represent the variation of $\cos\theta$ on nanostructured surfaces, with pitch 4.0 and 1.05 μm, respectively, both of which display a step change in $\cos\theta$ at a certain value of voltage. (With permission from T.N. Krupenkin et al., From rolling ball to complete wetting: The dynamic tuning of liquids on nanostructured surfaces, *Langmuir*, 20, 3824. Copyright 2004, American Chemical Society.)

TABLE 19.1

Influence of Electrowetting Number on Droplet States on Rough Electrowetted Surfaces

EW Number	Stable State	Comments
$\eta < -\cos\theta_0 - \left[(1-\phi)/(r_m - \phi)\right]$	Cassie	Finite energy barrier for Cassie–Wenzel transition. Barrier for reverse transition lower.
$\eta > -\cos\theta_0 - \left[(1-\phi)/(r_m - \phi)\right]$	Wenzel	Finite energy barrier for Cassie–Wenzel transition. Barrier for reverse transition higher.
$\eta > -\cos\theta_0$	Wenzel	No energy barrier for Cassie–Wenzel transition. Reverse transition disallowed.

Source: From V. Bahadur and S.V. Garimella, *Langmuir*, 23, 4918, 2007. With permission.

voltage, EW would be in the Wenzel state. An intermediate state occurs during transition from the Cassie–Baxter to the Wenzel state, where the droplet partially wets the grooves, and the contact angle is [43,44]

$$\cos\theta_c = R_1 f_1 \left(\cos\theta_0 + \frac{1}{2} \frac{\varepsilon_0 \varepsilon_d V^2}{d\gamma_{lv}} \right) - f_2. \tag{19.36}$$

The droplet state on a rough surface is related to the EW number Ew and the initial contact angle θ_0. The influence of these two factors is summarized in Table 19.1 [44].

19.3.2.2 Electrowetting on Curved Surfaces

Since EW on planar surfaces is a specific and relatively simple situation, structured surfaces with curved features are more common, such as liquid lenses with curved contact surfaces, flexible paper-like substrate, and lotus leaf. The L–Y equation is derived from and applicable for EW on planar surfaces under low voltage. However, it does not take the surface curvature effect into consideration.

Considering the surface curvature effect, by using the principle of energy minimization, an extended L–Y equation on spherical surfaces (Figure 19.8) was derived [48]:

$$\cos\theta(V) = \cos\theta_0 + \frac{\varepsilon V^2}{2\gamma_{lv}d} \cdot \frac{1}{1 \pm \xi}, \tag{19.37}$$

where $\xi = d/R_S$. Equation 19.37 means that for convex surface

$$\cos\theta(V) = \cos\theta_0 + \frac{\varepsilon V^2}{2\gamma_{lv}d} \cdot \frac{1}{1 + \xi}, \tag{19.38a}$$

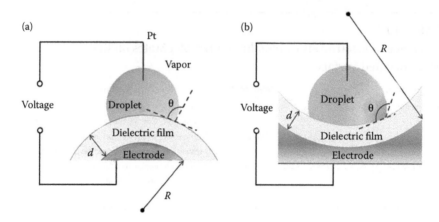

FIGURE 19.8 Schematic of EWOD on curved surfaces: (a) EWOD on a convex surface; (b) EWOD on a concave surface. A conducting liquid droplet is placed on a metal counter electrode coated with a dielectric film. The contact angle θ changes when a voltage is introduced. (Y. Wang and Y.P. Zhao, Electrowetting on curved surfaces, *Soft Matter*, 8, 2599, 2012. By permission of The Royal Society of Chemistry.)

and for concave surface

$$\cos\theta(V) = \cos\theta_0 + \frac{\varepsilon V^2}{2\gamma_{lv}d} \cdot \frac{1}{1-\xi}. \tag{19.38b}$$

Comparing with the EW number, a curvature-modified EW number was introduced:

$$Ew^* = \frac{\varepsilon V^2}{2\gamma_{lv}d} \cdot \frac{1}{1\pm\xi}. \tag{19.39}$$

The curvature-modified EW number can be expressed as a function of Ew and ξ:

$$Ew^* = Ew \cdot \frac{1}{1\pm\xi}. \tag{19.40}$$

The surface curvature has an influence on the contact angle; especially with decrease of the system size, the effect becomes even more significant. At the present experimental condition, ξ or the ratio of d to R_s is small, which makes the modified term of the EW number close to 1. In this case, the variation of the contact angle on curved surfaces is close to that on planar surfaces, and the surface curvature effect is not obvious. However, when R_s becomes very small or EWOD is on and below microscale, such as EW in carbon nanotubes (CNTs) or on a graphene substrate, the surface curvature effect cannot be neglected. Nevertheless, there are few experiments on and below microscale in the published work, and further investigation is necessary. Besides, since a curved surface can be either convex or concave, the variation of the contact angle is different. While the variation of the contact angle on convex surfaces decreases compared with that on planar surfaces, it increases on concave surfaces. This means that concave surfaces can enhance wettability better than convex surfaces when applying the same voltage.

As the capacitance of a spherical capacitor per unit area is $c = \varepsilon/d(1 \pm \xi)$, Equation 19.37 can be simplified as

$$\cos\theta(V) = \cos\theta_0 + \frac{1}{2} \cdot \frac{cV^2}{\gamma_{lv}}. \tag{19.41}$$

where c is the capacitance per unit area. Actually, Equation 19.41 is always applicable for EW on surfaces with various curvatures. For EW on a planar surface, substituting the capacitance of a parallel-plate capacitor per unit area $c = \varepsilon/d$ into Equation 19.41, we get $\cos\theta(V) = \cos\theta_0 + (cV^2/2\gamma_{lv})$. This is exactly the classical L–Y equation. Therefore, the L–Y equation can be seen as a degenerate equation considering the surface curvature effect in a sense. For EW on curved surfaces under low voltage, once the capacitance per unit area is known, the variation of the contact angle can be predicted. However, it should be noted that although the extended EW equation can be simplified as the form of capacitance, it is not just simply a problem of capacitor, since the mechanism of EW is still debatable.

Considering that the line tension works on and below microscale, a uniform expression of EW on various geometrical surfaces was given [48]:

$$\cos\theta(V) = \cos\theta_Y - \frac{\tau\chi}{\gamma_{lv}} + \frac{1}{2} \cdot \frac{cV^2}{\gamma_{lv}}, \tag{19.42}$$

where θ_Y is the Young contact angle, τ is the line tension, χ is the geodesic curvature, and c is the capacitance per unit area.

19.3.2.3 Electrowetting under High Voltage

According to the L–Y equation, complete wetting would be achieved at $V = \sqrt{2d\gamma_{lv}(1 - \cos\theta_0)/\varepsilon_0\varepsilon_d}$. However, complete wetting has never been observed experimentally [4]. Experiments [4,5,49,50] found that the contact angle saturated around 30°–80° under high voltages, and electrical breakdown of the dielectric film would happen when the voltage is high enough [50,51]. The factors that were identified as influencing critical voltage for breakdown include physical properties of the dielectric film, the contact time between the wire electrode and the droplet, and the electrode–dielectric film distance [51]. This means that the L–Y equation fails to describe EWOD under high voltages. New theories need to be established to explain the phenomenon. However, the mechanism of the phenomenon is still ambiguous and no consistent theory has been set up so far. Researchers have never stopped exploring the high-voltage phenomenon and various mechanisms have been proposed.

One of the most broadly accepted explanations was proposed by Verheijen and Prins [52]. They indicated that above a threshold voltage, almost all charges would get trapped in or on the insulating layer. By using the principle of virtual displacement, they derived the equation for EW:

$$\cos\theta = \cos\theta_0 + \frac{1}{2}\frac{\varepsilon_0\varepsilon_d\left(V-V_T\right)^2}{d\gamma_{lv}}, \tag{19.43}$$

where V_T is the voltage of the trapped charge, which is induced by the trapped charge $\sigma_T = \varepsilon_0\varepsilon_d V_T/d_T$ (see Figure 19.9). Below the threshold voltage, there is no trapped charge and the variation of the contact angle is proportional to V^2. When the applied voltage V exceeds the threshold value, trapped charges are generated so that the variation of the contact angle will be proportional to $(V-V_T)^2$. Equation 19.43 corresponded well with their experimental data. However, since the trapped charge is associated with the insulator properties, the model fails to establish the relation between the threshold voltage and the known material properties.

Another modified L–Y equation considering the saturation effect was proposed by Berthier et al. [53]. They used the Langevin function $L(\cdot)$ that is defined as

$$L(X) = \coth(3X) - \frac{1}{3X}, \tag{19.44}$$

and verified that the function fit well with the experimental results:

$$\frac{\cos\theta - \cos\theta_0}{\cos\theta_S - \cos\theta_0} = L\left[\frac{\varepsilon_0\varepsilon_d V^2}{2d\gamma_{lv}\left(\cos\theta_S - \cos\theta_0\right)}\right], \tag{19.45}$$

where θ_S is the saturation angle.

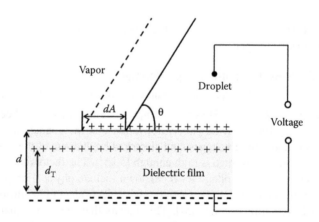

FIGURE 19.9 Schematic representation of contact angle saturation phenomenon. A sheet of trapped charge does not generate until the voltage is above a threshold value. d is the thickness of the dielectric film, and d_T is the distance between the sheet of trapped charge and the bottom of the dielectric film. (With permission from H.J.J. Verheijen and M.W.J. Prins, Reversible electrowetting and trapping of charge model and experiments, *Langmuir*, 15, 6616. Copyright 1999, American Chemical Society.)

Electrical breakdown of the dielectric film will happen when the voltage is high enough. It was pointed out [50,51,54] that the critical voltage for breakdown was proportional to the thickness of the dielectric film. A whole process of breakdown was recorded with a high-speed camera in 200 Hz by Feng and Zhao [51]. In this experiment, a spark was seen as the start of breakdown, at which stage the dielectric film was destroyed and there was an instant current between the wire and the counter electrodes. The heat generated by the instant current generated bubbles by electrolysis of the electrolyte, which swelled dramatically, like an explosion, and electrical energy was converted to dynamic energy of conductive liquids. Droplet ejection is a common phenomenon occurring under high voltages, which has also been observed and reported by other groups [38,55].

Many other studies have been carried out trying to explore EW under high voltage [38,55–58]. However, none of these existing explanations can fully explain the phenomenon and the mechanism is still unclear. More intensive work is needed to fully understand and explain EW under high voltage.

19.3.2.4 Electrowetting at Micro- and Nanoscales

Owing to a variety of applications, such as micro- and nanofluidic manipulation and LOC, EW at a small scale has attracted considerable attention. However, the L–Y equation was derived from and is applicable for EW in macroscopic systems. It neglects some effects such as line tension, which play a major role and cannot be neglected at a microscopic scale. Besides, the assumptions in the derivation may fail due to scale effect. For instance, the size of a nanodroplet is well below the Debye screening length, in which case the treatment of the charge distribution in droplets will be different from macroscopic droplets. Therefore, the study for EW at micro- and nanoscales is of great importance, both for understanding the mechanisms and for application.

19.3.2.4.1 Line Tension Effect

It has been recognized for a long time that the Young equation does not account for three-phase molecular interactions at the contact line. A number of studies have been conducted to consider this effect [33,59,60]. A modified Young equation that takes into account line tension is given by [61]

$$\cos\theta_{ac} = \cos\theta_Y - \frac{\tau H}{\gamma_{lv}} - \frac{\nabla\tau \cdot \nabla H + (\nabla\tau \times \nabla w) \cdot (\nabla w \times \nabla H)}{R\gamma_{lv}\sqrt{|\nabla H|^2 + |\nabla H \times \nabla w|^2}}, \quad (19.46)$$

where τ is the line tension, χ is the geodesic curvature at a certain point, H is the thickness of the liquid at each point, and w is the elevation of the solid surface. Generally, the last term in Equation 19.46 can be neglected and simplifies to

$$\cos\theta_{ac} = \cos\theta_Y - \frac{\tau\chi}{\gamma_{lv}}. \quad (19.47)$$

Especially, $\chi = 1/R$ for planar surfaces (R is the radius of the three-phase contact line) and $\chi = \sqrt{1 - (R/R_S)^2}/R$ for spherical surfaces (R_s is the radius of the solid surface). Values of reported line tension [34,62,63] range from 10^{-10} to 10^{-6} N in order of magnitude, which means it will apply when the system is at and below microscale.

While a great deal of work has been done to modify the Young equation considering the line tension effect, far fewer studies exist about the line tension correction for EW. Using the Gibbs–Johnson–Neimann (GJN) approach, which assumes that a free liquid drop automatically takes the shape that minimizes the total system energy, Digilov considered the line tension effect and the presence of charges on the three-phase contact line and gave the modified equation [64]

$$\cos\theta = \cos\theta_Y - \frac{\tau}{\gamma_{lv}R} + \chi_{slv}U_{slv}, \tag{19.48}$$

where χ_{slv} is the excess line charge and U_{slv} is the excess energy of the three-phase contact line. Equation 19.48 can be rewritten as

$$\left(\frac{\partial\cos\theta}{\partial\Phi_{slv}}\right)_{T,\mu} = \frac{1}{\gamma_{lv}R}\chi_{slv}, \tag{19.49}$$

where Φ_{slv} is the potential at the three-phase contact line. Equation 19.49 illustrates that the change in the cosine of the contact angle under an applied potential is governed by the electric charges on the three-phase contact line. This result is similar to the explanation that it is the Maxwell tension on the contact line that changes the contact angle to some extent.

19.3.2.4.2 Electrowetting (EW) at Nanoscale

In macroscopic EWOD systems, the conducting liquid will spread under a weak electric field, or the apparent contact angle will decrease with an applied voltage. The diameter of the droplet L is at millimeter level, which is much larger than the Debye screening length λ_D, that is, $L \gg \lambda_D$. For pure water, whose ionic concentration is on the order of 10^{-7} M, an EDL, on the order of a few nanometers that follows the Helmholtz model, will generate at the insulator–liquid interface. The electric field will then be confined in this interface layer. In this case, the contact angle can be described by the L–Y equation $\cos\theta = \cos\theta_0 + \langle\varepsilon\varepsilon_1|E|^2\rangle d_E/2\gamma_{lv}$, where the brackets $\langle\cdot\rangle$ denote the average over d_E (d_E is the thickness of the EDL) and ε_1 is the dielectric constant of the liquid. Note that in conventional EWOD, since the thickness of the EDL is much less than that of the dielectric film (on the order of several µm), the EDL effect can be neglected.

Unlike macroscopic drops, nanodroplets with size well under the Debye screening length when placed on a parallel-plate capacitor will essentially behave as conductors [37]. The electric field permeates the whole droplet, and polarization is strongest at the liquid surfaces, which may cause the dipolar molecules to be attracted to the interface [37,65]. The average interaction w between a free dipole $\boldsymbol{\mu}$ and an applied electric field $\boldsymbol{E}$ is [37]:

$$w \approx -\mid \boldsymbol{E} \parallel \boldsymbol{\mu} \mid L\left(\frac{\mid \boldsymbol{E} \parallel \boldsymbol{\mu} \mid}{k_{\mathrm{B}}T}\right) \tag{19.50}$$

where $L(\cdot)$ is the Langevin function and k_{B} is the Boltzmann's constant. Under a weak electric field

$$w \approx -\frac{\mid \boldsymbol{\mu} \mid^{2} \mid \boldsymbol{E} \mid^{2}}{3k_{\mathrm{B}}T}. \tag{19.51}$$

When the electric field gets stronger, the problem will become different and more complicated.

Here, we discussed only the line tension and dipole effects. Actually, while a continuum picture is still applicable, EW at micro- and nanoscales may become different from the macroscopic situation due to scale effect. Forces or other effects that are less important in macrofluidics may play critical roles in microfluidics, for instance, van der Waals (vdW) forces or double-layer forces [66] will become significant and should be considered as the system becomes smaller. At present, relevant research is not far enough, further studies are needed.

19.3.3 CONTACT ANGLE HYSTERESIS AND DROPLET ACTUATION

19.3.3.1 Contact Angle Hysteresis

EWOD has been widely used to manipulate and control microliter or nanoliter quantity of liquids in micro-total-analysis systems (μ-TAS) and LOC. It was found that the droplet did not move immediately when the voltage was applied, but started to move when the voltage exceeded a threshold value [67–69]. Different threshold values were required to initiate movement of the droplet [69,70] in different systems. Once the threshold was exceeded, movement would be both rapid and repeatable. The contact angle hysteresis was believed to be the mechanism responsible for the threshold effect [69].

The contact angle of a droplet advancing on a solid surface is usually different from that receding on the surface. This difference between advancing and receding angles is known as contact angle hysteresis. It can be caused by such factors as surface roughness, surface physical, and chemical heterogeneity [71–73]. The mechanism of contact angle hysteresis is not fully clear. Theoretical [73] and experimental [67,74] studies have been conducted to measure contact angle hysteresis. Figure 19.10a shows the experimental results for a microdroplet of deionized water immersed in silicone oil and placed on a SiOC substrate [67]. The vertical shift between the two curves in Figure 19.10b defines the EW contact angle hysteresis. Contact angle hysteresis is undesirable in EWOD devices due to its influence

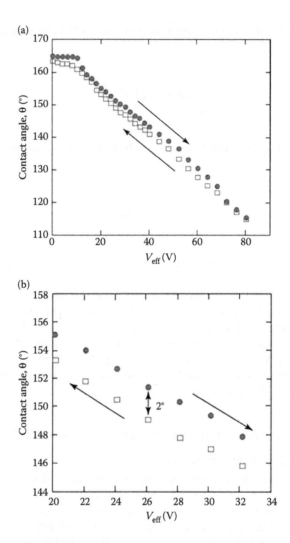

FIGURE 19.10 (a) Experimental results for a microdrop of deionized water immersed in silicone oil and placed on a SiOC substrate. (b) Local amplification of (a). The vertical shift between the two curves in (b) is the EW contact angle hysteresis. (From *Sens. Actuators A: Phys.*, 134, J. Berthier et al., Actuation potentials and capillary forces in electrowetting based microsystems, 471. Copyright 2007, with permission from Elsevier.)

on the minimum actuation potential. Different measures [68,73] were proposed to minimize hysteresis, such as using known-good dice (dice that show good repeatability after preliminary EWOD testing) to circumvent the inconsistent coating effect [68]. As contact angle hysteresis is closely dependent on the properties of the substrate, such studies will provide indications for optimizing the design of EWOD devices.

19.3.3.2 Minimum Actuation Voltage

Prior knowledge of the minimum actuation potential is important for the estimation and design of an EW system. Berthier et al. [67] established a model to estimate the minimum actuation voltage, assuming that the static advancing and receding contact angles were $\theta_0 + \alpha$ and $\theta - \alpha$, where θ is the actuated contact angle, θ_0 the nonactuated contact angle, and α is the hysteresis angle (see Figure 19.10b). This assumption stemmed from the Hoffman–Tanner law [75], which states that the advancing and receding contact angles were, respectively, larger and smaller than their Young values. The total capillary force was obtained:

$$F_x = R_e\gamma_{lv}\left(\cos\theta - \cos\theta_0\right) - R_e\gamma_{lv}\alpha\left(\cos\theta + \sin\theta - \cos\theta_0 + \sin\theta_0\right), \quad (19.52)$$

where R_e is the width of the electrode. Without hysteresis ($\alpha = 0$), the droplet would move immediately even when an infinitely small voltage is applied. Taking into account the hysteresis ($\alpha \neq 0$), the minimum electric potential is given by

$$\frac{c}{2\gamma_{lv}}V_{min}^2 = \frac{\alpha}{1-\alpha}\left[\sin\theta_{V_{min}} + \sin\theta_0\right], \quad (19.53)$$

where c is the capacitance of the dielectric film per unit area. Equation 19.53 is an implicit equation for calculating V_{min} due to the fact that θ depends on V. It can be solved iteratively, considering that $\theta_{V_{min}}$ can be obtained from the L–Y equation.

Berthier's model only considered the EW force F_{elec} and the contact angle hysteresis F_f. Other forces such as the viscous resistance $F_{viscous}$ and the drag force applied by the Pt electrode F_{drag} may also influence the actuation of the droplet. To include these effects, another model was proposed for the actuation force [70]:

$$F_a = F_{elec} - (F_{viscous} + F_f + F_{drag}). \quad (19.54)$$

The horizontal component of EW force is

$$F_{elec} = 2R\gamma_{lv}\left(\cos\theta_a - \cos\theta_r\right), \quad (19.55)$$

where R is the radius of the contact area of the droplet and the lotus leaf, and θ_a and θ_r are the advancing and receding angles with applied voltage, respectively. In the case of an open EWOD system [15], the viscous resistance $F_{viscous}$ is described by

$$F_{viscous} \approx \pi R^2 \tau_w, \quad (19.56)$$

where τ_w is the shear stress which can be approximately expressed as

$$\tau_w \approx \frac{5\eta v_{open}}{2h}, \quad (19.57)$$

in which h is the maximum height of the droplet, η is the viscosity of the liquid, and v_{open} is the actuation speed. Another resistive force is the force resulting from contact angle hysteresis F_f, which is given by

$$F_f \approx 2R\gamma_{lv}\left(\cos\theta_{r0} - \cos\theta_{a0}\right) = 4R\gamma_{lv}\sin\left(\frac{\theta_{r0} + \theta_{a0}}{2}\right)\sin\left(\frac{\theta_{a0} - \theta_{r0}}{2}\right), \quad (19.58)$$

where θ_{a0} and θ_{r0} are the advancing and receding angles without applied voltage, respectively. The drag force applied by the Pt electrode can be estimated by

$$F_{drag} \approx \pi D_{Pt}\gamma_{lv}, \quad (19.59)$$

where D_{Pt} is the diameter of the Pt electrode. The actuation force F_a should be larger than zero in order to fulfill the droplet actuation, and then the minimum voltage can be obtained. The droplet velocity can be estimated by

$$V_{open} = \int_0^{t_\tau} \frac{F_a(t)}{m}dt, \quad (19.60)$$

in which t_τ is the actuation time and m is the mass of the droplet.

19.3.4 MARANGONI CONVECTION

In EW, the thermal behavior and convection in droplets in both stable and unstable situations are of interest [50,76]. In this case, Marangoni convection, which results from temperature nonuniformity on the surface, concentration gradient, or even electric field, may need to be taken into consideration.

Marangoni convection is associated with surface tension gradient [76]. It occurs when the variation of the surface tension force dominates the viscous force. The Marangoni number Mg, which is dimensionless, determines the strength of the convection motion:

$$Mg = \frac{\Delta\gamma R}{\eta\zeta}, \quad (19.61)$$

where R is the radius of the droplet spherical cap, η is the dynamic viscosity, ζ is the thermal diffusivity, and $\Delta\gamma$ is the surface tension difference between the cooler and warmer parts of the surface. The critical value of the Marangoni number is $Mg_c = 80$ [77,78]. Generally, Marangoni convection cannot be ignored if it is larger than the critical value.

Surface tension is a function of surface temperature, which means that if the temperature varies from one region to another on the surface of a sessile droplet, the surface tension will be nonuniform on the surface. As a result, the liquid will flow from the region with lower surface tension to that with higher surface tension. This kind of fluid flow is exactly the thermal Marangoni convection. According to the Guggenheim–Katayama model [15], the surface tension is related to temperature as follows:

$$\gamma = \gamma^* \left(1 - \frac{T}{T_C}\right)^q,$$

(19.62)

where γ^* is a constant for a given liquid, T_C is the critical temperature in Kelvin, and q is an empirical factor. Considering that q is close to 1 and using a measured reference value γ_0 (at $T = 0°C$), Equation 19.62 can be rewritten as a linear approximation [15]:

$$\gamma = \gamma_0(1 + kT),$$

(19.63)

where k is the thermal coefficient, for water it has a value of $-0.15/°C$.

For many systems, the values of ρ, η, ζ, and γ under certain temperature have been reported, and can be used to estimate the Mg number using Equation 19.61.

19.3.5 PRECURSOR FILM

19.3.5.1 Huh–Scriven Paradox

The movement of the contact line over a solid surface was studied by Huh and Scriven [27]. By imposing the no-slip boundary condition, they constructed a hydrodynamic model of flow near an MCL, giving viscous stress components

$$\tau_{r\theta} = \frac{2\eta}{r}(g\cos\theta - j\sin\theta)$$

(19.64)

and the pressure field

$$p - p_0 = -\frac{2\eta}{r}(g\sin\theta + j\cos\theta),$$

(19.65)

where η is the fluid dynamic viscosity, p_0 is the hydrostatic datum, and g and j are undetermined coefficients. Owing to the no-slip boundary condition, r is set to zero at the contact line, making the total force exerted on the solid surface logarithmically infinite. The energy dissipation is also logarithmically diverging [79]:

$$D_{visc} \approx \frac{\eta v^2}{\theta} \ln\left(\frac{r_{out}}{r}\right),$$

(19.66)

where D_{visc} is the dissipation per unit time and unit length of the contact line, v is the constant moving velocity of the bottom fluid, and r_{out} is an appropriate outer length scale. From Equation 19.66, one can infer that energy dissipation is infinite at the contact line where r equals zero. However, no matter that the total force or the energy dissipation abhors local infinities virtually, and "not even Herakles could sink a solid" is absurd, which lead to the paradox.

About the 1971 paper, Professor Chun Hum, Department of Petroleum and Geosystems Engineering, University of Texas at Austin, recalled in an email to Professor Yapu Zhao on July 6, 2011 "My Ph.D. advisor Prof. Skip Scriven (U. Minnesota, Chem. Eng.; who passed away two years ago) was a highly respected fluid mechanism expert who is also known for his early work on the interfacial fluid mechanics ('Marangoni effects'). The 'corner flow' problem was a simple exercise problem that he assigned to me; and even when I reported to him the strange solution, I did not of course recognize its meaning. Prof. Scriven however immediately realized its significance and let other people know. I was fortunate enough to piggy back along."

19.3.5.2 Possible Explanations for Huh–Scriven Paradox

Huh and Scriven pointed out that the most obvious culprit leading to stress and energy dissipation singularities at the contact line was the no-slip boundary condition [27]. Later research [59,60] shows that the Huh–Scriven paradox arises from three other ideal assumptions besides the no-slip boundary condition: incompressible Newtonian fluid, smooth solid surface, and impenetrable liquid–solid interface. A list of mechanisms that were proposed to solve the Huh–Scriven paradox is shown in Table 19.2 [79].

19.3.5.3 Molecular Kinetic Theory and Properties of Precursor Film

The PF, usually a single molecular layer propagating ahead of the nominal contact line, may be one answer to the Huh–Scriven paradox [66,79,88] by introducing atomic details to eliminate the infinite dissipation. Hardy's pioneering work [89] predicted the existence of the PF when a droplet spreads. His results have been confirmed by numerous relevant theoretical [66,90] and experimental [91] studies. In order to explain the physical mechanism behind the PF phenomena, MKT [88, 92], which was proposed by Gladstone et al. [93], was adopted for instance in the EW case.

TABLE 19.2
Mechanisms Proposed to Solve the Huh–Scriven Paradox

Ideal Assumptions	Mechanisms to Relieve the Singularity
	Precursor film [80,81]
No-slip boundary condition	Slip boundary condition [27,82]
Smooth rigid solid walls	Surface roughness [83]
Incompressible Newtonian fluid	Shear thinning [84]
	Normal stresses [85]
Impenetrable fluid–fluid interface	Diffuse interface [86]
	Evaporation and condensation [87]

19.3.5.3.1 Molecular Kinetic Theory

The behavior of the contact line is determined by the statistical dynamic behavior of molecules within its vicinity. At equilibrium (see Figure 19.11a), each molecule has the same probability for jumping left or right, and the jump frequency v^0 can be expressed as

$$v^0 = \frac{k_B T}{h_P} \exp\left(\frac{-\Delta G_m}{k_B T}\right),$$
(19.67)

where k_B is the Boltzmann's constant, T is the absolute temperature, h_P is the Planck constant, and ΔG_m is the barrier energy. Once an external force (e.g., surface tension, electric field, disjoining pressure) is applied, the potential energy surface will tilt (see Figure 19.11b). The jump frequencies in forward and backward directions will be different:

$$v^+ = \frac{k_B T}{h_P} \exp\left(\frac{-\Delta G_m}{k_B T} + \frac{\lambda F_{\text{driving}}}{2 k_B T}\right),$$
(19.68)

$$v^- = \frac{k_B T}{h_P} \exp\left(\frac{-\Delta G_m}{k_B T} - \frac{\lambda F_{\text{driving}}}{2 k_B T}\right),$$
(19.69)

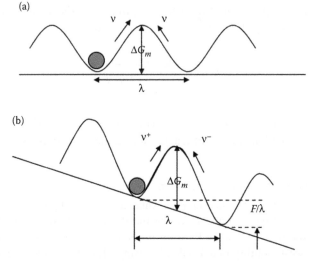

FIGURE 19.11 Schematic of the molecular kinetic theory (MKT) depiction of molecular jump to a neighboring site before (a) and after (b) application of external force. Note that application of external force results in different barriers for jump of molecule in the forward versus backward directions.

where v^+ is the forward frequency, v^- is the backward frequency, F_{driving} is the driving force, and λ is the distance between two neighboring surface sites. The velocity of the liquid molecules can then be written as

$$v = \lambda(v^+ - v^-) = 2\lambda \frac{k_B T}{h_P} \exp\left(\frac{-\Delta G_m}{k_B T}\right) \sinh\left(\frac{\lambda F}{2k_B T}\right). \qquad (19.70)$$

If $\lambda F_{\text{driving}} \ll 2k_B T$, which is true in most cases, then Equation 19.70 simplifies to

$$v \approx \frac{F_{\text{driving}}}{\lambda \upsilon}, \qquad (19.71)$$

where $\upsilon = (h_P/\lambda^3)/\left[\exp(\Delta G_m/k_B T)\right]$ denotes the friction coefficient per unit length of the contact line. υ has the same dimension as dynamic viscosity with unit of Pa·s. Equation 19.71 predicts that to a first approximation, velocity is directly proportional to the driving force, which is similar to a macroscopic friction.

Without considering the PF, Blake and Haynes [94] took $w = \gamma_{lv} (\cos\theta_0 - \cos\theta)$ for a spreading droplet, that is, wetting process. Here, $w = F_{\text{driving}}/\lambda$ is the work per unit area performed by the driving force, θ_0 and θ represent static and dynamic contact angles, respectively. When the droplet is small, w becomes complicated. By considering the PF, the driving work on the spreading droplet can be written as $w = w_V + w_P + w_S$ [95], where w_V is the work per unit area arising from vdW interactions, w_P is from the polar interactions between water molecules, and w_S is from the different structures of the PF from the bulk liquid. In the case of EW, an additional average electric energy $w_E \sim \Sigma_i\left[-|\boldsymbol{E}||\mu_i|L(|\boldsymbol{E}||\mu_i|/k_B T)\right]$ contributes, where $\boldsymbol{\mu}$ is the dipole moment vector. Then, the governing equation for EW can be expressed as

$$v = 2v_0\lambda \sinh\left(\frac{w_V + w_P + w_S + w_E}{2nk_B T}\right). \qquad (19.72)$$

For $w_V + w_P + w_S + w_E \ll nk_B T$, $v \sim (w_V + w_P + w_S + w_E)/\zeta_0$, where ζ_0 represents a friction coefficient per unit length, having the same dimension as dynamic viscosity.

19.3.5.3.2 Properties of Precursor Film

Previous studies considered that the PF advanced adiabatically much faster than the liquid above it, and hence, concluded that the PF behaved diffusively. However, a recent study [88] showed that the PF itself is not diffusive, but no-slip and solid-like and has the lowest mobility in a droplet. By using MD simulations, the propagation of the PF in wetting was obtained. The results showed that the molecules which finally formed the PF came almost from the surface in the initial state of the droplet. That is to say, the continuous and fast diffusion of surface water molecules to the front of the PF allows for fast propagation of the PF. Figure 19.12 shows the

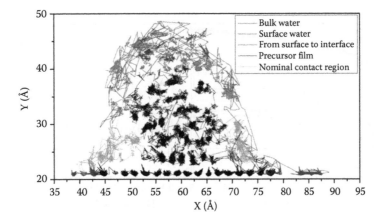

FIGURE 19.12 MD simulations of the movement of water molecules belonging to different regions of the droplet. (With permission from Q. Yuan and Y.P. Zhao, Precursor film in dynamic wetting, electrowetting, and electro-elasto-capillarity, *Phys. Rev. Lett.*, 104, 246101. Copyright 2010 by the American Physical Society.)

movement track of each water molecule during spreading. It can be seen that some water molecules move very fast at the surface, but once diffused to the region of the PF, they will be pinned by the surface and damped quickly, and then act as PF molecules with the lowest mobility. Besides, MD simulation of the EW process showed that the PF propagates even faster in EW compared with that in wetting.

Equation 19.72 derived from MKT is too complicated to have an analytical solution. Therefore, the power law $R \sim t^{n(E)}$ [88,96] was used to fit the relationship between R (spreading radius) and t (spreading time). The fit gave $n = 0.1554$ in wetting ($E = 0$ V/nm) and $n = 0.1819$ in EW ($E = 0.245$ V/nm), see Figure 19.13. The data of R versus t fits the power law very well. A critical electric field E_c (about 0.175 V/nm) and a saturated field E_s (about 0.625 V/nm) were obtained, which was in accordance with the experimental observation [4] in Figure 19.13c.

19.3.6 Spontaneous Electrowetting

Spontaneous EW is one of the variants of the classical EW configuration [97]. Unlike classical EWOD configuration, where the conducting droplet is placed on a dielectric film-coated planar electrode, spontaneous EW adopts a parallel line electrode configuration and the droplet is of high permittivity. Both configurations are shown in Figure 19.14 for comparison [40]. While in classical EW, a change in contact angle arises and the droplet will be stabilized at a certain apparent contact angle by applying a voltage, in spontaneous EW, a macroscopic finger of liquid film is pulled out ahead of the droplet and this front-running EW film maintains a constant contact angle. In this section, a theoretical analysis of the spontaneous EW will be presented first. Then, a detailed comparison between spontaneous EW and classical EW will be given.

A detailed theoretical analysis of spontaneous EW that combines electrodynamics with hydrodynamics was proposed by Yeo and Chang [40,98,99]. The electrode

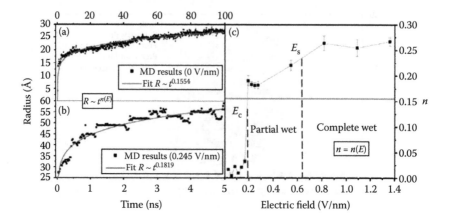

FIGURE 19.13 Propagation of the radius R of the precursor film (PF) as a function of time t, that is, $R \sim t^{n(E)}$ (n is a parameter relevant to E). (a) The spreading of the droplet (the external electric field $E = 0$). (b) Electrowetting (EW) of the droplet when $E = 0.245$ V/nm. (c) The change in n with respect to E. The horizontal line is $n = 0.1554$ when the droplet spread ($E = 0$). The left and right dashed lines represent the critical electric field E_c and the saturated field E_s in EW, respectively. (With permission from Q. Yuan and Y.P. Zhao, Precursor film in dynamic wetting, electrowetting, and electro-elasto-capillarity, *Phys. Rev. Lett.*, 104, 246101. Copyright 2010 by the American Physical Society.)

configuration in which the electric field is predominantly tangential to the TCL is shown in Figure 19.15. Given the absence of free space charge, the electrostatic potential Φ satisfies the Laplace equation

$$\nabla^2\Phi_i = 0, \tag{19.73}$$

where $i = v, l$ represents vapor and liquid, respectively. The boundary conditions are stipulated by continuity of the normal and tangential fields across the liquid–vapor interface Γ as

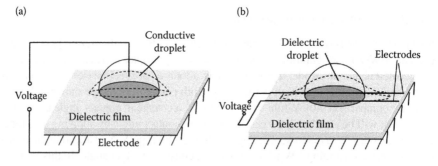

FIGURE 19.14 Schematics of classical versus spontaneous electrowetting (EW). (a) Classical configuration of EW, where a conducting droplet is deposited above an electrode coated with a dielectric film. (b) Spontaneous EW configuration, where a high permittivity polar dielectric droplet is placed on a parallel line electrode.

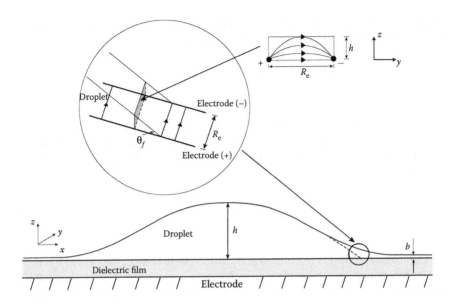

FIGURE 19.15 Schematic of the spreading droplet on a horizontal substrate for the case of spontaneous electrowetting. The inset is an enlarged view of the region near the three-phase contact line. The upper right depicts a y–z cross-section of the electric field in the droplet within the contact line region.

$$\left[\varepsilon_0\varepsilon_i\frac{\partial\Phi_i}{\partial n}\right]_v^l = \left[\frac{\partial\Phi_i}{\partial t}\right]_v^l = 0 \quad \text{on } \Gamma, \tag{19.74}$$

in which the square brackets $[\cdot]_v^l$ indicate a jump in the inner quantity across the interface. The normal continuity of the electric field can be expressed as

$$\varepsilon_1 E_{nl} = \varepsilon_v E_{nv} \quad \text{on } \Gamma. \tag{19.75}$$

As was mentioned above, the droplet is of high permittivity, thus the permittivity of the droplet is much large than that of the ambient vapor phase, that is, $\varepsilon_1 \gg \varepsilon_v$. Since E_{nv} is of finite value, the value of E_{nl} is small and can be neglected, and $E_1 \approx E_{tl}$. In this special case, the electric field in y–z plane will be dominant due to the polarity of the electrodes, and the Laplace equation for the electrostatic potential in the liquid phase can be rewritten as

$$\frac{\partial^2\Phi}{\partial y^2} + \frac{\partial^2\Phi}{\partial z^2} = 0. \tag{19.76}$$

The boundary condition at the solid–liquid interface is given by the potential of the electrodes

$$\Phi = \pm V \quad \text{at } y = \mp R_e/2, \tag{19.77}$$

where R_e is the electrode separation distance, and the boundary condition at the liquid–vapor interface in the limit $\varepsilon_l \gg \varepsilon_v$ is given by

$$E_{nl} = \frac{\partial \Phi}{\partial z} = 0. \tag{19.78}$$

By using the method of images (or mirror images, which is a mathematical tool used in electrostatics to simply calculate the distribution of the electric field), the solution to Equation 19.76 with the boundary conditions in Equation 19.77 and 19.78 was given [40] as

$$E_{tl} = \frac{4V}{\pi R_e}\left(1 - \frac{8h^2}{R_e^2}\right). \tag{19.79}$$

The tangential field is maximum at the three-phase contact line where $h = 0$, and decays linearly along the interface away from the contact line. Unlike classical EW, in spontaneous EW, the field is not singular at the contact line. The fluid pressure can then be obtained:

$$p = \gamma_{lv}\frac{\partial^2 h}{\partial x^2} - p_M = \gamma_{lv}\frac{\partial^2 h}{\partial x^2} - \frac{8\varepsilon_0\varepsilon_l V^2}{\pi^2 R_e^2}\left[1 - \frac{16\tan^2\theta_f}{R_e^2}(x_f - x)^2\right], \tag{19.80}$$

where p_M is the interfacial Maxwell pressure:

$$p_M = \frac{\varepsilon_0\varepsilon_l}{2}\left(E_n^2 - E_t^2\right), \tag{19.81}$$

and x_f is the position of the three-phase contact line, θ_f is the contact angle or slope of the capillary ridge (see Figure 19.16). The Maxwell stress is not singular and is responsible for negative capillary pressure in the contact line region, thus pushing out a thin spontaneous EW film ahead of the macroscopic spreading drop. This spontaneous EW film was first observed by Jones et al. [100] and Ahmed et al. [101]. It advances much faster than the macroscopic spreading drop and behaves in a self-similar manner. In particular, it should not be confused with the molecular PF, for it is macroscopic with thickness of several µm while the molecular PF is microscopic with only a few Å in thickness.

The governing equations for the two cases of EW (classical and spontaneous EW) are the same, while the boundary conditions are different, thus leading to different phenomena and properties. Comparisons between the two cases are given in Table 19.3 [40].

At the end of this section, we try to establish a more systematic framework of the EW problem. The method based on the phase-field model can be a broader way to solve EW. It can be used not only for obtaining the static contact angle, but also for exploring the dynamic details. In the phase-field model, which is

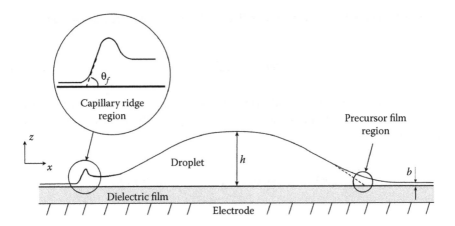

FIGURE 19.16 Spontaneous electrowetting film generated between the parallel planar line electrodes. The inset indicates the enlargement of the capillary ridge region. θ_f is the apparent static contact angle.

TABLE 19.3
Comparisons between Classical and Spontaneous Electrowetting

	Classical (Static) EW	Spontaneous EW
Configuration	Plate electrodes coated with dielectric film	Planar parallel line electrodes
Electric field	Normal to the vapor phase	Tangent to the liquid phase
	Singular and confined to small region ~d (d is the thickness of the dielectric film)	Nonsingular and vanish when $h \sim R_e$ (h is the local thickness of the droplet, R_e is the width of the two electrodes)
Maxwell pressure gradient	Point force at contact line	Body force within contact line region
	Can be balanced by surface forces	Net force and cannot be balanced by surface forces
Spontaneous EW film	Have no spontaneous EW film	Have self-similar spontaneous EW film
Phenomenon	Static change in apparent contact angle	Dynamic spreading with spontaneous film

based on the basic principles of thermodynamics, the free energy function is defined as [102]

$$F(\phi, \nabla\phi) = \int_V \left(\Psi(\phi) + \frac{1}{2}\kappa \, |\nabla\phi|^2 \right) dV + \int_S \varphi(\phi) dS, \qquad (19.82)$$

where ϕ is an order parameter used to distinguish different fluids, that is, $\phi = 1$ for one fluid (water droplet in the EW case) and $\phi = -1$ for the other fluid (air in the EW case). In Equation 19.82, the first term $\Psi(\phi)$ is the bulk free energy density and takes the form

$$\Psi(\phi) = a(\phi^2 - 1)^2, \tag{19.83}$$

where a is a constant. The second term is the interfacial energy density with κ being another constant. The last term in the surface integral is the surface energy density, which can be taken as

$$\phi(\phi) = -\gamma_{lv} \cos\theta \frac{\phi(3 - \phi^2)}{4} + \frac{1}{2}(\gamma_{sl} + \gamma_{sv}), \tag{19.84}$$

where $\cos\theta$ is obtained by the L–Y equation. The order parameter ϕ is governed by the Cahn–Hilliard equation:

$$\frac{\partial\phi}{\partial t} + (\mathbf{u} \cdot \nabla)\phi = \nabla \cdot (M\nabla\mu), \tag{19.85}$$

where μ is the chemical potential defined as the variation of free energy with respect to the order parameter:

$$\mu = \frac{\delta F}{\delta\phi} = \frac{d\Psi(\phi)}{d\phi} - \kappa\nabla^2\phi = 4a\phi(\phi^2 - 1) - \kappa\nabla^2\phi. \tag{19.86}$$

In phase-field models, interfaces between different phases are "diffuse" interfaces with a small, but finite, thickness. The EW problem can be solved by combining Equations 19.82 and 19.85 with specific boundary conditions. It is hard to have an analytical solution, and numerical simulations are often used [102,103]. However, since using a phase-field model for EW is still in its initial step, further studies are needed.

19.4   EXPERIMENTS AND MOLECULAR DYNAMICS SIMULATIONS

In this section, we will give a brief overview of EW-related experiments and MD simulations. While some of them were conducted to verify the existing theories, a number of experiments and MD simulations were also aimed at exploring new phenomena and mechanisms.

19.4.1   ELECTROWETTING EXPERIMENTS ON SURFACES WITH VARIOUS CONFIGURATIONS

Conventional EW experiments are focused on ideal planar surfaces, which means that the surfaces are planar and smooth. However, ideal planar surfaces are relatively simple to model, while surfaces with different geometries and rough properties are more common. Wetting can be seen as a specific state of EW, that is, EW at $V = 0$. The study of wetting on structured surfaces has been conducted for tens of years, both theoretically and experimentally. In contrast, the theoretical and experimental research in EW is not that flourishing. In this section, typical experiments of EW on surfaces with various configurations, including planar, rough and curved surfaces, are presented.

19.4.1.1 Electrowetting on Planar Surfaces

The L–Y equation is the basis of EW or EWOD, which has been verified by many experiments [5,6,38,52,54]. Poly(dimethylsiloxane) (PDMS) is widely used as a base material for bio-MEMS/NEMS devices [104,105] because of its excellent properties of being optically transparent, chemically inert, and flexible. However, since the surface of PDMS is inherently hydrophobic, it is difficult to transfer and spread aqueous solutions [106]. EW is an effective way to hydrophilize the PDMS surface. Therefore, the study of PDMS as the dielectric film in EWOD is of great significance.

Dai and Zhao [54] conducted a series of EWOD experiments using PDMS as the dielectric film. Figure 19.17 shows the images of a droplet on the PDMS film of 9.1 µm thickness. During the experiment, the voltage was increased from 0 to 360 V and back to 0. As shown in the three images in Figure 19.17, the contact angles changed from 110.1° to 46.9° and back to 104.6°, respectively. The surface converted from hydrophobic to hydrophilic and the decrease in the contact angle was 63.2°, so the effect of EW on a thin PDMS film was remarkable. They also studied the effect of PDMS film thickness on contact angle variation. Their results showed that the change in the contact angle increased with the thinning of the PDMS films, which was in good accord with the L–Y prediction.

Nanoscale EW effects were studied using atomic force microscopy by Guan et al. [107]. In their study, poly(methyl methacrylates) (PMMA) was used as the bare material. Adhesive interaction between the atomic force microscope (AFM) tip and the dielectric surface reflected the change of surface tension under the influence of applied voltage, thus leading to the EW effect. The results were in good accord with the Y–L prediction under a low electric field between the tip and the dielectric surface. Their study of nanoscale EW behavior provided complementary insights into macroscopic contact angle studies.

19.4.1.2 Electrowetting on Rough Surfaces

Surface roughness can increase the hydrophobicity of solid surfaces. EW can make the surface more hydrophilic. Both roughness and EW can physically modify the effective apparent contact angle by affecting the surface tension balance without altering the chemical properties of the surface or fluid. A combination of these two mechanisms will result in a large change in contact angle, or even change the surface from superhydrophobic to completely wetting.

EW on nanostructured superhydrophobic surfaces was investigated by Krupenkin et al. [46]. The surfaces were constructed by etching a microscopic array of cylindrical nanoposts into the surface of a silicon wafer. Each post had a diameter of about 350 nm and a height of about 7 µm, and the distance between posts varied from

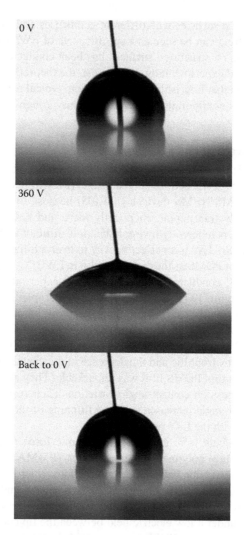

FIGURE 19.17 Images of water droplet on PDMS film with thickness of 9.1 μm. The three images are under the voltage of 0 V, 360 V, and (back to) 0 V, respectively.

1 to 4 μm. Several types of liquids were investigated, including water, alkanes, alcohols, ionic liquids, and various mixtures of these. In their experiment, the transition between a rolling ball droplet state and an immobile droplet state with the application of voltage was observed. In the rolling ball state, the droplet did not penetrate the nanoposts, which corresponded to the Cassie–Baxter model; in the immobile droplet state, the droplet completely wet the grooves, which corresponded to the Wenzel model. A transition from the rolling ball to the immobile droplet state was observed at a critical voltage of 22 V.

EW on superhydrophobic SU-8 (an epoxy-based negative photo-resist) patterned surface was investigated by Herbertson et al. [47]. The scanning electron microscopic image of the completed structure consisting of cylindrical pillars is shown in Figure 19.18a. The SU-8 pillars provided necessary surface roughness. The electrolyte solution was deionized water with 0.01 M KCl. In this experiment, voltages were applied from 0 to 130 V and back to zero. The change in the cosine of the contact angle as a function of voltage is shown in Figure 19.18b. A high hysteresis in the contact angle was observed. EW on an ideal planar surface can be described by the L–Y equation and a reversible change in the contact angle is expected. However, in their experiment, reversibility was not observed, as the starting contact angle was 152° but the ending was 114°. This phenomenon can be explained as discussed in Section 19.3.2.1, by change of wetting behavior from the Cassie–Baxter to the Wenzel regime due to applied voltage, which is an irreversible process. It is also observed that there is a

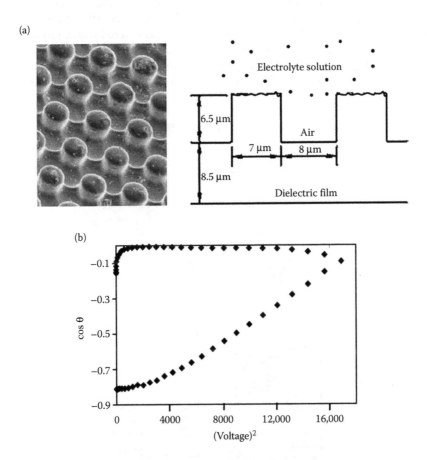

(a)

Electrolyte solution

6.5 µm

Air

7 µm   8 µm

8.5 µm

Dielectric film

(b)

$\cos\theta$

−0.1

−0.3

−0.5

−0.7

−0.9

0   4000   8000   12,000   16,000

(Voltage)^2

FIGURE 19.18 (a) Left: The scanning electron microscope (SEM) image of a superhydrophobic patterned surface (SU-8) consisting of cylindrical pillars. Right: The schematic of the microstructures. (b) Cosine of contact angle of electrolyte fluid on (SU-8) as a function of the square of applied voltage (from 0 to 130 V and back to 0 V). (From *Sens. Actuators A: Phys.,* 130, D.L. Herbertson et al., Electrowetting on superhydrophobic SU-8 patterned surfaces, 189. Copyright 2006, with permission from Elsevier.)

critical voltage, where there is a change in the slope of the data line in the EW process. From Figure 19.18b, the critical voltage is near 45 V, before which the contact angle changes little whereas after which the contact angle changes appreciably. It can also be explained by the transition from the Cassie–Baxter to the Wenzel regime, at the critical voltage. The Wenzel model was used to fit the experimental data for the voltage range from 45 to 130 V, giving a roughness factor of 1.92 ± 0.1. For voltage below 45 V, the liquid was considered unable to wick into the grooves and the Cassie–Baxter model was employed to explain the observation [43]. Both theoretical explanations were in good accord with the experimental data.

19.4.1.3 Electrowetting on Curved Surfaces

Kim and Steckl [26] studied the feasibility of using paper as a cheap and flexible substrate for e-paper, electronic display, and other EW devices. In their experiments, the paper was completely rolled into a cylinder, which was covered with a ground electrode, a dielectric film, and a fluoropolymer top layer, as depicted schematically in Figure 19.19b. Several categories of paper were investigated. The contact angle variations on these substrates as a function of voltage 0–60 V are shown in Figure 19.20.

(a)

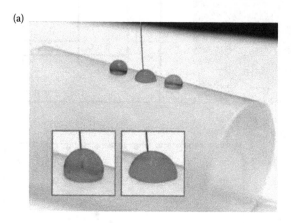

(b)

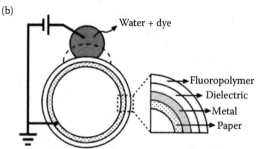

FIGURE 19.19 (a) Photographs of electrowetting on rolled-paper substrate. Insets show the contact angle with (right) and without (left) applied voltage. (b) Schematic of the EW structure. (From D.Y. Kim and A.J. Steckl, Electrowetting on paper for electronic paper display, *ACS Appl. Mater. Interfaces*, 2, 3318. Copyright 2010, American Chemical Society.)

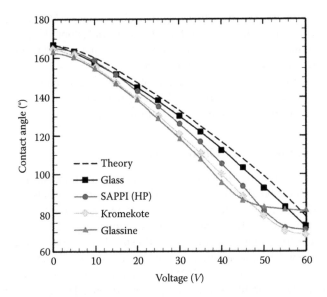

FIGURE 19.20 Contact angle (θ) versus applied voltage (V) on rolled substrate with different materials. All substrates were covered with a 1.0-μm-thick parylene-C dielectric film. The dashed line is the theoretical prediction using the Lippmann–Young (L–Y) Equation 19.6. The black line is the experimental data on a glass substrate. Three different materials of paper are tested, that is, SAPPI (HP), Kromekote, and Glassine. (From D.Y. Kim and A.J. Steckl, Electrowetting on paper for electronic paper display, *ACS Appl. Mater. Interfaces*, 2, 3318. Copyright 2010, American Chemical Society.)

The dotted line is the predicted CA value using the L–Y equation on planar surface ($\cos\theta = \cos\theta_Y + (\varepsilon V^2/2d\gamma_{lv})$). It can be seen that the experimental values deviate from the theoretical prediction, and the deviations are different for different categories of paper. It may arise from different properties of the paper. Possible variations of paper properties could be surface coating, roughness, thickness, and water uptake. Variations in surface curvature may also contribute to the variation of measured contact angles.

Fan and coworkers [22] studied "droplet-on-a-wristband" based on the study of EWOD in curved devices. They conducted a series of experiments on wrist-band-like curved surfaces whose curvatures (reciprocal of radius) were 0, 0.02, 0.04, and 0.06 mm⁻¹, and found that the curvature had no noticeable influence on the required driving voltages at the center point.

In Section 19.3.2.2, the theory of EW on curved surfaces has been presented. It predicts that at the present experimental condition (ξ or the ratio of d to R is small), the variation of the contact angle on curved surfaces is close to that on planar surfaces, and the surface curvature effect is not obvious. The experiments above accord well with the theoretical prediction. However, according to the theory, when R becomes very small or EWOD is on and below microscale, such as EWOD in CNTs or on graphene substrate, the influence of the surface curvature cannot be neglected.

Chen [108] studied EW in carbon nanotubes by means of MD simulations. The studies looked into the influence of surface curvature and adopted Lippmann's model of EW for a cylindrical capacitor. The results showed that the MD simulations were in good accord with the theoretical predictions.

Compared with the flourishing development of EW experiments on a macroscopic level, experiments or MD simulations of EW on and below microscale are rare in the published work. More studies are needed in future work.

19.4.2 BREAKDOWN OF THE DIELECTRIC FILM

Electrical breakdown of the dielectric film is a common failure form in EW, so how to reduce the critical breakdown voltage is of great importance. In typical EW experiments, the wire electrode is inserted into the droplet and maintains direct contact with it throughout the whole EW process. This method may lead to droplet contamination by the electrode, and the critical breakdown voltage is low. Feng and Zhao [51] put forward a new EW contact mode, aimed at effectively protecting the dielectric film from electrical breakdown. A PDMS membrane (thickness: 7.6 μm) was used to demonstrate the instability process in their experiment. The distance between the wire electrode and the PDMS membrane had influence on the EW behavior of the PDMS membrane. The EW process in contact mode can be divided into three steps (see Figure 19.21):

Step 1. The DC voltage is increased to a certain value (350 V in the experiment). In this step, the wire electrode does not contact the droplet.
Step 2. The electrode-membrane distance is reduced until the conductive droplet is in contact with the wire electrode. In this step, the contact angle will decrease and reach equilibrium, after which the wire electrode is separated from the droplet.
Step 3. The moving stage is adjusted to make the droplet contact the wire electrode again. As a result the contact angle will continue to decrease until it reaches a new equilibrium.

Step 2 and step 3 are unstable processes whose characteristic times are on the order 10 ms. The droplet will oscillate at its natural frequency of about 100 Hz. It was observed that the unstable process in fact started when the wire electrode was still a short distance above the droplet. When the voltage was increased to a high value, electrical breakdown of the dielectric film occurred. The whole process was recorded with a high-speed camera at 200 Hz (see Figure 19.22, voltage: 800 V, PDMS thickness: 4.8 μm). A spark can be seen in Figure 19.22b and this was taken as the start of an electrical breakdown. The whole process of EW from instability to electrical breakdown is shown in Figure 19.23. It can be seen that bubbles can still be observed after breakdown, which means that the dielectric film has been destroyed (see Figure 19.23d).

In their contact mode EW, the contact time between wire electrode and droplet was less than 5 ms, which dramatically lowered the chance of breakdown. In their experiment, voltage even higher than 800 V did not induce breakdown in the PDMS

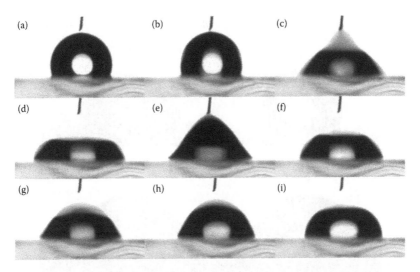

FIGURE 19.21 Droplet instability process of first contact with an electrode in EW contact mode. (a) The wire electrode does not contact the droplet. The DC voltage is increased to a certain value (350 V in the experiment). (b) The electrode–dielectric film distance is reduced until the conductive droplet is in contact with the wire electrode. (c) The contact angle decreases. (d) The contact angle reaches equilibrium and the wire electrode is separated from the droplet. (e) The moving stage is adjusted to make the droplet contact the wire electrode again. (f)–(i) The contact angle continues to decrease until it reaches a new equilibrium. (From J.T. Feng and Y.P. Zhao, Experimental observation of electrical instability of droplets on dielectric layer, *J. Phys. D: Appl. Phys.*, 41, 052004, 2008. With permission.)

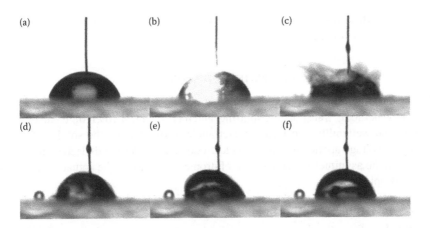

FIGURE 19.22 Breakdown process in electrowetting (EW) contact mode. The whole process was recorded with a high speed camera at 200 Hz (voltage: 800 V, PDMS thickness: 4.8 μm). The whole process is about 30 ms and the interval between each picture is 5 ms. (a) Initial state. (b) A sparkle can be seen. This is taken as the start of an electrical breakdown. (c) Electrical breakdown. (d)–(f) Electric energy is turned to dynamic energy of conductive liquids and a sputtered spray can be seen. (From J.T. Feng and Y.P. Zhao, Experimental observation of electrical instability of droplets on dielectric layer, *J. Phys. D: Appl. Phys.*, 41, 052004, 2008. With permission.)

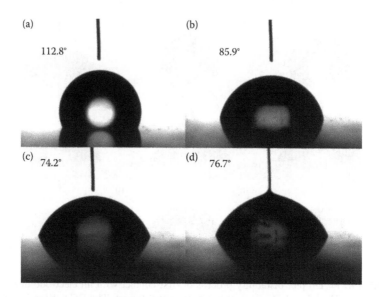

FIGURE 19.23 Images depicting the complete process in electrowetting (EW) contact mode (from instability to electrical breakdown, voltage 800 V, thickness of the PDMS film 4.8 μm). (a) Initial state. (b) First contact. The contact angle changes from 112.8° to 85.9°. (c) Second contact. The contact angle changes from 86.2° to 74.2°. (d) Bubbles can still be observed after breakdown. The contact angle is 76.7°. (From J.T. Feng and Y.P. Zhao, Experimental observation of electrical instability of droplets on dielectric layer, *J. Phys. D: Appl. Phys.*, 41, 052004, 2008. With permission.)

membrane (thickness: 4.8 μm) during the first two steps, while in normal mode the average critical voltage is 228 V.

19.4.3 VOLTAGE-INDUCED DROPLET ACTUATION

EW is an important technique for carrying out elementary operations on droplets, such as generating, transporting, splitting, and merging [4,21,109,110]. It can directly change the wettability and local contact angle of droplets on the solid surface by changing voltage applied to the microelectrode array under the dielectric film. This results in the asymmetric deformation of droplets and allows the actuation and control of droplets.

19.4.3.1 Droplet Transportation

Enclosed configuration is more common in digital microfluidics devices [69,111,112]. Figure 19.24 [69] shows the schematic of an EW microactuator. A droplet is sandwiched between two sets of planar electrodes whose two surfaces are coated with a thin dielectric film. The upper plate consists of a single continuous ground electrode, whereas the bottom one has an array of independently addressable control electrodes. The droplet volume is controlled such that it contacts the upper ground electrode and its footprint overlaps with at least two adjacent control electrodes. The

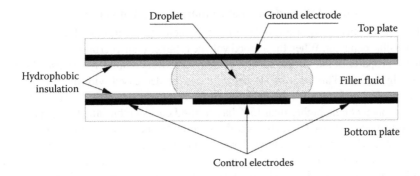

FIGURE 19.24 Schematic cross-section of an electrowetting microactuator. (With permission from M.G. Pollack, R.B. Fair and A.D. Shenderov, Electrowetting-based actuation of liquid droplets for microfluidic applications, *Appl. Phys. Lett.*, 77, 1725. Copyright 2000, American Institute of Physics.)

surrounding medium can be either air or a fluid that is immiscible to prevent evaporation of the droplet. All electrodes are initially grounded, so that the initial contact angle is the equilibrium contact angle described by the Young equation. Then, the potential on the adjacent electrodes is increased until motion is observed. This threshold potential, which is caused by the contact angle hysteresis, is dependent on the system. A prototype device consisting of a single linear array of seven interdigitated control electrodes at a pitch of 1.5 mm was fabricated and tested by Pollack et al. [69]. In their experiments, 30–40 V was required to initiate movement of the droplet. Sustained droplet transport over thousands of cycles at switching rates of up to 20 Hz was demonstrated. This rate corresponded to an average droplet velocity of 3.0 cm/s, which cannot be achieved with thermocapillary systems because it will require temperature difference between the ends of the drop in excess of 100°C [113]. Their results demonstrated the feasibility of EW as an actuation mechanism for droplet-based microfluidic systems. Several other possible microactuator electrode arrangements are illustrated in Figure 19.25.

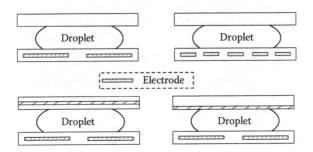

FIGURE 19.25 Several possible microactuator electrode arrangements investigated for moving droplets between electrodes.

19.4.3.2   Voltage-Induced Droplet Actuation on a Lotus Leaf

The general operations of a droplet using EW techniques can often be limited by an irreversible behavior. When the applied voltage is larger than a certain value, the hydrophobic property of the sample surface could not recover completely because of contact angle hysteresis. This irreversible behavior can be decreased with the use of superhydrophobic surfaces such as dielectric films. A superhydrophobic surface [114,115] with a contact angle higher than 150° has aroused considerable attention. It is of great interest to study droplet actuation on the superhydrophobic surface owing to its two outstanding characteristics: a large contact angle and a low contact angle hysteresis.

EW on lotus leaf (EWOL) was studied [70] because it is superhydrophobic (contact angle of greater than 150°) and a weak conductor. The scanning electron microscopy (SEM) images of a fresh lotus leaf (on both sides) are shown in Figure 19.26. There

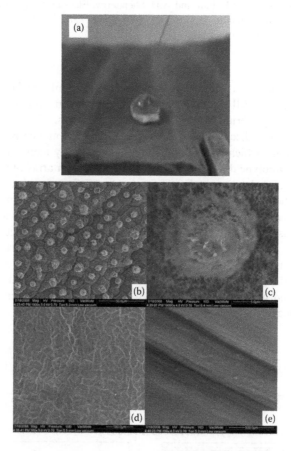

FIGURE 19.26   Images of lotus leaf under various magnifications. (a) An image of water droplet deposited on a lotus leaf. (b)–(e) Scanning electronic microscopy (SEM) images of the lotus leaf: (b) top side of lotus leaf, (c) papilla, (d) backside of the lotus leaf, and (e) the vein. The lotus surface has many papillae with radius of about 4 μm. (With permission from J.T. Feng, F.C. Wang and Y.P. Zhao, Electrowetting on a lotus leaf, *Biomicrofluidics*, 3, 022406. Copyright 2009, American Institute of Physics.)

are many papillae with radius of about 4 μm on the lotus surface. The contact angles of water on lotus and dorsal leaves are 159° and 141°, respectively. Experiments with applied voltage of 150 V were conducted to demonstrate the process of droplet motion induced by applied voltage. It was found that the droplet climbed up the vein (a height of 1.3 mm) of the leaf. This is because the lotus leaf is not an equipotential plate. There is a gradient of the electrical potential between two sides of the droplet, resulting in an asymmetric deformation of droplets, which allows actuation and control of droplets. The voltage-induced droplet motion induced in the top side is different from that in the bottom side of the lotus leaf. In these experiments, applied voltage on the top side of the leaf caused the advancing contact angle of the droplet to jump from 154° to 114°, and then recover from 114° to 124° when the droplet totally settled down. When the voltage was applied to the bottom side of the leaf, however, the advancing contact angle jumped from 141° to 79°, and recovered from 79° to 96° when the droplet stopped moving. The recovery of the contact angle may be a result of residual charges in the droplet moving to the leaf gradually. Moreover, the experiments showed that when the applied voltage was more than 300 V, a lot of heat would be generated and the lotus leaf would be dehydrated by the heat. High voltage also caused the appearance of the lotus leaf to change significantly (see Figure 19.27).

In this experiment, the actuation potentials were greater than 100 V, the actuation speed of droplet motion was on the order of 10 mm/s, and the actuation time was on the order of 10 ms. Figure 19.28 shows the variation of the contact angles when 150 V was applied on the top side of the lotus leaf. It can be seen that the change of the advancing contact angle is different from that of the receding contact angle. Toward the end of the motion, the advancing contact angle still increased while the receding contact angle was approximately constant. This is because the droplet separated from the Pt electrode toward the end of the motion, which opened the electrical circuit, and with release of the residual charges in the droplet the advancing angle increased. The capacitance of the lotus leaf can be estimated from the actuation criterion [67]:

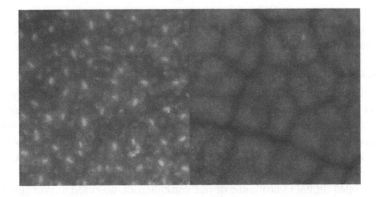

FIGURE 19.27 The lotus leaf before (left) and after (right) application of 400 V. (With permission from J.T. Feng, F.C. Wang and Y.P. Zhao, Electrowetting on a lotus leaf, *Biomicrofluidics*, 3, 022406. Copyright 2009, American Institute of Physics.)

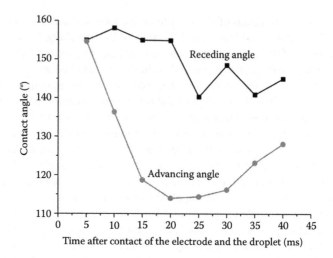

FIGURE 19.28 Advancing and receding contact angles as a function of time after contact of the droplet and the electrode (the applied voltage is 150 V). (With permission from J.T. Feng, F.C. Wang and Y.P. Zhao, Electrowetting on a lotus leaf, *Biomicrofluidics*, 3, 022406. Copyright 2009, American Institute of Physics.)

$$\frac{c}{2\gamma_{lv}} V_{min}^2 = \frac{\alpha}{1-\alpha} \left(\sin\theta_{V_{min}} + \sin\theta_0 \right), \tag{19.87}$$

where c is the capacitance of the lotus leaf per unit area and V_{min} is the minimum electric potential for actuation. In this case, $\theta_{V_{min}} \approx \theta_0$ and the contact angle hysteresis α is sufficiently small; Equation 19.87 can be simplified as

$$C \approx \frac{4\gamma_{lv}\alpha\sin\theta_0}{V_{min}^2}. \tag{19.88}$$

Equation 19.88 was used to estimate the capacitance of the lotus leaf per unit area to be about 0.63 μF/m^2 [70].

19.4.3.3 All-Terrain Droplet Actuation

Conventional study of digital microfluidics was mainly concentrated on droplet actuation on a single plane. As digital devices develop fast nowadays, conventional planar devices are not sufficient, and flexible and portable devices are in strong demand. Abdelgawad et al. [116] studied all-terrain droplet actuation (ATDA), showing that ATDA can be used to manipulate droplets across a wide range of geometries, including inclined, declined, vertical, twisted, and upside-down configurations. In their study, an open digital microfluidic device configuration was used to take full advantage of its flexibility. ATDA devices were formed by patterning arrays of copper electrodes and coating them with PDMS and Teflon-AF. Droplets were actuated by applying potentials to sequential pairs of electrodes. This enabled droplet movement on surfaces with different configurations. In all the geometries investigated, droplet movement was found to be facile and fast, with no significant differences relative to conventional planar microfluidics.

Since droplets have never before been manipulated on all-terrain surfaces, the researchers developed a model for predicting droplet motion. Unlike conventional planar situation, where the actuation force F_{EWOD} and the resistive friction force F_f work together to determine droplet motion, gravity effect was also considered in the ATDA situation and an additional force F_g was introduced. For example, for droplet movement up an inclined plane, F_g was calculated simply as the projection of droplet weight down the inclined plane. The researchers gave the maximum device inclination angle at which the actuation force (for a given droplet volume) was predicted to be larger than or equal to the resistive force, that is, $F_{EWOD} \geq F_f + F_g$. The experimental results showed good agreement with the theoretical prediction using $F_{EWOD} \approx \frac{1}{2}\pi R\gamma_{lv}(\cos\theta_a - \cos\theta_r)$ and $F_f = e^{(0.4084 \ln V + 0.4153)}$ [116]. The curve predicted that a droplet with volume less than ~7.3 µL can be driven up a 90° incline, whereas droplets with larger volumes can only be driven up inclinations with reduced angle. Moreover, it was observed that for an inclination angle far below the maximum, lower actuation force (and lower applied potential) was required to drive the droplet up the inclined plane.

19.4.4   IN SITU OBSERVATION OF THERMAL MARANGONI CONVECTION

An experiment on *in situ* observation of thermal Marangoni convection in sessile droplets was conducted by Wang and Zhao [117]. Infrared (IR) thermal imaging was applied to visualize transient temperature distribution on the surface of the droplet, while microparticle image velocimetry (micro-PIV) was used to illustrate the fluid field inside.

In the experiment, a droplet of aqueous KCl solution was trapped between a PDMS hydrophobic layer and an indium tin oxide (ITO) conductive glass hydrophilic layer (see Figure 19.29). When a platinum electrode with a diameter of 100 µm touched the apex of the droplet (60 V DC voltage applied), current passed through

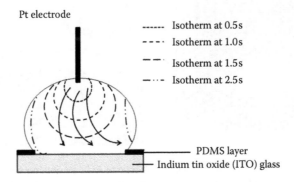

FIGURE 19.29 Schematic of surface flow in a droplet of aqueous KCl. The isotherms at different times are marked to show the development of the temperature field on the surface. The arrows indicate the direction of surface flow.

and heated the droplet. With the occurrence of the temperature gradient, the surface tension of the droplet between the bottom and the apex became uneven. The warmer region at the apex had smaller surface tension than the cooler region at the bottom. Consequently, the liquid on the surface flows downward when the liquid inside the drop flows upward. The heated surface region developed isotherms as shown by the dashed line profile in Figure 19.29. Since the cross-section was small at the apex, the current density was high and most of the heat was generated there (see Figure 19.30a). In the picture, the maximum surface temperature achieved was 100°C at the top region of the droplet. In this experiment, the Mg number of the droplet was calculated using Equation 19.61 to be 4.24×10^4 ($R = 1$ mm, $a = 0.14 \times 10^{-6}$ m^2/s, $v = 8.94 \times 10^{-7}$ m^2/s, and $\Delta T = 20$°C). This value is of the same order as that reported by Evren-Selamet et al. [118]. The thermal evolution was recorded in Figure 19.30a–d. The four thermal images were taken in series with a constant time interval of 0.5 s. The results agreed well with the theory of Marangoni convection.

The interior flow field, which was detected by the micro-PIV method, is shown in Figure 19.31. Two symmetrical vortices appeared as soon as the electrode touched the droplet. In the center of the droplet, the particles in the solution moved swiftly from the bottom ITO glass electrode to the top Pt electrode. The flow velocity in the center reached the maximum of about 17.5 mm/s; in the center of the vortex near the surface, the particles did not move and the velocity dropped to zero. Note that, in this experiment, gravity-driven flow was negligible because of the small variation in density. Also, the influence of liquid evaporation was ignored since the duration of the observation was short, and the flow was entirely due to Marangoni convection.

In ideal EWOD devices, the Mg number is less than the critical value ($Mg_c = 80$), so Marangoni convection can be neglected. However, if the dielectric film under the droplet developed a hole, just like the case in this experiment, the Mg number would be much larger than the critical value, and Marangoni convection must be taken into account.

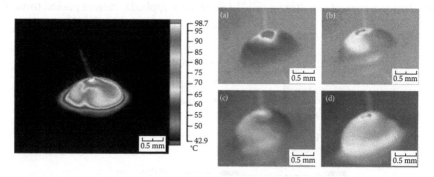

FIGURE 19.30 Infrared thermal image of a droplet of KCl solution trapped between PDMS and indium tin oxide (ITO) conductive glass. Left image is the temperature distribution on the droplet surface when the Pt electrode contacts the droplet. The temperature is highest at the top region and lowest at the bottom region. Right images (a)–(d) show the evolution process of temperature distribution. The time interval between every two neighboring pictures is 0.5 s.

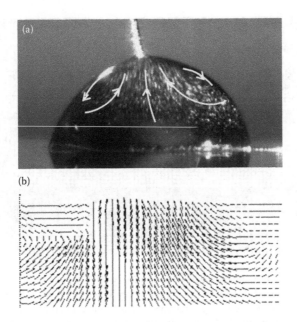

FIGURE 19.31 PIV images of the droplet heated by an electrode. (a) Vortices formed with an upward stream in the center. (b) Visualized flow field.

19.4.5 Electro-Elastocapillarity: Electrowetting on Flexible Substrate

Ever since self-assembled three-dimensional microstructures have been realized using elastocapillarity (EC) [119], studies on EC phenomenon have attracted considerable attention because of its realized and potential applications in NEMS/MEMS [120]. Since wrapping of droplet with flexible substrate in EC is spontaneous, it is irreversible, which restricts its application. The PF may be the key to solving this problem due to its unique transport properties. Besides, voltage is applied in order to produce reversible wrapping and unwrapping, and this phenomenon is called electro-elastocapillarity (EEC) [88]. In the case of EW, a typical electric energy $E \cdot \mu$ (E is the electric field vector and μ the dipole moment vector) in the PF cannot be neglected relative to thermal energy $k_B T$ (k_B is the Boltzmann's constant and T the absolute temperature). Polarization is strongest in the PF and dipolar water molecules may therefore be attracted to the PF [88]. Hence, the PF plays a critical role in EW.

An EEC process was simulated, using graphene to wrap the droplet [88]. The graphene automatically wraps the droplet when the radius of the droplet is larger than the EC length $L_{EC} = (B/\gamma)^{1/2}$ [121], where B is the flexural stiffness (for graphene its value is about 0.2 nN·nm [122]). Then, the wrapped droplet is unwrapped on a gold substrate under $E = 0.544$ V/nm along the $-y$ direction (see Figure 19.32). To avoid interfering with the process, the vdW force between the graphene and the substrate is reduced to 0.1% under the electric field. As the propagation of the PF is faster than the liquid above it and the PF actually acts solid-like, the PF pushes the graphene and unwraps it with a force on the order of 1 nN/nm. Thus, reversible wrapping and

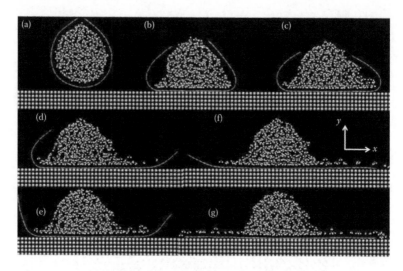

FIGURE 19.32 The dynamic electroelastocapillary (EEC) process of a droplet on gold substrate. The wrapped droplet in (a) progressively unwraps (b–g) on the gold substrate under $E = 0.544$ V/nm along the $-y$ direction. The line under the droplet represents the graphene. (From Q. Yuan and Y.P. Zhao, Precursor film in dynamic wetting, electrowetting, and electro-elasto-capillarity, *Phys. Rev. Lett.*, 104, 246101. Copyright 2010 by the American Physical Society.)

unwrapping of the droplet with graphene is realized. The dynamic process of EEC is a promising candidate for drug delivery application at a micro- or nanoscale.

19.4.6 ELECTROWETTING IN A HYDROPHILIC INTERIOR CORNER

Liquid transport in an interior corner is a crucial requirement for a variety of applications, such as aerospace, micro/nanofluidics, biology, fuel cell, and so on. It is of great interest to improve the transport velocity in these applications. The topological structure of the interior corner leads to significant changes in the dynamics of wetting. A droplet will spread until its dynamic contact angle reaches an equilibrium value θ_0 (Young's equation) when placed on a hydrophilic surface. The question is, what will happen when the two contact lines of the droplet on wedge surfaces encounter at a hydrophilic interior corner with an interior angle of 2α?

Dynamic wetting in a hydrophilic interior corner was simulated by Yuan and Zhao [123]. The atomic details of a water droplet in a hydrophilic gold interior corner with different interior angles in dynamic wetting were explored. The droplet was pulled along the bisector of the interior angle toward the interior corner by an attractive vdW force F. Then, the droplet spread on the substrate driven by the disjoining pressure. According to the MD results, the wetting transition is controlled by the interior angle. When $2\alpha < 135°$, an apparent PC advances along the interior corner, while the PF advances on the wedge surface, as shown in Figure 19.33c and d for $2\alpha = 45°$. The PC propagated much faster than the PF and completely wetted the interior corner. When $2\alpha \geq 135°$, only the PF was observed and partially wetted the interior corner, while

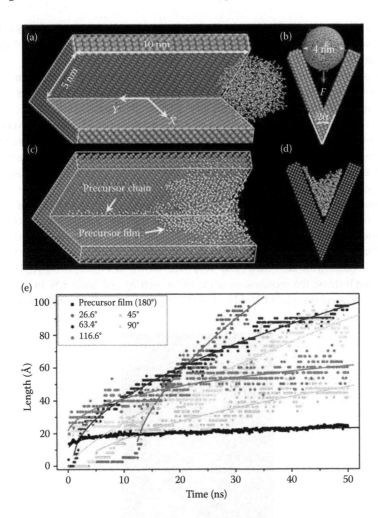

FIGURE 19.33 Formation mechanism and propagation characteristics of precursor chain (PC): (a) Initial ($t = 0$) oblique view. (b) Initial ($t = 0$) cross-section of the simulation domain. (c) Final ($t = 50$ ns) oblique view. (d) Final ($t = 50$ ns) cross-section of the simulation result. The interior corner consists of gold atoms. (e) The propagation length of the PC as a function of time (points are molecular dynamics simulations; lines are fits to power law). PC: red (26.6°), green (45.0°), blue (63.4°), cyan (90°), and purple (116.6°). PF: black. (From Q. Yuan and Y.P. Zhao, Topology-dominated dynamic wetting of the precursor chain in a hydrophilic interior corner, *Proc. R. Soc.* A, 468, 310, 2012. By permission of The Royal Society of Chemistry.)

no apparent PC could be found. The propagation lengths of the PC as a function of time for different interior angles are shown in Figure 19.33e. The data were fitted using a power law function similar to the treatment of the PF [88]. The velocity of PC v_{PC}, which is the slope of the length–time curve, is about 10^{-1} m/s. This is one order of magnitude higher than v_{PF} (~10^{-2} m/s). When 2α was small ($2\alpha = 26.6°$, $45.0°$), the propagation length initially remained zero for a characteristic time of about 10 ns, during which the droplet was pulled to the interior corner. The existence of the PC brings atomic details to eliminate stress singularity in the interior corner.

The advance velocity of the PC was one order of magnitude faster than the PF. MKT was used to investigate the transport properties of the PC. The results showed that (1) the PC was driven by the disjoining pressure, whose energy could be obtained from theoretical derivations and MD results (2) because water molecules were confined in the interior corner, the potential surface near the interior corner was lower and smoother than that on the bare surface, and quickly decayed over a distance of about the size of one water molecule. Hence, the PC could generate, exist stably, and propagate fast. In the PC, a one-dimensional (1D) hydrogen-bond (H-bond) chain generated the pressure to drive the PC to slip like ice. With an increase of 2α, the potential surface near the interior corner became high and rough, and the transport properties of the PC gradually changed and became the same as those of the PF.

EW in a hydrophilic interior corner was also explored. Models of EW in hydrophilic interior corners with $2\alpha = 26.6°$ and $90.0°$ are shown in Figure 19.34 [124]. Figure 19.35a and b shows the propagation of the PC in wetting and under EW in the interior corner with interior angles of $26.6°$ and $90.0°$, respectively [124]. Owing to the effect of

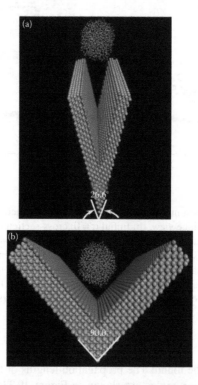

FIGURE 19.34 Configurations of EW in interior corners; surface gold atoms are charged ($q = 0.02e$): (a) opening angle, $26.6°$, (b) opening angle, $90°$. (Adapted from Q. Yuan. The electro-mechanical coupling property of topological wetting interface dynamics at nanoscale: [D]. Beijing: Graduate University of Chinese Academy of Sciences, 2011.)

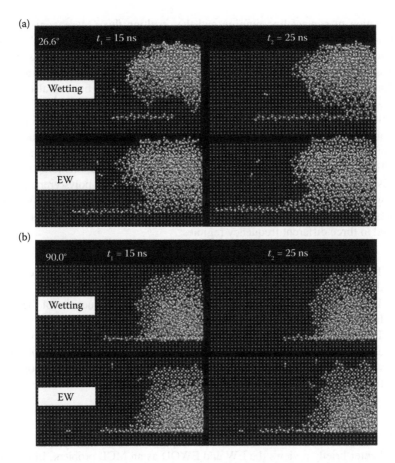

FIGURE 19.35 Comparison of precursor chain (PC) propagation in wetting versus electrowetting (EW) in an interior corner: (a) opening angle of 26.6°, (b) opening angle of 90°. (Adapted from Q. Yuan. The electro-mechanical coupling property of topological wetting interface dynamics at nanoscale: [D]. Beijing: Graduate University of Chinese Academy of Sciences, 2011.)

the electric driving work in EW, the advance speed of PC was one order of magnitude (i.e., 10^{-1} m/s) faster than that in the wetting process. As the interior angle becomes smaller, the charge density within its vicinity becomes larger so that the propagation velocity further increases (compared with that in the wetting process) more obviously.

19.4.7 ELECTROWETTING AS A TOOL FOR SUPPRESSING THE COFFEE STAIN

Ring-like solid residues along the contact line remain, after evaporating drops of colloidal suspensions or solutions of nonvolatile species. This is known as the coffee stain effect, which is undesired in applications such as coating and printing. Eral et al. [125] proposed a method to solve this problem by applying EW with an alternating (AC) voltage to an evaporating droplet. Coffee stains are formed because of the combination of contact line pinning and the resulting flux (Marangoni flows and electroosmotic flows) of solvent and solute toward the contact line. EW can be a particularly effective tool to counteract the coffee stain effect because of its ability to address both aspects of the problem simultaneously. It can not only set pinned contact lines in motion and weaken the contact angle hysteresis, but can also generate internal flow fields within the drops.

In the experiments, a transparent ITO glass, which was covered with a 5-µm-thick layer of SU8, was used as the electrode. Voltage with a fixed root-mean-square amplitude of $U_{rms} = 200$ V was applied throughout at frequencies varying from 6 Hz to 100 kHz. Different aqueous solutions were prepared and tested, including the colloidal suspensions of fluorescently labeled carboxyl-terminated polystyrene particles, ranging from 0.1 to 5 µm in diameter, as well as DNA solutions. The experimental suppression of the coffee stain effect using EW was given in Ref. [125]. The behavior referred to three different frequency regimes:

1. The quasistatic regime, at frequencies below the lowest eigenfrequency of the drop, that is, typically up to about 10 Hz for millimeter drops.
2. The hydrodynamic regime, in the range of the eigenfrequencies of the drop from several tens of Hz to a few kHz.
3. The electrothermal regime, at frequencies allowing for penetration of the electric field into the liquid bulk. It can be seen that smallest residue spots were achieved in the hydrodynamic regime or by a combination of high frequencies with low-frequency amplitude modulation.

19.5 SUMMARY AND CONCLUSIONS

This chapter briefly reviews the EW and EWOD as an MCL problem, in which the "Huh–Scriven paradox" is of main scientific concern. Systematic research is needed to focus on the elimination of the singularities for electrical and mechanical fields at the TCL, and a unified framework needs to be constructed. PF and PC are proven to be two effective physical mechanisms for the nano-EW of droplet spreading on a flat substrate [79] and in an interior corner [112]. Are slip boundary, diffuse liquid–vapor interface, or other mechanisms still applicable in the EW and EWOD cases as those in the classical MCL problem? This is still an open question.

EWOD has been widely used as a tool to manipulate microdroplets. Unlike macrosystems, surface tension effect will be significant due to the increase in surface-to-volume ratio in such microsystems. Therefore, a good understanding of the relevant characteristic time and length scales will assist the study of surface effects and help with the design of new micro- and nanosystems. In Section 19.2, some characteristic time and length scales are discussed. Capillary characteristic time and Lord Rayleigh's period are used to measure the strength of the droplet mass compared with surface tension, whereas viscous characteristic time is the ratio of the liquid viscosity to surface tension. These characteristic timescales may have a direct or indirect effect on the actuation of droplets. Then, the EW number is defined and used to measure the strength of the electrostatic energy relative to surface tension.

Considering size effects, other two characteristic length scales are also given, representing the line tension and Debye screening length, respectively.

The L–Y equation is the fundamental of EWOD. It can be derived using different approaches, three of which are given as examples: the thermodynamic approach, the energy minimization approach, and the electromechanical approach. Since it is derived under ideal assumptions that cannot be always satisfactory in reality, it needs to be modified in various situations. Two different models, that is, the Cassie–Baxter model and the Wenzel model, are used to modify and extend the L–Y equation for use on rough surfaces. It was found that rough surfaces can attain better hydrophobicity than smooth surfaces. Then, considering the surface curvature effect, an extended L–Y equation on curved surfaces is obtained and a curvature-modified EW number is introduced. The surface curvature effect will become even more significant with decreasing system size, so it cannot be neglected on and below microscale. Besides, it is found that concave surfaces can enhance wettability better than convex surfaces when applying the same voltage. Next, EWOD under high voltage is discussed. The experiments showed that the L–Y equation was only applicable under a low voltage. Contact angle saturation and dielectric film breakdown may happen under high voltage and the phenomenon cannot be explained by the L–Y equation. The mechanism is still not clear and possible explanations are proposed. EWOD at micro- and nanoscales are then discussed. The line tension works at or under microlevel, in which case the L–Y equation should be modified by adding this effect. As was presented in Section 19.2, the droplet is conductive at nanoscale because its characteristic length is less than the Debye screening length. An expression for EWOD is given for such situations. The L–Y equation or its extensions discussed above are focused on static cases. However, for many practical applications, however, dynamic properties are of great interest. Actuation of the droplet is one of the most common forms of dynamic movement. Here, the problem is concerned with minimum actuation voltage, which is relevant to contact angle hysteresis. A mechanical analysis is presented and an expression for minimum actuation voltage is proposed. Next, the Marangoni convection, which is related to uneven heat distribution on the droplet, is taken into consideration. The value of the Marangoni number can be a criterion for deciding whether the Marangoni convection should or should not be considered. Finally, the molecular PF, which may be one of the answers to the Huh–Scriven paradox, is discussed. At the end of Section 19.3, compared with the conventional EWOD, another form that is known as spontaneous EW is discussed.

In Section 19.4, relevant experiments and molecular dynamics simulations corresponding to the theories in Section 19.3 are presented, including the conventional EWOD experiments, breakdown of dielectric film under high voltage, voltage-induced droplet actuation, *in situ* observation of thermal Marangoni convection, wetting and EW in a hydrophilic interior corner, and the EEC phenomenon. The voltage-induced droplet actuation experiments are particularly discussed in relation to its application in devices and other potential uses. An experiment on droplet with actuation time on the order of 10 ms is presented. An all-terrain droplet actuation is shown, demonstrating the feasibility of future flexible devices. The feasibility is further verified by the electro-elastocapillarity phenomenon (both in experiments and simulations), which is a good candidate for drug delivery at the micro- or nanoscale. At the end of Section 19.4, EW as an efficient tool for suppressing the coffee stain effect is presented.

While considerable studies have been conducted and great progress has been made in recent decades, there are still many unknowns about the EW phenomenon, and more researches are needed. Since EW at micro- and nanolevels performs differently from that at macroscopic level due to scale effect, the study of EW at small scales is of great interest and significance. Several applications based on EW have already entered the market while lots of potential applications are under study. We hope that this review will stimulate further studies and progress in EW.

ACKNOWLEDGMENTS

This work was jointly supported by the National Natural Science Foundation of China (NSFC, Grant No. 11072244, 60936001, and 11021262), the Key Research Program of the Chinese Academy of Sciences (Grant No. KJZD-EW-M01) and the Instrument Developing Project of the Chinese Academy of Sciences (Grant No. Y2010031).

SYMBOLS

A	droplet base area (m^2)
c	capacitance per unit area (F/m^2)
Ca	capillary number (dimensionless number)
d	thickness of the dielectric film (μm)
d_E	thickness of the electrical double layer (m)
D	droplet diameter (m)
D_{visc}	energy dissipation per unit time and unit length of the contact line ($J \cdot m^{-1} \cdot s^{-1}$)
e	charge of an electron (C)
E	electric field intensity (V/m)
Ew	EW number (dimensionless number)
f_1, f_2, R_1	surface roughness factor
F	free energy (J)
F_d	electrostatic force (N)
F_{drag}	drag force applied by the Pt electrode (N)
F_e	net electrostatic force acting on the liquid–vapor surface (N)
F_{elec}	electrowetting (EW) force (N)
F_f	contact angle hysteresis force (N)
$F_{viscous}$	viscous resistance (N)
ΔG_m	barrier energy (J)
H	local thickness of the droplet (m)
h	maximum height of the droplet (m)
h_P	Planck constant (J·s)
k	thermal coefficient (/°C)
k_B	Boltzmann's constant (J/K)
l	characteristic length (m)
m	droplet mass (kg)
Mg	Marangoni number (dimensionless number)
M_i	molar concentration (mol/m^3)

p	pressure (Pa)
R	radius of the contact line (m)
R_e	width of the electrode (m)
R_S	radius of the substrate (m)
t	characteristic time (s)
T	absolute temperature (K)
$\boldsymbol{T}$	Maxwell stress tensor
T_{LR}	Lord Rayleigh's period (s)
T_c	capillary characteristic time (s)
T_{vis}	viscous characteristic time (s)
U	electric energy (J)
v	velocity (m/s)
V	voltage (V)
W_B	work that the voltage source performs (J)
z_i	valency
α	contact angle hysteresis (°)
γ_{lv}	liquid–vapor surface tension (N/m)
γ_{sl}	solid–liquid interfacial tension (N/m)
γ_{sv}	solid–vapor interfacial tension (N/m)
δ	average distance between the liquid and the solid molecules (m)
ε	dielectric constant (F/m)
ζ	thermal diffusivity (m²/s)
η	droplet dynamic viscosity (Pa·s)
θ	contact angle (°)
λ_D	Debye length (m)
μ	chemical potential (J/mol)
$\boldsymbol{\mu}$	free dipole (C/mol)
ν	molecule jump frequency (Hz)
ξ	order parameter representing for d/R_S
ρ	droplet mass density (kg/m³)
σ	surface charge density (C/m²)
τ	line tension (N)
τ_w	shear stress (N/m²)
Φ	electrostatic potential (V)
χ	geodesic curvature at the contact point (m⁻¹)
ϕ	an order parameter used to distinguish different fluids

REFERENCES

1. G. Lippmann, Relations entre les ph"nomenes "lectriques et capillaires, *Ann. Chim. Phys.*, 5, 494, 1875.
2. L. Minnema, H. Barneveld, and P. Rinkel, An investigation into the mechanism of water tree-ing in polyethylene high-voltage cables, *IEEE Trans. Dielectr. Electr. Insul.*, 6, 461, 1980.
3. G. Beni and S. Hackwood, Electro-wetting displays, *Appl. Phys. Lett.*, 38, 207, 1981.
4. F. Mugele and J. C. Baret, Electrowetting: From basics to applications, *J. Phys.: Condens. Matter*, 17, 705, 2005.

5. B. Berge, Electrocapillarity and wetting of insulator films by water, *CR Acad. Sci. II*, 317, 157, 1993.
6. C. Quilliet and B. Berge, Electrowetting: A recent outbreak, *Curr. Opin. Colloid Interface Sci.*, 6, 34, 2001.
7. V. Srinivasan, V.K. Pamula, and R.B. Fair, An integrated digital microfluidic lab-on-a-chip for clinical diagnostics on human physiological fluids, *Lab Chip*, 4, 310, 2004.
8. J.Y. Yoon and R.L. Garrel, Preventing biomolecular adsorption in electrowetting-based biofluidic chips, *Anal. Chem.*, 75, 5097, 2003.
9. A.R. Wheeler, H. Moon, C.J. Kim, J.A. Loo, and R.L. Garrell, Electrowetting-based microfluidics for analysis of peptides and proteins by matrix-assisted laser desorption/ionization mass spectrometry, *Anal. Chem.*, 76, 4833, 2004.
10. I. Barbulovic-Nad, Digital microfluidics for cell-based assays, *Lab Chip*, 8, 519, 2008.
11. R.B. Fair, A. Khlystov, T.D. Tailor, V. Ivanov, R.D. Evans, P.B. Griffin, S. Vijay, V.K. Pamula, M.G. Pollack, and J. Zhou, Chemical and biological applications of digital-microfluidic devices, *IEEE Des. Test Comput.*, 24, 10, 2007.
12. C. Karuwan, K. Sukthang, A. Wisitsoraat, D. Phokharatkul, V. Pattnanasettakul, W. Wechsatol, and A. Tuantranont, Electrochemical detection on electrowetting-on-dielectric digital microfluidic chip, *Talanta*, 84, 1384, 2011.
13. D.M. Ratner, E.R. Murphy, M. Jhunjhunwala, D.A. Snyder, K.F. Jensen, and P.H. Seeberger, Microreactor-based reaction optimization in organic chemistry—Glycosylation as a challenge, *Chem. Commun.*, 5, 578, 2005.
14. M. Pollack, A. Shenderov, and R. Fair, Electrowetting-based actuation of droplets for integrated microfluidics, *Lab Chip*, 2, 96, 2002.
15. J. Berthier, *Microdrops and Digital Microfluidics*, William Andrew Publishing, New York, 2008.
16. B. Berge and J. Peseux, Variable focal lens controlled by an external voltage: An application of electrowetting, *Eur. Phys. J. E*, 3, 159, 2000.
17. T. Krupenkin, S. Yang, and P. Mach, Tunable liquid microlens, *Appl. Phys. Lett.*, 82, 316, 2003.
18. F. Krogmann, W. Monch, and H. Zappe, A MEMS-based variable micro-lens system, *J. Opt. A, Pure Appl. Opt.*, 8, 330, 2006.
19. M.A. Bucaro, P.R. Kolodner, J.A. Taylor, A. Sidorenko, J. Aizenberg, and T.N. Krupenkin, Tunable liquid optics: Electrowetting-controlled liquid mirrors based on self-assembled janus tiles, *Langmuir*, 25, 3876, 2009.
20. N. Chronis, G.L. Liu, K.H. Jeong, and L.P. Lee, Tunable liquid-filled microlens array integrated with microfluidic network, *Opt. Express*, 11, 2370, 2003.
21. N. Verplanck, E. Galopin, J.C. Camart, and V. Thomy, Reversible electrowetting on superhydrophobic silicon nanowires, *Nano Lett.*, 7, 813, 2007.
22. S.K. Fan, H. Yang, and W. Hsu, Droplet-on-a-wristband: Chip-to-chip digital microfluidic interfaces between replaceable and flexible electrowetting modules, *Lab Chip*, 11, 343, 2011.
23. J.A. Rogers, Z. Bao, K. Baldwin, A. Dodabalapur, B. Crone, V.R. Raju, V. Kuck et al., Paper-like electronic displays: Large-area rubber-stamped plastic sheets of electronics and microencapsulated electrophoretic inks, *Proc. Natl. Acad. Sci. USA*, 98, 4835, 2001.
24. R.A. Hayes and B. Feenstra, Video-speed electronic paper based on electrowetting, *Nature*, 425, 383, 2003.
25. Q. Cheng, Brief in electronic paper and electronic ink, Printed Circuit Information, 2010.
26. D.Y. Kim and A.J. Steckl, Electrowetting on paper for electronic paper display, *ACS Appl. Mater. Interfaces*, 2, 3318, 2010.

27. C. Huh and L.E. Scriven, Hydrodynamic model of steady movement of a solid-liquid-fluid contact line, *J. Colloid Interface Sci.*, 35, 85, 1971.

28. L. Rayleigh, On the instability of jets, *Proc. London Math. Soc.*, 10, 4, 1879.

29. R.R.A. Syms, E.M. Yeatman, V.M. Bright, and G.M. Whitesides, Surface tension-powered self-assembly of microstructures—The state-of-the-art, *J. Microelectromech. Syst.*, 12, 387, 2003.

30. G.I. Barenblatt, Micromechanics of fracture, *J. Theor. Appl. Mech.*, Elsevier, Amsterdam, 25, 1993.

31. K.R. Sreenivasan and G. Stolovitzky, Self-similar multiplier distributions and multiplicative models for energy dissipation in high-Reynolds-number turbulence, *J. Theor. Appl. Mech.*, Elsevier Science, 1993.

32. Q. Yuan and Y.P. Zhao, Hydroelectric voltage generation based on water-filled single-walled carbon nanotubes, *J. Am. Chem. Soc.*, 131, 6374, 2009.

33. A. Marmur, Line tension and the intrinsic contact angle in solid-liquid-fluid systems, *J. Colloid Interface Sci.*, 186, 462, 1997.

34. A. Marmur and B. Krasovitski, Line tension on curved surfaces: Liquid drops on solid micro-and nanospheres, *Langmuir*, 18, 8919, 2002.

35. D.C. Brydges and P.A. Martin, Coulomb systems at low density: A review, *J. Stat. Phys.*, 96, 1163, 1999.

36. D. Li, *Electrokinetics in Microfluidics*, Elsevier Academic Press, London, 2004.

37. C.D. Daub, D. Bratko, K. Leung, and A. Luzar, Electrowetting at the nanoscale, *J. Phys. Chem. C*, 111, 505, 2007.

38. M. Vallet, M. Vallade, and B. Berge, Limiting phenomena for the spreading of water on polymer films by electrowetting, *Eur. Phys. J. B*, 11, 583, 1999.

39. K.H. Kang, How electrostatic fields change contact angle in electrowetting, *Langmuir*, 18, 10318, 2002.

40. L.Y. Yeo and H.C. Chang, Static and spontaneous electrowetting, *Mod. Phys. Lett. B*, 19, 549, 2005.

41. C. Neinhuis, and W. Barthlott, Characterization and distribution of water-repellent, self-cleaning plant surfaces, *Ann. Bot.*, 79, 667, 1997.

42. G. Palasantzas, and J.T.M. De Hosson, Wetting on rough surfaces, *Acta Mater.*, 49, 3533, 2001.

43. W. Dai and Y.P. Zhao, An electrowetting model for rough surfaces under low voltage, *J. Adhes. Sci. Technol.*, 22, 217, 2008.

44. V. Bahadur and S.V. Garimella, Electrowetting-based control of static droplet states on rough surfaces, *Langmuir*, 23, 4918, 2007.

45. A.W. Adamson and A.P. Gast, *Physical Chemistry of Surfaces*, Wiley, New York, 1997.

46. T.N. Krupenkin, J.A. Taylor, T.M. Schneider, and S. Yang, From rolling ball to complete wetting: The dynamic tuning of liquids on nanostructured surfaces, *Langmuir*, 20, 3824, 2004.

47. D.L. Herbertson, C.R. Evans, N.J. Shirtcliffe, G. McHale, and M.I. Newton, Electrowetting on superhydrophobic SU-8 patterned surfaces, *Sens. Actuators A: Phys.*, 130, 189, 2006.

48. Y. Wang and Y.P. Zhao, Electrowetting on curved surfaces, *Soft Matter*, 8, 2599, 2012.

49. W.J.J. Welters and L.G.J. Fokkink, Fast electrically switchable capillary effects, *Langmuir*, 14, 1535, 1998.

50. H. Moon, S.K. Cho, and R.L. Garrell, Low voltage electrowetting-on-dielectric, *J. Appl. Phys.*, 92, 4080, 2002.

51. J.T. Feng and Y.P. Zhao, Experimental observation of electrical instability of droplets on dielectric layer, *J. Phys. D: Appl. Phys.*, 41, 052004, 2008.

52. H.J.J. Verheijen and M.W.J. Prins, Reversible electrowetting and trapping of charge model and experiments, *Langmuir*, 15, 6616, 1999.
53. J. Berthier, Ph. Clementz, O. Raccurt, D. Jary, P. Claustre, C. Peponnet, and Y. Fouillet, Computer aided design of an EWOD microdevice, *Sens. Actuators A: Phys.*, 127, 283, 2006.
54. W. Dai and Y.P. Zhao, The nonlinear phenomena of thin polydimethylsiloxane (PDMS) films in electrowetting, *Int. J. Nonlinear Sci. Numer. Simul.*, 8, 519, 2007.
55. F. Mugele and S. Herminghaus, Electrostatic stabilization of fluid microstructures, *Appl. Phys. Lett.*, 81, 2303, 2002.
56. A. Quinn, R. Sedev, and J. Ralston, Contact angle saturation in electrowetting, *J. Phys. Chem. B*, 109, 6268, 2005.
57. D. Klarman, D. Andelman, and M. Urbakh, A model of electrowetting, reversed electrowetting, and contact angle saturation, *Langmuir*, 27, 6031, 2011.
58. A. Papathanasiou and A. Boudouvis, Manifestation of the connection between dielectric breakdown strength and contact angle saturation in electrowetting, *Appl. Phys. Lett.*, 86, 164102, 2005.
59. C. Miller and E. Ruckenstein, The origin of flow during wetting of solids, *J. Colloid Interface Sci.*, 48, 368, 1974.
60. L.R. White, On deviations from Young's equation, *J. Chem. Soc., Faraday Trans.*, 1, 73, 390, 1977.
61. G. Wolansky and A. Marmur, The actual contact angle on a heterogeneous rough surface in three dimensions, *Langmuir*, 14, 5292, 1998.
62. D. Li and A. Neumann, Determination of line tension from the drop size dependence of contact angles, *Colloids Surf.*, 43, 195, 1990.
63. J. Drelich and J.D. Miller, The effect of solid surface heterogeneity and roughness on the contact angle/drop (bubble) size relationship, *J. Colloid Interface Sci.*, 164, 252, 1994.
64. R. Digilov, Charge-induced modification of contact angle: The secondary electrocapillary effect, *Langmuir*, 16, 6719, 2000.
65. V. Ballenegger and J.P. Hansen, Dielectric permittivity profiles of confined polar fluids, *J. Chem. Phys.*, 122, 114711, 2005.
66. P.G. de Gennes, Wetting: Statics and dynamics, *Rev. Mod. Phys.*, 57, 827, 1985.
67. J. Berthier, P. Dubois, P. Clementz, P. Claustre, C. Peponnet, and Y. Fouillet, Actuation potentials and capillary forces in electrowetting based microsystems, *Sens. Actuators A: Phys.*, 134, 471, 2007.
68. S.K. Cho, H. Moon, and C.J. Kim, Creating, transporting, cutting, and merging liquid droplets by electrowetting-based actuation for digital microfluidic circuits, *J. Microelectromech. Syst.*, 12, 70, 2003.
69. M.G. Pollack, R.B. Fair, and A.D. Shenderov, Electrowetting-based actuation of liquid droplets for microfluidic applications, *Appl. Phys. Lett.*, 77, 1725, 2000.
70. J.T. Feng, F.C. Wang, and Y.P. Zhao, Electrowetting on a lotus leaf, *Biomicrofluidics*, 3, 022406, 2009.
71. R.E. Johnson, Jr. and R.H. Dettre, Contact angle hysteresis I. Study of an idealized rough surface, *Adv. Chem. Ser*, 43, 112, 1964.
72. R.E. Johnson, Jr. and R.H. Dettre, Contact angle hysteresis. III. Study of an idealized heterogeneous surface, *J. Phys. Chem.*, 68, 1744, 1964.
73. N.A. Patankar, On the modeling of hydrophobic contact angles on rough surfaces, *Langmuir*, 19, 1249, 2003.
74. H. Tavana and A. Neumann, On the question of rate-dependence of contact angles, *Colloids Surf., A*, 282, 256, 2006.
75. M. Fermigier and P. Jenffer, An experimental investigation of the dynamic contact angle in liquid-liquid systems, *J. Colloid Interface Sci.*, 146, 226, 1991.

76. S.H. Ko, H. Lee, and K.H. Kang, Hydrodynamic flows in electrowetting, *Langmuir*, 24, 1094, 2008.

77. J.J. Hegseth, N. Rashidnia, and A. Chai, Natural convection in droplet evaporation, *Phys. Rev. E*, 54, 1640, 1996.

78. J. Pearson, On convection cells induced by surface tension, *J. Fluid Mech.*, 4, 489, 1958.

79. D. Bonn, J. Eggers, J. Indekeu, J. Meunier, and E. Rolley, Wetting and spreading, *Rev. Mod. Phys.*, 81, 739, 2009.

80. H. Hervet and P.G. de Gennes, The dynamics of wetting: Precursor films in the wetting of "dry" solids, *C. R. Acad. Sci., Ser. II: Mec., Phys., Chim., Sci. Terre Univers*, 299, 499, 1984.

81. M.H. Eres, L.W. Schwartz, and R.V. Roy, Fingering phenomena for driven coating films, *Phys. Fluids*, 12, 1278, 2000.

82. P.A. Thompson and S.M. Troian, A general boundary condition for liquid flow at solid surfaces, *Nature*, 389, 360, 1997.

83. L.M. Hocking, A moving fluid interface on a rough surface, *J. Fluid Mech.*, 76, 801, 1976.

84. D.E. Weidner and L.W. Schwartz, Contact-line motion of shear-thinning liquids, *Phys. Fluids*, 6, 3535, 1994.

85. A. Boudaoud, Non-Newtonian thin films with normal stresses: Dynamics and spreading, *Eur. Phys. J. E*, 22, 107, 2007.

86. P. Seppecher, Moving contact lines in the Cahn-Hilliard theory, *Int. J. Eng. Sci.*, 34, 977, 1996.

87. P.C. Wayner, Spreading of a liquid film with a finite contact angle by the evaporation/condensation process, *Langmuir*, 9, 294, 1993.

88. Q. Yuan and Y.P. Zhao, Precursor film in dynamic wetting, electrowetting, and electro-elasto-capillarity, *Phys. Rev. Lett.*, 104, 246101, 2010.

89. W. Hardy, The spreading of fluids on glass, *Philos. Mag*, 38, 49, 1919.

90. T.W. Poon, S. Yip, P.S. Ho, and F.F. Abraham, Equilibrium structures of Si (100) stepped surfaces, *Phys. Rev. Lett.*, 65, 2161, 1990.

91. H.P. Kavehpour, B. Ovryn, and G.H. McKinley, Microscopic and macroscopic structure of the precursor layer in spreading viscous drops, *Phys. Rev. Lett.*, 91, 196104, 2003.

92. F.C. Wang and Y.P. Zhao, Slip boundary conditions based on molecular kinetic theory: The critical shear stress and the energy dissipation at the liquid–solid interface, *Soft Matter*, 7, 8628, 2011.

93. S. Gladstone, K. Laidler, and H. Eyring, *The Theory of Rate Processes*, Princeton University, New York, 1941.

94. T. Blake and J. Haynes, Kinetics of liquid/liquid displacement, *J. Colloid Interface Sci.*, 30, 421, 1969.

95. B. Derjaguin and N. Churaev, Structural component of disjoining pressure, *J. Colloid Interface Sci.*, 49, 249, 1974.

96. L. Tanner, The spreading of silicone oil drops on horizontal surfaces, *J. Phys. D-Appl. Phys.*, 12, 1473, 1979.

97. F. Mugele, Fundamental challenges in electrowetting: From equilibrium shapes to contact angle saturation and drop dynamics, *Soft Matter*, 5, 3377, 2009.

98. L.Y. Yeo and H.C. Chang, Electrowetting films on parallel line electrodes, *Phys. Rev. E*, 73, 011605, 2006.

99. H.C. Chang and L.Y. Yeo, *Electrokinetically Driven Microfluidics and Nanofluidics*, Cambridge University Press, New York, 2010.

100. T.B. Jones, M. Gunji, M. Washizu, and M.J. Feldman, Dielectrophoretic liquid actuation and nanodroplet formation, *J. Appl. Phys.*, 89, 1441, 2001.

101. R. Ahmed, D. Hsu, C. Bailey, and T.B. Jones, Dispensing picoliter droplets using DEP micro-actuation, Proc. First International Conference on Microchannels and Minichannels, Rochester, New York, 2003.

102. J.J. Huang, C. Shu, J.J. Feng, and Y.T. Chew, A phase-field-based hybrid Lattice-Boltzmann finite-volume method and its application to simulate droplet motion under electrowetting control, *J. Adhesion Sci. Technol*, 26, 1825, 2012.

103. M.A. Fontelos, G. Grun, U. Kindelan, and F. Klingbeil, Numerical simulation of static and dynamic electrowetting, *J. Adhesion Sci. Technol*, 26, 1805, 2012.

104. J.C. McDonald and G.M. Whitesides, Poly (dimethylsiloxane) as a material for fabricating microfluidic devices, *Accounts Chem. Res.*, 35, 491, 2002.

105. T. Fujii, PDMS-based microfluidic devices for biomedical applications, *Microelectron. Eng.*, 61, 907, 2002.

106. D. Bodas and C. Khan-Malek, Formation of more stable hydrophilic surfaces of PDMS by plasma and chemical treatments, *Microelectron. Eng.*, 83, 1277, 2006.

107. L. Guan, G. Qi, S. Liu, H. Zhang, Z. Zhang, Y. Yang, and C. Wang, Nanoscale electrowetting effects studied by atomic force microscopy, *J. Phys. Chem. C*, 113, 661, 2008.

108. J.Y. Chen, Electrowetting in carbon nanotubes, *Science*, 310, 1480, 2005.

109. F. Mugele and J. Buehrle, Equilibrium drop surface profiles in electric fields, *J. Phys.-Condes. Matter*, 19, 375112, 2007.

110. N. Verplanck, Y. Coffinier, V. Thomy, and R. Boukherroub, Wettability switching techniques on superhydrophobic surfaces, *Nanoscale Res. Lett.*, 2, 577, 2007.

111. Y.H. Chang, G.B. Lee, F.C. Huang, Y.Y. Chen, and J.L. Lin, Integrated polymerase chain reaction chips utilizing digital microfluidics, *Biomed. Microdevices*, 8, 215, 2006.

112. A.R. Wheeler, H. Moon, C.A. Bird, R.R. Ogorzalek Loo, C.J. Kim, J.A. Loo, and R.L. Garrell, Digital microfluidics with in-line sample purification for proteomics analyses with MALDI-MS, *Anal. Chem.*, 77, 534, 2005.

113. M. Washizu, Electrostatic actuation of liquid droplets for micro-reactor applications, *IEEE Trans. Ind. Appl.*, 34, 732, 1998.

114. S. Ren, S.R. Yang, Y.P. Zhao, T. Yu, and X.D. Xiao, Preparation and characterization of an ultrahydrophobic surface based on a stearic acid self-assembled monolayer over polyethyleneimine thin films, *Surf. Sci.*, 546, 64, 2003.

115. S. Ren, S.R. Yang, and Y.P. Zhao, Nano-tribological study on a super-hydrophobic film formed on rough aluminium substrates, *Acta Mech. Sin.*, 20, 159, 2004.

116. M. Abdelgawad, S.L.S. Freire, H. Yang, and A.R. Wheeler, All-terrain droplet actuation, *Lab Chip*, 8, 672, 2008.

117. Z.Q. Wang and Y.P. Zhao, In situ observation of thermal marangoni convection on the surface of a sessile droplet by infrared thermal imaging, *J. Adhesion Sci. Technol.*, 26, 2177, 2012.

118. E. Evren-Selamet, V. Arpaci, and A. Chai, Thermocapillary-driven flow past the Marangoni instability, *Numer. Heat Transf.*, 26, 521, 1994.

119. R. Syms and E. Yeatman, Self-assembly of three-dimensional microstructures using rotation by surface tension forces, *Electron. Lett.*, 29, 662, 1993.

120. J. Bico, B. Roman, L. Moulin, and A. Boudaoud, Elastocapillary coalescence in wet hair, *Nature*, 432, 690, 2004.

121. C. Py, P. Reverdy, L. Doppler, J. Bico, B. Roman, and C.N. Baroud, Capillary origami: Spontaneous wrapping of a droplet with an elastic sheet, *Phys. Rev. Lett.*, 98, 156103, 2007.

122. N. Patra, B. Wang, and P. Kra, Nanodroplet activated and guided folding of graphene nanostructures, *Nano Lett.*, 9, 3766, 2009.

123. Q. Yuan and Y.P. Zhao, Topology-dominated dynamic wetting of the precursor chain in a hydrophilic interior corner, *Proc. R. Soc. A*, 468, 310, 2012.

124. Q. Yuan. The electro-mechanical coupling property of topological wetting interface dynamics at nanoscale: [D]. Beijing: Graduate University of Chinese Academy of Sciences, 2011.

125. H.B. Eral, D.M. Augustine, M.H.G. Duits, and F. Mugele, Suppressing the coffee stain effect: How to control colloidal self-assembly in evaporating drops using electrowetting, *Soft Matter*, 7, 4954, 2011.

Index

A

Ablation pit, 458
Abrasive index (AI), 485
Abrasive wear, 485
  at different pH, 485, 486
  effect of surfactant on, 486
Abrasiveness test, 478
ABS, *see* Alkylbenzene sulfonate (ABS)
AC, *see* Ammonium chloride (AC)
Acetylenic glycols, 168
Acid hydrolyzable surfactants, 156; *see also*
        Hydrolyzable surfactants
  acyclic acetals, 157
  cyclic acetals, 156–157
  ketals, 157–159
  ortho esters, 159–160
  surfactants containing N=C bond, 160
Acyclic acetal, 157; *see also* Acid hydrolyzable
        surfactants
  alkyl glucoside, 157
  surfactants synthesis, 158
Adsorption, 466; *see also* Adsorption isotherm
  contact angle and wetting, 470
  equilibrium contact angle, 470
  equilibrium time, 471
  factors in surfactant, 469
  free energy of, 397
  hemimicelle, 472
  interfacial tension, 470, 471
  of ionic surfactant on polar surfaces, 473
  mechanisms, 469
  model, 63
  models of surfactant, 471
  nonionic surfactant, 474
  physiosorption, 466
  surfactant, 468
  surfactant mixture, 474–475
  and wettability changes on solid surfaces,
    470
Adsorption isotherm, 396, 467
  Brunauer-Emmett-Teller isotherm, 468
  determination, 476
  Freundlich isotherm, 468
  Langmuir isotherm, 467–468
  pH effect on, 479–481
AFM, *see* Atomic force microscopy (AFM)
AFM tip on C_{60}/graphite interface, 3, 26;
       see also Carbon hybrid interface
    superlubricity

  frictional force map, 11
  results, 12
AFM tip on graphite surface interface, 3, 26;
       see also Carbon hybrid interface
    superlubricity
  C_{60} bilayer on graphite, topographs of, 11
  C_{60} monolayer on graphite, topographs of, 11
  frictional force maps, 7, 10, 11, 12
  in frictional force microscopy, 7
  graphite surface model, 7
  graphite topographs, 11
  long-range capillary force, 10, 12
  potential energy, 7
  simulation, 7–9
  static Tomlinson model, 8
Aggregation and cloud point, 398
AI, *see* Abrasive index (AI)
Air bearing, 457
Alcohol; *see also* Dialkyl carbonates
  ethoxylate with carbon dioxide, 154
  ethoxylates, 169
  fatty, 164
  fatty alcohol ethoxylates, 169
  short-chain, 397
Alginates, 161
Alkali hydrolyzable esters, 152; *see also*
        Hydrolyzable surfactants
  betaine esters, 153–154
  monoalkyl carbonates, 154
  normal quaternary esters, 152–153
  sugar esters, 155–156
  surfactants with Si-O bond, 154–155
Alkylbenzene sulfonate (ABS), 164; *see also*
        Surfactants
Alkylcarbonates; *see also* Dialkyl carbonates;
        Dimethyl carbonate (DMC); Glycerol
        carbonate (GLICA)
  as cleaning agents, 359
  commercial production, 360
  dibutyl carbonate, 367
  diethyl carbonate, 367
  as friction modifiers, 358
  historical perspective, 359
Alkyl glucosides, 163; *see also* Sugar-based
        surfactants
  in detergent production, 172
  long-chain, 172
  mixing behavior, 392
  surfactants, 157
N-Alkyl imidazolium-derived ionic liquid, 113